UDC № 669 (082)

TREATISE ON POWDER METALLURGY

VOLUME I

Technology of Metal Powders and Their Products

TREATISE ON POWDER METALLURGY *in three volumes*
by Claus G. Goetzel

Volume I:
Technology of Metal Powders and Their Products

Volume II:
Applied and Physical Powder Metallurgy
Part 1. Applied Powder Metallurgy — Industrial Materials and Products
Part 2. Physical Powder Metallurgy — Practical Evaluations and Theoretical Analyses of the Materials, Products, and Processes

Volume III:
Classified and Annotated Bibliography
Part 1. Literature
Part 2. Patents

TREATISE ON
POWDER METALLURGY

By
CLAUS G. GOETZEL, Ph.D.

VICE-PRESIDENT AND DIRECTOR OF RE-
SEARCH, SINTERCAST CORPORATION OF
AMERICA, NEW YORK, N. Y.
ADJUNCT PROFESSOR OF CHEMICAL ENGI-
NEERING, NEW YORK UNIVERSITY

Volume I

Technology of Metal Powders and Their Products

1 9 4 9

INTERSCIENCE PUBLISHERS, INC., NEW YORK
INTERSCIENCE PUBLISHERS LTD., LONDON

INTERSCIENCE PUBLISHERS, INC.
215 Fourth Avenue, New York 3, N. Y.

For Great Britain and Northern Ireland:
INTERSCIENCE PUBLISHERS LTD.
2a Southampton Row, London

PRINTED IN THE UNITED STATES OF AMERICA BY
MACK PRINTING CO., EASTON, PA.

PREFACE

Powder metallurgy has come of age. This means that it has established itself as a legitimate branch of metallurgy, that a huge amount of theoretical, empirical, and technological material has been accumulated in the journals and patent literature, and that the application of metal powders is ever increasing in all branches of metal technology.

The author feels, therefore, that the time is ripe to survey, digest, and present the whole wealth of our present knowledge of the field in a systematic way in the form of an exhaustive treatise. Its aim is to serve as an introduction into the field for the novice who wants detailed information, as a reference book for the initiated, and as a source of stimulation for the inventor and engineer who has to work out problems and new developments and to find new ways of application.

Although there is a voluminous literature pertaining to this subject, it has been restricted mainly to papers read before the learned societies, articles in the technical and daily press, letters patent issued in the United States and abroad, monographs on restricted phases of the subject, symposia and conferences. Several outstanding men in this field have written books on the subject, but none has attempted to organize the mass of present-day knowledge of powder metallurgy in a methodical manner. Among these authors are:

Dr. W. D. Jones, who wrote the first prewar book, *Principles of Powder Metallurgy*, in 1937.

Professor John Wulff of the Massachusetts Institute of Technology, who organized two conferences on the subject (1940 and 1941) under the sponsorship of M. I. T.; the proceedings were recorded in one volume, *Powder Metallurgy*, in 1942 by the American Society for Metals.

Dr. P. Schwarzkopf, who in his recently published book, *Powder Metallurgy*, made a valuable contribution to the subject. However, according to Dr. Schwarzkopf, his book is intended to be not a text, but a record of his own experiences, reflections, and theoretical studies, which, because he has been active in the field for more than thirty years, will be stimulating and inspiring to everyone.

Drs. R. Kieffer and W. Hotop, who in 1943 and 1948 compiled two books in German, *Pulvermetallurgie und Sinterwerkstoffe* and *Sintereisen und Sinterstahl*, that may be considered to be the most thorough and valuable contributions made to date.

It is my hope that *Treatise on Powder Metallurgy* will be a useful and comprehensive text for the student and technical worker in the many engineering industries related in one way or another to the subject. Care has been taken to base the presentation as much as possible on established facts—based either on published information or on research carried out by the author himself.

It is my further hope that the reader will not only become familiar with the present status of the art and its various technological, industrial, and theoretical aspects, but also become aware of the great potentialities of this special branch of metallurgical engineering and thus be in a position to develop it further.

The text has been subdivided into five parts, printed in three separate volumes. *Volume I* (twenty chapters) deals with the technology of powder metallurgy processes, including a brief description of the underlying principles and a historical review. The production, properties, and testing of powders, as well as their conditioning for powder metallurgy operations, are discussed and a survey of the currently available commercial grades of powders is made. This feature is believed unique—and necessary—for the beginner entering the field.

Aware of the controversial nature of the subject, the author has tried to give a fair appraisal of the most commonly used powders, mills, presses, and furnaces on the basis of personal experience and of catalogs and advertisements by the various producers, with no intention of criticizing or favoring any particular product. Should a particular brand or a particular manufacturer have been omitted, this has been done unintentionally and may be attributed solely to the rapidly changing commercial picture.

The subject of pressing of powders is given considerable space, and coverage of both the theoretical and the practical aspects is attempted. It may be considered most fortunate that, just before closing the manuscript, two valuable scientific contributions in the field of pressing became available: the work of Hermann Unckel of Finspång, Sweden, published in one of the last issues of the German periodical, *Archiv für Eisenhüttenwesen,* before the end of the Second World War; and the work of John Wulff and collaborators of the Massachusetts Institute of Technology, presented at the 75th Anniversary Session of the American Institute of Mining and Metallurgical Engineers, March, 1947.

Pressing at elevated temperatures is intentionally treated before a discussion of sintering because the view has been taken that hot-pressing is merely a mechanism parallel to ordinary molding. It is hoped that the detailed description of the various hot-pressing apparatus and techniques used for experimental purposes during the last two decades will serve to stimulate further development.

The process of sintering is covered at great length in a theoretical review in Chapter XIV of Volume I and is also summarized at the end of Part Two of Volume II. The practical aspects of the sintering operation are separated from the theoretical discourses in individual chapters on operations, furnaces, and atmospheres.

The technological review constituting Volume I is concluded with the individual treatment of such operations as working, heat treating, and finishing. Finishing is believed to represent another unique feature, since it refers to a subject not yet treated in any other powder metallurgy publication. Although it belongs, strictly speaking, to the border region of powder metallurgy, much space is devoted to this highly important practical subject.

Volume II consists of two parts, the first dealing exclusively with applied powder metallurgy, and the second with some theoretical aspects. The various fields of industrial application of powder metallurgy products are presented in ten chapters. Here, the author has gone considerably beyond the contemporary work of Kieffer and Hotop and of Schwarzkopf. The chapter on ferrous powder metallurgy products has been developed to a large extent on the basis of the author's own experience.

The final chapter of Part One covers briefly the many uses for metal powders that are somewhat beyond the sphere of interest of the powder metallurgist, though of general interest. In writing a treatise on this subject one is faced with the difficulty of limiting the interpretation of the term. There are a number of bordering fields that can be included if the widest sense of the term is accepted or that can be excluded if the field proper as we know it today is surveyed. Since this text is already extremely voluminous in its presentation of the topics directly connected with the field, the bordering subjects are merely indicated. For instance, the subject of paints and pigments is only mentioned briefly despite the fact that by far the largest quantities of metal in powder form, such as aluminum or brass flake powders, are used in the pigment industry. The same holds true for powder uses in the field of pyrotechnics and explosives, which during the war overtook the paint industry in its consumption of light metal powders.

The material of Part Two of Volume II is presented under the broad heading of "Physical Powder Metallurgy" in lieu of a more suitable term. The subjects treated here do not constitute a complete survey of all the important phases of the physical metallurgy of the metal powders, their processing, and their products. Instead, the intention is to present a collection of specific, but loosely connected, subjects that appear to the author either to be helpful to the reader in his attempt to obtain an impression of the applicability of powder metallurgy to his particular problem, or to throw light on some of the fundamental principles of the process. In this collection of individual topics are incorporated: a survey of new alloy combinations—according to the periodic system—for potential applications; a critical discussion of the testing methods as applicable to powder metallurgy products; an analysis of the internal stresses prevailing in porous bodies; and a summary of the basic theories of bonding and sintering. The idea of the survey of the periodic system has been adopted from the work by Kieffer and Hotop, but the survey has been considerably broadened. The chapter on stress analysis has been largely based on the concepts expressed by the originator of the theory, Paul Schwarzkopf, in his contemporary book.

Perhaps the chief purpose in presenting this second part of Volume II with its somewhat unsystematic collection of topics of specific interest, has been to provide a basis of thought, discussion, and criticism, out of which earnest theoretical study and scientific research might grow, so that a true *science* of powder metallurgy may develop that will promote the field in its entirety.

A classified and annotated literature and patent reference collection, which may serve as a quick reference file for initial information pertaining to any of the important phases of the subject, is presented in the two parts of *Volume III*.

ACKNOWLEDGMENTS

The task of writing this book would hardly have been possible without the teachings and the immeasurable aid of a number of persons. It is unfortunately not possible to mention them all, but I would like to take this opportunity to express to them my sincere appreciation and thanks.

It appears appropriate to begin this acknowledgment in memory of my first teacher in powder metallurgy, the late Charles Hardy, who succeeded in awakening in me a deep and lasting interest in the field, and in laying the foundation for my work and endeavors during the past dozen years.

My most sincere appreciation is expressed to Dr. Paul Schwarzkopf for his kindness in placing at my disposal data, drawings, and photographs from his own publications and from his company files, for his advice and generous assistance in the preparation of drawings and in reviewing the manuscript, and lastly, for the many years of fruitful experiences during my association with him and with his firm.

In the preparation of the first part of this book, I have had the closest cooperation of Dr. Werner Leszynski, editor of the *Powder Metallurgy Bulletin*, who has been instrumental in reshaping the text and who has subjected it to critical editorial review. My appreciation is extended also to Mr. Warren B. Austin for reading the first part.

Others who contributed their efforts in bringing various phases of the work to completion include: Dr. John T. Norton, Dr. Robert Steinitz, Richard P. Seelig, Wilson N. Pratt, Dr. Myron Coler, and Dr. Werner H. Simon.

I should like also to acknowledge the help and encouragement received from many of my friends and former associates in the American Electro Metal Corporation: George Stern for reading the chapters on powders; Mrs. Helen Friedemann for typing the manuscript; and Miss Louise Toppi for aid in the photographic work.

I should like to thank Messrs. Joseph Bednar and Frank J. Loeffler, and Miss Elsa Kaufmann, for their relentless effort in preparing the drawings and photographs.

My sincere thanks go to my associates in the Sintercast Corporation of America for their encouragement and decisive aid in the final phases of the writing and organization of the book: Messrs. Erwin Loewy, John L. Ellis, Hans Eyck, and the Misses Gabriele Reichmann and Rita T. Trask.

A special acknowledgment is due to Interscience Publishers, Inc., especially to Mr. Allen Kent, for their assistance, and their indulgence and never-ending willingness to encourage and cooperate through the genesis and many years of tribulation of this book.

My everlasting gratitude goes to my family for their forbearance, and to my dear wife in particular for her material aid in collecting and classifying the references presented in Volume III.

C. G. G.

New York
December, 1948

CONTENTS

TERMINOLOGY

A glossary of the most commonly used terms in powder metallurgy is presented in this section both as an aid to the novice and as a guide to the general reader. The terminology is based chiefly on a set of standardized definitions released March 1, 1946, by the Powder Metallurgy Committee of the American Society for Metals, and is enlarged by a number of terms and interpretations used in the industry or literature. The standardized definitions are denoted by an asterisk (*).

Acicular powder. Needle-shaped particles.

Agglomerate. An assembly of powder particles of one or more constituents knitted closely together by settling from a solution or suspension, or by pressing or heating to subsintering temperatures.

Air classification. The separation of powder into particle size ranges by means of a controlled air stream.

Alloy powder. A powder, each particle of which is composed of the same alloy of two or more metals.

Amorphous powder. A powder which substantially consists of particles which are noncrystalline in character.

Apparent density. The weight of a unit volume of powder, usually expressed in grams per cubic centimeter, determined by a special method of loading.

Arborescent powder. See *Dendritic powder.*

Atomization. The dispersion of a molten metal into particles by a rapidly moving gas or liquid stream.

Ball milling. Grinding or mixing in a receptacle of rotational symmetry that contains balls of a metal or nonmetal harder than the material to be milled.

Batch sintering. Pre-sintering or sintering in such a manner that the products are furnace-treated in individual batches.

Binder. A cementing medium: either a substance which is added to the powder to increase the strength of the compact and is dispelled during sintering, or a substance (usually a relatively low-melting constituent) added to a powder mixture for the specific purpose of cementing powder particles which, alone, would not sinter into a strong body.

Blank. A pressed, presintered or fully sintered compact, usually in the unfinished condition and requiring cutting, machining, or some other operation to give it its final shape.

Blending. The thorough intermingling of different batches of powder of the same substance for the purpose of adjusting the physical characteristics.

Bridging. The formation of arched cavities in a powder mass which may result in voids or uneven density in a compact, or which may result in stoppage or interruption of flow of powder through a funnel or other feeding device.

Briquette. See *Compact.*

Cake. A coalesced mass of unpressed metal powder.

Carbonyl powder. Particles prepared by the thermal decomposition of a metal carbonyl.

Chemical deposition. The precipitation of one metal from a solution of its salt by the addition of another metal or reagent to the solution.

Classification. Separation of a powder into fractions according to particle size.

Cloth. Metallic or nonmetallic fabric used for screening or classifying powders.

Coated particles. Metal particles whose surfaces have been covered with another metal.

Coining. See *Sizing.*

Cold pressing. The forming of a compact at room temperature.

Comminution. See *Pulverization.*

Compact. An object produced by the compression of metal powders with or without the inclusion of nonmetallic constituents; synonymous with *Briquette.*

Compactibility. The ability of a powder to be formed into a compact of well-defined contours and structural stability at a given pressure; a measure of the plasticity of the powder particles.

Composite compact. A metal powder compact consisting of two or more adhering layers of different metals or alloys, with each layer retaining its original identity.

Composite structure. A composition consisting of a physical mixture of two or more metals, or one or more metals with one or more nonmetals, all in finely divided form, whose inability to form alloys or compounds results in retention of the properties of the individual components; synonymous with *Compound compact.*

Compound compact. A metal powder compact consisting of mixed metals, the particles of which are joined by pressing and/or sintering, with each metal particle retaining substantially its original composition.

Compressibility. The ability of a powder to be formed into a compact of a predetermined, small volume at a given pressure.

Compression ratio. The ratio of the volume of the loose powder to the volume of the compact made from it by the application of a specific pressure in a specific die at a specific pressing speed.

Core rod. A member of a die used in producing a hole in a compact.

Cored bar. A compact of bar shape heated by its own electrical resistance to a temperature high enough to melt its interior.

Crystalline powder. A powder that consists substantially of particles having a crystalline structure; most metal powders are crystalline.

Cut. See *Fraction.*

Dendritic powder. Particles, usually of electrolytic origin, having a typical "pine tree" structure; synonymous with *Arborescent powder.*

Density ratio. The determined density of a green or sintered compact, divided by the absolute density of metal of the same composition; usually expressed as a percentage.

*Die. The part or parts making up the confining form in which a powder is pressed. The parts of the die may include some or all of the following: die body, punches, and core rods. Synonymous with *Mold.*

Die assembly. See *Die set.*

Die barrel. A cylindrical liner for a die.

*Die body. The stationary or fixed part of a die.

Die cavity. The regular or irregular opening of the die into which the powder is fed, and in which the powder is compacted.

*Die insert. A removable liner or part of a die body or punch.

*Die set. The parts of a press that hold and locate the die in proper relation to the punches.

*Disintegration. The reduction of massive material to powder.

*Electrolytic deposition. The production of a metal from a solution of its salts by the passage of an electric current through the solution.

Expansion. Synonymous with *Growth.*

Feed hopper. A container used in storing the powder prior to compacting in a press.

Feed shoe. A member of a press connecting the feed hopper and the die cavity which has a channel or hole through which the powder flows by gravity or vibration.

*Fines. The portion of a powder composed of particles which are smaller than a specified size, currently less than 44 microns.

*Flake powder. Flat or scalelike particles whose thickness is small compared with the other dimensions.

*Flow rate. The time required for a powder sample of standard weight to flow through an orifice in a standard instrument according to a specified procedure.

*Fraction. That portion of powder sample which lies between two stated particle sizes or two stated screen-mesh sizes; synonymous with *Cut.*

Fritting. Sintering of a compact in the absence of a liquid phase.

*Galling. The impairment of the surface of a compact or of die parts caused by adhesion between the die wall and the metal powder; synonymous with *Seizing.*

*Gaseous reduction. The conversion of metal compounds to metallic particles by the use of a reducing gas.

*Globular powder. Particles having approximately spherical shape.

*Granular powder. Particles having approximately equidimensional, nonspherical shapes.

*Granulation. The production of coarse metal particles by pouring the molten metal through a screen into water, or by violent agitation of the molten metal while solidifying.

*Green. Unsintered.

Green density. The determined density of a green compact; usually expressed in grams per cubic centimeter.

Green strength. The ability of a powder to form compacts of sharply out-lined contours at a given pressure, and to maintain projections, angles, corners, edges, undercuts, or curved faces during storing or handling prior to the presintering or sintering operation.

*Growth. The increase in dimensions of a compact which may occur during the sintering operation.

High sintering. Synonymous with advanced sintering at high temperatures; usually, the final sintering close to the melting point of the material.

*Hot pressing. The simultaneous forming and heating of a compact.

*Hydrogen loss. The loss in weight of a metal powder or of a compact caused by heating a representative sample for a specified time and temperature in a purified hydrogen atmosphere—broadly, a measure of the oxygen content of the sample when applied to materials containing only such oxides as are reducible with hydrogen and no other hydride-forming element.

Impregnation. The process of filling the pores of a sintered compact, usually with a liquid, such as a lubricant; also the process of filling the pores of a sintered or unsintered compact with a metal or alloy of lower melting point. Also the process of mixing particles of a nonmetallic substance in a powder metal matrix as in diamond-impregnated tools. Also the process of coating a sintered compact with another metal by burying and heating the compact in a powder of the second metal.

Infiltration. The process of filling the pores of a sintered or unsintered compact with a metal or alloy of lower melting point by capillary action; usually accomplished by submersion of the compact in a liquid metal bath.

Interlocking particles. Mechanical intertwining of particles.

Interlocking porosity. A system of intercommunicating voids in a sintered article; synonymous with *Intercommunicating porosity.*

*Intercommunicating porosity. That type of porosity in a sintered compact in which the pores are connected so that a fluid may pass from one to the other, completely through the compact.

*Irregular powder. Particles lacking symmetry.

Knockout. Ejecting of a compact from a die cavity.

Knockout punch. A punch used for ejecting compacts.

Liquid disintegration. The dispersion of a stream of molten metal or alloy into particles by means of a gas blast, a liquid jet, or a set of mechanical knives, either one of which crosses the metal stream while revolving at very high speed.

*Loading. The filling of the die cavity with powder.

Loading sheet. The part of a die set used as container for a predetermined amount of powder used for feeding into the die cavity.

Loading weight. Synonymous with *Apparent density.*

*Lower punch. The lower member of a die which forms the bottom of the die cavity. It may or may not move in relation to the die body.

*Lubricating. Mixing with or incorporating in a powder some agent to facilitate pressing and ejecting the compact from the die body; also, applying a lubricant to the die walls and/or punch surface.

Matrix metal. The continuous phase of a polyphase alloy or mechanical mixture; the physically continuous metallic constituent in which separate particles of another constituent are embedded.

Mesh. The screen number of the finest screen of a specified standard screen scale through which substantially all of the particles of a powder sample will pass. Frequently, but not necessarily, a part of the sample may pass finer screens; also called "mesh size."

Micron. A linear distance of 0.001 mm.

Milling. The mechanical tratment of material—as in a ball mill—to produce particles, alter their size or shape, or to coat one component of a powder mixture with another.

Minus mesh. The portion of a powder sample which passes through a screen of stated size.

Mixing. The thorough intermingling of powders of two or more substances.

Mold. See *Die.*

Needles. Elongated, rodlike particles.

Nodular powder. Irregular particles having knotted, rounded, or similar shapes.

Oversize powder. Particles coarser than the maximum permitted by a given particle size specification.

Packing material. Any material in which compacts are embedded during the presintering or sintering operations.

Particle. A minute part of matter; a metal powder particle may consist of one crystallite or more.

Particle size. The controlling linear dimension of an individual particle as determined by analysis with screens or other suitable instruments.

Particle size distribution. The weight percentages of the fractions into which a powder sample has been classified according to particle size.

Plates. Flat powder particles having considerable thickness. (See also *Flake powder.*)

Plus mesh. The portion of a powder sample retained on a screen of stated size.

Pore. A minute cavity in a compact; synonymous with *Void.*

Porosity. A multiplicity of pores in a compact.

Pore-forming material. A substance included in a powder mixture which volatilizes during sintering and thereby produces a desired kind of porosity in the finished compact.

Powder. Particles of matter characterized by small size, usually within the range 1 to 1000 microns.

Powder metallurgy. The art of producing metal powders and objects shaped from individual, mixed, or alloyed metal powders, with or without the inclusion of nonmetallic constituents, by pressing or molding objects which may be simultaneously or subsequently heated to produce a coherent mass, either without fusion, or with the fusion of a low-melting constituent only.

Preforming. The initial pressing of a metal powder to form a compact which is subjected to a subsequent pressing operation other than coining or sizing; also, the preliminary shaping of a refractory metal compact after presintering and before the final sintering.

Presintering. The heating of a compact at a temperature below the normal final sintering temperature, usually to increase the ease of handling or forming the compact or to remove a lubricant or binder prior to sintering.

Press. A machine used to form or size a compact by pressure.

Pressed density. The weight per unit volume of an unsintered compact; sometimes called "green density."

Pressing. The application of pressure to a powder or compact.

Pressing crack. See *Slip crack*.

Puffed compact. A compact expanded by internal gas pressure.

Pulverization. The reduction of matter to powder by mechanical means; a specified type of disintegration; synonymous with *Comminution*.

Punch. Part of a die assembly which is used to transmit pressure to the powder in the die cavity.

Q-value. A "figure of merit" commonly used to designate the ratio of the inductive reactance to the effective resistance of electrical cores made from iron and other powders, determined by means of a Q-meter according to a specified procedure.

Rate of oil flow. The rate at which a specified oil will pass through a sintered porous compact under specified test conditions.

Relative density. Synonymous with *Density ratio*.

Relative volume. The determined volume of a green or sintered compact divided by the volume of the massive metal of the same composition for the identical weight.

Repressing. The pressing of a previously pressed and sintered compact for the purpose of altering physical properties or size.

Resintering. Sintering of a repressed compact under conditions identical or different from those prevailing at the first sintering operation.

Roll compacting. The progressive compacting of metal powders by the use of a rolling mill.

Rotary press. A machine fitted with a rotating table carrying multiple dies in which a material is pressed.

Screen analysis. Particle size distribution expressed in terms of the weight percentage retained upon each of a series of standard screens of descending mesh size and the percentage passed by the screen of finest mesh.

Screening. Separation of powder according to particle size, by passing it over cloth having the desired mesh size.

Screen classification. The separation of powder into particle size ranges by the use of a series of graded sieves.

Segment die. A die made of parts which can be separated for the ready removal of the compact; synonymous with *Split die*.

Seizing. See *Galling*.

Shrinkage. The decrease in the dimensions of a compact which may occur during sintering.

Sintering. The bonding of adjacent surfaces of particles in a mass of powder or a compact by heating.

Sintering treatment. The heat treatment of a mass of powder or a compact to produce the bonding of the adjacent metal surfaces, usually under protective atmosphere conditions; also, the formation of alloy powders from mixtures of the individual powders by interdiffusion below the melting point of the alloy so formed.

Size fraction. A separated fraction of a powder whose particles lie between specified upper and lower size limits.

*Sizing. A final pressing of a sintered compact to secure desired size; synonymous with *Coining.*

Sizing die. A die used for the sizing or coining of a sintered compact.

Sizing knockout. An ejector punch used for pressing or ejecting a sintered compact during the sizing operation.

Sizing punch. A punch used for pressing of the sintered compact during the sizing operation.

Sizing stripper. A stripper punch used during the sizing operation.

*Slip crack. A rupture in the pressed compact caused by the mass slippage of a part of the compact; synonymous with *Pressing crack.*

Specific pressure. The pressure applied to a powder or sintered compact of unit cross-sectional area.

*Specific surface. The surface area of the particles in one gram of powder, usually expressed in square centimeters.

*Split die. See *Segment die.*

*Sponge iron. The material produced by the reduction of iron with carbon, without melting.

*Spongy. A porous condition in powder particles usually observed in reduced oxides.

Standard 200-mesh screen. A woven-wire screen having square openings 74 microns across and 200 wires per linear inch.

Standard sizing scale. A scale of sizes based on the standard 200-mesh screen forming a series in which the ratio of consecutive sizes is $\sqrt{2}$.

*Stripper punch. A punch which, in addition to forming the top or bottom of the die cavity, later moves further into the die to eject the compact.

*Subsieve fraction. Particles all of which will pass through a 325-mesh screen.

*Superfines. The portion of a powder composed of particles which are smaller than a specified size, currently less than 10 microns.

*Sweating. The appearance on the surface of a compact of some or all of the low-melting constituents.

Swelling. Synonymous with *Growth* or *Expansion.*

*Tap density. The apparent density of a powder obtained when the volume receptacle is tapped or vibrated during loading under specified conditions.

Thermal sintering. Molding by sintering without prior compaction.

Ultrafine powder. A powder composed of particles which are 2 microns or smaller in size.

*Upper punch. The member of a die assembly which moves downward into the die body to transmit pressure to the powder contained in the die cavity.

*Void. See *Pore.*

CHAPTER I

Principles

THE POWDER METALLURGY PROCESS

Definition

Powder metallurgy has been defined[1] as the "art and science of manufacturing useful articles from metal powders and of producing those powders." Such articles may be in the form of ingots, semifinished, or finished shapes. The technique employed to produce them consists essentially of subjecting the metal powders to pressure and heat, the heat treatment being performed at some temperature below the fusion point of the products or at least of their main constituents. The metal powder can, in some instances, be replaced by an oxide powder or the powdered form of another metallic compound; the mixture of nonmetallic components with metal powders is also practicable. Powder metallurgy, thus, permits the production of metallic or metal-like bodies of simple or complicated shapes without the use of orthodox metallurgy practices, such as melting, casting, ingot solidification, metal working, and shaping by machine operation.

Technique

In its most common form, the technique consists of pressing the powders to form coherent masses and heating (sintering) the resulting compacts. The pressing operation may be carried out at ordinary or elevated temperatures. In the latter case, it is, under certain conditions, possible to eliminate the subsequent heating operation. The application of heat during or after the compression of the powder into a compact produces a structure with physical characteristics comparable to those of similar materials obtained by fusion. Although these characteristics are frequently adequate for the finished product, it is often necessary to improve the properties by further operations, *e.g.*, additional compression and heat treatments. The additional compression may again be conducted at room or elevated temperatures, and the heat treatment must be carried out at

[1] Report of the Powder Metallurgy Committee, *Inst. Metals Div., Am. Inst. Mining Met. Engrs.*, Oct. 17, 1945.

such a temperature that the shape of the body is not altered by fusion of a major part of the material. Frequently, a supplementary heat treatment serves merely to increase the ductility or hardness. An annealing treatment often has remarkable results, especially if the body has been work-hardened by the additional compression operation.

Quite distinct from the foregoing is a special technique which combines powder metallurgy with fusion practices, and often results in remarkable and unusual products. Following the pressing of the powder into a porous skeletonlike compact, the latter is heated and brought into contact with a liquefied lower melting metal which, in turn, becomes absorbed by the pore system of the skeleton body with the aid of capillary forces. In this manner all original pores may be completely filled by the liquid metal, and solid and sound structures may be obtained upon cooling. Depending on the composition and constitution of the resulting composite structure, products of very high strength and hardness or of appreciable ductility and impact resistance are obtainable.

Many metals or alloys produced by the powder technique are just as readily workable as their fused and cast counterparts; there are even technically important materials which can be worked from ingots to sheets or wires only by this technique. Hot working of heat-treated powder compacts is usually more practicable, and forging, hot hammering, hot swaging, extruding, and hot rolling or drawing of ingots made from powders are common practice in a number of applications. There are, however, instances when cold working, especially sheet rolling and wire drawing, of powder metals is feasible.

It is thus apparent that powder metallurgy covers a wide variety of processing methods. It impinges upon the field of plastics, since a number of applications call for products in which metallic particles with certain characteristics are embedded in a plastic matrix. The techniques, moreover, are similar in many respects to those employed in the manufacture of ceramics; combinations of ceramics with metal powders are also in practical use. Methods and equipment that have been accepted in the plastics and ceramics industries have also been used in powder metallurgy, and experience gained in these allied fields has been most valuable in solving many engineering problems.

Powder metallurgy techniques can be divided into two broad sections: (1) preparation and conditioning of the powder, and (2) processing of the powder.

POWDER SOURCES, PREPARATION, AND CONDITIONING

Metals in powder form can be obtained from ores, from salts and other compounds, or from bulk metals or alloys. The methods of production include crushing, milling, machining, graining, atomizing, evaporation, condensation, reduction, precipitation, displacement, electrodeposition, alloying, and alloy disintegration.

Powders produced by any one of these methods are prepared for processing by proper conditioning and by certain advance treatments. They are cleaned, dry stored, and tested; frequently they are mixed with addition agents to facilitate pressing. Powders for alloys or composite structures are prepared by blending desired proportions of the ingredients. This blending operation is carried out under various conditions, with dry or wet powders; in certain cases blending is combined with ball milling. Frequently, precautions must be taken that the powders do not deteriorate mechanically or chemically during the blending operation, in order to retain the desired qualities.

PROCESSING

Metal powders can be processed in any of the following ways: (a) by application of pressure alone, $e.g.$, in the production of briquettes for blast or cupola furnaces or in the manufacture of magnetic cores; (b) by application of heat alone, $e.g.$, in the production of very porous articles or crude ingots which are to be worked subsequently; (c) by simultaneous application of pressure and heat, $e.g.$, in the manufacture of special hard metals; and (d) by application of pressure and heat in two successive operations, the procedure most frequently employed in the manufacture of articles by the powder technique. Precautions must be taken against mechanical damage during molding and heating and, by the use of controlled atmospheres, against chemical contamination (especially from the reaction with oxygen or water vapor).

ADDITIONAL OPERATIONS

Supplementary operations which are used include: (a) cold-working the compact after the first application of heat,—by a second press operation, known as coining or repressing, or by any of the common cold-working techniques; (b) additional heat treatments, usually after cold deformation of the compact is completed—these may be analogous to the first sintering, or may represent an annealing, hardening, or precipitation treatment; and (c) hot-working the compact after the first application of heat—by a second press operation, but at elevated temperatures, or by any of the customary hot-working techniques. Any of these supplementary operations may be repeated if the characteristics desired for the end product require repetition.

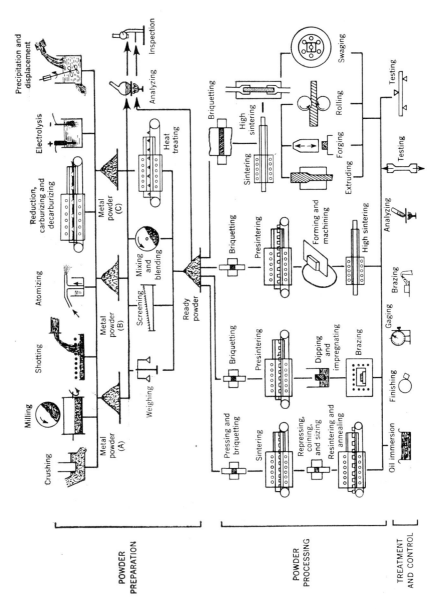

Fig. 1. Flow chart of the different powder metallurgy techniques.

MODIFICATIONS

There are also a number of modifications of the original processing steps in use, of which the most important are: (*1*) Interruption of the heat treatment at some intermediate temperature. The first heating, often referred to as presintering, may yield compacts having certain transitory properties, such as softness, good machinability, etc., which permit the use of operations not feasible after sintering is completed. (*2*) Division of the heat treatment into two or more stages at different temperatures to facilitate closer control of the dimensions or properties of the product. (*3*) Addition of a second constituent in the form of a liquid instead of as a powder. This method may result in more uniform products, but is applicable only if the second constituent has a melting point considerably below that of the main constituent and can infiltrate, *e.g.*, by dipping, into the porous structure of the sintered compact.

FINISHING OPERATIONS

Listing of the most important processing steps would not be complete without reference to the finishing operations required for most products made from metal powders. These operations include machining undercuts or re-entrant angles, as well as plating, buffing, sanding, grinding, or barrel tumbling of flat faces to remove fins and burs. A number of applications require joining operations; brazing, soldering, or welding of these parts onto other metal bases is common practice in the hard-metal and refractory-metal base contact industries. Finishing and joining operations must frequently be adapted to the specific properties of these articles: thus, care must be taken in machining because of the marked porosity of most sintered metals; plating methods must be adjusted to account for easy corrosiveness; and special fluxes or inert or reducing gaseous media must be used for brazing or welding at high temperatures in order to prevent excessive oxidation.

A flow chart representing some of the most common techniques used at present in the powder metallurgy field is shown in Figure 1. The powder preparation section indicates the most important industrial methods of powder manufacture, as well as methods of conditioning and testing raw material. The powder processing section illustrates the four principal methods used today: in the first column, the customary procedure is given for production of porous or dense parts from ferrous and nonferrous metals and alloys; in the second, a procedure used in the manufacture of refractory-metals base contacts; in the third, production methods used in the hard-metal field; in the fourth, refractory-metal manufacturing prac-

6 PRINCIPLES OF POWDER METALLURGY

tices. Final operations, such as oil impregnation, finishing, joining, and testing, can refer to all processing methods.

INDUSTRIAL APPLICATIONS

The use of metal powders as starting material can be considered for any of the following applications.

HIGH-MELTING METALS

The refractory metals have such high melting points that conventional melting and casting is impossible. The powder process has provided the first practical solution to the problem of producing tungsten, molybdenum, and other metals of the same group. By elimination of the coarse grain size inherent in fused metals and of contaminations originating from crucibles and molds, pure metals with extraordinary properties have been developed with the aid of special chemical equipment, milling and grinding apparatus, powerful presses, high-amperage electric furnaces, and a variety of metal-working machines.

COMPOSITE METALS

Metal combinations in which the characteristics of each constituent are retained are of particular interest for certain electrical applications. Heavy-duty contacts and welding electrodes combine a skeleton of refractory metal, highly resistant to abrasion and arcing, with a second low-melting metal of high conductivity. Alloying between the constituents is negligible, so that the original properties of the individual metals are preserved in the finished article, while surface bonding between the constituents is sufficient to give considerable strength to the material. Similar composite materials are employed for balancing parts, radium containers, or other products requiring especially high specific gravities. Composite metals offer a wide range of other engineering applications when special combinations of properties, such as hardness and wear resistance, are desired. For instance, composite structures of iron and steels permeated with copper and copper alloys have lately found applications as highly stressed machine and engine components.

METAL NONMETAL COMBINATIONS

The incorpation of nonmetallic constituents in a metallic matrix is one of the most intriguing applications of the powder metallurgy method which cannot be achieved by conventional engineering practices. The manufacture of cemented carbide high-speed cutting tools—the most striking example—has grown into a vital industry. Cobalt is used to bond the very hard particles of tungsten carbide into a coherent article with substantially metallic characteristics, but which is notable for hardness, strength, and excellent cutting properties. Diamond-impregnated grinding wheels, drill core bits, and dressing tools are of similar composite structure; they consist of diamonds embedded in cemented carbides or in more plastic metals or alloys. Friction elements, such as clutch facings and brake linings, are characterized by their content of abrasives (*e.g.*, emery or silica) embedded in metal, usually bronze. Metal–graphite combinations find widespread applications in copper current-collector brushes and in porous bronze and iron bearings. Combinations of metal powders with plastics are used in the manufacture of magnetic articles.

POROUS METALS

The development of metallic bodies of closely controlled porosity has led to an entirely new industry, engaged in making bearings, gears, and filters. Porous bearings impregnated with oil are self-lubricating and are therefore used in places which are inaccessible to outside lubrication. The mechanism of self-lubrication depends on the interconnection of the pores. Similarly, porous ferrous metal is used effectively for oil-pump gears where antifriction qualities and accurate contours of the teeth are more essential than strength. Metal filters and diaphragms compare favorably with their ceramic counterparts, particularly when heat or mechanical shock is encountered.

LAMINATED PRODUCTS

The previously mentioned current-collector brushes belong in this class. By the use of flake copper and flake graphite powders, articles of distinctly lamellar structure can be pressed. High electrical conductivity is thus obtained in the direction parallel to these laminations. Bimetals, which are modifications of this type of structure, consisting of only a few layers, such as one of iron and another of nickel, are used fo rthermostats.

ALLOYS OF UNUSUAL COMPOSITION

Closely related to the production of composite materials is the manufacture of alloys from metals in proportions which are not miscible in the liquid or solid state. The resulting articles can be of solid structure, with the second constituent more uniformly dispersed in the matrix than is possible from centrifugal casting or other metallurgical methods. Copper–lead bearing compositions are an outstanding example. Metals having widely different fusion temperatures can be alloyed by low-temperature sintering without contamination by addition agents which are necessary to facilitate the melting or casting process.

METALS OF HIGH PURITY

Elimination of these addition agents and prevention of reactions with gases and refractory linings lead to very closely reproducible analyses of the powdered·metal products, and make possible, through careful selection of the initial materials and close control of processing conditions, the manufacture of very pure metals, e.g., nickel, iron, precious metals, etc.

PRODUCTION OF PARTS

Since many metallic parts can be formed by powder metallurgy methods to finished dimensions, expensive and time-consuming machining operations are completely eliminated or greatly reduced. Technically dense parts of iron, brass, or bronze, sometimes requiring as many as twenty machining operations, can be produced efficiently by molding, sintering, coining, and a few finishing and cleaning operations. Savings in time and in reduced scrap metal losses and liberation of vitally needed tool machine facilities have in several instances been the deciding factors in use of the powder method. Molding of small gears, levers, cores, cams, sprockets, and other parts from iron and steel powders has become one of the most interesting and promising applications of the art. Pressing and sintering of small permanent magnets containing aluminum, cobalt, nickel, and iron has advantages over the casting process; machining, necessary to obtain smooth finished surfaces on the hard cast products, is very difficult and can be accomplished only by tedious grinding, while the powder products can be readily molded to accurate size. In addition, a finer grain structure and greater strength is obtained in the sintered magnets.

OTHER APPLICATIONS

Chemical reagents and catalysts containing aluminum, copper, iron, nickel, zinc, or other metal powders are widely used in the chemical and metallurgical industries. Aluminum and magnesium powders are used in explosives and pyrotechnics, and the former is irreplaceable in Thermit welding because of its powerful reducing qualities and easy oxidation by heat. Tin and lead powders are useful as solders, and copper and silver powders can be used as effective brazing agents, sometimes applied with the aid of a liquid carrier or flux. Copper, aluminum, and sponge iron have been added to cements and ceramics to control porosity, weight, thermal properties, or moisture resistance. Metal powders applied to the surface of nonconducting materials may render them satisfactory for electroplating, or may produce metallic appearance, greater strength, or higher conductivity in plastic articles. Other applications include certain dental alloys, amalgams, and a great variety of ornamental objects, *e.g.*, coins, medallions, and dies for engraving.

COATINGS AND PAINTS

Steel can be coated by applying metal powders to the surfaces at elevated temperatures. Zinc coatings are made by the Sherardizing process, and aluminum coatings by calorizing. Diffusion produces an intimate bond between the coating and the base metal. Calorized and Sherardized steels are used where corrosive conditions prevail. Coatings can also be obtained by spraying fused low-melting metal powders and allowing them to solidify on the surface of the base metal; zinc coatings are made extensively by this method.

The use of metal powders as paint pigments is probably the largest single use; the "bronze" powders are the best known. These copper, copper-alloy, aluminum, or nickel bronze powders have flakelike particle structures which distinguish them from the more equiaxed powders generally used in powder metallurgy. The flake form is desirable because it creates metallic luster, due to light reflection from the flat polished faces, and yields a large surface per unit weight. These characteristics are accompanied by good leafing properties which cause the flakes to arrange themselves in the liquid varnish in planes parallel to the coated surface.

CRITICAL EVALUATION

We have now reached the point at which we may compare the results achieved by applying powder metallurgy techniques with those obtained by applying orthodox metallurgical methods. Such a comparison will provide a balance sheet enumerating advantages and limitations, both technical and economical, and may serve as a guide for the engineer and manufacturer confronted with the problem of selecting the most suitable production method for any particular job.

Advantages

For a number of materials and applications, the metal powder technique has turned out to be the only practical manufacturing method.

Refractory metals and cemented carbides are examples of materials whose high melting points prohibit the application of fusion methods. If objects of unusual structure, composition, or properties are desired, the orthodox methods of metal production may be inadequate. Fusion and casting processes have not yet produced metal structures containing a pattern of *uniform* dispersal of pores or of foreign, nonmetallic matter. Even the most modern methods of centrifugal casting have failed to produce metallic bodies in which two or more metals, insoluble in each other in the liquid and solid states (such as copper and lead), are combined in a structure of uniformly distributed constituents. Such structures can be obtained only by the use of powder metallurgy technique.

Adjustments or changes in the components, control of the pressing operation and selection of the proper heating cycle have resulted in the development of many engineering products which cannot be produced by other methods. Typical examples are cemented carbide tools, refractory-metal base contacts, self-lubricating bearings, porous filters, friction clutches, or diamond-impregnated tools. In recent years, another application of powder metallurgy has achieved considerable practical importance. The production of metallic parts to exact dimensions has been made possible by molding the powders in accurate dies and controlling the shapes of the compacts during heat-treatment. Complicated shapes have been formed in this manner, and time-consuming and expensive machining has become unnecessary, or at least been greatly reduced. At the same time, excessive scrap losses due to removal of surplus material have been eliminated. The ability to produce complicated shapes within very close tolerances by a process which is much simpler and faster than machining and which requires equipment that can be handled by less skilled workers, and lower heating temperatures than casting, has become one of the greatest assets of powder metallurgy—the chief reason, in fact, for its impressive expansion during the war.

In the foregoing review are listed, without claim to completeness, the most important technical applications of metal powders, including their uses as chemical reagents and catalysts, as coating materials and pigments —that is to say, uses that are beyond the scope of this book. According to the definition given at the beginning of this chapter, only the manufacture of the powder material for these applications is considered as part of the art and science of powder metallurgy. As far as the processing of powders into compact metallic products is concerned, it has been indicated that the advantages of the method are the achievement of results which are impossible with ordinary methods, or products of better quality or lower cost. Extensive research and development work will undoubtedly extend the applicability of powder metallurgy techniques and will, in a good many cases, lead to the conversion from orthodox manufacturing practices to the powder method. Accessibility of raw material may have a determining influence on the question of whether a particular article can be produced more efficiently by melting or by sintering. Large-scale production of technically dense parts of more or less complicated design from iron, brass, and other metal and alloy powders has particularly promising aspects.

Limitations

That powder metallurgy is, however, no panacea, and that its application is subject to certain principal limitations, will be shown in the following. According to Patch,[2] these limitations are based on three main factors: material, men, and machines. The limitations may be of a technical or an economic nature.

MATERIAL LIMITATIONS

Heading the list of these limitations is the difficulty of securing suitable powder material, for not every powder on the market is suitable for molding and sintering. It is of prime importance that the powder be available in satisfactory quality and quantities.

The characteristics of the powder determine, to a large extent, the properties of the finished product. The purity and density of the powder are of utmost importance in controlling the quality of the end products. It has been found experimentally that compacts of high density and great purity exhibit physical properties which are at least equal to those of fused metals. As a matter of fact, iron and steel compacts, as well as refractory metals, have been developed in the laboratory, for which the tensile strength, elongation, electrical conductivity, and other properties have exceeded normal values by as much as 25%. The results obtained by Balke[3] on iron and steel compacts demonstrate the importance of density and purity for the attainment of maximum physical characteristics. The presence of traces of oxygen lowered elongation values as much as 50%, ·and highest tensile values were obtained only if extremely high pressures insured maximum density values.

These optimum conditions have not as yet been realized on an industrial scale in the case of iron and other important materials for which the daily production is on a tonnage basis. Since the daily output of the purest iron powders, e.g., electrolytic, carbonyl, and especially refined hydrogen-reduced iron, falls far short of meeting these demands, extreme purity

[2] E. S. Patch, *Iron Age, 146,* No. 25, 31 (1940).
[3] C. W. Balke, in *Symposium on Powder Metallurgy,* ASTM, Philadelphia, Pa., 1943, p. 11.

must be sacrificed. Since very high molding pressures cause excessive die wear and require frequent replacement of pressing tools in mass production, extreme density must likewise be sacrificed. These sacrifices in purity and density result in inferior physical properties. In the case of gears from iron powder, for instance, the tensile strength usually does not exceed 30,000 psi, but it is, at least in some instances, possible to increase the tensile strength to about 50,000 psi by controlling the carbon content, by special pressing and sintering procedures (resulting in a density increase), or by subsequent operations. However, in this case, tool costs are increased and manufacturing operations become more expensive.

Another major limitation inherent in the powdered material lies in the fact that powders do not follow the laws of hydrodynamics when being compressed in a die. Thus it is impossible with simple molding and press tools, to press parts with re-entrant angles, threads, steep bevels, or lateral protrusions.

Certain economic limitations can also be connected with the powdered raw material. Especially in the highly competitive field of parts production, the cost of the powder plays an important role. Although powder cost is usually insignificant in the case of production of very large quantities of small-sized parts, it becomes significant for parts weighing more than an ounce. When the weight per unit reaches one-half or one pound the cost becomes one of the most predominant economic factors. Of course, there are other points that must be considered in conjunction with the powder cost, such as the amount of machining that can be saved, *i.e.*, the ratio of the weights of the finished part and the rough blank weight of the cast iron, steel or brass stock, and the number of pieces involved. The more complicated the machining of the piece by the conventional method, and the larger the quantity of parts to be produced, the less significant become the raw material costs as well as the large expenses for equipment and tools for the powder process.

MANPOWER LIMITATIONS

The importance of proper education of the engineer, designer, and draftsman must not be underestimated. There are a large number of parts used in industry which can be produced efficiently from powders, but which have so far been neglected for such simple reasons as lack of knowledge, experience, or foresight. Only in recent years have several major engineering colleges instituted courses in which students are made familiar with powder metallurgy techniques and applications. At the same time, many engineers have learned that the design can be modified in many instances in such a manner as to allow the use of powdered metal articles of lower strength, ductility, and toughness without impairing the usefulness of the article. Of course it must be understood that a progressive mind and some imagination, as well as the desire to see the job through, are necessary on the part of the manufacturer for the successful conversion of a conventional manufacturing process. Frequently initial reverses, time-consuming development work, and, particularly, rigid specifications as to mechanical properties and dimensional tolerances cause discouraging results and bring discredit on the art. The situation is not improved by well-meaning enthusiasts who propagate exaggerated claims of extraordinary and sometimes almost incredible laboratory results which fail to materialize on the production line.

Because of the efforts of many engineering schools and technical societies, much information has been presented in recent publications which may serve as the basis for more extensive education of engineers and metallurgists. Studies dealing with theoretical and practical problems made by scientific institutions and by industrial laboratories have formed the basis for realistic and fruitful expansion. It is therefore probable that limitations imposed by lack of understanding of what to expect and what not to expect from powder metallurgy will be overcome in the future.

EQUIPMENT LIMITATIONS

The equipment used in the powder metallurgy industry imposes certain severe restrictions, both technical and economic. Powders are introduced into suitably constructed dies and presses. They are pressed at pressures varying from about 5 to 100 tsi, depending on the specific application and desired density. It is obvious that large, powerful, and therefore expensive, presses are necessary. The capacity, both in power and number of strokes per minute, decides the cost of the press; frequently a compromise must be made between tonnage and output. The fact that economical production of small- and medium-sized parts is possible only if the molding presses are fully automatic also increases the investment costs. Present limitations in tonnage for fully automatic presses with strokes up to 50 per minute, generally of the mechanical eccenter or cam type, are about 100 tons, thus limiting drastically the size of the part. Hydraulic presses with tonnages ranging in the thousands of tons have found use in the pressing of large parts, but the production rates are generally reduced. Even quick-acting presses of relatively low tonnage usually operate only semiautomatically and require the continuous attendance of an operator, thus limiting the output to a maximum of six to eight pieces per minute.

Other limitations are imposed by the construction of dies and punches. Strength and wear resistance of the die and toughness and fatigue resistance of the punches involve restrictions of both size and quantities of the molded part. Although porous bearings as large as 48 in. in diameter have been produced, high production rates can be obtained only for parts up to 3–4 in. in diameter (these rates, on the other hand, may reach the remarkable figure of 3,000 per minute for certain very small cores pressed in rotary pill presses using some twenty odd dies). The life of the dies depends on many factors, mainly pressure, plasticity and lubrication of the powder, composition of the die steel, and construction details. Die-wear figures have been published several times. The average die life for hardened alloy steel dies used in pressing porous bearings at about 20 tsi has been variously given as between 200,000 and 500,000 compressions. For higher pressures, e.g., 50 tsi, these figures range between 50,000 and 100,000 compressions. The life of the die was doubled or tripled by regrinding the cavity. It has also been found that special hard-metal inserts prolonged the die life twentyfold. The endurance of the punches is another matter of concern. Replacement figures for punches are not generally known, but it was recently ascertained, in one particular application involving coining of a plastic iron part, that alloy steel punches withstood not more than an average of 25,000 compressions at 70 tsi average pressure. The use of cemented carbide punches was found prohibitive because of unsatisfactory shock resistance, which resulted in almost immediate chipping of the corners of the punches.

The need for high production rates restricts the number of strokes of most presses, thus imposing limitations on the height of the compacts. This height is also affected by certain powder characteristics, such as flow and interparticle friction. Lubrication of the die cavities improves the uniformity of the compact by equalizing the density, but in the case of thick compacts (in excess of one-fourth of the diameter of the shorter rectangular dimension) uniform density parallel to the axis pressure can be obtained only by a double-action press movement or by a complicated floating die arrangement. Experience has shown that there is a maximum ratio of length to diameter which is obtainable in practice; with ordinary pressing equipment this ratio cannot go much beyond 4 : 1. The unit pressure necessary for molding dense parts can be considerably reduced by hot-pressing, thus taking advantage of the increased plasticity of the powders at higher temperatures. The hot strength of the die, however, imposes a drastic limitation on the temperature and pressure. High-speed steel and other alloy steel dies have withstood several thousand compressions at temperatures up to 800°C. (1470°F.) and at pressures up to 30 tsi, but the lack of knowledge and experience in this particular field gives no conclusive data on the performance of the molding tools. Hot-pressing imposes another severe restriction—it reduces production rates sharply because of time lost in heating the compacts prior to molding.

Other equipment limitations are imposed by the complicated and expensive equipment for producing powders, for preparing and conditioning them, and for storing large quantities without contamination by oxygen or moisture. Although large investments are necessary to air- and moisture-condition a powder metallurgy plant, such conditioning of the preparation and press departments is essential in many instances in order to assure reproducible results on a production scale. Other restrictions are imposed by the sintering equipment. Most commercial furnaces provide for continuous transport of the charge, *via* either a mesh conveyor belt or a roller hearth. Creep of these transport members restricts the operating temperature to about 1100°C. (2000°F.) with higher temperatures requiring frequent replacements and disturbances during production. Hence a moderate sintering temperature must often be used, although a higher temperature would lead to more desirable results. Extension of the furnace bed, resulting in longer sintering times, may compensate for the lower temperature but requires higher investment costs. The necessity for controlling the sintering atmosphere also leads to handicaps and temperature limitations, particularly in the sintering of ferrous metal parts.

Overcoming the aforementioned limitations may frequently call for knowledge which can be gained only by experience and trial-and-error methods. This is inevitable since both the practice and the theory of powder metallurgy are still in their infancy. A compilation of the results of such practical experiences on the broadest possible basis (as attempted in this book), and, of course, where possible, in conjunction with the more scientific revelations available, will constitute a sound foundation for new thought and progress in the development of better or more economical materials, processes, and products.

CHAPTER II
History

EARLIEST DEVELOPMENTS

Although the art of molding and firing useful or decorative objects of definite dimensions from clay and other ceramic materials is among the most ancient known to man, similar practices were only occasionally applied to metals during the early stages of history. Sintering of metals was entirely forgotten during the succeeding centuries, but was revived briefly in the early part of the nineteenth century. Another period of disuse of the method ended with a modern renaissance of powder metallurgy at the turn of the century.

Metal powders, especially gold, copper, bronze, and many powdered oxides (particularly iron oxide and other oxides used as pigments), were used for decorative purposes in ceramics, as bases for paints and inks, and in cosmetics, as far back as our historic record goes; powdered gold was used to adorn some of the earliest manuscripts. Little is known of the methods used by the ancients to produce these powders, but it is possible that some of the powders were obtained by granulating their melts. Low melting points and resistance to oxidation or tarnishing favored such procedures, especially in the case of gold powder. The use of these powders for pigments and ornamental purposes cannot truly be called powder metallurgy, since the essential features of the modern art are the production of the powder and its consolidation into a strong and solid form with precise contours by the application of pressure and heat at a temperature below the melting point of the major constituent.

Powder metallurgy principles, however, were used 5000 years ago to make solid objects from iron, long before furnaces were devised which could even approach the melting point of the metal. Egyptian implements were made by such a process as early as 3000 B.C. Pure iron oxide was heated in a charcoal fire which was intensified by air blasts from bellows and was thus reduced to a spongy metallic iron. A coherent metallic mass was then obtained by hammering the porous metal while still hot and final usable shapes were obtained by simple forging operations. The resulting metal was of uneven quality, ranging from unsound material contaminated with holes or inclusions to remarkably solid and sound structures, as evidenced by photomicrographic studies made by Carpenter and Robertson[1] of old Egyptian iron implements. The direct reduction of iron oxide without fusion was certainly carried out to an impressive extent about 300 A.D. in the famous Delhi Pillar, which weighs more than six tons, and in other similar monuments of even larger size.

Long before the discovery of the New World, the predecessors of the Incas, and later the Incas themselves, employed powder practices in making platinum metal, which would otherwise have required unobtainable temperatures for fusion. The technique which was used is most interesting since it was based on the cementing action of a lower-melting binder, a process closely resembling the present practice of making sintered hard carbides. According to Bergsøe,[2] the technique consisted of cementing of platinum grains (separated from the ore by washing and selection) by a metal of lower melting point which wet them, thus drawing the platinum grains together by surface tension and allowing the formation of a raw ingot suitable for further handling. The cementing material was an alloy of gold and silver of a fairly low melting point and great resistance to oxidation which wet the platinum surface readily without the aid of fluxes or a protective atmosphere. A change of color from the yellow of the sintered material to the whitish platinum pallor of the final metal was caused by diffusion during heating prior to working. Since all samples found by Bergsøe were very small (the largest hardly exceeding one-quarter inch across), it is believed that heating was accomplished by means of primitive charcoal fires fanned by blowpipes. Analyses of these alloys vary considerably, the platinum content ranging from 26 to 72%, and the gold content ranging from 16 to 64%. Silver additions were found to vary from 3 to 15%, and amounts of copper up to 4% were traced.

These ancient methods of direct production of metallic iron, by reduction of the oxide in the solid state and by subsequent coalescence of the grains by hot hammering and the production of metallic platinum, by careful selection of the native grains followed by a similar consolidating treatment facilitated by low-melting cementing agents, were eclipsed in the seventeenth century by processes based on melting and made possible by the introduction of new methods of heating which produced higher temperatures. Wrought iron was obtained by the indirect process of oxidation of cast iron, which was later to be replaced by melting low-carbon steels. At a later date heating methods were improved to such an extent that even the melting of platinum could be accomplished.

[1] H. C. H. Carpenter and J. M. Robertson, *J. Iron Steel Inst. London, 121,* 417 (1930).

[2] P. Bergsøe, *Ingeniørvidenskab. Skrifter, A* No. 44 (1937).

POWDER PRODUCTION METHODS

An account of the chronological development of powder metallurgy is incomplete without due reference to the early endeavors to make metallic powders. It must be understood, however, that the powders were sought primarily as a base for pigments and the process, therefore, belongs to the borderline applications of the field.

The earliest reference to the manufacture of metallic powders is found in an Egyptian-Greek manuscript dating from 300 A.D. and translated by Berthelot.[3] Powdered gold and silver, suitable for paints, inks, and lacquers, were obtained by mechanical disintegration or by heating gold amalgams. Eraclius[4] described similar methods in 994 A.D. The first full descriptions were given in a famous Latin manuscript written by the monk Theophilus[5] in the eleventh century. Theophilus gives a detailed account of the mill used for making powdered gold, silver, copper, brass, and tin. It consisted substantially of a mortar and a pestle whose shank was rotated by means of a leather strap and a flywheel which continued the motion and rewound the strap. A high-tin bronze was used for casting the mortar and the head of the pestle. Gold powder was obtained by mixing gold fillings with water and grinding for two hours, after which the suspended powder was poured off and the remainder ground further. To prevent adhesion of the powder to the mortar and pestle, grinding was done gently. The gold powder suspension was finally heated over a charcoal fire to drive off the water and restore the bright gold color of the rubbed particles. The same writer also describes the amalgam process. The gold amalgam was ground with dried salt, heated to drive off the mercury, and washed. In another process gold was made brittle and suitable for pulverizing by havings its melt poured into water saturated with lead; this was a first approach to the granulation method. Embrittlement may have been facilitated by minute additions of lead.

Allessio Piemontese[6] published in 1555 a method for making gold and silver powders from their amalgams. Gold amalgams were obtained by heating gold and mercury or by rubbing mercury and gold leaf together in the cold, with the aid of vinegar or lemon. Excess mercury was removed by staining; the amalgam was then ground with sulfur and heated to volatilize the sulfur and mercury.

[3] M. Berthelot, *Introduction a l'Étude de la Chimie des Anciens et du Moyen Age.* Paris, 1889.
[4] Eraclius, *De Coloribus et Artibus Romanorum,* in A. Ilg, *Quellenschriften für Kunstgeschichte und Kunsttechnik des Mittelalters und der Renaissance.* 1873, p. 4.
[5] Theophilus, *Schedula Diversarum Artium.* R. Hendrie, ed., London, 1847; W. Theobald, ed., Berlin, 1933.
[6] Allessio Piemontese (pseudonym), *Secreti.* 1555. Translated in London, 1559–60.

Granulation of gold and silver was described by Ercker[7] in 1574. In his book on assaying, Ercker indicates the way in which granulation was accomplished by pouring the molten metals into brooms kept under water. The same source indicates the practical application obtained by sintering these powders. Spongy gold obtained after parting was raised to red heat in order to give it sufficient strength for handling in weighing and also to restore its bright color. Cramer[8] reported (in 1739) the production of metal powders by mechanical disintegration in the temperature range between solidus and liquidus. The method was adopted for obtaining powdered lead for assaying.

In his review of the early development of powder metallurgy, Smith[9] points out correctly that it should not be overlooked that in the nineteenth century more metallic elements were produced in powder form than in any other form. Many of the refractory metals were produced by electrolysis or by reduction with hydrogen, carbon, or sodium. Sometimes amalgams were used which were later distilled; molten alloys were formed, from which the refractory metal solidified and was subsequently extracted by dissolving the mass. The products were almost exclusively in powder form and were then consolidated into a compact metal mass. Smith enumerates some 31 metals which were first produced (or within a short time after their discovery) in the form of powders. Nickel and cobalt were marketed throughout the nineteenth century in the form of briquets made by low-temperature reduction of their oxides by carbon, followed by sintering. In the nineteenth century metal powders became increasingly useful to dentists, who had used gold leaf filings for teeth for a long time. Toward the middle of the century, it was found that gold sponge could be made to cohere into a solid mass at the low temperatures prevailing in the human mouth. The use of powdered tin-silver alloys for amalgam fillings is another example of early cementing processes. Bessemer's[10] interest in making bronze powders is stressed by Smith, who concludes with a special reference to the fact that Bessemer's classic experiments on steel production were largely financed by the profits derived from his thriving powder manufacture.

The history of the metal pigment powders was detailed by Theobold[11] in 1913, and has been covered more recently by Chaston.[12]

[7] L. Ercker, *Beschreibung Allerfürnemisten Mineralischen Ertzt und Berckwercksarten*, Prague, 1574–1580.

[8] J. A. Cramer, *Elementa Artis Docimasticae*. Leyden, 1739. Translated in London, 1741.

[9] C. S. Smith, in J. Wulff, *Powder Metallurgy*. Am. Soc. Metals, Cleveland, 1942, chapt. 2.

[10] H. Bessemer, *An Autobiography*. Office of Engineering, London, 1905.

[11] W. Theobold, *Dinglers Polytech. J., 328*, 163 ff. (1913).

[12] J. C. Chaston, *J. Birmingham Met. Soc., 17*, 9 (1937).

POWDER METALLURGY OF PLATINUM

Development of the Art

The metallurgy of platinum as practiced during the second half of the eighteenth and the first half of the nineteenth centuries in Europe may be designated as the most important stepping stone on the road to the modern development of powder metallurgy. For the first time, complete records were available which gave some of the details of the various methods of powder production, but which dealt mainly with the processing of these powders into solid, useful articles.

The manufacture of platinum from its powder was accomplished by methods similar to those mentioned above as having been employed at an earlier date in America. It was won in the form of grains of the native metal, frequently contaminated with other metals of the platinum group or with iron oxide. In 1755 Lewis[13] found that when a lead–platinum alloy was oxidized at very high temperatures a spongy, workable mass was left after the lead oxide impurities had been driven off by volatilization. Scheffer,[14] at the same time, found that when platinum was heated with arsenic, the former metal showed signs of melting. This was confirmed by Achard[15] in a publication in 1781, which described the production of a fusible alloy of platinum and arsenic, probably by forming the eutectic containing 87% platinum and melting at the very low temperature of 597°C. (1107°F.). Achard obtained solid Pt-metal by hot-hammering a sponge, thus welding the individual particles into a massive body. (The sponge was obtained by high-temperature treatment of the platinum–arsenic alloy, which resulted in volatilization of the arsenic.) Toward the end of the eighteenth century,[16] Jannetty in Paris produced the first platinum vessels for chemical uses by this method.

Somewhat later, in a method employed by Mussin-Puschkin,[17] the process was modified by replacing arsenic with mercury; in similar processes palladium was obtained with the aid of sulfur and iridium, with the aid of phosphorus. In 1816 Ridolfi[18] proposed production of malleable platinum for chemical vessels on the basis of such a volatilization treatment, employing sulfur or lead. In 1786 Rochen[19] carried out his first successful experiments in producing solid platinum without the aid of

[13] W. Lewis, *Phil. Trans. Roy. Soc. London, 48,* 638 (1755).

[14] H. T. Scheffer, *Kgl. Svenska Vetenskapsakad., Handl. 13,* 269 (1752).

[15] H. F. Achard, *Nouveaux Mém. Acad. Roy. Sci. (Akad. Wissenschaften Berlin) 12,* 103 (1781).

[16] C. S. Smith, in J. Wulff, *Powder Metallurgy.* Am. Soc. Metals, Cleveland, 1942, chapt. 2.

[17] A. Von Mussin-Puschkin, *Allgem. J. Chem., Scherer 4,* 411 (1800).

[18] C. Ridolfi, *J. Sci. Arts Roy. Inst. Gr. Br. 1,* 259 (1816).

[19] A. Rochen, *J. phys., chim., hist. nat., arts, 37,* 33 (1798).

arsenic by simply welding together small pieces of platinum, preferably by wrapping them in sheets. He obtained malleable platinum by uniting native platinum grains, which had been purified with the aid of fluxes. Even more important were the observations made by Knight,[20] in 1800; he produced solid platinum metal by forging a coherent mass, which was obtained by heating chemically precipitated platinum powder at a high temperature in a clay crucible and compressing the hot powder. This process included the basic processing operations of powder metallurgy: powder production, powder annealing, mechanical consolidation (by compressing the hot powder), and final densification during hot-working. Here, for the first time, the principles of hot-pressing were clearly applied.

Carrying the diffusion treatment still further, Tilloch[21] in 1805 placed platinum powder in tubes bent from rolled platinum sheet and heated and forged the "sandwich" into a solid metal block. In 1813 Leithner[22] described another approach: he produced thin, malleable platinum sheets by drying out successive layers of the powder suspended in turpentine and subsequently heating the resulting films at a very high temperature, without using any pressure. Thus, a metal was made by applying a high-temperature sintering operation to its powder.

In 1822 a French process was described by Baruel[23] in which as much as 30 pounds of platinum powder was fabricated into a solid ingot by a great number of operations. Again, platinum was precipitated in powdered form and, after slight compression in a platinum crucible, was heated to white heat. The powder was then placed in a steel matrix and pressure was applied by means of a screw coining press. The compact was reheated in a platinum crucible and pressed again; this was repeated until a solid ingot was obtained, except that the last heat-treatments were made in a charcoal fire at lower temperatures. Since the platinum powder was introduced into the steel die while still very hot, this process was also based on the hot-pressing principle.

Sobolewskoy[24] reported a basically different process operating in 1826 in Russia. Here, for the first time in powder metallurgy practices, a high-temperature sintering operation was applied to a previously compressed powder compact on a commercial scale, in contrast to earlier methods based on hot-pressing. Platinum powder was compressed in a mold made of a cast iron cylinder with steel punches fitted closely in the bore. Pressure

[20] R. Knight, *Phil. Mag.*, *6*, 1 (1800).
[21] A. Tilloch, *Phil. Mag.*, *21*, 175 (1805).
[22] Leithner, letter quoted by A. F. Gehlen, *Beiträge zur Chemie and Physik, 7*, 309, 514 (1813).
[23] M. Baruel, *Quart. J. Lit. Sci. Arts, (Roy. Inst. Gr. Br.) 12*, 246 (1822).
[24] P. Sobolewskoy, *Ann. Physik Chem.*, *109*, 99 (1834).

was applied by a heavy screw press. The compacts were stable enough to be handled; they were annealed for more than a day at the high temperature used for firing porcelain. Great malleability was achieved in ingots prepared from carefully screened, washed, and purified powder. Considerable shrinkage, ranging from 20 to 30%, was caused by the high-temperature sintering.

A similar method of producing malleable platinum in Russia was reported by Marshall[25] in 1832. Fine, dry platinum powder was placed in a ring-shaped iron mold and then pressed by a screw press. After heating to a red heat, pressing was repeated. Finally, the resulting discs were worked in a rolling mill and used for coins.

The Wollaston Method of Producing Platinum

Basing his work on information obtained from some of the earlier publications, Wollaston[26] perfected a process for producing compact platinum from platinum sponge powder. At least sixteen years prior to his publication of 1829,[27] describing the manufacture of a product much superior to that of contemporary manufacturers, Wollaston laid down the classic foundations for modern powder metallurgy technique. In fact, it may be said that powder metallurgy had its first truly scientific enunciation in his work. Aside from leaving complete records of his work for the benefit of posterity, Wollaston made a twofold contribution of great merit. He was the first to realize all the difficulties connected with the production of solid platinum ingot from powdered metal and, accordingly, paid much attention to the preparation of the powder. He found that pressing the powder while wet into a hard cake (to be baked subsequently at red heat) was best done under considerable pressure. Second, since available screw presses were not sufficiently powerful, Wollaston developed a horizontal toggle press of the ingeniously simple construction shown in Figure 2.

Wollaston's process has been aptly summarized by Streicher,[28] who divides the method into nine steps:

(1) Precipitation of ammonium platinum chloride from dilute solutions.

(2) Slow decomposition of the finely divided and carefully washed ammonium platinum chloride precipitate into loose sponge powder by igniting and heating at not too high a temperature.

[25] W. Marshall, *Phil. Mag., 11,* 321 (1832).
[26] W. H. Wallaston, *Trans. Roy. Soc. Londin, 119,* 1 (1829).
[27] Leithner, Letter quoted by A. F. Gehlen, *Beiträge zur Chemie und Physik, 7,* 309, 514 (1813).
[28] J. S. Streicher, in J. Wulff, *Powder Metallurgy.* Am. Soc. Metals, Cleveland, 1942, p. 16.

(*3*)   Gentle grinding of this sponge powder in a wooden mortar, without applying pressure to the particles; this prevented burnishing of the powder particles, preserving all their surface energies essential for subsequent interparticle cohesion.

(*4*)   Sieving the metal powder to remove foreign matter.

(*5*)   Washing the black sponge platinum powder with water, thus carefully eliminating all volatile matter retained from the precipitation treatment.

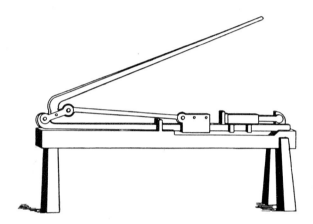

Fig. 2.   Toggle press used by Wollaston in the early
part of the nineteenth century.

(*6*)   Separating coarse from fine particles by sedimentation and eliminating the coarser sponge particles for use in the process.

(*7*)   Pressing the powder, suspended in water and containing only the finest platinum particles, into a slightly tapered cylinder (approximately one and one-quarter inches in diameter and six and three-quarter inches long); a brass mold closed at the bottom with an iron plug was used. The powder was first pressed in by hand with a wooden plug, then covered with a copper plug, and pressed in the toggle press.

(*8*)   Carefully drying the resulting stable, wet cake by heating it very slowly to a red heat on a charcoal fire. Thus, moisture and oil were removed and harmful absorbed gases, which caused cracks and blisters in competitive products, were eliminated. Heating was continued to yellowness in a coke-fired wind furnace for about twenty minutes, after which the compact was placed on pure quartz sand and covered by a refractory pot. This heat treatment resulted in a strong bonding of the individual particles into a coherent metallic blank.

(9) Hammering and forging the platinum cake while it was still hot. Consolidation was first achieved by striking the blank squarely on top with a heavy hammer; further forging was done in a conventional manner. Wollaston in using his special features succeeded not only in producing compact platinum, but also in producing large pieces of metal yielding thinly rolled sheets which were practically free of gas blisters. Crucibles made from this sheet were the best platinum products of his time. The inferior quality of competitive platinum articles was caused chiefly by the modified and simplified methods employed by his contemporaries. These simplified methods, less meticulous in the selection of a fine powder and the elimination of all volatile impurities and harmful absorbed gases before the high-temperature sintering, failed particularly to utilize the wet treatment and wet handling of the platinum powder before and during its briquetting, as exemplified by the St. Petersburg method described by Marshall.[29] Hence, Wollaston's process was successfully used for more than a generation and became obsolete only with the advent of the platinum fusion procedure in the eighteen-fifties.

OTHER DEVELOPMENTS IN THE NINETEENTH CENTURY

In giving an account of the historical developments of powder metallurgy in the nineteenth century, authorities often show a tendency to emphasize Wollaston's process to such an extent as to overshadow important contributions of his contemporaries. Particular reference is made to the work on copper in 1830 by Osann,[30] who by 1841 had clearly suggested the formation of articles of complicated shapes by pressing and sintering of metal powders—the basis of our modern practices. Smith,[31] in his review of early developments in the field, gives due credit to this work and, in consideration of its important implications, his excellently detailed description is substantially given in the following description.

During atomic weight determinations on copper, Osann observed that the metal, reduced from its oxide by hydrogen, sintered to a compact mass. He thereupon developed a process for making impressions of medals from copper powder, produced by the reduction of precipitated copper carbonate. He found that the reduction was best done at the lowest possible temperatures which could be used to produce a metal powder of the fineness known in platinum manufacture. High reduction temperatures resulted in granular masses which did not sinter well. Contaminations by

[29] W. Marshall, *Phil. Mag.*, *11*, 321 (1832).
[30] G. Osann, *Ann. Physik Chem.*, *128*, 406 (1841).
[31] C. R. Smith, in J. Wulff, *Powder Metallurgy*. Am. Soc. Metals, Cleveland, 1942, chapt. 2.

the atmosphere were prevented by immediate use of the powder after reduction or by careful storing in properly closed glass bottles. Before use, the powder was separated into three grades. In making an impression of a coin, the finest powder was sprayed on the surface and was then backed up with the coarser grades. Compression was obtained by placing die and powder in a ring-shaped mold and exerting pressure by hammer blows on a punch or by using a knuckle press. The copper powder used must have been very light, as the powder volume was reduced to one-sixth during compression. Sintering was carried out at temperatures close to the melting point of copper, after the compacts were placed in copper packets sealed with clay to exclude air. A shrinkage of 20%, but without distortion, was observed and the sintered copper was found to be harder and stronger than cast copper.

The same process was used by Osann to make medals of silver and lead. Advancing his method chiefly as substitute for the customary electrotype method of copying coins and medallions, he did not miss its possibilities for initial production of such objects as coins, medals, printing type, and concave- or convex-shaped backings for mirrors. He also suggested that measurement of the shrinkage of pressed copper powder compacts could be used for temperature indication, in the same manner as the shrinkage of clay cylinders was utilized in the Wedgewood pyrometer.

Toward the middle of the nineteenth century, powder metallurgy suffered a complete relapse. The superiority of the fusion method for platinum and iron, aided by perfection of high-temperature refractories and more powerful heating equipment, suceeded in eliminating the Wollaston process. An entire generation completely neglected all experiences and practices of the art and, except for some sporadic experimenting with the cohesion of minute metal surfaces in about 1880, Wollaston's method was revived only toward the turn of the century, when its principles were applied to the manufacture of ductile tungsten.

Among the few efforts made during the second half of the nineteenth century to carry the method further belong the noteworthy attempts by Gwinn[32] to develop bearing materials from metal powders. In fact, it was his work, as disclosed in a patent issued in 1870,[33] which marked the conception of the important and fundamental idea of a self-lubricating bearing, thus becoming the forerunner of a whole series of developments in the field some fifty years later. Gwynn employed a mixture of 99 parts of powdered tin, prepared by rasping or filing, and 1 part of petroleum-still residue. The two constituents were intimately stirred while being

[32] U. S. Pats. 101,864; 101,866; and 101,867.
[33] U. S. Pat. 101,863.

heated. A solid form of desired shape was then produced by subjecting the mixture to severe pressure while enclosing it in a mold. The patent specifically states that journal boxes made by this method or lined with material thus produced would permit shafts to run at high speeds without the need of any outside lubrication.

In 1880 Spring[34] published a report on studies of the effect of pressure on the interparticle cohesion of metallic powders. He also investigated the production of alloys by pressure alone.[35] Employing very high pressures, he found that the compressed compacts were similar to the alloys obtained by ordinary methods. Experimenting with mixtures of metallic filings, he was successful in producing a number of alloys. He obtained true brass by mixing copper and zinc filings, pressing them into a compact, and refiling and repressing them as often as six times. Similarly, he produced Wood's metal when experimenting with filings of a number of low-melting metals, notably lead, tin, cadmium, bismuth, etc. Although Spring failed in all these studies to consider the important effects of sintering upon his compressed materials, his success in producing fusible alloys of low melting points must be attributed to diffusion (facilitated by conditional heat-treatments of the compacts) caused by the heat of friction during pressing or filing.

Toward the end of the nineteenth century, fine powders, which presented the greatest surface area, became of great interest for the manufacture of chemical catalysts, the effectiveness of which depends in large part upon the surface area which can be utilized for reaction. Nickel and cobalt powders especially became the focus of much experimental work; Edison[36] developed an electrolytic process for making nickel flake powders, in which a composite sheet of nickel and copper or zinc is electrodeposited in alternate layers, the sheet is subsequently stripped from the cathodes, and the copper or zinc is leached out to leave a mass of fine nickel flakes.

REFRACTORY METALS AND THEIR CARBIDES

Edison's discovery of the incandescent electric light encouraged the search for efficient materials for filaments and resulted in the successive development of osmium, tantalum, and tungsten filaments from their powders. For this purpose, methods were used which were derived from the procedures used by Edison, Swann, Westinghouse, and others, from 1878

[34] W. Spring, *Bull. classe sci., acad. roy. Belg., 49,* 323 (1880).
[35] W. Spring, *Ber. deut. chem. Ges., 15,* 595 (1882).
[36] U. S. Pats. 821, 626; 865,688; and 936,525.

through two decades, in making carbon filaments by the extrusion and subsequent sintering of carbonaceous materials.

Osmium filaments were used for a short time in 1898–1900, and Auer von Welsbach[37] described the production of filaments of osmium by chemical precipitation of the powder and formation of a mixture with sugar syrup, which served both as binder and, if osmium oxide powder was used instead of the metal, as reducing agent as well. The mixture was squirted through fine dies and the resulting fine threads were subsequently fired in protective atmospheres to carburize and volatilize the binder, to reduce the oxide, and to sinter the metal particles into a coherent metallic wire capable of functioning as an electrical conductor.

The osmium electric lamp was soon succeeded by tantalum filament lights, which were used widely from 1903 to 1911. The general procedure[38] employed was similar to that used for osmium, with the exception that tantalum had to be purified by a vacuum treatment in order to become ductile. Similar techniques were used for the production of filaments from zirconium, vanadium, and tungsten; with the last metal, especially, the extruded wires were bent into hairpin shapes before sintering to obtain them in a form suitable for use as filaments. Since, however, lack of ductility was the paramount shortcoming of these filaments, attempts were made to improve these conditions by the addition of few per cent of a lower melting, ductile metal. Tungsten powder was mixed with two to three per cent of nickel, pressed into a compact, and sintered in hydrogen at a temperature slightly below the melting point of nickel. The resulting bars could be drawn and the final filaments were then freed from nickel by a vacuum heat-treatment at a high temperature.[39] Although this process was not commercially successful, it must be considered as an important step toward the industrial development of cemented hard metals and composite materials. The early work on metallic filaments is recounted in a number of patents by Lodyguine[40] and von Bolton;[41] Kuzel[38] discloses the manufacture of alloys by slowly adding a solution of a salt, oxide, or hydroxide of a metal such as aluminum, titanium, or platinum to a colloidal solution of a refractory metal, next making the solution alkaline with caustic soda, and then precipitating the colloid with sodium chloride. A hydrogen heat-treatment at a very high temperature resulted in a crystalline metal powder.

[37] U. S. Pat. 976,526.
[38] U. S. Pats. 899,875 and 912,246.
[39] C. R. Smith, in J. Wulff, *Powder Metallurgy*. Am. Soc. Metals, Cleveland, 1942, chapt. 2.
[40] U. S. Pats. 575,002 and 575,668.
[41] U. S. Pats. 927,935 and 930,723.

Because of certain important properties of tungsten, it was soon realized that this metal would be the most nearly ideal material for lamp filaments. The problem was, however, to evolve an economical procedure for producing these filaments in large quantities. A number of procedures to produce powdered tungsten had been worked out earlier. In 1783 the D'Elhujar brothers[42] first produced tungsten powder by heating a mixture of tungstic acid and powdered charcoal and then, after cooling, removing the small cake, which crumbled to a powder of globular particles. The purification of tungsten powder by boiling, scrubbing, and skimming to remove soluble salts, iron oxide, clay, and compounds of calcium and magnesium, was disclosed by Polte.[43]

At the beginning of the twentieth century, Coolidge[44] made the important discovery that tungsten could be worked in a certain temperature range and would retain its ductility at room temperature. His process consisted essentially in briquetting the tungsten powder in the form of small ingots and sintering in a protective atmosphere below the melting point of tungsten. The sintered ingots, although very brittle at room temperature, were workable at very high temperatures (only slightly below the sintering temperature), and could thus be reduced in cross section by swaging. With greater reduction in cross section, the annealing temperatures could be lowered, until a stage was finally reached at which the metal was worked down to a thin rod which was ductile at room temperature. Subsequent drawing produced extremely thin and ductile filaments. This process is still being used, since there is no satisfactory method of preparing tungsten by the fusion process common to most other metals. Certain modifications and improvements were presented in later disclosures by Coolidge,[45] the von Bolton patents,[41] the Kitse patent,[46] and the Just and Hanaman patent,[47] and in the years that followed the patent literature on sintered tungsten became voluminous. Lederer,[48] Krüger,[49] Farkas,[50] Pfanstiehl,[51] and many others described inventions closely related to the subject.

The procedures disclosed for the production of tungsten were fre-

[42] A. W. Deller, in J. Wulff, *Powder Metallurgy*. Am. Soc. Metals, Cleveland, 1942, chapt 51.
[43] U. S. Pat. 735,293.
[44] U. S. Pat. 963,872.
[45] U. S. Pats. 1,026,429; 1,077,674; and 1,082,933.
[46] U. S. Pat. 971,385.
[47] U. S. Pat. 1,018,502.
[48] U. S. Pats. 1,034,018; 1,035,833; 1,047,540; 1,071,325; and 1,132,523.
[49] U. S. Pat. 1,154,701.
[50] U. S. Pat. 1,188,057.
[51] U. S. Pats. 1,282,122; 1,305,975; and 1,315,859.

quently adaptable to the manufacture of molybdenum. Lederer[52] developed a method of making molybdenum by starting with powdered molybdenum sulfide. The sulfide, mixed with amorphous sulfur and kneaded into a paste, was formed into a filament. When exposed to air, the filaments became strong enough to be placed in a furnace. Heating in hydrogen resulted in formation of hydrogen sulfide and sintering of the metal into solid filaments. A similar process was patented by Oberlander[53] who used molybdenum chloride and other halides as starting materials. When the chloride was treated with a reducing agent such as ether, a paste was obtained.

In addition to tungsten, molybdenum, and tantalum, the three most important refractory metals used today in the lamp, electronics, x-ray, and chemical industries, other refractory metals of minor significance were developed by the powder metallurgy method in the early part of the century, notably columbium, thorium, and titanium. However, at the same time another development, originating in refractory metal processing, took form and rapidly grew to a subject of such importance that it far overshadows the parent field: the cemented carbide field has become one of the greatest industrial developments of the century.

In drawing tungsten wires and filaments it was found that the usual drawing dies were unsatisfactory. The need for a very hard material to withstand the greater wear became pressing and, since it was known that tungsten granules combined readily with carbon at high temperatures to give an extremely hard compound, this material was used as the basis for a very hard and durable tool material, known as cemented carbides. The tungsten carbide particles, present in the form of finely divided hard and strong particles, are bonded into a solid body with the aid of a metallic cementing agent. Early experiments with a number of metals established that this cementing agent had to possess certain properties to permit solidification of the hard metal body. Close affinity for the carbide particles, a relatively low melting point, limited ability to alloy with the carbide, and great ductility (not to be impaired by the cementing operation) were the most important requirements of this metallic binder. It was found that cobalt satisfied these requirements most closely. The early work was mostly carried out in Germany by Lohmann and Voigtländer[54] in 1914 by Liebmann and Laise[55] in 1917, and by Schröter[56] in 1923 to 1925. Krupp[57] perfected the process in 1927 and marketed the first product of

[52] U. S. Pat. 1,079,777.
[53] U. S. Pat. 1,208,629.
[54] German Pats. 289,066; 292,583; 295,656; and 295,726. Swiss Pats. 91,932 and 93,496.
[55] U. S. Pats. 1,343,976 and 1,343,977.
[56] German Pat. 420,689. U. S. Pat. 1,549,615.
[57] Brit. Pats. 278,955 and 279,376. Swiss Pat. 129,647. U. S. Pat. 1,757,846.

commercial importance, "Widia." In 1928 this material was introduced to the United States and the General Electric Company which held the American patent rights, issued a number of licenses. The process consists essentially of carefully controlled powder manufacture, briquetting a mixture of carbide and metallic binder (usually 3–13% of cobalt), and sintering in a protective atmosphere at a temperature high enough to allow fusion of the cobalt and partial alloying with the tungsten carbide. The molten matrix of cobalt and partly dissolved tungsten carbide serves as a bond, holding the hard particles together and giving the metallic body sufficient toughness, ductility, and strength to permit its effective use as tool material.

COMPOSITE METALS

The next development in the field was the production of composite metals used for heavy-duty contacts, for electrodes, and for balancing parts and radium containers. All of these composite materials contain refractory metal particles, usually tungsten, and a cementing material with a lower melting point, present in various proportions. Copper, copper alloys, and silver are frequently used; others are cobalt, iron, and nickel, while some combinations also contain graphite. The first attempt to produce such materials was recorded in the patent of Viertel and Egly,[58] issued shortly after 1900. The procedures used were either similar to the one developed for the hard metals[59] or called for introduction of the binder in liquid form by dipping, infiltration, etc. Gebauer developed such a procedure in 1916[60] and the process was carried further by Baumhauer[61] and Gillette[62] in 1924. Pfanstiehl[63] in 1919 obtained patent protection for a "heavy" metal, consisting of tungsten and a binder which contained copper and nickel.

OTHER MODERN DEVELOPMENTS

Simultaneously with the development of the refractory metals and their carbides, another important branch of powder metallurgy came to maturity during the early part of this century. Porous metal bearings were an innovation in metallurgy since, for the first time, metals could be produced with a structure which displayed uniformly distributed, closely interconnected porosity, with controlled pore sizes and quantities. Special

[58] U. S. Pat. 842,730.
[59] U. S. Pats. 1,418,081; 1,423,338; and 1,531,666.
[60] U. S. Pat. 1,223,322.
[61] U. S. Pat. 1,512,191.
[62] U. S. Pat. 1,539,810.
[63] U. S. Pat. 1,315,859.

types of these porous bearings are known as self-lubricating and, as mentioned previously their origin dates back to the last quarter of the nineteenth century. [64-66] The modern type of bearings, usually made of copper, tin, and graphite powders and impregnated with oil, were first developed in processes patented by Loewendahl[67] and Gilson.[68,69] Gilson's material was a bronze structure throughout which finely divided graphite inclusions were uniformly distributed. It was produced by mixing powdered copper and tin oxides with graphite, compressing the mixture, and heating it to a temperature at which the oxides were reduced by the graphite and the copper and tin could diffuse sufficiently to give a bronze-like structure. Excess graphite (up to 40% by volume) was uniformly distributed through this structure. The porosity was sufficient to allow for the introduction of at least 2% oil. The process was later improved by Boegehold and Williams,[70] Claus,[71] and many others, primarily by utilization of the metal powders rather than the oxides. Metallic filters were the next stage in the development of these porous metals and patents date back as far as 1923,[72] when Claus patented a process and machine to mold porous bodies from granular powder. Similarly, the manufacture of laminated products, such as current collector brushes, was derived from the porous bearing process, the main difference being the incorporation of flaky particles of copper and graphite. Early procedures for this type of product were described by Loewendahl[67] and Gilson.[73]

Another development of the last two decades has been the utilization of powdered iron for cores in electric circuits. In about 1927, Polydoroff[74] introduced powdered iron cores made from hydrogen-reduced iron, which provided efficient operation in radio tuning devices at frequencies as high as 1500 kilocycles. In 1930, Vogt in Germany introduced powdered iron tuning inductances, which consisted of powdered iron particles sprayed on a sheet of paraffined paper which was rolled into a cylinder to form the core of a radio inductance. Later carbonyl iron powder was introduced which, due to its fineness and low losses, was particularly suitable for tuning devices operating at frequencies as high as 40,0000 kilocycles.

[64] U. S. Pats. 101,863; 101,864; 101,866; and 101,867.
[65] U. S. Pat. 189,684.
[66] U. S. Pats. 304,500 and 313,916.
[67] U. S. Pat. 1,051,814.
[68] U. S. Pat. 1,177,407.
[69] E. G. Gilson, Gen. Elec. Rev., 24, 949 (1921).
[70] U. S. Pats. 1,642,347; 1,642,348; 1,642,349; and 1,766,865.
[71] U. S. Pat. 1,648,722.
[72] U. S. Pat. 1,607,389.
[73] U. S. Pat. 1,093,614.
[74] A. Crossley, J. Applied Phys., 14, 9, 451 (1943).

Among the latest developments are cores from powdered iron–nickel alloys (Permalloy), with and without small amounts of molybdenum.

This brief historic review of powder metallurgy developments is not complete without reference to the two most recent developments: magnets and parts, especially from iron powder. In both cases the powder method has been introduced when the conventional methods of casting, working, or machining were difficult or uneconomical. An example is the production of small permanent magnets of irregular shapes and certain small iron parts of complicated configurations, but tolerant physical requirements, for the electrical and machine industries.

CHAPTER III

Methods of Powder Production

Any discussion of metal powders as raw materials for the manufacture of powder metallurgy products must necessarily stress the close connection between the manner of production and the desired properties and characteristics of the powder. Generally speaking, each of the many methods of powder production yields a specific product with definite qualities, which may or may not make it suitable for a particular application. Sometimes, several entirely different methods of production result in powders equally well adapted to processing the same type of industrial article, e.g., electrolytic and reduced copper powders for porous bronze bearings. Usually, however, each application requires a powder with special properties, which can be obtained only by a particular method of powder production.

For this reason it has been suggested that the preparation of a metal powder should be considered as an integral part of the industry which finally processes it into the finished product. The powder producer must meet the requirements of the user, whose specifications depend on the particular application and method of manufacture. This has often led to the development of special powders for particular applications.

A general classification of industrial methods of powder manufacture is presented in Table 1.[1] A recent comprehensive survey of the various methods of metal powder manufacture has been published by Miller.[1a] The following discussion will deal only with the few processes which are actually being used on an industrial scale and which are, therefore, particularly interesting to the powder metallurgist. The methods for producing metal powders may be arbitrarily grouped into two general classes: (a) mechanical processes, and (b) physicochemical and chemical processes.

MECHANICAL PROCESSES

There are six important methods of disintegrating metals and alloys by mechanical means: (1) machining, (2) crushing, (3) milling, (4)

---

[1] J. C. Chaston, *J. Birmingham Met. Soc.*, 17, 9 (1937).
[1a] G. L. Miller, *Symposium on Powder Metallurgy*, The Iro nand Steel Institute Special Report No. 38, London, 1947, p. 8.

TABLE 1   Production of Metal Powder (Chaston[1])

Raw material	State	Process	Principle involved	Product	Average particle size, μ	Uses
Metal		Machining Bessemer process Screening beaten foil	Tearing	Mg Cu and Al alloys Au, Cu, and alloys	50–150 0.8–50 0.8–50	Flashlight powders Paints Paints
		Stamp mills Hametag impact mills		Al, Cu, and alloys Al, Cu, and alloys	0.8–50 0.8–50	Paints Paints
	Solid	Eddy mills	Severe working	Fe Cu	75 75	Magnetic cores, commutator brushes Magnetic cores
		Grinding sponge		Fe	100	Chemicals
		Grinding Cleavable metals	Fracturing of cleavage planes	Bi, Sb, etc.	150	Magnetic cores
		Grinding brittle electrolytic metals	Intercrystalline fracturing	Fe	180	Magnetic cores
		Grinding brittle metals made fine by hot wax		Ni-Fe alloys	75	Magnetic cores
	Molten	Atomization by air blast or steam	Spraying	Al Pb Pb alloys	200 8 15	Thermit process Paints Solders
		Granulation by stirring	Graining	Al Pb alloys	250 —	Thermit process Solder
	Vapor	Condensation at normal or low pressure	Condensation	Zn, Mg	—	Chemicals
	Solid	Reduction by hydrogen or other gases at temperatures below melting point	Reduction	W, Mo Ni, Co Fe Fe, Cu	0.5–50 — — 0.5–150	Lamp filaments Sintered carbides Magnetic cores Porous bearings Sintered products
Chemical	Solution	Chemical precipitation	Precipitation	Pt, Pd Sn	—	Catalysts Condenser paper
	Solution	Electrodeposition as a powder	Electrolysis	Cu, Fe, etc.	0.5–150	Porous bearings Sintered products
	Gas	Carbonyl process	Thermal decomposition	Fe Ni Ni-Fe alloys	3 —	Magnetic cores Sintered products

graining, (5) shotting, and (6) atomizing. The first three are carried out at room temperature and apply to solid metals; the last three require elevated temperatures and liquefaction of the metal.

Comminution of Solid Metals

MACHINING

The machining of metals to produce powders suitable for processing into compacts is employed only in a few special cases, since the product is invariably coarse, with many bulky particles. Because of the sharply outlined, shell-shaped or flaky particles this type of powder is difficult to mold and usually produces very porous compacts, with a low green strength and a tendency to fracture laterally. Further grinding of the powder in a ball mill may improve the molding properties; this may, however, result in increased work-hardening, which may necessitate annealing prior to briquetting.

One of the chief applications of the machining method is in the manufacture of magnesium powder for pyrotechnic purposes, in which case the malleability and explosiveness of the powder prohibit the use of other methods. The danger is considerably lessened by the use of the relatively coarse chips. When a finer powder is produced by ball milling, special precautions are necessary to prevent explosions. The introduction of an inert atmosphere into the ball mill to avoid explosive dust–air mixtures is most effective. During removal of the charge from the mill, however, the powder must not be brought into sudden contact with air, which would cause spontaneous combustion. Instead, the gradual formation of oxide layers around the particles must be encouraged, with sufficient time allowed to permit the heat of reaction to be dissipated without appreciably raising the over-all temperature. Mixtures of an inert atmosphere and small amounts of air can also be used effectively, if the oxygen content is kept high enough to allow the formation of protective surface films, yet not high enough to cause an explosion. If milling must be carried out in air, careful choice of mill linings and ball charges must be made in order to avoid sparking.

Silver solders and certain dental alloys are also produced by machining, especially when the product must be free from fine particles. Iron powders for use as chemical reducing or precipitating agents or as hardeners in concrete are often produced by this method; in the case of Permalloy (80% nickel, 20% iron), sulfur must be added to facilitate machining. Machining costs are high and the method is practicable only when there is no other possible method which can be used, e.g., magnesium powder, or when the cost of the metal itself is so high, *e.g.*, dental alloys, that cost of machining is only incidental.

CRUSHING

Crushing of metals into small particles is very similar to milling, since it also provides comminution of bulk metal. The apparatus used depends largely on the malleability of the metal. To facilitate the operation, certain metals or alloys are first subjected to an embrittlement treatment, for example, certain ferrous stampings, which are hardened before being pulverized into the fine mesh sizes used for sandblasting. Figure 3 shows the

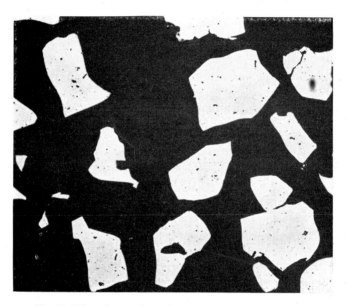

Fig. 3. Microshape of crushed steel particles (×200).

shape of crushed steel powder (in a photomicrograph). There are many types of crushers suitable for this kind of work, ranging from heavy drop hammers for crushing ores, brittle metal lumps, scrap, and steel stampings into coarse granules, to jaw-crushers and gyratory crushers used for further grinding of the metal particles into fine sizes. Frequently crushing forms only an intermediate step and is succeeded by other milling or grinding operations.

MILLING

Milling comprises many methods of disintegrating metals, all of which can, however, be classified as comminution of brittle, friable metals and pulverization of malleable metals. By milling or grinding, brittle and hard materials can be reduced to powder of practically any degree of fineness. Metals such as tungsten, molybdenum, titanium, manganese, silicon, antimony, bismuth, and many of their alloys and compounds, notably the refractory carbides, can be ground to closely controlled size and size distribution, thus making them useful for a variety of applications. Because of their brittleness and irregular particle shape these metal powders are not generally suitable for direct processing into pressed and sintered objects; in certain cases, e.g., carbides, the powder can be used only if softer, lower melting material is added as a binder. During grinding the powder may also be contaminated by oxides or other impurities, which add to the difficulties of molding. For these reasons, large quantities of finely pulverized cast iron and steel powders, now available at very low prices, have not yet been utilized for powder metallurgy processing.

The equipment used for milling these hard and brittle materials is of many kinds, varying with the toughness of the metal. It may consist of one mill or a combination of several, and may utilize rod mills, ball mills, impact mills, disk mills, stamps, crushers, and rolls. To minimize contamination of the powder, the wearing parts of such equipment are provided with hard and durable facings. Undue heating and ductilization of the powder is sometimes prevented by water-cooling the grinding surfaces.

As mentioned previously, alloys may be embrittled to facilitate mechanical disintegration.[2] It is possible in the case of aluminum–magnesium alloys to produce a powder which is brittle enough for pulverizing and then to adjust the composition to that of the standard alloy by addition of more metal. Nickel–iron alloys have sufficient intercrystalline brittleness for easy comminution. A particularly fine grain is, however, necessary because the size of the powder particles will be identical with the grain size of the alloy before pulverizing. Such nickel–iron alloys, suitable for magnetic materials, can be made more brittle by omitting the deoxidizer or by the addition of ferrous oxide or ferrous sulfide to the molten metal. After casting, the grain size is coarse, and must be diminished by hot-rolling, starting at temperatures above 1300°C. (2370°F.), to assure malleability. Rolling is continued at decreasing temperatures until fragmentation occurs[3] and the fragments are pulverized further by ball milling. Oversize particles can be reground only after recrystallization, as the result of an annealing treatment at 100°C. (212°F.).[4]

The milling or stamping of soft, malleable metals produces particles which are flaky or disk-shaped, rather than granular and equiaxial, and

[2] U. S. Pats. 1,669,649 and 1,739,052.
[3] J. C. Chaston, *Metal Treatment*, *1*, No. 1, 3 (1935).
[4] Brit. Pat. 327,419.

is the method employed in making aluminum and bronze powders for pigments. With the exception of flakelike copper powder, which is used primarily for current collector brushes, this type of powder is generally unsuited for use in sintering. The particles, of leaflike or laminal shape, are extremely thin (0.3 to 1.3 μ); their area is much larger (25 to 50 μ) and can be controlled to a certain extent.[5–7] Lubricants, such as lard, stearin, etc., tend to prevent interparticle cohesion and facilitate spreading of the metal into flakes during milling.

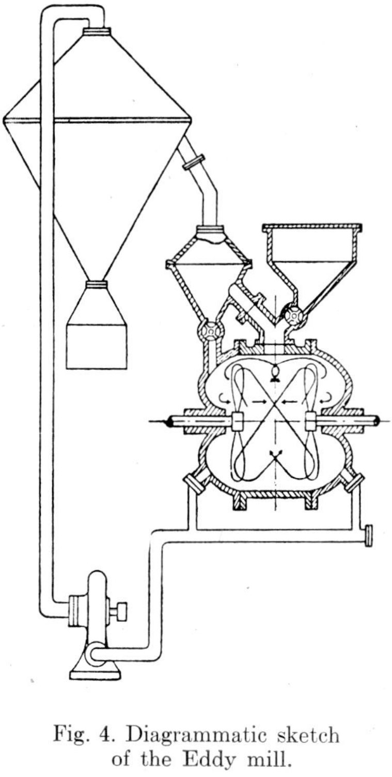

Malleable metals have been pulverized in Germany by the Hametag process which has been described in many patents.[8] Based on the principle of confining the impact blows to the particles themselves, it avoids contamination from grinding balls or from the walls of the mill, and in the case of flaking of aluminum prevents dust explosions by the introduction of an inert atmosphere in the mill. The Hametag process has found wide application in Germany during the War. According to Comstock,[9] the German industry in 1944 produced about 3500 tons of moldable steel powder per month, and approximately 85% of this production was obtained from Hametag mills.

In Figure 4 a diagrammatic illustration of the Eddy mill is reproduced from Jones' description.[10] The mill consists of a housing in which two fans are mounted at opposite ends which rotate in opposite directions at

Fig. 4. Diagrammatic sketch of the Eddy mill.

[5] H. H. Mandle, in J. Wulff, *Powder Metallurgy*. Am. Soc. Metals, Cleveland, 1942, chapt. 10.

[6] J. D. Edwards and R. B. Mason, *Ind. Eng. Chem., Anal. Ed., 6*, 159 (1934).

[7] J. D. Edwards, in J. Wulff, *Powder Metallurgy*. Am. Soc. Metals, Cleveland, 1942, chapt. 9.

[8] German Pats. 395,075; 459,595; 459,695; 471,310; and 479,337. U. S. Pats. 1,832-868; 1,930,684, and 1,932,741.

[9] G. J. Comstock, *Communication at Powder Metallurgy Colloquim*, New York University, April 26, 1946.

[10] W. D. Jones, *Principles of Powder Metallurgy*. Arnold, London, 1937, pp. 177-178. See also U. S. Pat. 1,573,017.

high, but individual speeds. Thus two opposing streams of gas of high velocity are produced which serve as powder carriers. The particles are pulverized by collision. Before being introduced into the mill, the material must be reduced to a fairly small size, *e.g.*, small pieces of wire, $1/_{64}$ in. diameter. The mill has provisions for collecting and grading the powder which is produced. The Eddy mill is useful for pulverizing malleable metals into particles of various shapes. It is claimed that close control can be exercised over particle size and shape; angular, flake, pebble, and spherical powders have been produced, many exhibiting the peculiar characteristic of a saucerlike depression on the surface. Jones in describing this process reports that this depression can be closed up if necessary, yielding hollow particles, and sintered compacts of very fine and highly dispersed porosity can be obtained with incompletely closed particles. If

Fig. 5. Microsection through aluminum flake
powder particles (×200).

the mill is filled with an inert gas, the particles remain substantially free of oxide films and form a powder particularly suitable for the manufacture of sintered products.

The Micronizer, which is a jet pulverizer, operates on a similar principle. Air or superheated steam, at pressures of 100 to 500 psi, is injected into an annular chamber. The direction of the jets is tangential to a common circle concentric with the chamber. For certain nonmetallic materials, *e.g.*, organic chemicals and pigments, extremely fine powders have been obtained with this grinder but work with metals has not yet been done on an industrial scale.

The best-known process for producing flake metal powders is that of Hall,[11] by which almost any malleable metal or alloy can be comminuted by ball milling. The metal to be flaked is suspended in a solution of a lubricant, e.g., stearic acid in alcohol, so that a sludge is formed which effectively protects the mixture from dust explosions and at the same time improves flaking. Another advantage is that the sludge can be used directly for paint pigments, since simply filtering off the excess solvent produces a paste suitable for this purpose. Dry flakes can be obtained by distillation of the solvent after filtering (Figure 5).

Another type of mill, operating on the impact principle, is suitable for both brittle and malleable materials, and is known as the Micro-pulverizer.[12,13] This mill consists chiefly of a group of small impact hammers fastened to a rotating disk by small arms swinging on pivots. Disintegration of the powder particles is effected by the terrific impact forces created by the very high velocity of the rotating disk. This method has been found especially useful for breaking up spongy particles of malleable metals, such as copper and iron; the reduction in size of the particles is accomplished without changing the equiaxial particle shapes, and without substantially work-hardening the individual particles. Powders thus treated have been found to be most suitable for molding and sintering.

Pulverization of Molten Metals

GRAINING

The formation of solid metal granules by the graining process has only very limited applications for the powder metallurgist. The success of this method is dependent on the formation of an oxide surface on the individual particles, achieved by mechanically agitating molten metal (in contact with air) when it is in the semisolid (or *mushy*) state. Forceful stirring or shaking provides rather coarse granules whose coalescence is inhibited by the surface oxide films. Since the oxide content of such powders usually amounts to several per cent, their use for powder metallurgy is excluded. The method has applications, however, in the manufacture of grained aluminum (by the Thermit process) and granulated brass for brazing powders.

A slightly modified process has been described by Rees[14] for the

[11] U. S. Pats. 1,569,484 and 2,002,891.
[12] U. S. Pats. 1,622,849 and 1,711,464.
[13] Canadian Pat. 281,198.
[14] R. W. Rees, *J. Inst. Metals, 57*, 193 (1935).

manufacture of low-melting powders, *e.g.*, lead and tin alloys. By using an impeller-type pulverizer, which revolves at 1500 r.p.m. at a temperature a few degrees below the solidus point, powders as fine as 200 mesh have been reported being produced on a commercial scale.

Attempts to modify the oxide content of grained powder are described in a recent British patent,[15] disclosing a process in which the molten metal is agitated at the lowest possible temperature in gases containing only a small per cent of oxygen. However, it is not known whether this powder has found applications in powder metallurgy.

SHOTTING

Shotting is a method of producing minute metal particles by pouring molten metal into air or a neutral atmosphere. Practically all metals and alloys can be shotted, the size and character of the resultant shot depending on details of the operation. The shot may consist of large granules which can hardly be classified as powders, or may be of a very fine grade comparable with the common grades of metal powders. Frequently, shotted metal is an intermediate product which is later subjected to further comminution. The optimum temperature of the molten metal is slightly above the melting point of the metal or alloy. To obtain a uniform product, a fine, continuous stream is maintained. The metal may be poured or forced through screens or orifices into water. The particle shape of the shot is considerably affected by variations in the quenching procedure. If the metal is still liquid (due to overheating the molten metal), by placing the crucible close to the quench bath and using cold water, feathery particles are obtained. On the other hand, heating the molten metal to a temperature just above the melting point, placing the crucible considerably above the bath (sometimes requiring the construction of a *shot tower*), and using hot water, yields solid, spherical particles, which are formed even before they reach the bath. For the production of very fine shot a blast of compressed air, steam, or gas, or a jet of water is used instead of the quench bath. The stream of molten metal can also be poured onto a metal surface, level with the water-bath surface, or onto a rotating disk which discharges the particles into the bath.

One method involves pouring on a rotating surface, the cooling being accomplished by water.[15a] The apparatus, known as a *liquid disintegrator*, is shown in Figure 6. In this case the direction of the liquid jet is identical with that of the metal stream; the coolant surrounds the metal when impacting the rotating surface and the atomizing effect is enhanced by a

[15] Brit. Pat. 440,768.
[15a] U. S. Pats. 2,271,264; 2,304,130; 2,305,172; and 2,306,449.

44

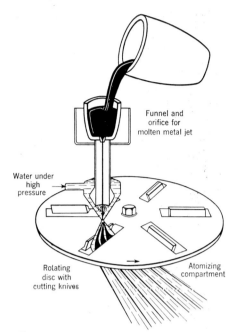

Funnel and
orifice for
molten metal jet

Water under
high
pressure

Rotating
disc with
cutting knives

Atomizing
compartment

Fig. 6. Diagrammatic sketch of liquid
disintegrator of Degussa type.

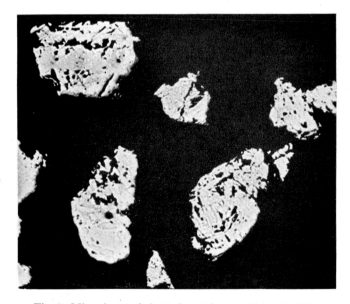

Fig. 7. Microshape of shotted cast iron particles (×300).

series of blades. In the so-called Degussa or DPG process, developed in Germany and employed to a large extent in England during the war, the cooling is performed by water under high pressure and the disk is rotated at 10,000 r.p.m. A serious difficulty encountered in the operation of rotating-disk atomizers was the freezing of metal on the blades, which set the rotating mechanism off balance and enforced an immediate interruption of operation. Attempts to replace the rotating disk by a system of rotating arms were only partly successful. Recent developments aim at the complete elimination of the disk and its replacement by a rotating, horizontal water curtain. The rotating-disk method and its modifications are of special interest for the production of alloy powders, particularly those only miscible in the liquid state (e.g., copper–iron, copper–lead, silver–lead).

Metals commonly shotted are tin, lead, zinc, aluminum, copper, silver, and gold, and many alloys, notably cast iron (Fig. 7) and brass (Muntz metal). Operating conditions must be varied with the particular properties of the metal or alloy, e.g., melting point, surface tension, etc., if particle size and shape are to be controlled. Special alloying additions are sometimes made to influence these properties. In addition to being an intermediate product in the manufacture of many powders, shot is used widely in the metallurgical field to facilitate weighing of alloying additions and, because of the increased surface area, to aid dissolution in fusion or chemical processes.

ATOMIZATION

The atomization method is most frequently employed for the production of powders from metals having relatively low melting points, and is occasionally referred to in the literature as *spraying*.

Atomization, closely related to shotting, consists essentially in forcing a thin stream of molten metal through a small orifice and then bombarding it with a stream of compressed gas, which causes the metal to disintegrate and solidify into finely divided particles. Usually the gas stream is directed through a nozzle, partly submerged in the molten metal, in such a manner as to draw the metal up through the nozzle to the tip. Solidification of the metal occurs instantaneously upon contact with the gas stream. The product is then removed by means of a suction system and collected in baghouses or cyclone dust collectors.[16,17]

The degree of fineness and the particle size distribution of the atom-

[16] D. O. Noel, J. D. Shaw and E. B. Gebert, *Trans. Am. Inst. Mining Met. Engrs.*, *128*, 37 (1938).
[17] D. O. Noel, in J. Wulff, *Powder Metallurgy*. Am. Soc. Metals, Cleveland, 1942, chapt. 8.

ized product can be varied over a considerable range: (*1*) by proper design of the orifice, nozzle, and tip, (*2*) by controlling temperature, viscosity, and flow of the liquid metal through the orifice, and (*3*) by adjusting the pressure and temperature of the gas.

A thorough study of the effects of some of the process variables upon the important characteristics of the final powder product and the method of operation has recently been undertaken in England by Thompson.[17a] In producing atomized aluminum powder, special attention was paid to the rate of atomization, the ease and efficiency of operation, and the particle size distribution of the resulting powder. Of particular interest seems the finding that variables concerned with the nozzle design are apparently of relatively minor importance, while the rate of atomization simply appears related to such variables as metal head, air pressure, metal temperature, and area of metal orifice, when each is considered separately. These variables are interdependent, however, and when two or more are considered together the relationships become more complex.

Many different nozzle designs have been patented in recent years, some of them being special combinations of orifices and nozzles, and others embodying the conventional features of spray guns. The basic process was developed by E. J. Hall for the production of aluminum more than twenty years ago, and was patented in 1924.[18] Various modifications have been suggested in the patent literature.[19-24]

Steady, uniform feed of the molten mass is especially important if fine particle size is the object. The compressed gas must often be preheated to a temperature above the melting point of the metal, to prevent the metal from solidifying within or around the orifice. Air or steam is commonly used as the compressed gas, but inert gases may also be used when formation of oxide films around the particles is to be avoided. However, even the use of hot compressed air or steam causes only slight oxidation in many atomized powders, and Noel[17] reports the remarkable figures of 0.2 to 0.3% oxygen for such metals as atomized lead, tin, zinc, cadmium, and aluminum. This low oxygen content is not objectionable as it tends to prevent further oxidation during the process, especially because the metal particles normally cool very rapidly and remain for only a fraction of a

[17a] J. S. Thompson, *J. Inst. Met.*, *74*, No. 3, 101 (1947).
[18] U. S. Pat. 1,501,449.
[19] U. S. Pats. 1,351,865 and 1,356,780.
[20] U. S. Pat. 1,545,253.
[21] U. S. Pats. 2,271,264; 2,304,130; 2,305,172; and 2,306,449.
[22] U. S. Pat. 2,330,038.
[23] U. S. Pat. 2,341,704.
[24] Swiss Pat. 206,995.

second in a temperature zone at which oxidation proceeds at a rapid rate. If an inert gas is necessary for the atomizing process, the powder must stay in this atmosphere during the entire period of collection, and air must be admitted only at a very slow rate after the powder has been completely cooled, to prevent spontaneous combustion, caused by the heat of oxidation raising the temperature of the powder above the ignition point. The maintenance of a low temperature in the dust-collecting system and strict precautions against sparking through static discharges or mechanical friction are additional safeguards against explosion hazards.

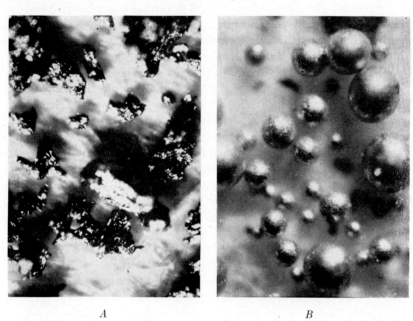

A   B

Fig. 8. Shape and surface characteristics of atomized 70–30 brass powders (×100): A, irregularly shaped powder; B, spherically shaped powder. (Courtesy of New Jersey Zinc Co.)

The commercial application of the atomization method is at present restricted to low-melting metals, chiefly because of limitations imposed by the corrosiveness of the nozzle material at high temperatures. Recent development work has been carried out with metals whose melting point is considerably above 900°C. (1650°F.), the temperature limit observed in commercial applications. Commercial batches of atomized copper and silver, and laboratory samples of nickel and iron powders have been produced, and their examination has suggested applications in which their special characteristics are useful.

Atomized powders generally contain particles of irregular or tear-drop-like shape, solid in structure, as shown for a fine powder in Figure 8A. The particles may also approach ideal spheres, as can be seen in Figure 8B. Low-melting metals, as, for instance, lead, tin, or cadmium, are basically soft and can readily be compacted even at very low pressures. The suitability of atomized aluminum powder for molding purposes depends on details of the atomizing technique which govern some of the powder characteristics, especially amount and type of oxide films, particle size distribution, etc. Although certain atomized aluminum powders are highly suitable for powder metallurgy purposes, other types cannot be used because of lack of purity and particle plasticity, or inability to form coherent bonds between the particles during cold-pressing, which result in compacts with cleavage fractures, lateral cracks, or other serious defects.

With increased basic hardness of the higher melting metals and alloys, such as copper and the commercial brasses and bronzes, the adaptability of their atomized powders for molding becomes even more difficult. The physical properties of parts made from these powders have been found to compare unfavorably with similar articles made from powders obtained by other methods. Sintering was found to be inhibited partially by a lack of contact areas, resulting from resistance of the individual particles to plastic deformation under pressure. Atomized nickel and iron powders, obtained on a laboratory scale, have been found to be even harder and more resistant to plastic molding, and have thus far resulted in compacts of unsatisfactory quality. However, the possibility of producing a low-priced powder having spherical particles of closely controlled fineness, by this method, has suggested a future use for atomized iron in the manufacture of powdered iron cores, in which case the poor compressibility and presence of oxide films might not be objectionable.

PHYSICOCHEMICAL AND CHEMICAL PROCESSES

This class includes the following methods: (1) condensation, (2) thermal decomposition, (3) reduction, (4) precipitation and replacement, (5) carburization and decarburization, (6) electrodeposition, (7) electrical dispersion, (8) diffusion alloying ("sintering"), and (9) alloy disintegration and intergranular corrosion. The first two may be considered essentially physical processes and the next three (3–5) chemical processes; the following two, (6) and (7), may be classified as electrochemical and electrical methods, while the last two, (8) and (9), belong to the group of miscellaneous chemical methods.

Physical Methods

CONDENSATION

Metals which can be vaporized at fairly low temperatures can be condensed on glass, metal, or other cool surfaces. The most important commercial application of this method is for the manufacture of zinc dust, of which tons are produced daily. This process, in fact, is a combination of two chemical processes—reduction and condensation—for the first step is the vaporizing of zinc oxide, followed by carbon monoxide reduction to zinc vapor; the final stage is the condensation either to molten zinc or zinc powder, depending on the prevailing conditions of temperature, gas concentration, composition, etc. In this process zinc oxide is mixed with powdered coal and heated until zinc vapor is formed by the reaction of the oxide with carbon monoxide. The vapor is condensed in a cooler extension of the retort. Reoxidation of the vapor during cooling, caused by the formation of carbon dioxide, results in oxide films around the droplets, preventing coalescence into a liquid and yielding *blue powder* as a by-product. Excess oxygen in the oxide–carbon mixture and a low temperature in the condenser tubes (causing rapid condensation) increase the amount of powder produced. Because of its high oxygen content, this type of powder is not suitable for general powder metallurgy purposes. The product can be considerably improved by altering certain conditions during condensation. By adding a second condenser and by reducing and closely controlling the amount of carbon dioxide in the condenser, a very fine powder can be obtained, with a relatively low oxygen content, although it is still higher than that of an atomized powder. The product is always fine and can be varied within fairly wide limits of particle size distribution, a relatively high proportion of extremely fine dust being always present. Commercial grades of zinc dust are fairly uniform and are used in large quantities for the pigment industry, as reducing agents, for Sherardizing, and for metal spraying of zinc coatings.

Extension of the process to include other, higher melting metals (*e.g.,* magnesium) has been reported,[25] and vaporization by means of an electric arc or by dropping the metal on a very hot plate, to allow the fine droplets of metal to vaporize quickly, was suggested by the same source.

THERMAL DECOMPOSITION

Thermal decomposition of a metal powder from its gaseous phase has found industrial application in the carbonyl process in which the decomposition is followed by condensation of the vapor. A number of metals

[25] W. J. Baëza, *A Course in Powder Metallurgy.* Reinhold, New York, 1943, p. 28.

form organic compounds which, under proper temperature and pressure conditions, decompose to yield a gas and a metallic powder; iron and nickel, however, are the only important metals which can be obtained in powder form by the decomposition of their carbonyls.

The carbonyl process had its origin in Germany where I. G. Farbenindustrie produced commercial quantities of iron and nickel powders having characteristics which greatly facilitated recrystallization and homogenization of the grain structure during sintering.[26-36] The demand for a pure iron powder with extremely fine and spherical particles to be used for the molding of high-frequency iron core coils has inaugurated the production of carbonyl iron in America in appreciable quantities and various grades, replacing the former German source.[26]

The manufacture of metal powders from their carbonyls has been described in many publications and patents. Detailed accounts were given by Mittasch, Schlecht, and others in Germany,[27-37] by Job and Cassal[38] in France, by Chaston,[39] and, more recently, Pfeil[39a] in England, and by Fieldauer and Jones[40] and Trout[41] in America. The first patents covering the carbonyl process were taken out by Mond[41a] as early as 1890; more recently many important basic patents have been taken out by I. G. Farbenindustrie,[42-45] by Mond,[46] by General Electric Co., Ltd.,[47,48] and by others.[49-51]

[26] General Aniline Works, New York, N. Y., Catalogs for 1943 and 1945.
[27] A. Mittasch, *Z. angew. Chem., 41,* 827 (1928).
[28] A. Mittasch, *Stahl u. Eisen, 48,* 979 (1928).
[29] L. Schlecht, W. Schubardt, and F. Duftschmid, *Z. Elektrochem., 37,* 485, (1931)
[30] F. Duftschmid and E. Houdremont, *Stahl u. Eisen, 51,* 1613 (1931).
[31] F. Stablein, *Z. tech. Physik, 13,* 532 (1932).
[32] L. Schlecht, W. Schubardt, and F. Duftschmid, *Stahl u. Eisen, 52,* 845 (1932).
[33] G. Hamprecht and L. Schlecht, *Metallwirtschaft, 12.* 281 (1933).
[34] W. Rüdorff and V. Hofmann, *Z. physik. Chem., B28,* 351 (1935).
[35] W. Hieber, *Angew. Chem., 49,* 463 (1936).
[36] E. K. Offerman, H. Buchholz and E. H. Schulz, *Stahl u. Eisen, 56,* 1132 (1936).
[27] L. Schlecht and G. Trageser, *Metallwirtschaft, 19,* 66 (1940).
[38] A. Job and A. Cassal, *Bull. soc. chim ind., 41,* No. 2, 1041 (1927).
[39] J. C. Chaston, *Elec. Commun., 14,* 133 (1935).
[39a] L. B. Pfeil, *Symposium on Powder Metallurgy,* The Iron and Steel Institute Special Report No. 38, London, 1947, p. 47.
[40] A. C. Fieldauer and G. W. Jones, *Am. Gas Assoc. Monthly, 6,* 439 (1924).
[41] W. E. Trout, Jr., *J. Chem. Education, 15,* No. 3, 113 (1938).
[41a] Brit. Pat. 12626/90; U. S. Pats. 455,227; and 455,228.
[42] U. S. Pats. 1,759,658; 1,759,661; 1,816,388; 1,825,241; 1,836,732; 1,852,541; 1,858,- 220 and 2,004,534.
[43] German Pats. 499,296; 500,692; 511,564; 517,831; 518,387; 518,781; 520,220; 520,221; 520,852; 523,601; 524,963; 531,479; 532,534; 535,437; 544,283; 545,710; 546,353; and 615,524.
[44] Brit. Pats. 334,976; 336,007; and 364,781.
[45] French Pats. 677,858; 690,991; 691,100; 691,243; 708,260; 715,206; and 717,568.
[46] French Pat. 709,381.

Carbonyls are obtained by passing carbon monoxide over spongy metal at specific temperatures and pressures. Iron pentacarbonyl, $Fe(CO)_5$, is a liquid at room temperature, boiling at 103°C. (217°F.), while nickel tetracarbonyl, $Ni(CO)_4$, boils at 43°C. (109°F.). When the pressure is reduced to one atmosphere and the temperature is raised correspondingly, both of these carbonyls dcompose to re-form the metal and carbon monoxide, the latter being recycled to form more carbonyl and to continue the process. In other words, the process constitutes a reversible reaction, the direction of which is determined by the prevailing conditions.

Nickel made by this method, known as the Mond process, is usually allowed to build up to shot of appreciable size, with commercial grades running between $1/_8$- and $1/_2$-inch grain diameter. Nickel carbonyl is led into a heated chamber where a cloud of microscopic condensed particles is first deposited. These particles, whose size has been found[52] to be about 0.01 μ, serve as nuclei on which more condensing metal is built up until quite large shot is formed. The build-up period necessary for $1/_2$-inch diameter shot covers many months. Particle growth follows a pattern of characteristic concentric structure, frequently referred to as *onionlike*. Annular discontinuities are probably caused by differences in the rate of decomposition or by the inclusion of carbon films. Although nickel can be collected as very fine powder, it has been used only to a limited extent, because of the high cost and the small quantities available. Iron, on the other hand, is usually collected as a very fine powder, its particles ranging from $1/_2$ to 10 μ in diameter. This fine size, the high purity, and spherical particle shape—and possibly the peculiar concentric structure—give the material qualities which are very advantageous for use in magnetic cores. Figure 9 shows the particle shape of carbonyl iron powder at high magnification.

For the manufacture of iron powder, iron carbonyl is obtained by passing carbon monoxide over powdered iron in a pressure chamber at pressures of about 2000 psi and temperatures near 200°C. (390°F.). Of the many methods of obtaining the powdered iron, the one usually used is the reduction of iron oxide with hydrogen, natural gas, or carbon monoxide. In the latter case carbon monoxide is obtained by the interaction of coke and the powdered oxide, and the reduction is followed up immediately by the formation of the carbonyl, the speed of gas flow through the pressure

[47] German Pat. 542,783.
[48] Brit. Pat. 300,691.
[49] U. S. Pat. 2,070,079.
[50] Brit. Pats. 244,895; 269,677; 284,087; 353,671; 367,996; and 374,560.
[51] French Pat. 765,893.
[52] R. Brill, Z. *Krist.*, 68, 387 (1928).

chamber being controlled to give a concentration of about 6% of iron carbonyl in the gas.[53]

During the decomposition phase, the temperature of the chamber is usually maintained at about 200°C. (390°F.), since a lower temperature would result in slow or incomplete decomposition, while a temperature above 400°C. (750°F.) would result in oxidation of the iron particles in the carbon monoxide atmosphere. Cooling conditions must be controlled to influence the final properties. The optimum conditions of reduction, of faces results in nodules, sponges, or smooth layers. If a pure, very fine product is to be achieved, the decomposition chamber must be of sufficient size to prevent decomposition taking place on the walls, or special cooling systems must be installed inside the walls. Precipitation of the metal as a fine powder in the hot space is facilitated by mixing the carbonyl with a stream of hot inert gas, such as nitrogen.[54] After decomposition, the

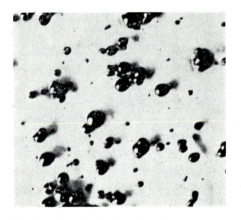

Fig. 9. Microshape of carbonyl iron powder
(×500). (Courtesy of W. D. Jones.)

powder can be mechanically ground if further reduction in size is desired. To prevent pyrophoric reactions, milling in inert atmospheres is recommended.

The high cost of carbonyl iron powder has prevented more general use, although there have been many reports of its advantages for general powder metallurgy applications. German research, notably by Offerman et al.,[55] has shown the excellent sintering qualities of carbonyl iron which

---

[53] French Pat. 715,206.
[54] Brit. Pat. 300,691.
[55] E. K. Offerman, H. Buchholz, and E. H. Schulz, Stahl u. Eisen, 56, 1132 (1936).

largely compensate for the poor compressibility of its rigid, spherical particles and for certain impurities, such as carbon and, to a lesser degree, oxygen. Very recently it has been stated that a few per cent of carbonyl iron powder added to commercial hydrogen-reduced iron improve the product considerably without materially affecting the cost. On a smaller scale the carbonyl process has also been used in the manufacturing of other metals, and methods covering the production of cobalt[56] and molybdenum[57,58] have been patented by I. G. Farbenindustrie. Joint decomposition of mixed carbonyls has led to the production of alloy powders; a number of patents cover alloys of iron, nickel, cobalt, tungsten, and others.[59] Iron–nickel alloy powders obtained by this method are preferentially annealed to allow diffusion to occur and to assure uniformity of the solid solution.

Major Chemical Processes

REDUCTION

Reduction of metal compounds, especially oxides, by the *action of a gas* or a gas-forming solid, is a thoroughly familiar process to the metallurgist, since it is one of the basic steps in the extraction of most metals from their ores. In applying this method to the production of metal powders, however, lower operating temperatures (usually well below the melting points of the metal and the oxide) are employed. The treatment of pulverized oxides at elevated temperatures in reducing atmospheres has become one of the most extensively used of all methods of powder production, since it provides a convenient and flexible means for controlling the properties of the product over a wide range. Oxides are generally very brittle and, since comparatively little energy is consumed in comminuting them to the desired degree of fineness, the process is economical.

By varying the raw material and the conditions of reduction, considerable control over the properties of the final product can be exercised. The nature, particle size, and distribution of the starting material all influence the form of the deposited particles. Decomposition on hot surcourse, differ for each metal. The temperature and rate of reduction are extremely important, the finest powders being produced at the lowest reduction temperatures. It is possible to obtain pyrophoric powders by grinding the oxide extremely fine and maintaining a low reduction tem-

[56] German Pat. 511,564.
[57] German Pat. 531,402.
[58] French Pat. 708,379.
[59] Brit. Pats. 284,087; 367,996; 374,560; and 423,823.

perature. On the other hand, an upper limit for the reduction temperature is imposed by the tendency of the powder to sinter, forming solid crusts which are impervious to the reducing gases, and which impede the rate of reduction. The time necessary for complete reduction depends on particle size of the raw material, temperature, and furnace conditions, e.g., type of furnace, method of suspension and transportation of material in the furnace, etc.

The selection of a suitable reducing agent is also very important, and must be made from the standpoint of composition, impurities (especially moisture content), and economy of operation. Hydrogen, carbon monoxide (obtained from a mixture of the powdered oxide and charcoal or coke), natural gas, partially combusted hydrocarbons, and dissociated ammonia have all been used for this purpose, the choice being frequently dictated by hydrocarbons, coal gases, or carbon monoxide, decarburization by hydrogen, or reoxidation by carbon dioxide.

The reduction process is open to many variations in order to produce a powder with specific properties. (The same results can be obtained by blending several powders produced by various methods.) One of the advantages of the method is that the particle shape of the final powder can, to a certain extent, be controlled. The individual particles are usually spongelike in structure and suitable for molding. The particle size distribution, apparent density, and flow of the powders can be modified over a wide range to meet industrial requirements. The fluffiness of the spongy particles facilitates plastic deformation of each particle during molding and considerably increases the contact areas between adjacent particles, thus aiding interparticle cohesion and resulting in compacts of high "green" strength, which can readily be handled in industrial operations and yield sintered metals of great strength. In Figure 10A the microstructures of reduced powder particles of different degrees of sponginess are shown. The method can also be adapted to control particle shape. Under suitable conditions the metal particles retain the shape of the particles of the starting material; in some cases the particles produced from oxides display distinctly angular contours, similar to those of pulverized brittle materials. The particles are, however, almost always slightly porous. If any oxide remains it is mostly some of the original oxide which has not been reduced and is, therefore, in the center of the particle (see Figure 10B), as distinguished from oxide films so frequently present in mechanically produced powders.

The reduction process is used for two classes of metals: (1) those for which the reduction can be carried out at temperatures very considerably below the melting point of the metal or oxide, making the process the only practicable method available, and (2) those for which the process is widely used in preference to other available methods because of economic considerations, comparative ease of reduction, or because the product has

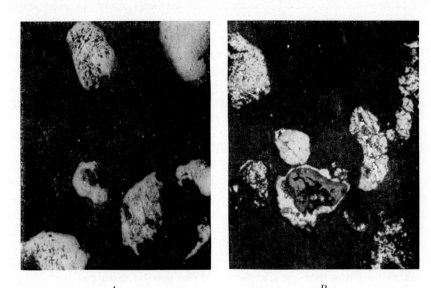

A   B

Fig. 10. Microsections through reduced iron powder particles (×500): A, completely reduced particles; B, particles containing unreduced oxides. (Courtesy of Metals Refining Co.)

certain characteristics which are difficult or impossible to obtain by other methods. Tungsten and molybdenum powders belong to the first group; iron, nickel, cobalt, and copper (by reduction of their oxides) are examples of the second group.

The manufacture of tungsten and molybdenum powders by the reduction method has been described comprehensively by Skaupy and other workers in a number of sources[60-63] and will be dealt with more fully in Chapter VI and also in Chapter XXI, Volume II. Cobalt powder,

[60] F. Skaupy, *Metallkeramik*. Verlag Chemie, Berlin, 1943.

[61] C. J. Smithells, *Tungsten*. 2nd ed., Chapman & Hall, London, 1936.

[62] H. W. Highriter, in J. Wulff, *Powder Metallurgy*. Am. Soc. Metals, Cleveland, 1942, chapt. 37.

[63] P. E. Wretblad, in J. Wulff, *Powder Metallurgy*. Am. Soc. Metals, Cleveland, 1942, chapt. 38.

important because of its use in sintered carbides, is produced almost exclusively by the reduction method. It is common practice to pulverize the oxide to a very fine state, mix it with powdered coal, and then reduce it with the resulting carbon monoxide at very low temperatures. Close control of the particle size is important, and much of the material produced has particles only a few microns in diameter. Only the production of copper and iron powders by the reduction method has reached a daily tonnage output. The direct reduction of copper oxide scale from wire-drawing and sheet-forming operations has resulted in products not considered pure enough for most applications.[64]   Oxides obtained from finely divided electrolytic copper or by chemical precipitation are the chief commercial sources of reduced copper powder. The same factors affecting the reduction of other metal oxides apply to copper, e.g., degree of oxidation, particle fineness, details of the reducing cycle, reducing agent, and the grinding procedure used to pulverize the sintered cake.

The manufacture of iron powder by the reduction process is probably the most interesting and, for the future industrial development of metal powder parts, the most promising application of the method. Numerous investigations on the direct reduction of iron ores to sponge iron will be outlined only briefly. Some of the individual processes will be discussed in greater detail in Chapter VI and also in Chapter XXV, Volume II.

Before the war Sweden was the source of most of the iron powder used for molding parts in this country. The powder, a by-product of the production of Swedish sponge iron,[65-69] was the only low-cost material available in larger quantities. It was obtained by the reduction of high-grade magnetite by charcoal or anthracite and was quite pure—being free from objectionable impurities, e.g., titanium, phosphorus, etc., which can not be removed during the low-temperature reduction.

In the last few years efforts to develop a low-cost domestic source of iron powder have led to the production of reduced iron powders on tonnage scales.[70-72]   Oxide ores of satisfactory purity, such as magnetites

[64] D. O. Noel in J. Wulff, *Powder Metallurgy*. Am. Soc. Metals, Cleveland, 1942, p. 119.

[65] E. Edwin, *Tek. Tids. Bergsvetenskap, 56*, 41 (1926); *Tek. Ukeblad, 73*, 225, 235 (1926).

[66] A. Johansson, *Jernkontorets Ann., 111*, No. 4, 51 (1927).

[67] M. Tigerschiöld, *Blad Berghandteringens Vänner, 21*, 219 (1932).

[68] E. Edwin, *Tijdschr. Kjemi Bergvesen, 22*, 23 (1942).

[69] E. Ameen, *Iron Age, 153*, No. 3, 55 (1944); No. 4, 56 (1944).

[70] A. T. Fellows, *Metals & Alloys, 12*, No. 3, 288 (1940).

[71] A. H. Allen, *Steel, 109*, No. 14, 58, 90 (1941).

[72] C. Hardy, *Metal Progress, 43*, No. 1, 62 (1943).

and hematites, are ground to desired fineness and reduced at temperatures ranging from 815° to 980°C. (1500° to 1800°F.) by charcoal or reducing gases. The purity of the product is affected by the choice of the ore, by the processing, and by the reducing agent. Appreciable amounts of carbon are usually introduced into the iron by carbonaceous reducing agents, and substantial residues of acid-insoluble matter (sometimes amounting to several per cent) are encountered in the products unless separation of the gangue during concentration is complete. The presence of carbon or silica is objectionable for many purposes because they impair the compressibility of the powder, cause excessive wear on the molding tools, and cause difficulties in controlling dimensions and properties in the sintered parts.

A recent method by which Swedish sponge iron powder is obtained as a by-product in a process for making a highly pure raw material for steel manufacture, has been described by Eketorp.[72a] A rich concentrate of magnetite ore is charged in cylindrical ceramic muffles in alternate layers with solid carbon, such as mixtures of bituminous coal and coke or anthracite fines. Heating is carried out in much the same manner as that used for firing refractory bricks. The iron sponge is obtained in form of rectangular cakes of a density of about 2 g. per cc. or in form of compressed cylinders of approximately twice the density. The resulting iron analyzes as 97.2% iron, 0.009% sulfur, and 0.11% phosphorus.

Additional details of the prevailing Swedish sponge iron process, known as the Höganas process, have also been recently revealed by Tigerschiöld.[72b,c] The crude ore from Gellivare, which contains 65–68% iron and about 1% phosphorus, is milled and concentrated into a magnetite containing 71.5% iron, about 0.008% phosphorus and practically no sulfur, with a total gangue content of less than 0.9%. Höganas sponge iron is produced in fire-clay saggers. The ore is packed in layers alternately with layers of a mixture of coke breeze and limestone. The charge is heated to a maximum temperature of about 1200° C. (2190°F.) and the reduction cycle including the heating and cooling periods requires as much as 180 hours. The silica content of the reduced sponge iron is about 0.4%, and can be reduced to less than 0.2% by magnetic concentration.

Substituting pure oxides for the ores (e.g., mill scales) and replacing carbon monoxide by hydrogen results in pure and very spongy iron powders, highly suitable for molding purposes. The cost is correspondingly

[72a] S. Eketorp, Jernkontorets Ann., 129, No. 12, 703 (1945), translated into English in Iron Steel Inst. London, Translation No. 275, July, 1946; Metal Powder Report 1, No. 1, 4 (1946).
[72b] M. Tigerschiöld, Metals Technol., 14, No. 5, T.P. 2188, 2 (1947).
[72c] M. Tigerschiöld, Jernkontorets Ann, 131, 295 (1947).

higher, although the product may still be economical if savings in tool wear, power, and fuel are considered. For many applications, iron powders which were subjected to a final brief treatment in hydrogen after the initial carbon monoxide reduction of the oxide ore have been found satisfactory. Since particles of such materials are distinguished from hydrogen-reduced mill scales by greater compactness, they usually require somewhat higher molding pressures and higher sintering temperatures.

Reduction by hydrides applies to cases where ordinary reducing agents, such as hydrogen or carbon monoxide, are not sufficiently effective.

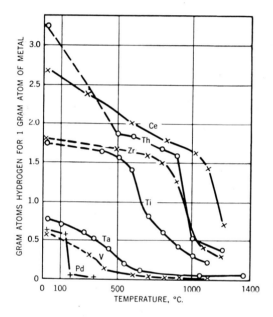

Fig. 11. Solubility of hydrogen in hydride-forming metals at atmospheric pressure, according to Sieverts.[73]

For refractory metal oxides (*e.g.*, titanium and zirconium oxides) and for the production of certain alloys which include one or more components that tend to form very stable oxides (*e.g.*, aluminum or chromium, which are vital components in sintered permanent magnets and sintered stainless steels, respectively), the use of atomic hydrogen, a very powerful reducing agent, has proved successful. Nascent hydrogen—*i.e.*, hydrogen in the atomic state—is evolved from certain metal hydrides which are commercially available as fine powders (300 mesh or finer). Under suitable

[73] A. Sieverts, *Z. Metallkunde, 21,* 41 (1929).

conditions, elements of the fourth and fifth groups of the periodic table form hydrides which are quite stable at room temperature, are nonhygroscopic, and can be preserved indefinitely in air, since they do not oxidize under such conditions. Evolution of hydrogen occurs upon heating these compounds under vacuum or in a nonoxidizing atmosphere above $350\,^{\circ}$C. ($660\,^{\circ}$F.); complete dissociation into hydrogen and the pure metal is obtained at red heat. The effect of temperature on the solubility of hydrogen in these metals can be seen from Figure 11.[73]

Alexander[74-77] developed the hydride process for producing refractory and other metals in powder form, using the hydride of an auxiliary metal such as calcium. Calcium hydride dissociates into calcium and nascent hydrogen, and the latter reduces the oxides of the metals to be prepared in powder form. The general reaction for the hydride process may be summarized by the following equation:

$$M_xO_y + y\,CaH_2 \rightleftharpoons x\,M + y\,Ca + y\,H_2O$$

For example, the reaction of chromium oxide and calcium hydride is as follows:

$$Cr_2O_3 + 3\,CaH_2 \rightleftharpoons 2\,Cr + 3\,Ca + 3\,H_2O$$

The metal hydride powders can be mixed readily with other metal powders, and during heating of the briquetted mixtures to sintering temperature, the hydrogen evolved—which is nascent at the instant of dissociation of the metal hydride—is powerful enough to reduce surface oxide films on the powder particles and form a protective atmosphere within the compact, which, under certain favorable conditions, can prevail throughout the sintering treatment.

By mixing a metal hydride with another metal powder, a eutectic alloy can be produced. Copper–titanium alloys are made in the following manner. Copper powder and titanium hydride powder are mixed in proportions corresponding to the eutectic composition; heat-treatment of the mixture, in hydrogen or under vacuum, to a temperature slightly above the melting point of the eutectic, and drawing at this temperature for a sufficient time to allow diffusion between the copper and titanium to be completed, result in liquefaction of the eutectic. After cooling, the eutectic alloy can be pulverized or can be used as a master alloy to replace addi-

[74] U. S. Pats. 2,038,402; 2,043,363; and 2,287,771.

[75] P. P. Alexander, *Metals & Alloys, 5*, No. 2, 37 (1934).

[76] P. P. Alexander, *Metals & Alloys, 8*, No. 9, 263 (1937); *9*, No. 2, 45 (1938); *9*, No. 7, 179 (1938); *9*, No. 10, 270 (1938).

[77] P. P. Alexander, in J. Wulff, *Powder Metallurgy*. Am. Soc. Metals, Cleveland, 1942, chapt. 12.

tions of pure titanium (which generally cause difficulties in alloying because of their lightness and tendency to float on top of a metal bath). In this connection, another method of interest may be mentioned, *i.e.*, the reduction of mixtures obtained from solutions of isomorphous salts.[78]

PRECIPITATION AND REPLACEMENT

The *precipitation* process offers means of producing commercially important powders, such as iron and copper, in good qualities and in very large quantities—permitting the production of these commodities at very low costs. It may therefore be expected that the process will become increasingly important with further expansion of the parts-producing industry.

The principle of precipitating a metal from solution by chemical means in a relatively insoluble form and subsequently reducing the precipitate to the metal in finely divided form, found its classical application in the Wollaston process, described in Chapter II, page 23. Wollaston precipitated platinum, as ammonium platinum chloride in powder form, from dilute aqua regia solutions and then heated the powder to obtain pure platinum. Substantially this method is still in use in the manufacture of platinum and palladium powders for use as chemical catalysts.

By the same principle, selenium and tellurium are precipitated from solutions derived from smelter fumes and sludges from electrolytic tanks in copper refineries.[79] Selenium powder is used in photoelectric cells; tellurium is employed as an alloying element for lead to increase its hardness and acid resistance. Both elements are also used for small additions to steels to improve their free-cutting properties.

The other well-known method of precipitating metallic powders is based on the principle of precipitating a metal from its aqueous solution by the addition of a metal which is higher in the electromotive series. Thus, tin powder is produced in quantity by precipitation by metallic zinc from solutions of stannous chloride. Similarly, silver powder is precipitated from a nitrate solution by copper or iron, and gold powder is precipitated by zinc and copper. The powders are very fine and of low apparent density, their typical particle structures consisting of crystalline aggregates. The tin powder is used for coating condenser paper; the silver powder is useful for molding purposes.

The addition of zinc scrap to a solution of cadmium sulfate results in the formation of zinc sulfate and a spongy precipitate of cadmium which

[78] Brit. Pat. 413,526.
[79] D. O. Noel, in J. Wulff, *Powder Metallurgy*. Am. Soc. Metals, Cleveland, 1942, chapt. 8, p. 121.

can readily be broken down to a fine powder. Similarly, the action of aluminum powder and an activating agent[80] is used to precipitate copper, nickel, and iron powders from solutions of their salts. In a process patented by Drouilly,[81] first copper and then nickel are precipitated in powder form from a solution of copper and nickel sulfates by the addition of aluminum powder. The process permits close control of the precipitate by regulating the temperature and the concentration of the metal in the sulfate solution. If aluminum powder of a certain mesh size is moistened with 10% hydrochloric acid (as an activating agent) and added to a solution of copper sulfate under certain temperature conditions, copper powder is precipitated which has exactly the same mesh size as the original aluminum. Dilute mercuric chloride or alkaline chlorides are also effective activating agents.

A very common method for producing copper powder is by precipitation from sulfate solutions or mine waters, flowing over scrap iron. This type of powder, suitable for certain chemical purposes, has not, however, proved to be of high enough quality to be useful in the manufacture of molded parts for the bearing or electrical industries. (It is generally known as "cement-copper" and contains up to 10% iron, as well as other impurities.)

The *chemical replacement* method is really a precipitation reaction with a limited material transfer into the solution. It is applicable for the production of composite powders whose individual particles consist of cores of a base metal and shells of a second metal. The best-known example is the copper-coated lead powder used for heavy-duty bearings (60–40, Cu–Pb).[82−84] In this process fine lead powder is introduced into an agitated copper acetate solution which is kept at a certain temperature. The reaction between the surface layers of the lead particles and the copper acetate solution results in superficial dissolution of the lead and precipitation of copper films on the particles. In other words, the surface layers of the lead are replaced by surface films of copper. The reaction is completed when all surface areas are covered with copper. A correct proportioning of the lead powder and the acetate solution results in the formation of lead acetate solution and complete precipitation of the copper, yielding the desired composite powder. The powder obtained after filtering, drying, and low-temperature reduction, consists of spherical lead particles completely surrounded by copper shells; it is ideally

[80] Brit. Pat. 403,469.
[81] U. S. Pat. 1,963,893.
[82] E. Fetz, *Metals & Alloys, 8,* 257 (1937).
[83] U. S. Pats. 2,033,240 and 2,234,371.
[84] Brit. Pat. 463,775.

suited for molding, since it permits the use of sufficiently low pressures to prevent impairment of the copper films, which retain the liquid lead droplets during sintering. The photomicrograph in Figure 12A shows a section through copper–lead particles made in the above manner. Although this process has been in the experimental stage for several years,[82] it is not known to the author whether it has reached a commercial production scale.

Figure 12B shows a section through copper-coated iron particles made by the replacement method.

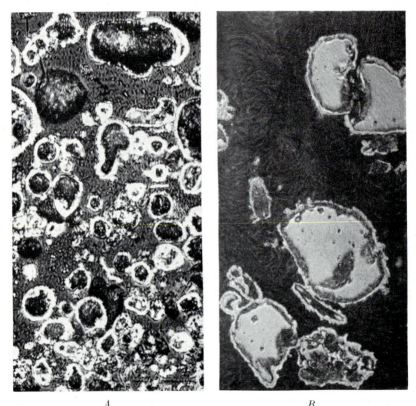

A   B

Fig. 12. Microsections through coated powder particles ($\times 500$): A, copper-coated lead particles; B, copper-coated iron particles. (Courtesy of Charles Hardy, Inc.)

CARBURIZATION AND DECARBURIZATION

The severe curtailment in the importation of Swedish sponge iron at the outbreak of the war resulted in numerous attempts to increase domestic sources of iron powder. One direction in which work proceeded was the partial (or complete) decarburization of granulated pig iron or of shotted cast iron. Comstock[85] in 1940 described a product made by such a process. Available in three grades, with carbon contents varying from practically carbon-free to the eutectoid composition, the powder compared favorably in many respects with the Swedish type. Although the powder equalled Swedish sponge iron powder in purity and compressibility, the physical properties of sintered compacts were even better.

The process for making this powder, designated as NY 100, was developed by Ekstrand and Tholand, on the basis of the process of Kalling and Rennerfelt for granulating pig iron.[86,87] It consists essentially of granulating molten pig iron by spraying it into a water bath, followed by a decarburization treatment in a rotary furnace in a carbon monoxide–dioxide atmosphere, without melting or oxidizing the powder. In a similar manner, cast iron in the form of comminuted particles, such as are used in sandblasting grits and fine shots, can be decarburized into iron. Several decarburizing agents have been used for this purpose, especially steam, hydrogen saturated with water, and carbon monoxide–dioxide gas mixtures. It has also been found that decarburization is greatly facilitated by mixing the carbon-rich iron with iron oxide powder, the resulting powder being a mixture of decarburized steel and reduced sponge iron.[88] Boegehold[89] has applied the decarburization method to the formation of a nickel–iron powder of low coercive force. After atomizing a molten composition containing 1–4% carbon and the desired amounts of iron and nickel, the powder is chilled and mechanically comminuted. Finally, the material is subjected to a decarburizing and alloying treatment which leaves the alloy powder with not more than 0.05% carbon.

During the last ten years or so many attempts have been made to produce steel powders suitable for molding parts. Poor molding properties resulting from increased hardness and resistance to plastic deformation have accounted for the relatively slow advance of steel powder metallurgy until recently. One of the methods used has been to prealloy the steel powders by carburizing iron powders with the aid of solid or gaseous reagents. Compressibility of such carburized steel powders has been considerably improved by a limited decarburization treatment, which causes the formation of soft, ferritic surface regions in the hypoeutectoid, eutectoid, or hypereutectoid steel particles.[90] Photomicrographs of such gas-

[85] G. J. Comstock, *Steel*, *106*, No. 22, 54 (1940).
[86] B. Kalling and I. Rennerfelt, *J. Iron Steel Inst. London*, *140*, No. 2, 137 (1939).
[87] U. S. Pat. 2,170,158.
[88] U. S. Pat. 2,315,302.
[89] U. S. Pat. 2,289,570.
[90] U. S. Pat. 2,175,850.

carburized steel particles before and after partial decarburization are shown in Figures 13A and 13B.

In a method by General Motors Corporation[91] for the production of iron powder, carburizing plays an important role. The process employs steel scrap as raw material. Since this type of material cannot readily be pulverized by mechanical means, the process provides first for an em-

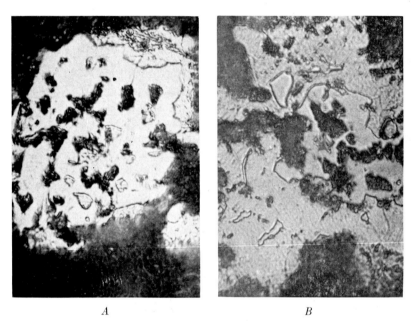

A   B

Fig. 13. Microsections (×1000) through gas-carburized iron powder particles: A, particle gas-carburized to complete cementite; B, particle gas-carburized and surface-decarburized showing core of cementite and case of ferrite. (Courtesy of Charles Hardy, Inc.)

brittlement treatment by carburization. After crushing and grinding, the carbon-rich iron powder can either be decarburized or used "as is," depending on the desired carbon content of the final product. It is distinguished from other iron powders by substantially solid particle structures and by smooth particle surfaces. When mixed with spongy iron powder in certain proportions, it forms compacts which, after sintering, are especially suitable for impregnating work.

Carburization plays its most important part in powder metallurgy in

[91] U. S. Pat. 2,164,198.

the manufacture of sintered hard metals. There are many ways of producing the raw materials for cemented carbides, as will be explained in detail in Chapter VI, and also in Chapter XXII, Volume II. Engle[92] recently summarized the carburizing phase by stating that tungsten powder is mixed with lampblack and heated in hydrogen for a few hours at 1400° to 1500°C. (2550° to 2730°F.). Tantalum and titanium carbides can be obtained by a somewhat similar process, but, because of difficulties in producing the metal powders, are more often made by heating mixtures of their oxides with carbon to much higher temperatures (i.e., 1800° to 2000° C.; 3270° to 3630°F.). Tungsten carbide can also be produced by heating a mixture of the oxide with lampblack. Although the process is rarely used, gas carburizing of the metal powder constitutes another method. After the carburizing treatment, the sintered cake is crushed and screened within narrow limits, the maximum size limit usually being 15μ.[93] The carbide powders are then ready for mixing and milling with the cobalt binder.

Electrochemical and Electrical Processes

This class of methods of powder production includes: (1) electrolytic deposition from solutions, (2) electrolytic deposition from fused salts, and (3) electrical dispersion (sputtering) and comminution. The last method finds only rare application in the manufacture of extremely fine, near-colloidal powders; the other two are among the most common and universally used processes for manufacturing metallic powders for industrial purposes.

ELECTROLYTIC DEPOSITION FROM SOLUTIONS

Electrolytic deposition of metals from solutions (coherent deposits) results—through control of certain factors—in the production of deposits serving as a base for a powder. Jones[94] classifies electrolytic processes into three distinctly different methods: (a) deposition of a hard, brittle metal, a powder being obtained by mechanical comminution, e.g., grinding; (b) deposition of a soft spongy substance, a powder being obtained by pulverizing, e.g., light rubbing; and (c) direct deposition of a powder from the electrolyte. Powders obtained by the first method are generally unsuitable for molding purposes, but large quantities of electrolytic powders ideally suited for many applications are manufactured by the last two methods.

[92] E. W. Engle, in J. Wulff, *Powder Metallurgy*. Am. Soc. Metals, Cleveland, 1942, chapt. 39.
[93] V. Fisher, *Metal Progress, 36,* 247 (1939).
[94] W. D. Jones, *Principles of Powder Metallurgy*. Arnold, London, 1937, p. 173.

Hard, brittle electrodeposits can be obtained with comparative ease, but the powder obtained after crushing the drying has characteristic needlelike particles, which are extremely resistant to compacting under ordinary conditions. The compressibility of these powders can, however, be improved to some extent by annealing under conditions favorable for the removal of hydrogen. Iron and nickel powders are obtainable by this process; electrolytic iron powder produced by this method was at one time used in the manfacture of magnetic cores.[95],[96]

The depositions of slimes of spongy or powdery nature, disastrous for ordinary electrolytic work (*i.e.*, plating), forms the basis of the majority of commercial processes for the manufacture of electrolytic metal powders. The first patents covering this process date back to the turn of the century, when Clark[97] described a process for producing spongy lead deposits, and Edison[98] patented his well-known process for producing cobalt, nickel, and nickel–cobalt flakes for the Edison storage battery. Since then there have been an increasingly large number of patents issued; Rossman[99] gives an extended abstract of patents, covering the period up to 1931. During the last fifteen years the methods for producing spongy or powdery electrodeposits have been used extensively in the production of powders of copper, iron, and, to a lesser degree, for silver, cadmium, zinc, tin, antimony, chromium, and nickel. Probably the most important advantage of this kind of powder over most of the other kinds is its extraordinary purity. Moreover, certain advantages of other processes, such as the large measure of control over apparent density, particle size, and particle shape made possible by the reduction method, are equally assured by the electrolytic process. Variables permitting close control of chemical and physical properties of the powder include: (*1*) composition of the electrolyte (concentration of metal and pH), (*2*) temperature of the electrolyte, (*3*) rate of circulation of the electrolyte, (*4*) current density, (*5*) size and type of anode and cathode and their distance from each other, (*6*) removal of deposits at the cathode, and (*7*) addition agents. The relationship between these variables and the important powder characteristics has been discussed in detail by Passer.[100]

In powder production higher current densities are used than are generally employed for refining or plating. The deposit is rendered more spongy by the evolution of hydrogen at the cathode, achieved by high

[95] D. J. MacNaughton, *J. Iron Steel Inst. London, 109*, 409 (1924).
[96] B. Speed and G. W. Elmen, *Trans. Am. Inst. Elec. Engrs., 40*, 1321 (1921).
[97] U. S. Pat. 598,313.
[98] U. S. Pats. 821,626; 865,688; and 936,525.
[99] J. Rossman, *Metal Ind. New York, 30*, 321, 396, 436, 468 (1932).
[100] M. Passer, *Kolloid-Z. 97*, 272 (1941).

acidity and low metal ion concentration. Addition agents may consist of suitable organic substances in colloidal form. The colloidal additions are believed to act as *crystallization centers*, favoring the formation of many small crystallites instead of a few larger grains. Soluble organic agents, such as sugar, glycerin, and urea, have been recommended by Hardy[101] for producing electrolytic iron powder. Drouilly[102] uses glucose in his basic process for producing electrolytic copper powder. Drouilly's method, which involves fuming of the sulfuric acid with glucose prior to the addition of the acid to the electrolyte, eliminates hydrogen evolution at the cathode and improves current efficiency. The method is particularly valuable since it permits production of powders to closely specified screen analyses and apparent densities. The size can be varied over a wide range, from granular material to powders finer than 325 mesh, and the apparent density can range from one-tenth to one-half of the actual density of copper.

Sulfates are most frequently used for the electrodeposition of copper and iron powders, but chlorides, cyanides, etc., have also given satisfactory results. Sometimes a combination of several salts has been used, as in the case of the above-mentioned hard deposits of electrolytic iron, where an electrolyte consisting of ferrous sulfate, ammonium sulfate, and ammonium chloride was used, which was acidified by the addition of sulfuric acid. Details of the process for producing electrolytic copper powders have not been disclosed, but it appears from a review of the patent literature that the electrolyte contains (as copper sulfate) 2 to 3% copper, and sulfuric acid additions of about 10% are made. Bath temperatures are above room temperature and may even exceed 50°C. (122°F.). Cathode current densities may vary over a very wide range, usually running above 10 amp. per ft.^{2} and sometimes as high as 300 amp. per ft.^{2}

In order to maintain control of the particle size and to prevent fluctuations in the yield, the adequate removal of the powder, either continuously or at frequent intervals, is of great importance. The effort has, therefore, been made to produce a substantially nonadherent deposit, which either drops to the bottom of the cell or can be readily removed by regular scraping of the cathode. The removal of the powder can also be assured by several other means, involving the use of rotating cathodes, rapid circulation of the electrolyte, or periodic interruption of the current.

All powders produced electrolytically must, after their removal from the cell, be subjected to cleansing and drying treatments. The powders must be carefully washed to remove all electrolyte clinging to the particles, because even the smallest amounts of salts left on the surface or in the

---

[101] U. S. Pat. 2,157,699.
[102] U. S. Pat. 1,799,157. German Pat. 520,835. Brit. Pat. 303,984. French Pat. 656,777.

cavities of the particles will cause rapid oxidation upon exposure to air. Washing in a centrifuge serves to eliminate the last traces of salts and acids, and drying in inert or reducing atmospheres further prevents oxidation.

Powders obtained by the electrolytic method are generally somewhat harder than those produced by reduction. This may not be objectionable if further milling to subsieve sizes is intended. Thus, it has been possible to grind a fine electrolytic copper powder (all passing through a 325 mesh

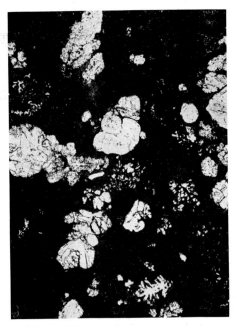

Fig. 14.   Microsection through electrolytic
copper powder particles (×150).

sieve) for a period of 30 hours in a laboratory ball mill, employing flint pebbles and small sections of copper rods. More than half of the original sample could be reduced by this dry milling operation to particles under 0.001 inch in size (as determined by elutriation and microscopic sizing). About 5% of the particles were as small as 0.0002 inch, with 25% about 0.0006 inch. The oversize material, averaging 0.002-inch particles, was distinctly flaky.[103]

---

[103] C. Hardy, *contribution to discussion*, Mass. Inst. of Tech. Powder Metallurgy Conference, 1940.

The structure of electrolytic powder particles is characteristically crystalline; their shape is generally *dendritic* or fernlike, but many types of powders, especially electrolytic iron, consist of particles of substantially solid, *nodular* structure. Each particle usually contains several individual grains, often spreading from the center to the circumference. Although dendritic structure is generally believed to give good molding properties because of the interlocking of the individual particles during compacting, it usually results in a powder of rather low apparent density and flow factor.[104] These characteristics are, however, much improved in the commercial powders by subsequent treatments. The dendritic structure is often altered to a more compact shape by a high-temperature drying and annealing treatment or by pulverizing of the sintered cakes. Figure 14 shows a photomicrograph of a variety of particle shapes included in one kind of electrolytic copper powder.

The electrolytic process of powder production has not been confined to the manufacture of pure metals. Alloy powders have been effectively produced by electrodeposition, with the simultaneous deposition of two metals.[105] Hardy,[103] pioneering in this field, has reported the successful production of electrolytic bronze and brass powders. He has also initiated development work on the electrodeposition of composite metal powders, consisting of metal particles of one kind enveloped by coats of another metal, listing in addition to copper-coated lead,[106] the following electro-coated powders: copper-coated silver, copper-coated graphite, nickel-coated iron, cobalt-coated tungsten, silver-coated molybdenum, silver-coated nickel, cadmium-coated nickel, and lead-coated nickel.

ELECTROLYTIC DEPOSITION FROM FUSED SALTS

By using a fused-salt bath below the melting point of the metal to be deposited, one can—with sufficiently high current densities—obtain deposits having very small crystallites or dendrites. Among metals obtained in this fashion are refractory metals, metals of the platinum group, iron, nickel, cobalt, chromium and manganese, silver, copper, beryllium, and aluminum. But only in the case of the high-melting metals—tantalum, thorium, and uranium—are powders made by fused-salt electrolysis suitable for sintering purposes.[107] The chief objection to this type of material lies in the difficulties of eliminating entrained or adherent fused salts. The mechanism of powder depositions from fused salts has

[104] U. S. Pat. 2,216,167.
[105] Brit. Pat. 312,441.
[106] U. S. Pat. 2,182,567.
[107] F. H. Driggs and W. C. Liliendahl, *Ind. Eng. Chem. Ed., 22,* 516, 1302 (1930); *23,* 634 (1931).

been studied in detail by Kroll[108] for the possible production of inexpensive lower melting metal powders.

In Figure 15 the apparatus for the fusion electrolysis of uranium is reproduced.[107,109] The metal is melted in a graphite crucible that serves as the anode; a strip of molybdenum serves as the cathode. The crucible is heated with Nichrome resistor elements, and is insulated by Sil-O-Cel.

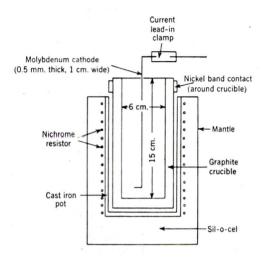

Fig. 15. Apparatus for the fusion electrolysis of uranium, according to Driggs and Liliendahl.[107,109]

Current is introduced with the aid of a nickel band. The electrolyte consists of a mixture of 250 g. calcium chloride, 250 g. sodium chloride, and 30 g. potassium uranium pentafluoride (KUF_5). Electrolytic deposition occurs at about 775°C. (1425°F.) at 30 amp. and 5 volts, with a cathode current density of approximately 15 amp. per ft.² Under these conditions, a dendritic deposit of about 1-in. thickness is obtained in 45 minutes. A surface coat of frozen salt protects the deposit from oxidation during cooling. Because of loss of the very fine particles during washing, the yield is only approximately 60%.

ELECTRICAL DISPERSION AND COMMINUTION

This method has been used in producing extremely fine powders by means of an electric arc formed between electrodes of the metal submerged

[108] W. J. Kroll, *Trans. Electrochem. Soc.*, *87*, 551 (1945).
[109] F. Skaupy, *Metallkeramik*. Verlag Cheimie, Berlin, 1943, p. 23.

under a liquid, *e.g.*, water or an organic solvent, enriched by small amounts of electrolytes. Kuzel,[110] in a patent obtained in 1908, refers to a colloidal tungsten powder obtained by the dispersion method in the presence of an alkali. Precious metals of the platinum group, as well as gold and silver, have been obtained in the form of hydrosols by this method; silver has also been produced by dispersion in alcohol.

By another process extremely fine powders can be produced by short-circuiting an electric current. The process involves forcing molten metal through a hollow electrode and spraying it toward a central electrode while an arc is formed. The current, which superheats the liquid metal as it is sprayed toward the central electrode, causes vaporization, thus interrupting the current and causing a short circuit. The rest of the metal volatilizes before reaching the central electrode and is then condensed in a hood. The resulting powder is so fine that its particles are all below one micron in size, some powders being actually in a colloidal state.

Miscellaneous Chemical Processes

DIFFUSION ALLOYING (SINTERING)

The principle of forming alloys from individual metal powders by *sintering* (depending on complete interdiffusion of the constituents at a temperature at which the compact is still solid) has been successfully employed to form nonferrous alloy powders from mixtures of the individual metal powders. The treatment may be carried out at a temperature below the melting point of the individual powders and the alloy formed, as in the case of copper–nickel alloys; or it may be used to form alloys in which one component melts below the temperature at which the diffusion treatment is carried out, as in the case of brasses and bronzes.

The method involves careful sizing and blending of the individual powders, placing them in vessels, and passing the mixture through a furnace with a neutral or reducing atmosphere. Temperature and time of the heat-treatment are governed by the diffusion laws, but are also affected by the characteristics and conditions of the powders and by their size distribution, which determines the closeness of packing and the interparticle contacts necessary for an effective diffusion of one metal into the other.

The cycle of the diffusion treatment must be carefully chosen, because otherwise great difficulties may be encountered in pulverizing the furnace product in attrition or impact mills. The tendency of the powders to sinter into a solid cake at temperatures above their recrystallization temperature limits the temperature and time. Very high temperatures may

[110] U. S. Pat. 899,875.

72

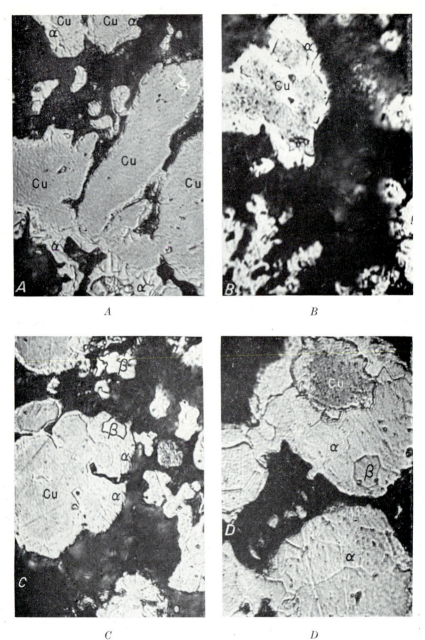

A   B

C   D

Fig. 16. Microstructure of particles of brass powders of following Cu–Zn
compositions produced by diffusion-alloying at 450°C. (840°F.) for 2 hours
(×1000): A, 85–15; B, 75–25; C, 65–35; D, 55–45.

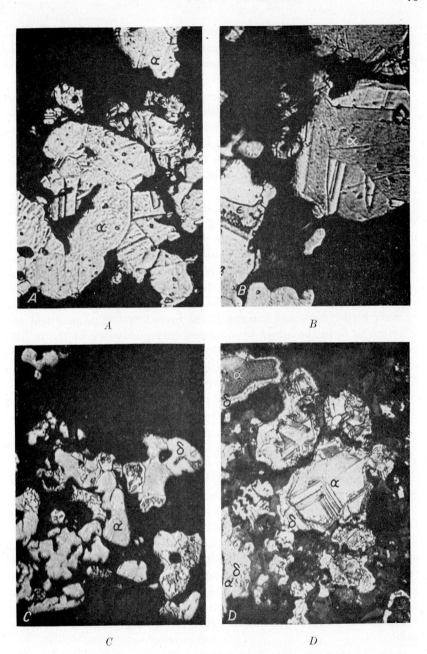

Fig. 17. Microstructure of particles of bronze powders of following Cu–Sn compositions produced by diffusion-alloying at 550°C. (1020°F.) for 1 hour: *A*. 95–5 (×1000); *B*, 90–10 (×1000); *C*, 85–15 (×500); *D*, 80–20 (×500).

be necessary if rapid diffusion is required. In these cases the sintered cakes frequently become so compact and malleable that their comminution into powder is only possible by drastic mechanical means which may partly, or completely, destroy important properties of the alloy powder particles, such as equiaxial shape, sponginess, or plasticity sufficient to permit molding. The percentage of the low-melting component may also have to be limited in those cases in which the diffusion temperature required is above the melting point of this conponent. Lower temperatures may be the only solution if a certain analysis has to be maintained, unless this increases the time necessary for complete diffusion to such an extent as to make the process impractical.

A considerable number of alloy powders have been produced by this method, especially binary alloys of copper with tin, zinc, aluminum, nickel, manganese, phosphorus, etc. Photomicrographs of several brass and bronze powder compositions made by the sintering method are shown in Figures 16 and 17; incomplete homogenization is apparent in most particles. Some of these alloys have found use in molding work, but their use is not comparable to that of the individual metal powders, because of their higher costs and because it has been found more practical and effective to form the alloys by diffusion during sintering of the molded compacts. Although the use of prealloyed powders may shorten the necessary sintering time, molding difficulties may be increased by the greater resistance to compression of the harder and stronger alloy particles. A poor bond between the particles will result, which may be reflected in the final physical properties of the molded part.

ALLOY DISINTEGRATION AND INTERGRANULAR CORROSION

Alloy disintegration involves a treatment with a solvent capable of dissolving only one constituent, the other remaining as finely divided particles or as a spongy mass which can then be pulverized. A classic example is the well-known assaying practice of treating silver–gold alloys with hot nitric acid, which dissolves the less noble silver and leaves the gold as a sponge readily susceptible to mechanical pulverization. Silver and gold powders can be obtained by amalgamation of these metals, grinding the amalgams with dry salt, and distilling off the mercury. Gold powder can also be obtained by alloying it with a small amount of bismuth, which contaminates the grain boundaries, thus facilitating chemical disintegration.

The method has also been adapted for the production of certain powders for use as catalysts; in this case, special alloys—sometimes of unusual composition—are prepared and subsequently decomposed by dissolving the secondary alloying constituent, leaving the primary metal behind in powder form. Thus, nickel powder is obtained by first alloying the nickel with aluminum or silicon and then dissolving the secondary metal with caustic alkali. Mercury is also used as secondary metal; after amalgamation, the mercury is distilled off. Powders produced by this method are usually extremely fine and active. They tend to be pyrophoric and must be stored in inert gases or solutions. The fact that they can be subjected to air only immediately before being used has been one of the chief reasons why this method has found little interest for powder producers supplying the powder metallurgy industries.

Fig. 18. Microsection through stainless steel powder particles produced by alloy disintegration (×200).

The method is also applicable to electrolytically deposited materials. If two metals are deposited in fine alternate layers, one layer can be dissolved, causing the other to disintegrate.

Intergranular corrosion, a form of alloy disintegration, has recently been successfully applied to the production of certain alloy powders. By promoting intergranular corrosion, by means of heat-treatment which makes the grain boundaries more susceptible to chemical attack than the grains, a complete disintegration of the bulk material may be accomplished. Wulff[111] has developed this ingenious process for the production of austenitic stainless steel powders of the 18–8 type. When stainless steel in the form of scrap, shotted metal, or bulk material is subjected to a carburizing heat-treatment between 500° and 750°C. (930° and 1380°F.) for a definite time, carbides will precipitate in the grain boundaries, and many corrosive substances will cause intergranular corrosion to the point of complete metal disintegration. Wulff suggests boiling Strauss solution (11% copper sulfate, 10% sulfuric acid, rest water) as a very effective corrosive agent but states that even sea water would give results. Copper, deposited on the grains during the corrosion treatment in the Strauss solution, is easily removed by a nitric acid etch. During the entire process less than 5% metal is lost, and the final carbon content is as low as 0.04%. The powders consist of particles having single crystals and can be used for molding and sintering under suitable conditions, i.e., high temperature and especially purified atmosphere. A photomicrograph of such a powder is shown in Figure 18.

The process of sensitizing materials and chemically disintegrating them into powder form by intergranular corrosion is not limited to stainless steels. Wulff reports interesting experiments with alloy steels containing titanium, columbium, or molybdenum, and with shotted nickel–base alloys of high sulfur content. Recently certain cobalt–base alloys have also been produced in this manner. Other experiments include the successful disintegration of various brasses, and even of pure iron, by analogous methods. Although relatively high losses of parent grain material and high cost of the required chemicals exclude the last two examples from commercial applications, the process for making stainless and alloy steel powders is today commercially employed by several concerns.

OTHER CHEMICAL PROCESSES

Other chemical methods, e.g., oxidation, are not included in this group, because they are not generally classified as processes used in manufacturing metal powders. Roasting and oxidizing are either operations preliminary to actual production methods or steps in specific processes, for example, in the manufacture of certain hard metals.

Summary

The various methods of producing metal powders discussed here are summarized in Table 2. The products are listed as metals, alloys, compounds, or composites, and the most important particle characteristics are noted.

[111] J. Wulff, in J. Wulff, *Powder Metallurgy*. Am. Soc. Metals, Cleveland, 1942, chapt. 11; also U. S. Pat. 2,361,443.

Production of Metallic Powders (Summarizing Tabulation)

Mode of production	Metals[a]	Alloys	Compounds	Composites	Particle characteristics
		Products			
Machining	(Mg, Fe)	(Dental alloys)			Coarse powders, flaky particles
Crushing	(W, Mo, Cr)	(Cast irons, steels)			Hard coarse powders, angular particles
Milling	W, Mo, Ti, Mn, Sb, Bi, (Cu, Fe)	Ni-Fe, Al-Mg, Al-Fe	Refractory metal carbides		Fine powders and spherical particles of brittle metals, leaflike particles of ductile metals
Shotting	Sn, Pb, Zn, Al, Cu, Ag, Au	Brasses			Granular or fine powders, solid particles
Graining	(Al, Pb)	(Brasses, solders)			Coarse granular powders, high in oxygen
Atomization	Sn, Zn, Pb, Cd, Al, (Ag, Cu, Ni, Fe)	Brasses, bronzes			Fine powders, particles are spheres, near spheres, or droplets
Condensation	(Zn, Mg)				Fine powders, spherical particles, high oxygen content
Thermal deposition	Carbonyl Fe, Ni	Fe-Ni			Very fine powders, uniform spherical particles
Gaseous reduction	W, Mo, Fe, Ni, Co, Cu				Powders of various grades, particles soft, plastic, and spongy
Hydride reduction	Ca, Ti, Cr	Cu-Ti			Very fine powders, fluffy particles

[a] Parentheses symbolize powders not generally suitable for powder metallurgical purposes.

(continued on page 78)

TABLE 2 (*Concluded*)

Production of Metallic Powders (Summarizing Tabulation)

Mode of Production	Metals[a]	Products			Particle characteristics
		Alloys	Compounds	Composites	
Precipitation	Pt, Pd, Ag, Cu, Ni, Fe, Sn, Cd, (Se, Te)	Cu–Ni			Very fine powders of low apparent density
Replacement				Cu-coated Pb	Fine powder, spherical particles
Decarburization		Cast iron, Ni–Fe			Hard powders, angular or rounded solid particles
Carburization		Steels	Refractory metal carbides		Fine, hard powders, particles solid or porous
Electrolytic deposition	Cu, Fe, Ag, Ni, Cd, Zn, Sb, Sn, Mo, Cr, Ta	Brass, bronzes		Cu-coated Pb	Fine powders, initially hard, particles dendritic or nodular, plastic, moldable
Electrical dispersion	(W, Au, Ag, Pt)				Extremely fine powders, near-colloidal powders
Sintering		Brass, bronzes, Cu–Ni			Medium-fine powders, particles spongy or solid, strained or plastic
Alloy disintegration	(Ni, Fe)	Stainless and alloy steels, (brass)			Fine or very fine (pyrophoric) powders, solid, single-crystal or spongy particles

[a] Parentheses symbolize powders not generally suitable for powder metallurgical purposes.

Three processes are most useful for powder metallurgy products:

(1) *Atomization.* The most important method; employed for the manufacture of low-melting metal powders, such as tin, lead, zinc, cadmium, aluminum, etc.

(2) *Gaseous reduction.* The most flexible and economical method; employed for the manufacture of commercial quantities of powders of the most common metals (iron and copper), of the less common metals (nickel and cobalt), and of the refractory metals (tungsten and molybdenum).

(3) *Electrolysis.* The method most suitable for the manufacture of extremely pure powders of a variety of metals, which include copper and iron in commercial quantities, and, to a lesser degree, silver, nickel, manganese, chromium, tantalum, etc.

Of the remaining methods only a few are of interest to the powder metallurgist: mechanical comminution of brittle metals (milling), thermal decomposition (metal carbonyls), chemical precipitation and gaseous reactions (carburization and decarburization), and formation of alloy powders by diffusion or by chemical disintegration.

CHAPTER IV

Characteristics and Properties
of Powders

In discussing the factors determining the usefulness of metal powders for powder metallurgy applications, we must distinguish between physical and chemical properties. The workability of a powder into a compact and the properties of the metallic products are affected to a marked degree by such physical characteristics as particle size, shape and structure, and surface conditions. Other characteristics, such as apparent density and flow, depend on the basic physical properties mentioned above. The recently concluded systematic study by Leadbeater, Northcott, and Hargreaves[1] in England of the relationship between these physical powder characteristics and the mechanical properties of sintered products, as applied to 28 different commercial iron powders, is ample proof that a definite correlation between powder characteristics and ultimate properties of the resulting compacts can be derived by statistical analysis (see also Chapter XXV, Volume II).

On the other hand, the chemical condition of a powder, *i.e.*, the degree of purity, is equally important, as solid or gaseous impurities in the form of inclusions, occlusions, or surface films also have a great influence on the performance of the powder during processing and on the final properties. The physical character of the particle and the particle shape and size are the principal factors determining the success or failure of a powder metallurgy process and their importance cannot be overemphasized.

PHYSICAL CHARACTERISTICS AND PROPERTIES

Order of Magnitude of Particle Size

Next to the molecules in order of size are the aggregates of atoms or molecules which are known as colloids. The behavior of these ultramicro-

---

[1] C. J. Leadbeater, L. Northcott, and F. Hargreaves, *Symposium on Powder Metallurgy*. The Iron and Steel Institute, Special Report No. 38, London, 1947, p. 15.

81

scopic particles is treated in colloid chemistry and is governed by the extremely high ratio of the surface to the volume. Beyond the particle size range covered by colloid chemistry are particles the limits of which extend from microscopic resolution to visibility by the naked eye. These particles, which behave differently from molecules and colloids, are frequently encountered in nature, for example, in the erosion of soils, silting of rivers, etc. An excellent treatise on the technology of these fine particles, dealing with particle measurement, size distribution, packing arrangements, physical properties, and industrial applications, has recently been published by DallaValle.[1a]

It is in this range of particle sizes that metal powders suitable for processing into solid metallic objects belong. The powder particles vary between the extreme limits of 10^{-1} and 10^3 μ, thus including submicroscopic, microscopic, and macroscopic sizes. The relationship of this range to colloid and molecular dimensions is shown in Figure 19, which clearly

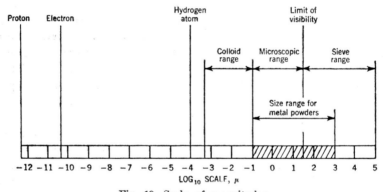

Fig. 19. Scale of magnitudes.

indicates the wide variation of powder particle sizes, all the way from colloidal particles to particles approximately $1/32$ in. in size. Accordingly, the limits of the volumes of these particles are 10^{-3} μ^3 and 10^9 μ^3, or 10^{-15} cm.^{3} and 10^{-3} cm.^{3} (1 mm.^{3}), respectively, based on the volume of cubes with sides of the above-mentioned dimensions.

It is customary to define the size of a particle by specifying a linear dimension, e.g., the diameter, and not by specifying the volume (for spherical particles the diameter is the true indication of size). However, most metal powders do not consist of spherical particles and the particle size is not a definite quantity which can be measured directly. As a matter of fact, different size values may be assigned to any given nonspherical

[1a] J. M. DallaValle, *Micromeritics*. Pitman, New York, 1943.

particle, according to the method of size determination employed. Large particle sizes may be defined as the cube root of the particle volume, as determined by weighing and counting. Within the sieve range, the particle size is measured as the size of the opening of a screen of standard woven square mesh which will just pass or retain the particle. For particles falling within the microscopic range, the size is determined by averaging several dimensions, such as length and width. The sedimentation method defines particle size as the diameter of an ideal sphere having the same specific gravity and settling velocity as the actual particle.

Metal powder particles spanning such a wide range of sizes behave according to different laws. The particles neighboring the lower limit of the range (approximating the size of tobacco smoke) diffuse almost like a gas and follow Cunningham's modification of Stokes' law in settling. This law applies without correction to larger particles, between 0.2 and 50 μ, the force retarding them being due entirely to the viscosity of the medium. Near the upper limit of the size range, viscosity loses its importance and resistance encountered in falling is caused substantially by difference in density between particle and fluid medium. The rate of fall for these large particles varies directly as the square root of the diameter, not as the square of the diameter, as in the case of the smaller particles.

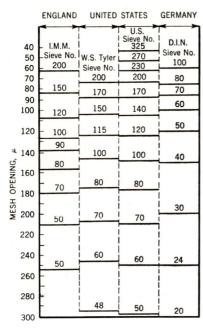

Fig. 20. Relationship of international sieve units (Skaupy).

For all practical purposes metal powders can be classified into sieve and subsieve size ranges. Powders generally used for molding of parts, magnets, bearings, etc., belong to the first class; powders used in the manufacture of refractory metals, cemented carbides, or magnetic cores belong to the second class. In both cases the prime requisite is that the particle size range and distribution remain constant within close limits to assure uniformity and reproducibility of the product. Specifications covering particle size are usually very strict, as a small change in average particle diameter or in distribution of sizes may have a considerable influ-

TABLE 3
Nominal Dimensions of Standard Sieves

Size, μ	Tylor standard screen scale Openings, mm.	Openings, in.[a]	Sieve No. mesh	U. S. standard screen scale Openings, mm.	Openings, in.	Sieve No. mesh	British standard scale[b] Sieve No.
26,670	26.67	1.050 (1)					
22,430	22.43	0.883 ($^7/_8$)					
18,850	18.85	0.742 ($^3/_4$)					
15,850	15.85	0.624 ($^5/_8$)					
13,330	13.33	0.525 ($^1/_2$)					
11,200	11.20	0.441 ($^7/_{16}$)					
9,423	9.423	0.371 ($^3/_8$)					
7,925	7.925	0.312 ($^5/_{16}$)	$2^1/_2$	—	—	—	—
6,680	6.680	0.263 ($^1/_4$)	3	—	—	—	—
5,660	—	—	—	5.66	0.233	$3^1/_2$	—
5,613	5.613	0.221 ($^7/_{32}$)	$3^1/_2$	—	—	—	—
4,760	—	—	—	4.76	0.187	4	—
4,699	4.699	0.185 ($^3/_{16}$)	4	—	—	—	—
4,000	—	—	—	4.00	0.157	5	—
3,962	3.962	0.156 ($^5/_{32}$)	5	—	—	—	—
3,360	—	—	—	3.36	0.132	6	—
3,327	3.327	0.131 ($^1/_8$)	6	—	—	—	5
2,830	—	—	—	2.83	0.111	7	—
2,794	2.794	0.110 ($^7/_{64}$)	7	—	—	—	6
2,380	—	—	—	2.38	0.0937	8	—
2,362	2.362	0.093 ($^3/_{32}$)	8	—	—	—	7
2,000	—	—	—	2.00	0.0787	10	—
1,981	1.981	0.078 ($^5/_{64}$)	9	—	—	—	8
1,680	—	—	—	1.68	0.0661	12	—
1,651	1.651	0.065 ($^1/_{16}$)	10	—	—	—	10
1,410	—	—	—	1.41	0.0555	14	—
1,397	1.397	0.055	12	—	—	—	12
1,190	—	—	—	1.19	0.0469	16	—
1,168	1.168	0.046 ($^3/_{64}$)	14	—	—	—	14
1,000	—	—	—	1.00	0.0394	18	—
.991	0.991	0.039	16	—	—	—	16
840	—	—	—	0.84	0.0331	20	—
833	0.833	0.0328 ($^1/_{32}$)	20	—	—	—	18

ence on the performance of the powder during molding or sintering, or on the physical properties of the finished product. Even a small change in size cannot always be compensated for by varying a cycle of operations.

Metal powders belonging to the sieve-size class are usually classified according to the finest screen through which all of the powder will pass. If, for example, all of the powder passes through a 100-mesh screen, it is designated as a minus 100-mesh powder; if it passes through a 200-mesh

TABLE 3 (*Concluded*)
Nominal Dimensions of Standard Sieves

Size, μ	Tylor standard screen scale			U. S. standard screen scale			British standard scale[b]
	Openings, mm.	Openings, in.[a]	Sieve No., mesh	Openings, mm.	Openings, in.	Sieve No., mesh	Sieve No.
710	—	—	—	0.71	0.0280	25	—
701	0.701	0.0276	24	—	—	—	22
590	—	—	—	0.59	0.0232	30	—
589	0.589	0.0232	28	—	—	—	25
500	—	—	—	0.50	0.0197	35	—
495	0.495	0.0195	32	—	—	—	30
420	—	—	—	0.42	0.0165	40	—
417	0.417	0.0164 ($^1/_{64}$)	35	—	—	—	36
351	0.351	0.0138	42	—	—	—	44
350	—	—	—	0.35	0.0138	45	—
297	—	—	—	0.297	0.0117	50	—
295	0.295	0.0116	48	—	—	—	52
250	—	—	—	0.250	0.0098	60	—
246	0.246	0.0097	60	—	—	—	60
210	—	—	—	0.210	0.0083	70	—
208	0.208	0.0082	65	—	—	—	72
177	—	—	—	0.177	0.0070	80	—
175	0.175	0.0069	80	—	—	—	85
149	—	—	—	0.149	0.0059	100	—
147	0.147	0.0058	100	—	—	—	100
125	—	—	—	0.125	0.0049	120	—
124	0.124	0.0049	115	—	—	—	120
105	—	—	—	0.105	0.0041	140	—
104	0.104	0.0041	150	—	—	—	150
88	0.088	0.0035	170	0.088	0.0035	70	170
74	0.074	0.0029	200	0.074	0.0029	200	200
62	—	—	—	0.062	0.0024	230	—
61	0.061	0.0024	250	—	—	—	240
53	—	—	—	0.053	0.0021	270	—
52	0.052	0.0021	270	—	—	—	—
44	0.044	0.0017	325	0.044	0.0017	325	—
37	0.037	0.0015	400	0.037	0.0015	400	—

[a] Fractions are approximate.
[b] British Standards Institution.

screen, it is designated as a minus 200-mesh powder, etc. Specifications may call for a certain percentage of minus 200-mesh, or minus 300-mesh powder. As will be shown later, these broad terms must frequently be modified to include a distribution of certain particle sizes within close limits. For these size specifications, the Tyler sieve series is generally used as the standard scale, although the U. S. sieve series has been adopted in special cases. In Table 3 a tabulation of the nominal dimensions of the Tyler and U. S. standard sieve series is given, and Figure 20 shows the relationship to international screen units.

Powders of the subsieve-size class all pass through a 325-mesh sieve, the smallest screen used in practice. Hence, their size specifications are based on different concepts, e.g., average particle size, maximum and minimum sizes, etc.

Particle Size Distribution

The effects of distribution of sizes in a given powder can be best understood by studying the packing volume of the particles under different conditions. It is obvious that an ideal packing with perfect and complete contact between the particles, as is possible with particles having plane surfaces (e.g., cubes), can neither be attained nor approached in practice with powdered materials. The degree of contact that can be obtained is, nevertheless, of great significance for all considerations relating to powder molding and sintering, since the result of incomplete contact between the particles is the appearance of interparticle voids and a packing volume larger than the massive volume of the metal fitting into the same space.

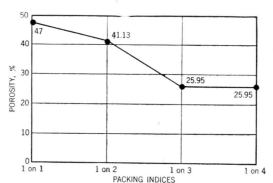

Fig. 21. Porosity of systematically packed equal-size spheres (Maldau and Stach).[2]

The discussion of this subject can be simplified by first studying the different degrees of porosity obtained by packing units of the simplest possible shape, namely spheres. Except in the case of the uniformly spherical particles of the carbonyl powders, little direct aid for the attainment of high-density packings of commercial types of powders can be derived from conclusions drawn from theoretical considerations. Such considerations will, however, give some indications of the effect of different size distributions on the relative density of the packing and will thus contribute to the understanding of the mechanism involved in the molding of metal powders.

Spheres of equal size may be packed into a given volume by various methods; the choice of method influences the volume of the space enclosed between the spheres. Meldau and Stach[2] have shown that this interparticle space has no relationship to the degree of symmetry with which the spheres are packed. In Figure 21 these data are plotted in a diagram showing porosity vs. packing indices, i.e., the manner in which the spheres are arranged. If spheres of uniform size are systematically packed one on top of another, the interparticle porosity will be 47%, regardless of the size of the spheres. If one sphere is packed on two others and the arrangement continued in that way, the percentage of interparticle voids will drop to 41%. A minimum of 26% is reached when the arrangement is such that one sphere is packed on top of three or four. Since no systematic packing mode other than these four is possible, the porosity value of 26% represents the minimum value that can be obtained by closest packing of spheres of equal size. It follows, therefore, that any chance packing of equal spheres will result in a porosity greater than 26%. Graton and Fraser[3,4] studied this subject closely, experimenting mainly with sands, and found that maximum porosities are obtained with particles which are all equal in size. The addition of larger or smaller particles has a tendency to reduce the porosity; the reduction is related to the number of different sizes added.

Considering again the case of ideal spheres, it is obvious that an increase in the packing density can only be achieved if the voids created by the closest packing of the spheres of one size are filled by spheres of smaller sizes. If, for example, a number of spheres of equal size are arranged in a systematic packing mode, corresponding to the closest packing, there is a certain diameter ratio between these spheres and smaller equal-sized spheres which allows the smaller spheres to fit exactly into the triangular interstices. Furthermore, the remaining triangular voids can be filled by spheres of still smaller diameters, thus reducing again the interparticle porosity. This method leads to the systematic packing of spherical particles of mixed sizes, making it theoretically possible to reduce the interparticle voids practically to zero by calculating the

[2] R. Meldau and E. Stach, Ber. deut. Tech. Wirt. Sach. d. Reichs Kohlenrats, 50 (1933); J. Inst. Fuel, 7, 336 (1934).
[3] L. C. Graton and H. J. Fraser, J. Geol., 43, 785 (1935).
[4] H. J. Fraser, J. Geol., 43, 910 (1935).

diameter ratios and by using definite amounts of spheres of the various diameters, that is, by using a definite particle *size distribution*. In practice, this is of course impossible, even with powders of spherical particle shape, and packing always remains haphazard, with increased porosities as a result. Even closest control of size distribution by screen analysis could not be correlated with porosity by Fraser, who arrived experimentally at a porosity of 33% (by volume) for four quite different screen analyses, as shown in Table 4. Attempts to obtain minimum porosi-

TABLE 4

Relationship between Screen Size and Porosity (Fraser[4])

Material	No. 1	No. 2	No. 3	No. 4
Screen size, %				
8.1 mm	79.86	29.04	28.55	45.34
2.3 mm	9.63	41.39	29.84	44.48
1.3 mm	10.51	29.57	41.43	10.17
Porosity, %	33.32	33.84	33.13	33.39

ties have been numerous, both by mathematical and experimental studies[5-7] but for all practical purposes the decision as to which size distribution is most desirable for a specific application is obtained empirically, based on experience and availability of possible raw materials. The following simple principles have been empirically established:

(*1*) A finer powder is generally preferred over a coarser powder, for the finer powders provide larger contact areas and thus usually result in better physical characteristics after sintering.

(*2*) Coarse powders are to be avoided because they tend to create larger pores and the maximum pore size is usually more important than the total porosity.

(*3*) The pore size decreases as fine powder sizes are added to a coarse powder.

(*4*) A large proportion of powders of any particular size is to be avoided, since it increases the total porosity.

(*5*) Too fine powders are not suitable for large compacts, as they require higher specific pressures and may result in appreciable variation in density along the longitudinal axis.

(*6*) A coarser powder has been found more suitable for hot- and cold-working of such sintered metals as copper or iron.

(*7*) Very small particle sizes are necessary for compressing very hard powders, such as tungsten, to assure sufficient adhesion for handling.

According to these principles, coarse powders (of particle sizes larger than 100 mesh) are not usually employed in the mass production of parts.

[5] C. C. Furnas, *Ind. Eng. Chem. Ind. Ed., 23,* 1052 (1931).
[6] F. O. Anderegg, *Ind. Eng. Chem. Ind. Ed., 23,* 1058 (1931).
[7] R. N. Traxler and L. A. H. Baum, *Physics, 7,* No. 1, 9 (1936).

Advantage is taken of the fact that the addition of fine powders to coarse material causes a progressive decrease in porosity as the smaller and smaller voids are occupied by ever smaller particles. Hence, most powder specifications provide for a certain amount of "fines"; powders suitable for molding usually have 30%, and sometimes even 50%, of their particles passing through a 325-mesh sieve.

Typical screen analyses of commercial powders used for the molding of parts are given in Tables 17–26 in Chapter VI. Specifications for these analyses are set up by agreement between the powder producer and the user; the producer is sometimes compelled to segregate the different size ranges of his products and then to mix the different sizes in the amounts given in the specifications. Subsieve sizes are conventionally segregated by centrifugal classifiers, cyclon separators, etc. Three brackets are significant in these commercial specifications: (*1*) all particles must pass a 100-mesh sieve; (*2*) the amount passing through a 325-mesh sieve is closely limited; and (*3*) the amount retained between the 200- and 250-mesh sieves is always considerably smaller than any other fraction. While the first two considerations are dictated by practical experience of the user, the last is significant for "natural" screen analyses as obtained by ball milling oxides or sinter cakes.

Small-Size Effects

The importance of the subsieve-size particles is based chiefly on the enormously increased amount of surface area, a factor of paramount importance for the creation of sufficient contact areas to permit action of the adhesive forces during molding and sintering. The increase in surface area upon subdivision of a material into very fine particles is evident in the following example: a solid cube of matter with a volume of 1 cm.^{3} has 6 cm.^{2} of surface; if the same mass is reduced in size to a powder with cubic particles that pass completely through a 325-mesh sieve, the total surface area becomes equal to 384,000 cm.^{2}, thus increasing 64,000 times.

This situation, incidentally, becomes even more aggravated if the particle shape is not isometric, as in the case of flaky powders. Such flaky particles constitute an extreme case because of their special shape—with the ratio of average particle breadth to thickness perhaps as great as 50 to 1 or even 100 to 1. These proportions are highly desirable in the pigment industry, where a maximum surface area is sought for greatest possible coverage. According to Mandle,[8] a pound of raw sheet material having a surface of less than 200 in.^{2} and a thickness of only $1/32$ in. can be

[8] H. H. Mandle, in J. Wulff, *Powder Metallurgy*. Am. Soc. Metals, Cleveland, 1942, chapt. 10.

stamped into a coarse 120-mesh powder with particles having an average breadth of about 50 μ and a thickness of 0.8 μ, the ratio of breadth to thickness being approximately 60 to 1. This pound of powder would contain close to 40,000,000,000 particles and the total expanded surface would be approximately 100,000 in.^{2}, an area 500 times as great as at the start. If, on the other hand, the same pound of raw material is reduced to a very fine lining powder, with all of the particles passing through a 400-mesh screen and having an average breadth of 25 μ at a thickness of about 0.4 μ, an expanded surface of approximately 220,000 in.^{2} (1100 times the original surface) is obtained and the number of particles would be in excess of 200,000,000,000!

Thus, it is understandable why a fine powder may be compressible into compacts of considerable cohesive strength, whereas a coarse powder of the same material consisting of a small number of large-size granules will not be compactible under the same conditions. In addition to the immense increase of the surface-to-surface contacts obtained automatically by the vast augmentation of the total surface area, other factors of no less importance come into play with these very fine particle sizes. The very large contact area causes a considerable increase in interparticle friction during molding, which in turn results in internal heating to various degrees. The heat improves the plasticity of the individual particles in certain cases, still further increasing the interparticle contact areas and adding to the strength of the pressed compact. This has been borne out in practice when compacting very fine tin and lead powders, where interparticle friction created during pressing has heated the compacts above room temperature, increasing the plasticity to such an extent that almost completely dense compacts with tensile strength values comparable to the massive metal resulted.

Friction promoted by the very small particle sizes may be caused otherwise than by interparticle contacts. In filling and packing these powders a considerable proportion of air is carried along—the larger the surface area of the particles, the larger the amount of air entrained. Friction among the air films and between the air and metal particles may thus add to the heat generated by interparticle rubbing.

In many cases the air entrapped among the myriads of tiny particles has detrimental effects, as will be shown in more detail later. It may be sufficient to note here that the larger the particle surface area in a given space, the more air is absorbed. During compression severe disturbances may be caused by the tendency of the air to escape in planes perpendicular to the axis of applied pressure, a tendency which frequently results in laminar sectioning of the compacts.

Next to friction effects, other influences—notably electrostatic and bridging effects—become of increasing importance for the smaller particle sizes. Although for comparatively large particles the size of the spheres has no effects on interparticle space for a given systematic manner of packing, this is not applicable to the very small sizes. The behavior of the particle is governed increasingly by surface phenomena as particle size diminishes, that is, as the ratio between surface area and volume increases. This increasing influence of surface properties for smaller sizes has been shown by a number of investigators, experimenting with both nonmetallic and metallic powders. Ellis and Lee,[9] working with sands in 1919, found the interparticle porosities to increase with finer sizes, from about 40% for coarse sand to about 52% for fine silt, as is illustrated in Table 5.

TABLE 5

Influence of Particle Size on Porosity (Ellis and Lee[9])

Material	Porosity, %	Material	Porosity, %
Coarse sand	39–41	Fine sand	44–49
Medium sand	41–48	Fine sandy loam	50–54

Similarly, Roller,[10] experimenting with such nonmetallic powders as gypsum, cement, and others, found that the relative volume per unit weight (the interparticle porosity) was constant for particle sizes above 14 μ, while a steady increase in volume was noted for progressively smaller sizes. Data on copper powders, published recently by Hardy,[11] show a similar trend of a steady decrease of the apparent density, or increase in the relative volume, with finer powders (see Table 22, Chapter VI).

In correlating these porosity figures with particle size, it must again be recognized that the maximum pore size is of considerable importance, as cohesion and shrinkage forces during sintering affect the closing of only small and uniformly shaped pores. Hence, preference for fine rather than coarse powders is also based on data accumulated from sintering work. Here, the actual adhesional forces between contiguous particles become of greatest importance and, if the bond between the particles after sintering is stronger than the cohesive strength within the particles themselves, the strength of the product increases proportionally to the specific surface of

[9] A. J. Ellis and C. H. Lee, *U. S. Geol. Survey Water Supply Papers, 446,* 12 (1919).

[10] P. S. Roller, *Ind. Eng. Chem. Ind. End., 22,* 1206 (1930).

[11] C. Hardy, *Symposium on Powder Metallurgy.* ASTM, Philadelphia, Pa., 1943. p. 1.

the powder, *i.e.*, to the ratio between surface and unit weight. This consideration is significant for sintering compacts using exclusively fine powders as initial material, *e.g.*, refractory metals or carbonyl metals, as we are dealing mainly with surfaces in these cases.

Particle Shape

The specific surface loses its importance in sintering coarser powders, for which the arrangement and shape of the particles are the predominant factors. If three shapes, *e.g.*, spheres, cubes, and plates, are compared in a systematic, as well as a haphazard, particle arrangement with regard to suitability for sintering, the spheres are more suitable for sintering when packed at random, despite having smaller specific surfaces than the two other shapes. If the particles are equal in size and systematically arranged, then the cubes or plates with their larger specific surfaces result in closer packing and a continuous network of strong interparticle boundaries.

From this consideration it becomes apparent that the shape of particles is as important as their size, although there are differences of opinion about the extent of the influence of particle shape on the behavior of the powders and on the final properties of the products. Most sources attribute good packing and flow characteristics of powders primarily to particular particle contours, but Baëza[12] disputes these claims and shows the predominant influence of size distribution by demonstrating that removal of certain size brackets can stop flow completely.

In discussing the influence of particle size and shape, Greenwood[13] has grouped the various metal powder shapes into seven classes as follows:

(1)   Uniform spherical particles (carbonyl iron and nickel).

(2)   Spheroids, near-spheres, droplets (atomized or sprayed copper, zinc, tin, aluminum, etc.).

(3)   Irregular spongy grains (reduced copper, iron, tungsten, etc.).

(4)   Dendritic or mossy grains (electrolytic copper, iron, silver, etc.).

(5)   Angular particles (crushed antimony, cast iron, etc.).

(6)   Round or oval plates, thick periphery (copper, iron, etc., of Eddy mill or Hametag process).

(7)   Flakes or leaves (stamped or ball-milled aluminum, copper, etc.).

In the photomicrographs of Figure 22, the different particle surfaces are shown for 12 typical metal powder products.

[12] W. J. Baëza, *A Course in Powder Metallurgy*. Reinhold, New York, 1943, p. 4.
[13] H. W. Greenwood, *Metal Ind. London*, *60*, 225, 242, 265, 279 (1942).

A, spherical silver (×160)

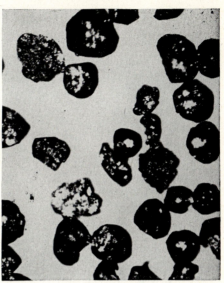

B, granular nickel (×160)

C, angular bismuth (×40)

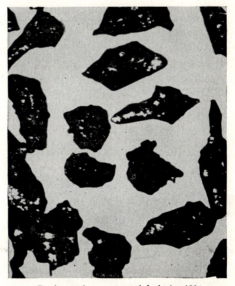

D, irregular cupro-nickel (×40)

Fig. 22 (Parts *A–L*). Typical particle shapes of different metal powders.
(Courtesy of Metals Disintegrating Co.)

(*Continued on page 94*)

94

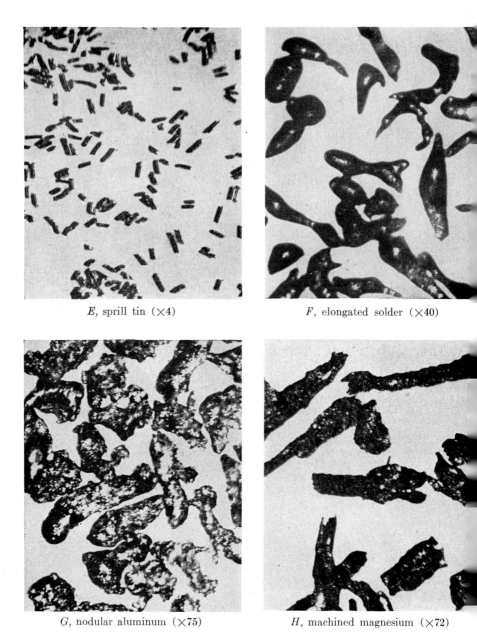

E, sprill tin (×4)   *F*, elongated solder (×40)

G, nodular aluminum (×75)   *H*, machined magnesium (×72)

Figure 22 (*Continued*)

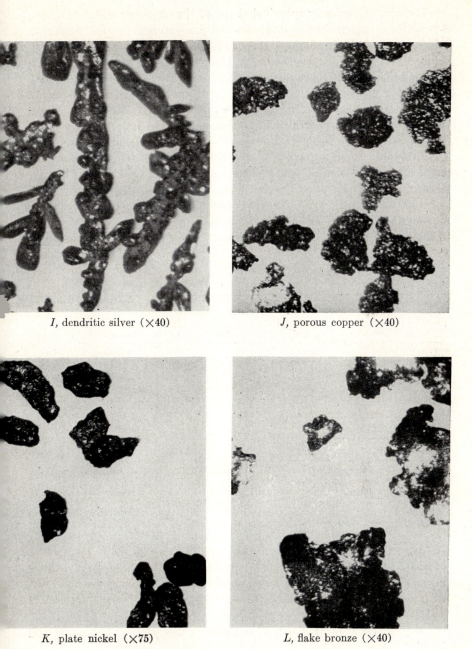

I, dendritic silver (×40)

J, porous copper (×40)

K, plate nickel (×75)

L, flake bronze (×40)

Figure 22 (*Concluded*)

The almost perfectly spherical shape of the carbonyl metal particles is considered extremely advantageous for the sintering process. Schlecht, Schubart, and Duftschmid,[13a] as well as Offerman, Buchholz, and Schulz,[13b] in the publications cited in Chapter III attribute the excellent sintering qualities of these powders and the uniform physical characteristics of the end product mainly to the uniformity and perfect spherical shape of the particles, which together with the extreme fineness lead to a very uniform porosity. Bridge formations (see page 100) are also more easily destroyed with spherical powders than with irregular shapes. Nonspherical and angular particles, obtained by pulverizing the carbonyl metal after deposition on heated surfaces, compared unfavorably in sintering. Fraser,[14] measuring the porosities of various shapes under dry and wet conditions, in the loose state and after compacting by tapping, arrives at somewhat similar conclusions. His results (partly reproduced in Table 6) show that: (1) wet materials pack more loosely than dry materials; and

TABLE 6

Influence of Particle Shape on Porosity (Fraser[14])

Material[a]	Specific g.avity	Porosity, %			
		Dry[b]		Wet[c]	
		Packed loose	Compacted	Packed loose	Compacted
Spherical lead shot....	11.21	40.60	37.18	42.40	38.89
Spherical sulfur shot...	2.024	43.38	37.35	44.14	38.24
Marine sand..........	2.681	38.52	34.78	42.96	35.04
Beach sand...........	2.658	41.17	36.55	46.55	38.46
Dune sand	2.681	41.17	37.60	44.93	39.34
Crushed calcite.......	2.665	50.50	40.76	54.50	42.74
Crushed quartz.......	2.650	48.13	41.20	53.88	43.96
Crushed halite........	2.180	52.05	43.51	—	—
Crushed mica	2.837	93.53	86.62	92.38	87.28

[a] All samples were of −18 to +35 mesh size; the average particle diameter of the shots and sands was 1.5 mm.
[b] The loose state was the particle arrangement as poured into the test container; the compacted state was obtained after tapping the container.
[c] The materials were first wetted and poured into the test container filled with water; they were measured after drying at 110°C.

(2) porosity values increase substantially from about 40% for shotted particles to above 90% for mica crushed into tiny lamellae. Irregularities in particle shape tend to increase porosity, the effect being more marked the more needlelike or the flatter the particles become. The high porosity

[13a] L. Schlecht, W. Schubardt, and F. Duftschmid, *Stahl und Eisen, 52,* 845 (1932).
[13b] E. K. Offerman, H. Buchholz, and E. H. Schulz, *Stahl und Eisen, 56,* 1132 (1936).
[14] H. J. Fraser, *J. Geol., 43,* 785 (1935).

figures for mica, even after compaction by prolonged jarring, are quite remarkable, although these results contradict somewhat Fraser's findings of a decreased porosity for disk-shaped particles, scarcely less angular than the mica flakes. The decreased porosity of the disk-shaped particles may be attributed to a favored orientation in a bricklike pattern.

In contrast with these observations, irregular spongy grains and feathery, dendritic, or nodular particles have been found superior to strictly spherical or tear-drop shapes for practical molding. Although claims that over-all porosities during packing or compaction of these fluffy particles with angular projections can be substantially reduced by an interlocking of the rough surfaces[15,16] can scarcely be accepted— especially in view of the intraparticle voids present in many types of electrolytic powder particles and in all reduced spongy particles—the vastly increased specific surface of these particles and the generally satisfactory plasticity of the metals made in this manner create an abundance of contact areas that yield compacts of considerable "green" strength at moderate pressures.

Flaky powders, such as aluminum flakes (shown in the photomicrograph in Figure 5, p. 41, are not generally suitable for molding as they result in a preferred orientation of the particles and in increased porosities, unless packing is arranged systematically. They have, however, been found to be suitable for laminated products, for which certain properties (e.g., electrical conductivity) are desired in a preferred direction, that is, in a plane to which the molding pressure is applied at right angles. The molding of such laminated bodies requires very careful technique, because the escape of air in these planes tends to disrupt the body by breaking it into many thin plates perpendicular to the axis of the applied pressure.

Structure of Particles

The structure of metal powder particles is closely related to the contour, but usually has little connection with the size. Thus, an electrolytic powder, containing substantially feathery particles in the larger size brackets (easily visible with a binocular microscope), has particles of the same character in the finer sizes, which may be resolved only by a high-power microscope.

In discussing metal powders, it is necessary to make a distinction between particles and grains or crystals: these are identical only in certain cases. Generally speaking, metal powder particles may be classified into two types of structure:

[15] C. Hardy, *Metal Progress, 22*, No. 1, 32 (1932).
[16] C. Hardy, *Eng. Mining J., 134*, No. 9, 373 (1933).

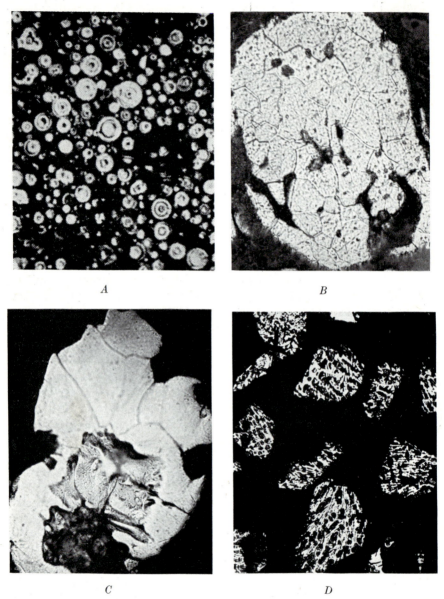

Fig. 23. Microstructure of particles of differently produced metal powders: A, carbonyl iron particles ($\times$500) (courtesy of W. D. Jones); B, reduced iron particle ($\times$1000) (courtesy of Charles Hardy, Inc.); C, electrolytic nickel particle ($\times$1000) (courtesy of American Electro Metal Corp.); D, crushed cast iron particles ($\times$300) (courtesy of American Electro Metal Corp.).

(a) Polycrystalline particles, which consist of a number of small grains or crystallites having either equiaxed or asymmetric shapes. In this group belong most of the powders of practical importance, such as reduced and electrolytic powders, as well as some powders obtained by atomizing, precipitation, condensation, or deposition.

(b) Monocrystalline particles, in which case each particle is a single crystal with identical grain and particle boundaries. In this class belong particles produced by liberating individual grains from a metal (*e.g.*, by the alloy disintegration method); also, all those powders in which each particle is a fragment of an original grain, with grain and particle boundaries becoming identical production, as in pulverizing.

Reduced powder particles usually consist of a large number of equiaxed crystallites with minute pores present throughout the particles, the number and size being dependent on the kind and history of the powder. Particles often display a lacelike fringe around more solid cores. If the reduction has not been complete, oxide inclusions take the place of some of these cavities, the inclusions being commonly found near the center of the particles. Electrolytic particles are also aggregates of crystallites, but the grains tend to be less symmetrical. Nodular particles often display columnar grains radiating from the center of the periphery. Carbonyl powders exhibit still another arrangement of crystallites and are characterized by concentric shells arranged in onion-skin fashion (Figure 23A), the discontinuity being attributed to intermittent deposition from the vapor phase. Similar structures can also be observed in other powders obtained by condensation of vapor phase. In Figures 23B and C, photomicrographs of (1) reduced and (2) electrolytic powder particles give evidence of the polycrystalline structure of these particles. Most powders obtained by atomization or condensation are representative of the second group. Mechanically comminuted powders have, in general, particles that are fragments of original grains (Figure 23D).

On the basis of a comparison of reduced and atomized copper and iron, it has been suggested that particles containing a large number of very fine crystallites tend to promote compressibility of the powder, whereas particles consisting of only one or a few grains are rather resistant to compacting. However, due consideration must be given to the nature of the metal, its basic ductility or brittleness, and the porosity or compactness of the particles in question. Atomized powders of the low-melting metals compare in compressibility with powders of the same metal produced by electrolysis or reduction.

A discussion of the particle structure is not complete without reference to crystal changes taking place in the course of powder manufacture.

Virgin metal powders are rarely ready for processing into parts. Powders whose production is based on mechanical comminution must be subjected to annealing to remove stresses and recover the true metal hardness; powders whose production is based on electrolytic deposition must be treated to release stresses inherent in the process; powders whose production is based on heat-treatment (such as reduction or sintering) must be subjected to subsequent cold-working to pulverize the sintered cakes. All these treatments, which follow the actual production of the powder and which are frequently termed "conditioning," may leave their marks on the internal particle structure. Twin bands (often found in electrolytic and pulverized powder particles) gives evidence of recrystallization, and prolonged annealing at sufficiently high temperatures sometimes causes transformation of the grains of small particles into single crystals. On the other hand, severe working during pulverization of the sintered cakes after reduction frequently results in distorted particles, in which the small crystallites take on a fibrous appearance and the pores are transformed into streaks or cracks.

In the case of many alloy powders obtained by sintering, the grain structure of the individual particles compares with that of normally recrystallized alloys, since the diffusion treatment is usually effectively performed only at temperatures above the recrystallization range. Of course, the internal particle structure may be obscured by the presence of transitory phases if the diffusion treatment has not been complete.

Packing of Particles

The packing characteristics of a powder are closely connected with particle size and shape. Discussing this subject at length, Jones[17] gives three processes which have a deciding influence on the mode of packing: (1) filling of gaps between larger particles by smaller ones, (2) breakdown of bridges or arches, and (3) conjunction of the particles by mutual sliding and rotation. The first process was discussed previously (p. 86 ff.), but the effects of the two other processes on the packing structure are of equal importance. Irregularities in packing due to the formation of bridges or arches contribute greatly to an increase in porosity of random packings. That such bridges are formed even in the case of spheres is illustrated by Jones,[18] who shows examples of bridge formation for lead spheres rolled at random in a single layer into a box with almost vertical sides. These bridges are arches in which the spheres support each other by thrusting against each other. Removal of the spheres forming the lower half of the

[17] W. D. Jones, *Principles of Powder Metallurgy*. Arnold, London, 1937, pp. 25, 26.
[18] W. D. Jones, *loc. cit.*, p. 22.

arches does not necessarily destroy the bridges. These conditions are not changed if spheres are arranged in three-dimensional packings where bridges can be formed in a variety of planes.

The formation of cavities due to bridge formation during packing of fine, irregularly shaped particles has been studied by Meldau and Stach.[2] The structure of the packings and the particle orientation were investigated for fine anthracite powders of particle sizes varying between 88 and 200 μ. The technique used in fixing the powder particles in their original positions prior to the preparation of sections of the packing structures involves the use of molten waxes, and was described by the same authors in an earlier publication.[19]

The mechanism of forming a dense packing from a loose one by settling, shaking, etc., entails a breakdown of the bridges, the reduction of the packing volume being caused by the collapse of the least stable bridges and the filling of the newly created craters with overlying particles. Collapse of the bridges is most pronounced at the beginning of the shaking operation, as the most stable bridges break down only after prolonged shaking, causing a general loosening of the component particles. It is even possible that new bridges are formed during this process. This mechanism of packing by promoting breakdown of the bridges occurs more rapidly for less stable bridges. Hence, spherical particles may lose many of these bridges after a short time, although angular particles may retain them even after a long shaking period.

Apart from the tendency to bridge, the orientation and conjunction of the particles by mutual movements during packing must be considered. A loosely filled powder is not uniform in density; during settling of the powder by shaking, certain denser regions function as nuclei and grow by agglomeration of nearby particles. This process is obvious when a powder containing some particles of a higher specific gravity is shaken, e.g., a mixture of anthracite and a small amount of pyrite. A tangential arrangement of the anthracite particles around the pyrite particles yields concentric layers which constitute a very dense form of packing.

Sliding and rotation of the particles during packing must be taken into consideration for curved particles. Closer conjunction of two particles, one having a concave face and the other a convex one, is often made possible by slight rotation to bring the corresponding faces into contact. Such movements obviously cannot occur as a result of simple pouring, but only by means of settling by some kind of agitation. Shaking brings the faces and edges of the particles into contact; lateral or rotary movements aid

[19] R. Meldau and E. Stach, Z. Ver. deut. Ing., 76, 613 (1932).

in matching corresponding faces, which may proceed until the densest possible packings are established. During settling of the powder by agitation, the particles tend to adjust their positions toward closest aggregation; the shape of the particles, however, often prevents the attainment of ideal conjunction.

Finally, Jones[20] reports time studies with loosely packed powders. Two reasons are given for the fact that a loose packing usually remains loose even after a long passage of time: first, the ratio between the total interparticle contact area and the total surface area is very small and, consequently, any cohesive forces are too small to have practical effects; and, second, surface films generally interfere with adhesion. However, an exception to the latter effect can be found in powders kept under vacuum: freshly prepared tin filings loosely packed in an evacuated stationary glass tube become slightly adherent after several days. On the other hand, similar effects noticed after several months even without evacuation, in the case of zinc and iron filings packed in sealed tubes, may be attributed to the moisture content of the air and the questionable effects of oxide films formed on the particle surfaces.

Apparent Density

The mechanism of packing has practical significance in loading die cavities with metal powders. The mass necessary to fill a given die cavity completely is usually determined as the weight of a loosely heaped quantity of powder and is generally referred to as *loading weight* or *apparent density*. Simple empirical measurements result in sufficiently accurate values for most practical purposes. Although this weight per unit volume refers to freely loaded powders, a similar value can be obtained for powders settled by tapping under specified conditions, frequently referred to as *tap density*. A list of apparent density ranges of a number of metal powders is given in Table 7. The apparent density is greatly influenced by a number of factors, some of which affect the packing mode primarily, while others are derived from the nature or condition of the metal. The effects of particle size have already been discussed and Roller,[21] working on nonmetallic materials, found that the apparent density decreased with decrease in particle size below a critical size (about 14 μ). Metal powders in general behave similarly, even in the coarser size range, as shown, for example, in the data for copper powders in Tables 22 and 23, Chapter VI. Of the other factors found to influence the loading weight, particle shape is most important. Thus, flaky powders have a very low

[20] W. D. Jones, *Principles of Powder Metallurgy*. Arnold, London, 1937, p. 27.
[21] P. S. Roller, *Ind. Eng. Chem. Ind. Ed., 22,* 1206 (1930).

density due to edgewise position and bridging tendency of many particles. Similarly very feathery or spongy particles have a low density because of their many projections.

Other factors influencing the apparent density include: (*1*) manner of packing the container, (*2*) shape and size of the container, (*3*) absolute unit incumbent weight of powder, (*4*) electrostatic forces, (*5*) surface conditions, and (*6*) specific properties of the powder, *e.g.*, hardness.

TABLE 7

Apparent Densities of Commercial Metal Powders

Material	Specific gravity	Apparent density	Material	Specific gravity	Apparent density
Aluminum	2.70	0.7–1	Molybdenum	10.2	3–6.5
Antimony	6.68	2–2.5	Nickel	8.9	2.5–3.5
Cadmium	8.65	3	Silicon	2.42	0.5–0.8
Chromium	7.1	2.5–3.5	Silver	10.50	1.2–1.7
Cobalt	8.9	1.5–3	Tin	5.75	1–3
Copper	8.93	0.7–4	Tungsten	19.3	5–10
Lead	11.3	4–6	Zinc	7.14	2.5–3
Magnesium	1.74	0.3–0.7	Iron and steel	7.85	1–4

Apparent density is of great importance for both the molding operation and sintering. In most briquetting operations, the dies are filled by volume measure and presses operate either to a definite stroke or pressure. If the press operates to a definite stroke, the pressure can be kept constant only if the apparent density does not change. On the other hand, if the press operates to a definite pressure, consistency in apparent density is necessary to ensure briquettes of equal height. Minor fluctuations in apparent density can usually be compensated by adjustments of pressure or stroke of the presses, but large-scale molding requires that the apparent density of the powder be controlled within close limits.

Another consideration is that actual values of the apparent density must fall within certain practical ranges; a low-density powder requires a longer compression stroke and deeper cavities to produce a compact of given size and density. This is best explained by means of an example: copper powder is available with apparent densities ranging from 0.75 to 4.5 g. per. cc., that is, with apparent densities of one-twelfth to one-half of the gross density of copper. If a briquette of gross density 8.9 is to be pressed from copper powder with an apparent density of 0.75, the ratio of the powder-fill depth to the thickness of the pressed compact—known as *compression ratio*—would be 8.9 to 0.75. If the desired briquette is to have a thickness of 1 in., a die cavity of approximately 12 in. is necessary, whereas a briquette 5 in. deep requires a die cavity 60 in. deep. The same comparative values apply to the plungers and also to the stroke and working space of the press; the total press opening must be in excess of this dimension. It is obvious that this involves extremely expensive equipment and operation, with a considerable loss of energy by friction, excessive abrasion of plunger and die walls, and very slow production rates because of the necessity of eliminating more than ten volumes of air for each volume of copper.

While such disproportionate densities are impractical for molding relatively thick parts, very light powders have found some use for very thin pieces, or in mixtures with very light nonmetallic powders, e.g., in the manufacture of current collector brushes from copper and graphite powders (see below, and also in Chapter XXIII, Volume II).

Because of the many difficulties encountered in the use of low-density powders, it would seem possible to use very heavy powders for all molding applications. If we return to the example of copper powder, a powder with an apparent density of 4.5 requires a die cavity and punch length only twice as great as the final compact height; the stroke and opening of the press and the travel of the press ram would be considerably reduced, thus insuring economy of operation. Unfortunately, there are certain limitations in producing and using such high-density powders. Of the methods in use for manufacturing metal powders on a large scale, only electrolysis and atomizing yield powders of high apparent density. Powders obtained by reduction or chemical precipitation are very light and fluffy, each particle containing a considerable amount of entrapped air. Only prolonged working, such as ball milling, would make the individual particles more dense, thus increasing the apparent density. Such working, however, always results in straining and work-hardening of the particles, rendering them less susceptible to plastic deformation during molding. Even annealing of such a powder does not always restore the compressibility of the original low-density powder.

There is still another factor to be considered in the use of high-density powders. It is evident that the average path which the particles must travel during compression from the fill volume to the completely densified compact is very short in the case of a dense powder, but very long in the case of a light powder. This difference in path is of little importance when objects with flat parallel faces are to be molded, as all particles will travel and settle in a general direction parallel to the pressure axis (disregarding local deflections caused by particle contacts, as previously described). On the other hand, the problem is very different when objects with curved or stepped faces perpendicular to the pressure axis are molded. A very short path would not permit sufficient deflection and by-passes of the particles to allow equalization of the relative density of the compact. Hence, powders with apparent densities below average have proved more satisfactory for these special applications.

Variations in apparent densities, moreover, have an indirect influence on the behavior of the material during sintering and on the physical properties of the finished product. Bal'shin[22] studied the effect of the apparent density of various metal powders on shrinkage, hardness, and strength figures of the sintered compact. The tendency of the compact to shrink during sintering is shown to decrease with increasing apparent density, a phenomenon generally observed in the production of most metal powder compacts. Bal'shin further shows an impressive decrease in bending strength, elongation, and hardness with increasing apparent density of the powder at constant relative volume of the compacts (i.e., relative density). He explains these results by the control of the specific surface area over the conflicting effects of a group of processes favoring motion of the particles toward each other and a group exerting the opposite effect. With decreasing specific surface area, the action of adhesive forces decreases; the ratio of internal to external changes is shifted in favor of the internal processes; finally, the absolute value of these internal changes caused by different processes increases.

Bal'shin's investigations leave no doubt about the importance of the apparent density, although the direct influence on the final physical properties must be considered in connection with particle size distribution and it is not clear whether he employed powders of constant particle size. In subsequent publications, Bal'shin[23] discusses the mathematical aspects of apparent densities, while extending his research to include an investigation of the effects of temperature and the blending of powders of different loading weights.

The effect of temperature on the apparent density is interesting; Bal'shin's results are reproduced in Figure 24. The diagram in Figure 25 shows the effect of temperature on the relative volume (packing volume) equal to the ratio of the specific gravity of the metal to the apparent density of the powder (indicating the number of times the volume of the loosely heaped powder is greater than a corresponding weight of massive metal). It is apparent that changes in apparent density (relative volume) commence only at a temperature of about 200°C. (390°F.), regardless of whether the powders were work-hardened or not. Further, all curves, irrespective of the degree of work-hardening, follow a similar course. The packing volume of coarse powders shows a continuous, though slight in-

[22] M. Yu. Bal'shin, *Vestnik Metalloprom.*, *16*, No. 17, 87 (1936).
[23] M. Yu. Bal'shin, *Vestnik Metalloprom.*, *16*, No. 18, 82, 91 (1936).

crease as the annealing temperature is raised, while in fine powders this increase appears only at lower temperatures, with a reversal at higher temperatures. The general trend of the curves can be explained on the

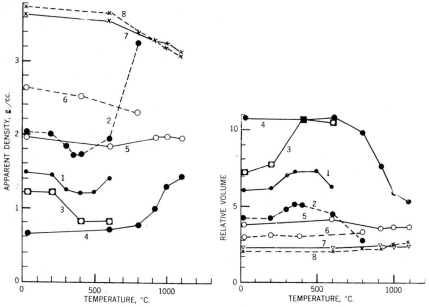

Fig. 24. Temperature effect on the apparent density of different metal powders (Bal'shin)[23]: 1, electrolytic copper, apparent density, 1.5 g./cc.; 2, same, pressed; 3, concentrate copper; 4, reduced iron, 0.67 g./cc.; 5, electrolytic iron, 2.11 g./cc.; 6, same, pressed; 7, electrolytic iron, 3.66 g./cc.; 8, same, pressed.

Fig. 25. Temperature effect on the relative (packing) volume of different metal powders (Bal'shin)[23]: 1, electrolytic copper, apparent density, 1.5 g./cc.; 2, same, pressed; 3, concentrate copper; 4, reduced iron, 0.67 g./cc. 5, electrolytic iron, 2.11 g./cc.; 6, same, pressed; 7, electrolytic iron, 3.66 g./cc.; 8, same, pressed.

basis of recrystallization effects. The expansion which all powders undergo during the first stage of the annealing treatment is caused by a number of factors, such as reduction of oxide films, evolution of adsorbed and absorbed gases, possible increase of particle surface, or tendency of the particles toward equiaxiality. The decrease in volume found in the fine powders at higher annealing temperatures can best be explained on the basis of particle growth phenomena as a result of atomic changes. The fact that such a decrease in volume was not observed in the coarser powders under investigation tends to support Bal'shin's explanation that the role played by external (collective) processes, such as growth of the particles, is only minor.

In order to determine the actual density of the individual powder particles, Hardy,[24] used methods standardized in the cement and pigment industries.[25,26] He found that certain of the following relationships exist between solid density, particle density, and apparent density for several types of iron, copper, and lead powders—as shown in Table 8.

TABLE 8

Relation between Solid, Particle, and Apparent Densities of Iron, Copper, and Lead Powders (Hardy[24])

Material	Solid density g./cc.	Average particle density g./cc.	Apparent density g./cc.
Electrolytic iron powder A	7.87	7.80	2.2
Electrolytic iron powder B	7.87	7.75	2.3
Reduced iron powder	7.87	7.49	2.1
Electrolytic copper powder	8.93	8.86	2.3
Atomized lead powder	11.34	11.36	4.5

It is questionable whether the figures given as particle density truly represent the over-all or *apparent* density of each particle, if the particle volume is taken to include pores or interstices. If they do, the figure of 7.49 for a spongy reduced iron particle appears to be very high. As the methods are generally used for nonmetallic particles of solid shapes, the results must rather be considered as deviations from the theoretical density of the material caused by inaccuracies in employing the method. The very close agreement between solid density and particle density in the case of atomized (*i.e.*, nonporous) lead particles is in agreement with this view. Since, however, the porosity of individual particles contributes to the compressibility of metal powders, further study of this subject is undoubtedly desirable.

Flow

In the plastics field the term "flow" designates a property of the powder which, under definite molding conditions, is equivalent to plasticity. Although comparisons have been made between plastic and powder metallurgy processes because of the similarity of some of the operations involved, it must be noted that in powder metallurgy "flow" has a different meaning. It generally designates the speed at which the powder flows over an inclined surface. Flow in powders is generated by a shear stress.

[24] Anonymous, *Iron Age, 154*, No. 11, 57 (1944).
[25] ASTM Standard C77–40.
[26] ASTM Standard D153–39.

If a powder is moved by gravity, the stress is proportional to the loading weight, provided that the cross section of the stream remains unchanged. The resistance against flow depends primarily on the regions in which one particle hampers the free movement of other particles, either by direct contact or indirectly. This is mainly determined by the coefficient of inter-particle friction. Particles may be prevented from moving separately by temporary adherence or interlocking. In this manner, clusters are formed which may cover considerable space. The phenomenon of cluster forma-tion depends on the movement and type of powder, the flow varying markedly with the size and structure of the particles. If the particles were all truly spherical, they would, in general, roll readily into a die cavity. This can never be achieved in commercial powders, in which differences in size and shape are unavoidable even in globular or pseudospherical powders, such as carbonyl iron. If the flow of iron and other ferromagnetic powders is considered, hindrance of the independent mobility of the par-ticles by magnetic forces must also be taken into account.

The rate of flow of a powder is of considerable importance wherever large-scale manufacturing processes depend on rapid filling of the molds, as, for instance, in the manufacture of porous bearings or machine parts. Since feeding is accomplished by volumetric measure, rapid and uniform filling with elimination of pockets in the die cavity or in the feeding device is one of the chief requirements. Poor flow characteristics of a powder en-force slow and uneconomical feeding and pressing and, at the same time, tend to make it very difficult to secure even fills of the die cavity.

The flow characteristics are dependent—in addition to the previously mentioned coefficient of friction—upon a large number of variables, in-cluding those which influence apparent density and packing capacity of the powder, i.e., particle size distribution, particle shape, etc., and the amount of absorbed moisture, air, or gases within the particles. In general, powders with too large a percentage of fine particles (less than 40 μ) have a poor flow rate. Freshly reduced, very fine powders usually have inferior flow as compared with other materials. For certain applications it has been possible to employ successfully powders with poor flow rates but otherwise desirable properties, by overcoming the poor flow factor with the aid of expedients, such as vibrators attached to feed hoppers or dies, or by de-humidifying or evacuating the powders as they enter the mold.[27,28]

During flow of a powder into a die cavity, segregation phenomena may occur, especially in free-flowing powders. In a one-component powder,

[27] C. Hardy, *Symposium on Powder Metallurgy.* ASTM, Philadelphia, Pa., 1943, p. 1.

[28] U. S. Pats. 2,198,612 and 2,259,465.

classification can usually be controlled by proper selection of particle sizes. With mixtures in which the components vary widely in density, control of segregation is more difficult, but can be managed by decreasing the apparent density of the heavier material to the point at which the apparent densities of the components are close together. For example, in the manufacture of current collector brushes, graphite powder is mixed with a copper powder of very low apparent density (less than 1 g. per cc.), to prevent excessive segregation. Another way of achieving the same purpose is to give the different components nearly the same weight range for the individual particles, *i.e.*, by mixing small or porous particles of the metal of greater density with larger or more solid particles of the lighter material. Surface-active agents are also employed to prevent segregation (see also Chapter VII).

Compactibility

Discussion of the various properties and characteristics of metal powders must include due reference to their performances during molding operations (see Chapter VIII). In contrast to conditions prevailing during the compression of fluids, where the pressure is distributed uniformly because of molecular mobility, the properties of compacted metal powders have frequently been observed to be anisotropic in different directions, especially when these compacts have been formed from powders consisting substantially of nonequiaxed particles. Microscopic examination and mechanical testing have in many instances disclosed that the particles show a preferred orientation in planes perpendicular to the axis of pressure. In addition, the deformation of the particles has often been found to be quite different in the longitudinal and transverse directions. In his important work, Bal'shin[23] has investigated this anisotropy with different materials and under different conditions, and has found that the coefficient of anisotropy increases with the irregularity of the particles, while it decreases with increasing density, especially in the initial stages of compaction.

It is obvious that the nature of the metal has a bearing on the compactibility and on the capacity to form sound and solid objects with a low coefficient of anisotropy. Skaupy and Kantorowicz[29] divided metal powders into two groups, based on the electrical conductivity of unsintered compressed briquettes: one group consists of *soft* powders, including copper, tin, etc.; the other consists of the *hard* powders, including the refractory metals. On the basis of this classification, iron powder belongs in the second group. In Figure 26 the microstructure and corresponding micro-

[29] F. Skaupy and O. Kantorowicz, Z. *Elektrochem.*, *37*, 482 (1931).

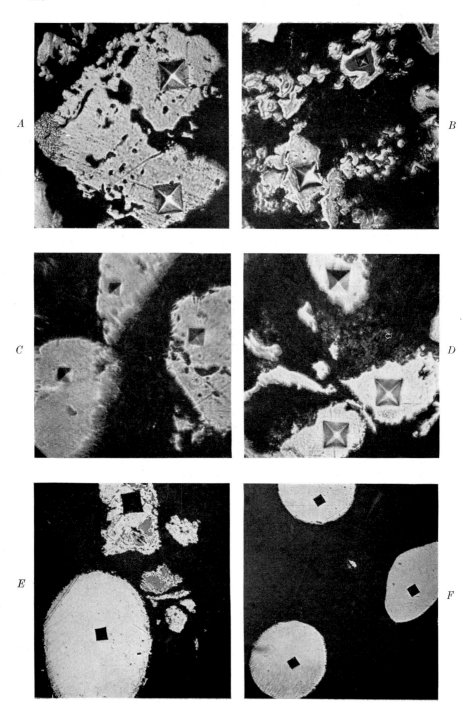

hardness numbers are given for a number of different iron and iron alloy powders, indicating a wide hardness range, and casting doubts on the correctness of the position of iron in the classification.

A similar division into hard and soft powders is also possible on the basis of plastic deformability of the particles during compaction with the following results:

Metal	Ductile and malleable	Brittle
Heavy refractory metals.............	W, Mo, Ta, Pt, Pd, Ir, Os	Ti, V, Cr, Zr
Heavy medium-melting metals......	Au, Ag, Fe, Ni, Co, Cu	Mn, As
Heavy low-melting metals..........	Zn, Sn, Pb	Sb, Bi, Cd
Light metals.......................	Al, Mg	Be

If the powder particles are soft and porous, and from metals which are tough and plastic by nature, they tend to flatten out upon compaction. When surface roughness is smoothed out, the particles become denser and the specific surface value is decreased together with the volume taken up by the powder. The tighter packing of the particles during compression results in an increase in apparent density, the increase being proportional to the pressure applied. The group of soft metals to which these principles apply includes tin, lead, aluminum, copper, etc. Reduced or electrolytic iron powder also belongs in this group and the apparent density can be closely controlled by applying these principles during the manufacture of such powders.[30]

As examples of hard powders, chromium, martensitic steel, and refractory metal carbides may be mentioned. The particles of these powders, being brittle and resistant to deformation, are not condensed or flattened in compaction, but tend to break down in fragmentation. Hence, the specific surfaces and the packing volumes of these powders do not necessarily decrease as a result of compaction—they may even increase under certain circumstances—and the apparent density may be reduced by such treatment. Other kinds of mechanical working, e.g., ball milling, effect changes in apparent density and packing volume similar to those which occur during compaction.

It must, of course, be realized that division into soft and hard powders is an oversimplification and that conditions differ from metal to metal,

[30] U. S. Pat. 2,306,665.

———

←——Fig. 26. Microstructure and microhardness of different iron and iron alloy powders: A, reduced iron (×470), microhardness 47–49; B, reduced iron with oxide inclusions (×470), microhardness of iron 60, oxide 348; C, electrolytic iron before annealing (×470), microhardness 147–232; D, electrolytic iron after annealing (×470), microhardness 52–59; E, cast and reduced iron (×200), microhardness of cast iron 750, reduced iron 168; F, quenched steel (×200), microhardness 750. (Courtesy of P. Schwarzkopf, R. Steinitz, and American Electro Metal Corp.)

with metals like nickel or soft steels constituting an intermediate group. Even powders of the same metal, differing only in size distribution or particle shape, may greatly differ in compactibility and susceptibility to plastic deformation. The behavior of two types of copper powder during compaction is indicated in Figures 27A and B. Figure 27A is a photomicrograph of a medium-sized electrolytic copper powder after compacting at 45 tsi, and Figure 27B, a photomicrograph of a somewhat finer grade of reduced copper powder after identical compression. Whereas the flattening of some of the larger particles of the reduced powder in planes perpendicular to the pressure axis indicates that this powder is highly susceptible to deformation, this tendency is less pronounced in the electrolytic powder. In the latter, signs of deformation can be detected only in several large particles which display deformed twin bands, originating from the electrolytic process and from the subsequent annealing treatment to which the powder was subjected. This deformation of the twin bands appears most impressive when several large particles form contact and friction prevents their dislodging during the compaction of the powder. On the other hand, individual particles of larger size show the least signs of deformation; they usually appear to be surrounded by small particles which evidently act as a protective cushion.

Other factors remain to be considered, notably the effect on the coefficient of the interparticle friction in conjunction with various lubricants. The work-hardening capacity of the metal and its recovery characteristics are both of considerable importance in connection with the compactibility of a powder. In certain cases it is possible to strain-harden basically soft powder particles by severe compaction to such an extent as to make them resistant to any further deformation or condensation at any pressure. Further compaction then becomes possible only after annealing above the recovery temperature releases the stresses sufficiently. In the case of very low-melting metals, the recovery from stresses may even occur at room temperature, as in the case of pure tin powder, which can be compacted to the specific gravity of tin at relatively low pressures (about 1 tsi) by alternate compaction and storage at room temperature for several days. The packing capacity of a powder, indicated (as shown above) by a tendency of the particles to slip and by-pass neighboring units during compaction, is frequently related closely to the plasticity of the particles. Both properties in concert can contribute to an increase in the total number of contact areas during compaction. As a result, the cohesive forces are increased, and the rigidity of the compact, known as *green strength*, is also increased. This is of particular significance for the molding of intricate shapes having curved faces, sharp corners, etc., when retaining solid corners and details is as important as equalization of the relative density throughout the entire cross section.

CHEMICAL PROPERTIES

In support of the claim that for many industrial applications (*e.g.*, structural parts) the chemical properties of metal powders are less critical than the physical, examples have been cited of the manufacture of metal powders, especially iron, into parts in the presence of a relatively high percentage of nonmetallic impurities, such as oxides or silica. Furthermore, development of the art has made possible the molding of many combinations of metal and nonmetal powders in various proportions, the nonmetal component varying in magnitude from a few per cent (*e.g.*, in friction materials) to a considerable proportion, as in the case of composites containing refractory materials, when the metal component is given the secondary role of a cementing material.

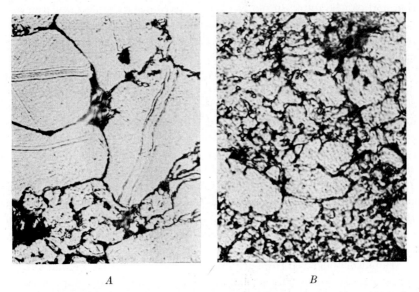

A   *B*

Fig. 27. Microsections through compacted powder particles (×1000): *A*, electrolytic copper powder particles after compaction at 45 tsi; *B*, reduced copper powder particles after compaction at 45 tsi.

Nevertheless, it must be emphasized that in a great majority of applications the chemical integrity of the initial powders is of the utmost importance. The presence of impurities creates many problems which cannot be easily solved by the user and frequently causes difficulties in processing or leads to products with inferior physical properties or distorted contours.

Various impurities may be present in metal powders, in solid form or included as dissolved gases. They may consist of foreign elements or compounds, or of the compounds from which the metal is derived; or they may consist of other metals which combine with the metal in question to form alloys or intermetallic compounds.

Solid Impurities

Four forms of solid impurities have been generally recognized: (*1*) individual particles of foreign matter, (*2*) occluded or semioccluded foreign substances within metal particles, (*3*) particle surface films of foreign matter, and (*4*) impurities combined with the metal within the particles.

Individual particles of foreign nature originate almost exclusively from raw material. Metal oxide particles are evidence of incomplete reduction and can be found in many commercial powders, especially iron and copper. As an example the microstructure of reduced iron powder particles is reproduced in Figure 28, in which completely unreduced oxide particles can be seen among the metal particles.

These oxide particles usually have detrimental effects on the processing of the powder and on the final properties; lack of plasticity has an effect on molding. It is possible to reduce the oxide particles by sintering in a reducing atmosphere, but the efficiency of this operation depends greatly on access of the gas to the oxide. Consequently, the more loosely the powder is compacted, the more effective is this reduction. The porosity of the compact is further increased by reduction of the oxide particles to metal particles with smaller volume and porous structure. If the oxide particles remain intact—as would be the case when very dense compacts are sintered or an inert atmosphere is used—they constitute islands of weakness throughout the structure, similar to contaminations in ordinary metals.

In powders derived from ores, residual particles of the original gangue are frequently found. Silica, alumina, magnesia, and other impurities resistant to acid treatment, which cannot be eliminated completely by customary methods of mechanical, magnetic, or chemical separation, are very harmful for several reasons. During molding, these impurities have an abrasive action on the die walls. Their removal during the sintering treatment is impossible as they remain stable under conditions prevailing during this heat-treatment. If present in amounts in excess of fractions of one per cent, they constitute an interfering phase which reduces metallic

cohesions; the effect of their presence on the final properties is as detrimental as that caused by residual particles of the initial metal oxide. In
powders derived from sources other than ores, particles of acid-insoluble
matter are present in lesser amounts. Such particles frequently adhere to
the raw materials during collection or storage, for example, in iron powders
derived from scrap metal or mill scale. Their presence in exceedingly
large proportions in many types of grinding scrap has so far prevented the
utilization of such materials for molding purposes.

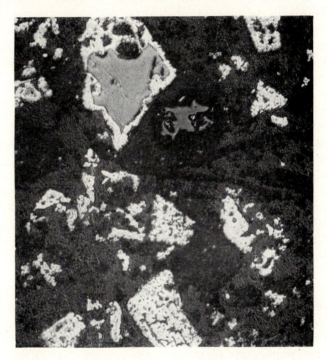

Fig. 28. Microsection through reduced iron powder particles
containing occluded and free oxides (×200).

Carbon, in the form of residual or free carbon or graphite, is usually
less objectionable than refractory compounds, since it acts as a lubricant
during molding. However, in some instances it may tend to combine with
the metal to form carbides, the presence of which has detrimental effects.
Organic compounds have a similar effect; their presence may even be
desirable and they may be added as lubricants or binders. Their removal
in the form of escaping gases at low-temperature levels during the early

stages of the sintering operation is most desirable so that only traces of residue remain in the sintered compacts. On the other hand, organic compounds which show a tendency to break down into active carbon (which may combine with the metal) or which leave a substantial residue must be considered objectionable.

Similar consideration must be given to occluded or semioccluded foreign substances within individual metal particles. Oxide impurities of this type are most often present in metal powders obtained by the reduction method. The reducing action proceeds from the outside of the particle toward the center and may not proceed far enough in some particles to remove the last traces of oxygen, thus leaving regions of unreduced oxide, usually in or near the center of the particle. This is illustrated in Figure 29, which shows a reduced and gas-carburized iron powder particle with a huge oxide occlusion.

Oxide areas completely enveloped by the reduced metal are often very stable. Processing of the powder offers little opportunity to remove these foreign cells, since the sintering treatment of the compacted powder, unless it is carried out at exceedingly high temperatures for long periods, constitutes a less severe reduction process than the initial manufacturing process. Generally speaking, it may be said that if the severe reducing treatment to which the loose powder is subjected during manufacture is insufficient to reduce these occlusions, it is obvious that the less favorable conditions for continued reduction during sintering will not do so. The oxide occlusions left untouched within the metal particles constitute a regular pattern of harmful solid inclusions in the final structure which contribute to diminished strength and ductility of the product.

Metal particles containing occlusions of the acid-insoluble type are frequently found in powders produced from ores. They constitute a definite problem since customary concentrating and separating equipment, based on magnetic, gravity, or flotation principles, is not very efficient when the two components are mechanically combined in different proportions in the individual particles. The novel method of electrostatic separation,[31] however, has solved this problem in several special cases involving ferrous powders from ore and scrap sources, and good results have been obtained in segregating these dual particles from ore particles and from metal particles.

Most metal powders contain particles which are enveloped by films of impurities, e.g., grease or oxide coats. Even powders brought to a high degree of purity during manufacture are subject to contamination during

[31] Product of the Separating Engineering Co., New York, N. Y.

shipment, storage, and handling because of the enormous surface area. Special precautions are sometimes taken by the manufacturer to keep oxidation at a minimum; high-purity powders are sometimes shipped in

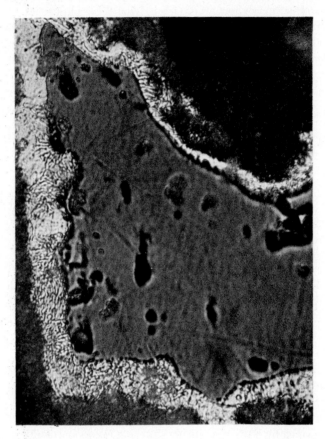

Fig. 29. Microsection through incompletely reduced and gas-carburized iron powder particle (×1000); core is unreduced iron oxide, case is eutectoid steel (Courtesy of Charles Hardy, Inc.)

evacuated containers or in containers filled with a neutral atmosphere. Commercial quantities of pure powders, e.g., iron or copper, are usually shipped with satisfactory results in hermetically sealed steel drums.

These particle surface films in most cases do not interfere noticeably with the processing of the powder. During molding they are punctured or partly displaced at the contact areas; the points at which the films have been mechanically removed offer fresh metallic surfaces which promote action of the adhesive forces. Less stable oxide films, as in iron, copper, etc., are readily removed during sintering under reducing conditions, and the final product may show no ill effects from these impurities.

In certain cases it has even been found that oxide films surrounding the particles (or small amounts of oxygen chemically bound within the metal particles) have beneficial effects. Tungsten powder, for example,

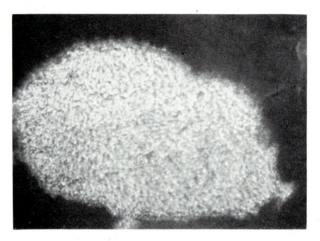

Fig. 30. Microsection through atomized aluminum powder particle displaying stable oxide surface film (×1000).

does not sinter very easily if absolutely free from oxygen, whereas sintering is considerably facilitated by a small oxygen content—especially if the oxide is present in the form of thin envelopes around the particles (sintering must, of course, be conducted under reducing conditions). The tungsten formed by reduction during sintering favors bonding and cements the particles together. Similarly, as reported by Sauerwald and Elsner,[32] oxide films have been found useful in the sintering of iron.

The situation is quite different, however, when metal particles form stable oxide films, as in the case of chromium, zirconium, aluminum, magnesium, lead, or tin. Figure 30 shows an aluminum particle with stable oxide film. Although these oxide envelopes can also be punctured during molding to permit sufficient adhesion of the particles in forming compacts, they cannot usually be removed by subsequent heat-treatments. The

[32] F. Sauerwald and G. Elsner, Z. Elektrochem., 31, 15 (1925).

structure of these compacts after sintering shows evidence of contaminated particle boundaries which often prevent the formation of a genuine crystalline structure. The effects of these surface films on the structure are not always so pronounced as to cause serious harm; sintered aluminum and aluminum alloys, for example, have been produced with rather clean structures and remarkable properties, in spite of clearly recognized oxide films on the initial particles.[33]

The abrasive effects of these stable oxide films on dies and punch faces are most undesirable and have constituted a great obstacle in the development of molded parts from aluminum and magnesium powders. In addition to attempts to lessen the oxide content of these powders by special manipulations during their manufacture, improvements in lubrication[33] have helped to diminish the abrasive effects during molding. Effective reduction of some of these films with the aid of metal hydrides as reducing agents has been claimed by Kalisher in the case of aluminum-nickel-iron alloys for magnetic materials.[34]

Impurities combined with the metal within the particles may be of different types. They may be of the metallic type (generally encountered in small quantities) such as manganese in iron, lead in copper, or nickel in cobalt. These impurities are not objectionable except for certain uses, e.g., in radio applications when electronic properties would be influenced.

Nonmetallic impurities in combined form, on the contrary, have profound effects on the metal. Combined carbon in iron, for example, has serious consequences: in certain types of carbonyl powders, the hardness of the particles is increased, while the plasticity is reduced. This, in turn, lowers the compressibility of the powder, requiring the use of higher unit pressures and yielding compacts of diminished green strength. Combined carbon in uncontrolled proportions influences the properties of the finished product and may be particularly objectionable when materials with definite magnetic, electrical, or chemical properties are being produced.

Sulfur and phosphorus in abnormal proportions are objectionable for obvious reasons. In addition to having detrimental effects on the strength and ductility of the final products, they create acid conditions during sintering which endanger the life of the furnace equipment. Gases formed by the reaction with oxygen or hydrogen may cause an expansion in volume of the compacts, thus increasing porosity and contributing to distortions of shape. Small amounts of these impurities, however, are sometimes not too objectionable, and may even be desirable, as Offerman,

---

[33] G. D. Cremer and J. J. Cordiano, *Trans. Am. Inst. Mining Met. Engrs.*, *152*, 152 (1943).

[34] P. R. Kalischer, *Trans. Am. Inst. Mining Met. Engrs.*, *145*, 369 (1941).

Buchholz, and Schulz[35] have shown in connection with experiments with carbonyl irons; powders completely free from sulfur and phosphorus gave unsatisfactory results in the production of sintered steels.

The reaction between carbon and oxygen in carbonyl iron or in steel powders is another example of internally produced gases. The theoretical oxygen-carbon ratio for complete removal of both elements can be calculated, but the actually required ratio is found only by experience, since variations in furnace atmosphere must be taken into account.

Gaseous Impurities

Much attention has been given to the interaction of gases and metals.[36] Jones[37] has given considerable space to the discussion of the volume effect, physical effect, and chemical effect of gases and vapors on metal powders. The topic will be discussed only briefly here, but some specific reactions caused by the evolution of these gases will be studied in a later chapter (XIV).

Practically all commercial powders contain gases, which cause contamination in different degrees. Absorption of gases in the powders (aided greatly by their vast surfaces) may originate from the method of manufacture or may take place during storage, shipment, or conditioning. In addition to dissolved gases, other sources of gas contamination are: (1) adsorbed films, (2) gases mechanically entrapped during powder production, mixing, or molding, (3) gases developed by chemical decomposition of solid impurities during heating, and (4) gases developed from intentionally added agents, such as lubricants.

Separately or in combination, these various sources may be responsible for large quantities of gas. Some gases may be dissolved in due time during heating of the metal, particularly if the powder has an initially low gas content. On the other hand, a metal which is saturated will not allow further gas dissolution; heating of the compact under pressure may even result in expansion due to an increase in volume of entrapped gases and liberation of sorbed gas films. The sorptive capacity of particle surfaces is also influenced by the conditions prevailing at these surfaces. Stresses and lattice distortions in the surface areas, caused, for example, by severe working or milling of the powder, affect the capacity to absorb gases at

[35] E. K. Offerman, H. Buchholz, and E. H. Schultz, *Stahl und Eisen, 56,* 1132 (1936).

[36] F. N. Rhines, in Symposium on Practical Aspects of Diffusion, *Trans. Am. Inst. Mining Met. Engrs., 156,* 335 (1944). F. J. Norton and A. L. Marshall, *ibid., 156,* 351 (1944).

[37] W. D. Jones, *Principles of Powder Metallurgy.* Arnold, London, 1937, pp. 38 ff.

room temperatures; the evolution of these impurities takes place in conjunction with stress releases at higher temperatures.

Recent studies in Germany of sintering processes in copper powders are of great interest in this connection. The sorptive capacity of these powders has been investigated for methanol vapor by Hüttig et al.,[38] and for dissolved dyes (Congo red, eosin, methylene blue, and rhodamine) by Hampel.[39] The sorption experiments were conducted on copper powders which were previously treated in a stream of pure dry hydrogen for one-half to four hours at temperatures ranging between 100° and 800°C. (212° and 1470°F.). Hampel's interesting results are reproduced in Figure 31, in which the sorption (in per cent) is shown as a function of the preheating temperature. The abscissa also indicates Tammann's temperature factor, a, the working temperature expressed in fractions of the melting point in absolute values.

The dye adsorption curves show a striking similarity in essential characteristics, independent of the dye used and of the time of preliminary heating within the range between one-half to four hours. Thus, it seems permissible to speak of the adsorptive capacity for dye solutions without specifying the dye, and of the influence of the preheating temperature without specifying the duration of heating. In the evaluation of the dye experiments, it is necessary to take a levelling effect of the solvents into account.

Hüttig's theory of sintering[40] is based on these adsorption experiments. This theory postulates different temperature stages of sintering (seeChapter XIV, page 518). In the interpretation of adsorption data it must be realized that the measured adsorption represents superimposition of two effects: that of the available surface area and that of the activity of this area. In agreement with Hüttig's theory are additional experiments of Hüttig and Bludar[41] on the degassing of iron powders and experiments of Hüttig and Arnestad[42] on the rusting of fritted iron powders. The former authors studied eight different iron powder samples and observed that the gas content varied with the method of preparation and with the previous history of the samples, but that, in all cases, the rate of gas evolution exhibited maxima at approximately 200°, 400°, and 710°C. (1390°, 750°, and 1310°F.). Hüttig and Arnestad studied the oxidizability and

[38] G. F. Hüttig, C. Bittner, R. Fehser, H. Hannawald, W. Heinz, W. Hennig, E. Herrmann, O. Hnevkovsky, and J. Pecher, Z. anorg. allgem. Chem., 247, 221 (1941).
[39] J. Hampel, Z. Elektrochem., 48, 82 (1942).
[40] G. F. Hüttig, Kolloid-Z., 97, No. 3, 281 (1941); 98, No. 1, 6 (1942); 98, No. 3, 263 (1942).
[41] G. F. Hüttig and H. H. Bludau, Z. anorg. allgem. Chem., 250, 36 (1942).
[42] G. F. Hüttig and K. Arnestad, Z. anorg. allgem. Chem., 250, 1 (1942).

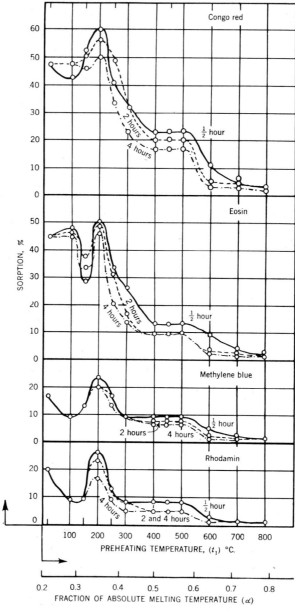

Fig. 31. Adsorptive capacity of copper powder, pre-heated to different temperatures for different lengths of time, with regard to dissolved Congo red, eosin, methylene blue and rhodamin (Hampel).[39]

the rate of reaction with ferric chloride solutions of electrolytic iron. Armco iron, and wrought iron powders. After identical heat-treatment in hydrogen, the powders showed maxima of oxidizability at 355°, 530°, and 850°C. (670°, 985°, and 1560°F.), respectively, and minima at 200°, 400°, and 780°C. (390°, 750°, and 1435°F.), respectively. Practically the same maxima and minima were observed in the curves representing the temperature dependence of the rate of reaction with ferric chloride.

TABLE 9

Air-Adsorption Experiments with Reduced Copper and Iron Powders
(Ruer and Kuschmann[43])

	Increase in weight	
Time of exposure to air, hrs.[a]	Mg. per 100 g.	Per cent
Copper		
1	5.44	0.005
3	6.26	0.006
23	8.07	0.008
70	9.57	0.010
Iron		
1	20.43	0.020
4	21.3	0.021
20	23.13	0.023

[a] At room temperature after evacuation at elevated temperatures.

Ruer and Kuschmann's[43] investigation of reduced copper and iron powders, subjected to evacuation and subsequent exposure to air, have shown that the amount of adsorbed gases may be appreciable even at room temperature. Their results are reproduced in Table 9. Dureau and Franssen,[44] producing copper powder by milling in high vacuum, found a small but definite adsorption at room temperature of hydrogen, nitrogen, carbon monoxide, and some hydrocarbons, and a large chemosorption of oxygen and carbon dioxide. Similarly, films of adsorbed water were observed when powdered gold and other metals were subjected to saturated water vapor.

Additional sources of gas include materials which may be added to a powder intentionally to facilitate lubrication or to promote porosity. Solid volatile materials may also be added deliberately to vaporize and

[43] R. Ruer and J. Kuschmann, Z. anorg. allgem. Chem., 154, 69 (1926); 166, 257 (1927); and 173, 233 (1928).
[44] F. Durau and H. Franssen, Z. Physik, 89, 757 (1934).

remove less volatile impurities. Certain stearates, camphor, and salicylic acid belong in the first group, while caustic potash, sodium, potassium, or ammonium chloride, and sodium or calcium silicate are examples of the second group.

With conventional molding and sintering conditions, entrapped air, carbon, and lubricants are probably the most important of all these sources of gas. The total amount of gas thus produced may be large; even the quantity caused by simple adsorbed films may be quite substantial. The apparent surface of a copper powder with strictly cubic particles, the sides of which are 1 mm. in length, is 0.67 m.^{2} per g. The true surface of an average powder may be a hundred times as large and, on the basis of an absorbed hydrogen film only one molecule thick, the gas content would be approximately 0.33 cc. per g.[45]

Although no perfectly gas-free powders are available for practical purposes, many obvious sources of contamination can be avoided. Milling and other types of comminution may be accomplished in protective atmospheres of gases which are not readily absorbed; packing, storing, and shipping in air-tight containers also aid in avoiding contamination. Processing of the powder with a minimum delay after manufacture is especially to be recommended in seasons when the humidity is excessive. Air and moisture conditioning have proved to be of great value in minimizing these difficulties; in modern plants the powders remain under controlled atmospheric conditions during storage, preparation, and molding.

Very pure or very fine powders (particularly electrolytic types) are sometimes shipped under protective layers of alcohol, carbon tetrachloride, etc., or while still wet. In the latter case, the powder is compressed while wet, the bulk of the water being removed during this compression. The rest of the water is volatilized at low temperatures during the early stages of sintering. However, compacts thus prepared are usually contaminated with oxide inclusions which cannot easily be reduced during sintering under conventional conditions.

The fact that most powders are likely to evolve large quantities of gas on heating *in vacuo* has led to attempts to liberate the gases before sintering, that is, while still at room temperature. Evacuation of powders both before and during molding has been recommended repeatedly in the literature.[27,28] Whereas the feasibility of such treatment has lately been established with vacuum operatable presses,[46,47] the effectiveness remains questionable, especially in view of the fact that heating for several hours

[45] W. Trzebiatowski, Z. physik. Chem., B24, 87 (1934).
[46] Products of the F. J. Stokes Co., Philadelphia, Pa.
[47] Anonymous, Mech. Eng., 69, 928 (Nov., 1947).

in vacuo is often necessary to remove all gaseous inclusions, the rate of evolution depending on the thickness of the powder layer. Removal of the bulk of entrapped air has also been attempted by simple vibration of the powder. Although vibration and settling of the powder in the die frequently involves mechanical difficulties, vibration of the powder in the feed hopper and feed shoe has been generally adopted to facilitate flow and to eliminate gross air pockets.

Summary

The properties of metal powders may be divided into physical and chemical characteristics. Of the physical characteristics, particle size distribution, particle shape and structure, and certain technological properties such as apparent density, flow, and compactibility, are of particular practical interest for the molding of metallic shapes.

The chemical characteristics of the powders are of equal importance since chemical contaminations may seriously influence molding and sintering. The character and the extent of these effects are dependent primarily upon the type of the contamination, which may be classified into two general groups, solid and gaseous constituents.

CHAPTER V

Methods of Testing Powders

For the purpose of controlling operations, as well as products, in the field of powder metallurgy, means of measuring or determining the various physical and chemical properties of the powders are very important. Many methods for testing these characteristics have been worked out in the laboratory and the plant.

SAMPLING

When the testing of certain properties is necessary for mere routine control of production, obtaining a true sample is important. Certain methods of securing representative samples have been detailed in individual specifications agreed upon by powder producers and users, these agreements generally being based upon standard procedures and specifications used in related fields, such as powdered lime, coal, etc.[1,2]   Only recently has the American Society for Testing Materials set up tentative standards directly related to metal powders, and provided a method of sampling finished lots of metal powders.[3]

The Metal Powder Association has also recently set up a tentative method for sampling finished lots.[4]   This is a method for removal of representative samples of finished lots of metal powders and has two parts: (1) a method for powders blended in mechanical blending equipment; and (2) a method for powders blended manually and which are adaptable to a second (or check) sampling.

In the first procedure, a sample of the entire cross section of the stream of thoroughly blended powder flowing from a blender is taken halfway during the filling of the first shipping container, another sample is taken in the same manner after one-half of the load of the blender has been discharged, and a final sample is taken after the last shipping container is halfway filled. 5000 grams of each of the three samples taken

[1] ASTM Standard C50–27.
[2] ASTM Standard D21–40.
[3] ASTM Standard B215–46T.
[4] Metal Powder Assoc. Standard I–45T, June, 1945.

during the blender discharge are then blended together. The resulting blend is finally reduced by passing it through a sample splitter of the W. S. Tyler type, until a sample of the desired size is obtained.

In the second procedure, a scoopful of powder is taken in rotation from each of the containers that holds the constituents of the blend and passed over a scalping screen. The relationship between scalping screen and mesh size of the powder is:

Mesh size of product	Mesh size of scalping screen
100	40
150	60
200	60
325	80

The "fines" discharge of the screen is connected to the shipping container and the containers are filled by this method until all of the constituents of the finished lot have been used. A *thief sample* is then removed from each of the final shipping containers at a point halfway between the center and the wall of the container. The individual samples are composited by blending and are then reduced in a sample splitter until a sample of the desired size is obtained.

Average samples from large powder shipments are frequently obtained by testing the individual batches or drums with the aid of a brass thief which permits withdrawal of sample cores throughout the entire length of the container or heap. Another conventional method involves sectioning a powder batch by quartering a cone made from the original batch, then forming a cone from the quarter selected and quartering it similarly, and repeating this procedure until the amount left is suitable for testing. Unless this procedure can be conducted in air-conditioned premises, it is not practical for large volumes, as it involves the exposure of a large surface area to the influence of the atmosphere or to contaminations originating from dust, etc. The method is useful, however, for smaller lots.

In sampling, care must be taken that the specimen remains truly representative of the entire batch with respect to all properties, especially size and impurities. Classification due to vibration, shock, etc., is frequently experienced during transport or storage. Thus, powders may show various concentrations of oxide in different sections of one container. depending upon the location of the section and the possibilities of access of air or moisture. Classification is common, moreover, when drums must travel by road or are handled roughly during loading or unloading.

CLASSIFICATION OF TESTING METHODS

There are many available methods for determining the properties of metal powders. They can be discussed only briefly here, but have been described in detail elsewhere.[5,6] The testing methods for powders can be classified into three types, for the determination of: (1) particle size, size distribution, and particle shape, (2) packing and molding characteristics, and (3) chemical properties.

Methods of Determining Particle Size and Shape

Methods of testing the particle size and size distribution of metal powders are quite numerous and are equally useful for research and production control work. The screen is the chief testing tool for coarse powders; powders containing a substantial proportion of subsieve sizes require more complicated methods of examination.[7] The microscope is exceedingly helpful for this kind of work, especially for purposes of calibration of other methods. Since microscopic examination is not limited to the subsieve particle sizes, the microscope is the most generally useful tool for determining the size and shape of powder particles.

The term *subsieve size* refers to all powder particles passing through a 325-mesh screen, although a 400-mesh screen is available which is used in the laboratory. Values from 45 to 60 μ have been given by various investigators[8,9] as the average diameter of a particle just large enough to pass through the 325-mesh screen; for metal powders an average value of 50 μ seems reasonable and agrees closely with values obtained by the correlation of screen analysis data with the results of other methods.

The conventional methods of measuring the size of fine powdered materials are given in the form of a diagram in Figure 32, according to Rigden.[10] The ranges of the different testing methods for particle size determination are shown in Table 10 according to data given by Baëza.[11] From the values given in this table it can be seen that these methods cover particles all the way from the one extreme of coarse granules about 10 cm. in diameter to the other extreme of fine dust having particles of molecular size.

[5] J. M. DallaValle, *Micromeritics*. Pitman, New York, 1943.
[6] R. Schumann, Jr., in J. Wulff, *Powder Metallurgy*. Am. Soc. Metals, Cleveland, 1942, chapt. 17.
[7] *Symposium on New Methods for Particle Size Determination in the Subsieve Range*. ASTM. Washington, March 4, 1941.
[8] R. N. Traxler and L. A. H. Baum, *Proc. ASTM, 35*, Pt. II, 157 (1935).
[9] S. S. Fritts, *Ind. Eng. Chem., Anal. Ed., 9*, 180 (1937).
[10] P. J. Rigden, *Chemistry & Industry, 21*, 393 (1943).
[11] W. J. Baëza, A Course in Powder Metallurgy, Reinhold, New York, 1943, p. 43.

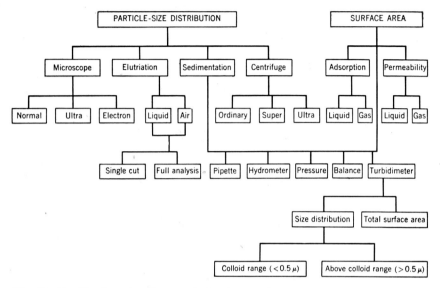

Fig. 32. Classification of various methods of measuring fine particles in the subsieve range, according to Rigden.[10]

TABLE 10

Range of Particle Size Testing Methods (Baëza[11])

Method of testing	Particle size μ	Method of testing	Particle size μ
Sieve	60–100,000	Turbidimeter	0.2 –50
Elutriation	5–100	Ultramicroscope	0.05 – 2
Sedimentation	1–100	Electron-microscope	0.005– 1
Microscope	0.3–100	Adsorption	0.001–10

These methods cannot be relied upon for the determination of absolute size, especially in the overlapping ranges which are covered by more than one method. Because of certain basic errors originating from the irregular shape of metal powder particles, each method can be used to obtain relative values which can be compared only with other values obtained by the same method. This brings up the problem of which dimension is to be measured or used as a basis for calculations of particle size. Conventionally, the assumption is made that the particle is a sphere; the particle size is, therefore, expressed in terms of the average diameter, from which the average volume, average surface, or average weight can be derived.

The measured magnitude is different for different methods: it is the true diameter for microscopic count, the average diameter for sedimentation, the smallest cross section for sieving, and the surface area for adsorption and permeability methods. It is, therefore, frequently necessary to combine the results of more than one method to obtain information regarding particle size and shape.

SCREEN TEST

Particle size distribution of most metal powders used for molding purposes is determined by conventional screen test methods. For many applications this type of test is sufficient to determine the suitability of the powder. In successive lots of powder made by the same method of manufacture, the amount passing through the finest sieve can thus be determined and is generally found to be about the same. The particle size distribution of this subsieve fraction is often substantially uniform, but this does not apply to all cases. There are applications in which even slight variations in the particle size distribution of the subsieve fraction have distinct effects on the processing of the powder. This is especially true when the majority of particles are of subsieve size and when it is no longer possible to depend on sieves as a means of classifying powders. In these cases the sieve method must be supplanted by other methods, such as direct microscopic count or sedimentation sizing tests. The results of sieve tests are usually given in the form of a screen analysis; the individual brackets designate the material which passes through a coarse screen (*minus*), but which is retained by a fine screen (*plus*). The brackets are given in weight per cent of the total test sample.

For practical purposes it has sometimes been found convenient to correlate these brackets with absolute size values (see Table 3, page 84) by assuming that the arithmetic mean of the two absolute size openings of the sieves is equivalent to the average particle size. Using a powder portion passed through a 250-mesh screen but retained on a 325-mesh screen as an example, the assumption can be made that the average particle diameter of this portion is 53 μ, as the 250-mesh screen possesses openings corresponding to 62 μ and the 325-mesh screen possesses openings corresponding to 44 μ. There is no actual basis for this assumption, since the same screen analysis would represent a powder sample with particles of uniform sizes anywhere within these two limits, *e.g.*, with particles of

61 μ, as well as with particles of 45 μ (about 27% smaller in absolute size). As in all other size testing methods, particle shape contributes materially to inherent errors in screen analyses. The error caused by the departure of the particle shapes from true spheres implies that the diameter of the particle's controlling cross section, *i.e.*, the cross section with an area smaller or larger than the sieve openings, is not identical with the average diameter of an ideal sphere. For example, in the case of flat disklike particles, any diameter measured will represent only one dimension and not the diameter of the theoretical sphere. In this case the proportion of the material retained on the sieve has a smaller volume than that indicated by the screen analysis on the basis of ideal spheres. If the particle shape is substantially needlelike, the short diameter will be the controlling factor in the number of particles passing through the sieve openings and the volume of the retained fraction will be much larger than that indicated by the screen test (based on ideal spheres). Since metal powders, though nonspherical, are neither disklike nor needlelike (with a few exceptions, *e.g.*, flaky aluminum or nodular steel grindings), the screen method has been found suitable for almost any application employing sieve-size powders.

The tendency of some powders to form agglomerates by interlocking introduces another source of errors. Breakdown of these agglomerates is usually secured after shaking the sieve for a sufficient time. Other errors are introduced, however, as prolonged agitation of the powder may in many instances cause comminution and attrition by a grinding action, resulting in gradual decrease in particle size. These errors have been brought under control, though not eliminated, by setting up standard specifications for the details of screen tests. Dry sieving is almost exclusively employed for metal powders; the period of agitation may vary between 10 and 45 minutes. One procedure, which has been widely accepted in the industry, specifies the screening of 100 grams of metal powder for 15 minutes. The Tylor Ro-Tap automatic sieve shaker, which combines a horizontal rotary motion with a vertical jarring motion obtained with the aid of a hammer, is used as standard shaking machine. A tentative method of determining a sieve analysis of granular metal powders has recently been standardized by the American Society for Testing Materials[11a]; this method is based on previously issued standard specifications for sieves used for testing purposes[11b] and on a standard definition for the term "screen (sieve)."[11c]

[11a] ASTM Standard B214–46T.
[11b] ASTM Standard E11–39.
[11c] ASTM Standard E13–42.

MICROSCOPIC METHOD

The various adaptations of the microscope—binocular, metallurgical, and electron[12]—make it the most dependable and most flexible tool for examination of size, shape, and structure of metal powder particles. Permanent records in the form of photographs and photomicrographs provide evidence of the nature of the powders, but the time, labor, and fine technique required to obtain a complete and reasonably reliable sizing analysis by the microscopic method virtually prohibit its use for routine work.

The direct microscopic method[13] for particle size determination has been standardized[14] and consists in actually counting the number of particles in each of several arbitrarily established micron ranges (size fractions) on a slide sample of the powder. Thus, for example, the number of particles between 1 and 5 μ appearing in a given microscopic field are counted and recorded; similarly, the number of particles in each of the other micron ranges are counted in the same microscopical field and recorded. An average size for all particles counted in each particular size range is assumed (the average size for all particles in the 1 to 5 μ range is assumed to be 3 μ) and, from the total number of particles found in the various micron ranges, the percentage by weight of the powder appearing in each micron range can be calculated. In evaluating count results, the fact that the weight fraction does not correspond to the number fraction must be taken into account.

For reliable results, the powder sample to be used on the slide must be very small and truly representative. Uniform dispersion over the slide is essential and classification, caused by vibration, must be prevented. This may be accomplished by preparing wet slides, *e.g.*, using mineral oil, benzene, diacetone alcohol, varnish dissolved in butyl acetate, etc. The error inherent in lack of uniformity in dispersion can be corrected for if a large number of fields on each slide are subjected to the count and the composite results of all the fields are added. It has been suggested that the procedure be standardized by preparing standard slides, counting 25 fields, and after calculating, repeating the process 4 times. If a reasonable check is obtained, the slide may be taken as standard. Errors in sampling may be eliminated by making a count of more than one slide.

[12] E. F. Burton and W. H. Kohl, *The Electron Microscope.* Reinhold, New York, 1942.
[13] C. R. Rogers, in J. Wulff, *Powder Metallurgy.* Am. Soc. Metals, Cleveland, 1942, chapt. 18, p. 216.
[14] *ASTM Tentative Standards*, 1614 (1938). See also *Proc. ASTM, 35*, Pt. I, 497 (1935).

Measurements may be made by direct examination through the microscope with the aid of standard grids in the eyepiece, by measuring the particles on photographs, or (as is generally preferred) by projecting the field onto a screen. Figure 33 is a photograph of the microscopic testing setup, and Table 11 shows the calculation of the size distribution of an iron powder sample, all of which passes through a 325-mesh sieve.

TABLE 11

Microscopic Particle Size Count of Iron Powder Sample (−325 Mesh)

Particles	Micron range	Average particle diameter (d), μ	Total No. of particles of 25 different fields	Frequency $F =$ average number of particles per field	Volume factor, $d^3 \times 10^{-8}$	Relative total volume, $F \times d^3 \times 10^{-8}$	Volume percentage of size fraction
These particles	0–5	2.5	3700	148	0.0156	2.31	1.04
counted only	5–10	7.5	7500	300	0.422	126.60	57.18
in the sixteen							
central 3-in.							
squares of							
screen							
These particles	10–20	15	300	12	3.370	40.44	18.27
counted in all	20–30	25	50	2	15.625	31.25	14.11
twelve 6-in.	30–40	35	10	0.4	42.875	17.15	7.75
squares of	40–50	45	1	0.04	91.125	3.65	1.65
screen							

Irregularity in particle shape presents one of the chief problems of microscopic analysis. Particles tend to fall with the largest axis parallel to the slide and with the flattest side down, thus presenting a classified diameter for examination. Although the apparent length may be considered to be the true one, the width is generally that of the flatter side. The thickness, on the other hand, can be determined only by focus adjustment manipulations, which introduce other errors. On the assumption that all three dimensions can be determined, an average diameter can be established, which is equal to the cube root of the product of length, width, and thickness. In most cases it is necessary to measure these three dimensions for several thousand particles in order to obtain fairly reliable results.

Because of the tediousness of the procedure, several short cuts have been suggested. One of them eliminates the insecure determination of the thickness by setting the diameter equal to one-half of the sum of length and width. Although this is not a true diameter, the result may not be much different from the result obtained by measuring all three dimensions, especially since the assumption that the third dimension is substantially the same as the two actually measured is within reason.

Another method of determining the diameter requires only a single measurement employing a filar eyepiece micrometer; the assumption is made that particles fall on each line in such a way that the line will be crossed by a statistically correct distribution of diameters. The value of the diameter at the point at which the filar line intercepts the particle is taken as representing the average diameter of the particle. In this method, however, the thickness of the particle is not taken into account.

Fig. 33. Microscope and wall screen for study of subsieve powder particle size distribution. (Courtesy of Metals Disintegrating Co.)

Although extremely small—and therefore possibly not truly representative—samples must be used, the microscopic method gives rather accurate and reproducible results if carefully carried out. As its use for size determination is extremely time consuming (even if one of the simplified methods is used), the value of the method consists primarily in its thoroughness as a research tool and as a means of standardizing more rapid methods.

The most obvious use of the microscope is for the examination of powders of various sizes for the purpose of investigating the structure and shape of the particles. Without exaggeration it can be said that the entire modern development of powder metallurgy has been made possible by the

136 METHODS OF TESTING POWDERS

intelligent use of the microscope for the examination of these properties.
The binocular microscope discloses surface conditions and characteristics
at moderate magnifications, while the standard metallurgical microscope
permits the study of internal particle structure in minute detail. Representative samples and cross sections are obtained in the same manner as
for particle size determinations; the powder samples, however, are most
conveniently fixed in mounting masses (e.g., Bakelite, Lucite) and the sections are polished and etched in accordance with common metallographic
practices. When the danger of distorting the particles exists (as in very
soft powders), pressure molding is sometimes replaced by a procedure
whereby the particles embedded in sodium silicate are fixed onto a base.
Lately special microscopes have come into use for investigating the shape,
structure, size, or geometric properties of very fine particles. Although the
ultramicroscope can be used to examine and classify particles only within
a relatively small range, the advent of the electron microscope[15,16] provides a tool of remarkable possibilities and is bound to give substantial
aid in clarifying many questions regarding particle structure and shape,
as well as particle surface conditions. The electron microscope permits
particle size measurements[17] in the range from 10 to 0.003 μ and may
even permit estimation of particles as small as 0.001 μ in diameter.[18]
According to Brubaker,[19] the average sizes obtained by electron microscopy are between 25 and 50% smaller than sizes obtained by light
microscopy.

SEDIMENTATION SIZING

Methods that classify powder particles according to their settling
velocities have found practical applications in powder metallurgy, both
for commercial sizing and for laboratory sedimentation work. The chief
merit of these methods lies in the fact that less time-consuming manipulations are involved than are needed for a direct microscopic count. However,
these methods cannot be depended upon for absolute values and the relation between particle size and settling velocity must be considered to be
empirical or at least semiempirical. Sedimentation methods depend in
principle on Stokes' law for falling particles, which states that under con-

[15] G. G. Harvey, in J. Wulff, *Powder Metallurgy.* Am. Soc. Metals, Cleveland,
1942, chapt. 20.
[16] J. Hiller, *Symposium on New Methods for Particle Size Determination in the
Subsieve Range.* ASTM, Washington, March 4, 1941, p. 90.
[17] T. F. Anderson, *Advances in Colloid Science.* Interscience, New York, 1942.
p. 353.
[18] M. V. Ardenne, *Z. physik. Chem., A189,* 1 (1940).
[19] D. G. Brubaker, *Ind. Eng. Chem., Anal. Ed., 17,* 184 (1945).

trolled conditions (*i.e.*, spherical particles falling at low velocity in a quiescent homogeneous fluid of infinite extent), the fall of the particles through a viscous medium is at a rate proportional to the square of the particle diameters. The law may be given by the formula:

$$v = [g(\rho_p - \rho_f)/18\eta] \times d^2$$

where v is the terminal velocity, g is the gravity constant, ρ_p is the density of the particle, ρ_f is the density of the fluid, η is the viscosity of the fluid, and d is the diameter of the spherical particle.

Unfortunately the idealized conditions on which Stokes' law is based do not prevail in practice, and a number of errors caused by the following factors must be taken into account: (*1*) irregular, nonspherical, particle shape of most metal powders; (*2*) viscosity effects; (*3*) incomplete dispersion; (*4*) excessive concentrations of the suspension; (*5*) wall effects, if container is too small; (*6*) effects caused by convection currents resulting from temperature gradients or from mechanical disturbances of the suspension; (*7*) tendency to form agglomerates of irregularly shaped particles due to interlocking; (*8*) effects of gas evolutions or bubble formation; (*9*) tendency of attrition caused by partcle impacts; and (*10*) effects of extremely small particle sizes, as compared with the inhomogeneities in the fluid.

To make possible the use of sedimentation methods based on Stokes' law, the assumption is generally made that the rate of fall of irregular particles is proportional to the square of their average diameters. For irregularly shaped particles, however, the viscosity effects are often quite marked and the Reynolds number should, therefore, not exceed 0.2 for consistent results. To extend the applicability of sedimentation methods to particle sizes of the sieve range, it has been recommended that two liquids of different viscosities be used. This may make possible the elimination of screen sizing, so that the data will not show the discontinuities so frequently found at the point at which a change is made from the screening to the sedimentation method; such discontinuities are encountered most frequently in the case of powder mixtures containing species of different specific gravities.

The attainment of complete dispersion is essential and must be assured by careful examination, preferably with the aid of the microscope. The wetting properties and settling behavior of sample mixtures are often the deciding factor in choosing the proper dispersing agent. In too concentrated suspensions the particles no longer move independently, falling more slowly than according to Stokes' law. Hence, it is practical to keep the suspension below approximately 0.5% solids by volume. Low concentrations also tend to keep the powder well dispersed in the liquid. The use of various dilutions is often helpful in running check sizing analyses.

The wall effects are usually negligible if the smallest inside dimension of the container is not less than 5 cm. Thermostatic control and shielding of the apparatus against external thermal effects (*e.g.*, sunlight, radiator heat, draft, etc.) eliminate convection currents caused by temperature gradients. Shielding from external mechanical influences (*e.g.*, shock, vibration, etc.) prevents convection currents caused by mechanical disturbances. Convection which may result from the fact that during settling the concentration gradient may be opposed to gravity is almost unavoidable.

Vigorous agitation is not always successful in preventing agglomeration of particles, particularly when the particles are oddly shaped and spongy, and display a natural tendency to interlock. In fact, vigorous agitation may aggravate this situation if gas bubbles are formed. Such gas bubbles, which act as nuclei for particle accumulations and may result in particle clusters, may also be caused by temperature gradients. Agglomerates and clusters do not fall either as single large particles or as independent small ones, and therefore deviate from Stokes' law.

If air is used as the suspension medium, errors due to attrition are also serious. Continuous agitation is caused by the air flow around and through the dry particles, resulting in separation of the finer particles. The impact force with which the individual particles are thrown against each other is considerable and may cause a grinding action more severe than that occurring during a sieve test. An increase of the amount of fines and a reduction of the large particle sizes are logical consequences; the sample under test can no longer be considered truly representative. Finally, there are the minor effects caused by extremely small particle sizes. If the particles are of the same order of magnitude as the intermolecular distances in the fluid they will fall more rapidly than anticipated by Stokes' law. In air sedimentations (page 141) this effect has been noted for particles less than 5 μ, but for liquid media it becomes noticeable only for particles below 0.1 μ. Furthermore, electrostatic forces may cause more pronounced wall effects for extremely fine particle sizes and deviations from Stokes' law must be expected.

Despite these restrictions, sedimentation methods based on Stokes' law have found widespread use. Generally speaking, these methods can be classified into two groups: (*a*) fractionation methods, when it is desirable or necessary to separate the various fractions for individual examination, and (*b*) nonfractionation methods, which are limited to measuring size distribution and estimating specific surfaces. A wide choice of techniques is available, with new techniques being developed constantly. In this survey only the most important methods, especially suitable for the testing of metal powders, will be described briefly. For further reference the reader should consult the standard work by DallaValles'[20] or Schuhmann's [21] paper on laboratory sizing methods.

FRACTIONATION METHODS

(1) *The sedimentation and decantation method* is based on a very simple principle, but the procedure is rather tedious. An initially uniform suspension in a certain size container (*e.g.*, 1-liter beaker) is allowed to settle for a certain time, after which it is decanted to a certain depth (10 or 20 cm.) by a large-bore (1-cm.) siphon. The residue contains mostly particles with high settling velocities and some with low settling velocities, whereas the decanted portion contains only particles with low settling

TABLE 12

Standard Size Scale (Ratio of Successive Sizes $= \sqrt{2})^a$

Size, μ	Tyler standard screen scale, mesh/linear in.
833	20
589	28
417	35
295	48
208	65
147	100
104	150
74	200[b]
52	270
37	400
26	
18.5	
13	
9.3	
6.5	
4.6	
3.3	
2.3	
1.6	
1.2	

[a] For sizes above this range see Table 3.
[b] Reference size used for sedimentation.

[20] J. M. DallaValle, *Micromeritics*. Pitman, New York, 1943.
[21] R. Schuhmann, Jr., in J. Wulff, *Powder Metallurgy*. Am. Soc. Metals, Cleveland, 1942, chapt. 17.

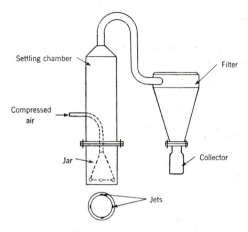

Fig. 34. Sketch of elutriator for sizing
small particles (DallaValle).[21a]

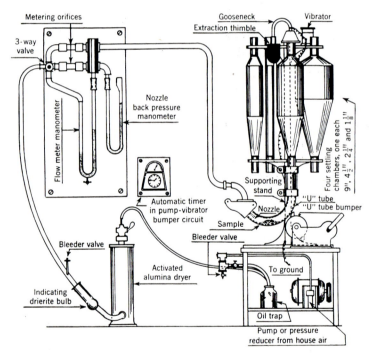

Fig. 35. Diagrammatical sketch of Roller air analysis assembly.
(Courtesy of Metals Disintegrating Co.)

[21a] J. M. DallaValle, *Micromeritics*. Pitman, New York, 1943, p. 87.

velocities. If more liquid is added to the residue and the whole operation is repeated several times, the residue finally contains substantially all the particles which settle quickly, while the decanted portion contains substantially all the particles which settle more slowly. The accuracy of the method increases with the number of repetitions. The end point is reached when the decanted portion is clear. The values of settling velocity at which the separations are made are calculated by Stokes' law to correspond to the sizes of the standard scale, and the velocities are doubled from each size to the next larger size in the $\sqrt{2}$ scale (see Table 12).

(2) *Elutriation in liquids or in air* is based on the principle that particles in a fluid rising vertically at a certain velocity are carried upward if their settling velocities are less than the velocity of the fluid current. High recovery of particles just below the separation size requires a long time; moreover, in the case of liquids (*e.g.*, water), wall effects influence the determination of the velocity of the liquid, thus making Stokes' law inaccurate. Elutriators can be readily constructed in the laboratory from standard materials (Fig. 34), although there have been several good commercial apparatuses developed, notably the one by Cooke[22] for liquid elutriation, and the one by Haultain[23] for air elutriation.

A well-known air-classifying apparatus is produced by Federal Classifier Systems, Inc.,[24] in laboratory and commercial sizes. Classification is effected by two opposed centrifugal currents of air; the method is both precise and flexible. The fractions are not clearly divided into units of definite size ranges; microscopic checking is necessary to control the orifice setting, which determines the air velocity. The setting then gives reproducible results for a particular material; but not for a powder of different apparent density.

An air-elutriation apparatus which is suitable for testing metal powders is the Roller air analyzer,[25–27] which may be found in many powder metallurgy laboratories and upon which many specifications are based. This separator, working on the basis of Stokes' law as applied to the fall of particles through a rising gas stream, is particularly suitable when separation of the various fractions for calibration purposes by microscopic examination is desirable. Figure 35 is a sketch of a Roller air analysis assembly as used for production control work.

[22] S. R. B. Cooke, *U. S. Bur. Mines Repts. Investigations.* No. 3333, 39 (1937).
[23] H. E. T. Haultain, *Trans. Can. Inst. Mining Met., 40,* 229 (1937).
[24] Federal Classifier Systems, Inc., Chicago, Ill., Bull. Nos. 5, 25, 26.
[25] P. S. Roller, *U. S. Bur. Mines Tech. Papers, No. 490,* (1931).
[26] P. S. Roller, *Proc. ASTM, 37,* Part II, 675 (1937).
[27] P. S. Roller, *J. Am. Ceram. Soc., 20,* 167 (1937).

NONFRACTIONATION METHODS

This group comprises several different methods in which measurements of various kinds are made at suitably spaced time intervals on powder suspensions settled under quiescent conditions. In all cases the uniformity of the suspension at the start of the test is essential.

(1) *The sedimentation balance* method is a frequently used tool, originally designed by Oden[28] about thirty years ago and later improved by Calbeck and Harner.[29] In this method the pan on one side of the balance is extended so that it can float freely in a suspension contained in a large cylinder. Perfect dispersion of the powder in water can be obtained by proper agitation. After the suspended pan is balanced, recordings of the increase in weight due to settling of some of the particles are taken at certain intervals, and from these data sedimentation curves can be drawn. The method can even be adapted to automatic recording of the sedimentation curve and so becomes a very simple device. The possibility of making direct weighings of the sediments rather accurately without disturbing the suspension is another advantage of this method. However, care must be taken to overcome one basic fault: systematic errors in the sedimentation data are caused by the fact that the pan sets up convection currents. A modification of this method uses centrifugal forces for the measurement of extremely fine particles.[30]

(2) *The pipette method* provides for setting up a series of test tubes marked off a certain distance from the base (*e.g.*, 10 or 20 cc.) and with another mark the same distance above. After a 1 or a 2% suspension is introduced into the tubes it is agitated and then allowed to settle. The fraction in the upper section is withdrawn by a pipette from the different tubes at certain intervals (*e.g.*, at 1, 2, 4, 8, 16, 32, and 64 minutes) and discarded. The suspension remaining in the lower sections, containing the powder settled during the interval, is filtered, dried, and weighed in the usual manner. Corrections are necessary for the amount of dispersing agent and other dissolved nonvolatile matter in the liquid. Slight variations of the method have been worked out by several investigators[31,32]; although somewhat more time consuming and not quite as simple, the pipette method has the advantage that only simple laboratory material is needed, and for this reason it is probably the best all-around nonfrac-

[28] S. Oden, *Bull. Geol. Inst. Univ. Upsala, 16,* 15 (1917).
[29] J. H. Calbeck and H. R. Harner, *Ind. Eng. Chem., Ind. Ed., 19,* 58 (1927).
[30] N. B. Nichols and H. Liebe, *Colloid Symposium Monograph Minnesota, 3,* 268 (1925).
[31] F. G. Tickell, *Examination of Fragmental Rocks.* Stanford Univ. Press, Stanford Univ., California, 1939, pp. 11–15.
[32] A. H. M. Andreasen, *Ingeniørvidenskab. Skrifter,* Raekke B. No. 3, (1939). (in English).

tionation method for general use. Several tests can be carried out at once, and in studying powder mixtures it is possible to determine the size distributions of all components from one sedimentation by chemical analyses of the initial mixture and of each pipette sample.

(3) *The hydrometer method*[33,34] is based on a similar principle, except that the concentration of the solids is measured indirectly by measuring the specific gravity of the suspension at certain time intervals (*e.g.*, 0.5, 1, 2, 5, 15, 45, 120, and 300 minutes). Special "streamlined" hydrometers have been developed which minimize agitation of the suspension during insertion and removal. But to obtain reproducible results various details of the technique must be given careful attention. Temperature control and cleanliness are especially important, and safeguards against contamination and disturbances of the suspension are necessary. Rogers[35] has reported two interesting, although rather empirical, deviations from the standard method which have worked satisfactorily in the determination of particle size for tungsten and tungsten carbide powders.

(4) *The manometric method* is based on the principle of observing density changes of the suspension by measuring the changes in pressure at a fixed height for varying times, or by measuring the pressure changes at varying heights for a fixed time. In the method as originally developed by Wiegner,[36] the pressure changes are measured manometrically with a small-diameter side tube. Goodhue and Smith[37] improved the method by employing differential manometry, using immiscible liquids. An automatic recording of the meniscus by photography is possible.

(5) *The turbidimeter,* one of the most useful instruments for quick and simple comparison work, uses measurements of the transmission of light to determine the rate of fall of particles in a liquid suspension. The apparatus best known to the powder metallurgist is the one originally developed by Wagner[38] based on the principle that the intensity of a light beam transmitted through a dilute suspension is related to the cross section of the suspended particles in the beam. The change in intensity of the light striking a photoelectric cell, as measured by a photelometer or microammeter, is an indication of the change in tur-

[33] A. Casagrande, *The Hydrometer Method for Mechanical Analysis of Soils and Other Granular Materials.* Soil Mechanics Lab., Mass. Inst. Tech., Cambridge, Mass., 1931.

[34] "Tentative Method of Mechanical Analysis of Soils," *Proc. ASTM, 35,* Part I, 953 (1935); ASTM Standard D422–35T.

[35] M. F. Rogers, in J. Wulff, *Powder Metallurgy.* Am. Soc. Metals, Cleveland, 1942, p. 173.

[36] G. Wiegner, *Landw. Vers. Sta., 91,* 40 (1919).

[37] L. D. Goodhue and C. M. Smith, *Ind. Eng. Chem., Anal. Ed., 8,* 46 (1936).

[38] L. A. Wagner, *Proc. ASTM, 33,* Pt. II, 553 (1933).

bidity of the suspension in the path of the light beam. Changes of the turbidity are assumed to be in direct proportion to changes in the surface area of the suspension at any level in the tank. By application of Stokes' law it is possible to calculate the time required for all particles of a given size to fall below the path of the light beam.

A number of errors are inherent in this method, apart from the necessity of obtaining a uniform suspension by proper agitation. Materials differ in the way they absorb and scatter light rays, and this difference affects the intensity reading. Deviation of the particle shape from the ideal sphere and variations in the density of the individual particles are also important factors which must be taken into account. As it is practically impossible to determine the exact configurations or the true density of the particles of some metal powders (as, for instance, in products obtained by the reduction of metal oxides), a calibration of the instrument for each specific powder is necessary. This must be accomplished by actually determining the rate of fall of the powder of a known size range in the particular liquid used for the suspension. Thereafter the instrument can be used for any sample of this particular powder and all sizes can be calculated on the basis of Stokes' law. If used in this manner, the method has given fairly reproducible results which check with those from other methods. In spite of the short time sufficient for the completion of an analysis (approximately 45 minutes) the method appears to be sensitive enough to indicate clearly significant changes in particle size distribution if used for a single material. However, serious errors are introduced when the data determined by the analysis of one material are applied to another with different physical characteristics. It must also be noted that a definite overlapping of particle sizes in the different ranges and a lack of sharp divisions into the different micron ranges are inherent in this method.

Wagner's original method has recently been improved by Stater[39] by using a photelometer of higher resolving power. Papers by Kalischer[40] and Steinour[41] describe in detail results of turbidimetric measurements on metal powders.

MISCELLANEOUS METHODS OF PARTICLE DETERMINATION

Of the remaining methods of size determination,[42] which include centrifugal analysis, photographic and pressure methods, and perme-

[39] M. N. Stater, *Proc. ASTM, 39*, 795 (1939).
[40] P. R. Kalischer, *Can. Metals Met. Inds.*, 7, 34 (1944).
[41] H. H. Steinour, *Iron Age, 155*, No. 20, 65 (1945).
[42] H. E. Schweyer and L. T. Work, *Symposium on New Methods for Particle Size Determination in the Subsieve Range*. ASTM. Washington, March 4, 1941, p. 1.

ability, adsorption and reaction tests, only the last three methods appear to have some significance in the metal powder field. In each case the specific surface rather than the size range is determined, and the methods are only suitable for establishing average size values. The physical structure of the particles (*e.g.*, their porosity) and particle surface conditions have, of course, a deciding influence. Thus the permeability method is useful only for purposes of comparison on the same material. The adsorption method, on the other hand, gives values for the surface area on either porous or nonporous particles; but it can be used for size determination only if the material is nonporous, while in all calculations of the particle size an uncertainty due to the shape factor exists.

The permeability method makes possible computation of the specific surface, for certain types of materials, from permeability data by empirical relationships. These relationships, for computing specific surface can be used only for materials similar to those used in the establishment of the relationships. But within this limiting scope the permeability method offers the advantages of saving of time and simplicity of operation. Various techniques have been developed for permeability testing with liquids,[43,44] and with gases,[45-48] and a Subsieve Tester based on the latter technique has recently been developed.[48a] The instrument measures the pressure loss of air caused by friction with particle surfaces, with the result that a packing of small particles offers more resistance to air flow than one of large particles, provided that the intraparticle porosity can be neglected. In its capacity as a quick control instrument for size variations of different lots of the same commercial powder, this instrument has found its place in the field.

Adsorption methods are ideally suited for the computation of the total surface area and the specific surface. The methods are based on the adsorption of gas films (*e.g.*, nitrogen or argon at temperatures close to their boiling points) on the surface of a powder[49,50] or on the adsorp-

[43]P. C. Carman, *J. Soc. Chem. Ind., 58,* 1 (1939).

[44] P. C. Carman, *Symposium on New Methods for Particle Size Determination in the Subsieve Range,* ASTM. Washington, March 4, 1941, p. 24.

[45] F. M. Lea and R. W. Nurse, *J. Soc. Chem. Ind., 58,* 277 (1939).

[46] R. L. Blaine, *ASTM Bull.,* No. *108,* 17 (1941).

[47] E. L. Gooden and C. M. Smith, *Ind. Eng. Chem., Anal. Ed., 12,* 479 (1940).

[48] E. L. Gooden, *Ind. Eng. Chem., Anal. Ed., 13,* 483 (1941).

[48a] Product of Fischer Scientific Co., Pittsburgh, Pa.

[49] P. H. Emmet, in J. Wulff, *Powder Metallurgy.* Am. Soc. Metals, Cleveland, 1942, chapt. 19; see also in *Advances in Colloid Science,* Vol. I., Interscience, New York, 1942, p. 1; *J. Am. Chem. Soc., 57,* 1754 (1935); *ibid., 59,* 310, 1553, 2682 (1937); *Trans. Electrochem. Soc., 71,* 383 (1937).

[58] J. P. Askey and G. G. P. Feachem, *J. Soc. Chem. Ind., 51,* 272 (1938).

tion of material from a solution (usually a dye)[51,52] after the powder has been mixed with the solution. If all other conditions are kept constant, the quantity of gas or material adsorbed by the powder is directly related to the surface exposed to the gas or solution. The method is consequently most useful for the studying of particle surface and shape and for the investigation of equilibrium adsorption relations for dyes and other surface-active materials. Although the principle of adsorption methods is comparatively simple, elaborate apparatus is usually required to give accurate results. It is also important to evaluate the results obtained by this method and comparison with some standard sample, whose surface has been determined by some other method, may be necessary in certain cases: when expressed in terms of surface, the results may define an entirely different surface from that obtained by other analytical measurements. In the case of a porous particle, for instance, the surface as measured by other methods is a spherical surface related to the average diameter of the particles, but excluding the surface presented by the pores. The adsorption method, on the other hand, also measures surfaces within the theoretical sphere, and the results refer to the total surface exposed to the testing medium. The efficiency of the adsorption depends upon the total surface exposed to the testing medium, which may depend on the character of the pores, their possible degree of interconnection, etc. For example, a dense particle with a diameter of 20 μ would present a considerably smaller figure for total surface than a porous particle with the same diameter, while a particle with interwoven pores would give the largest figure. Hence, adsorption methods cannot generally be used as a size indication for metal powders, except in the rare cases in which powders with absolutely dense particles are to be tested (e.g., flaky or spherical particles).

The rates of chemical reaction and dissolution of powder particles are also dependent on the exposed surface and could be utilized for the estimation of relative surface areas. At present they are not used for any practical purpose except in the control of the reactivity of powders, such as for spontaneously combustible pyrophoric powders. It has been found, however, that the ignition temperatures of aluminum and magnesium powders are a function of the surface area, and may vary for aluminum between 100° and 250°C. (212° and 480°F.), depending on the grade of powder.[53] The rate of dissolution of metal powders in acids

[51] G. F. Hüttig, C. Bittner, R. Fehser, H. Hannawald, W. Heinz, W. Hennig, E. Herrmann, O. Hnevkovsky, and J. Pecher, *Z. anorg. allgem. Chem.*, 247, 221 (1941).

[52] J. Hampel, *Z. Elektrochem*, 48, 82 (1942).

[53] A. Haid, *et al.*, *Jahresber. Chem.-tech. Reichsanst.*, 8, 136 (1930).

is also a function of the exposed surface area and therefore of the particle size. A method developed by Martin[54] and improved by Gross and Zimmerly[55] for the surface determination of quartz by the measurement of its rate of dissolution in hydrofluoric acid should in principle be applicable to metal powders, with certain adjustments.

The measurement of the surface area is of importance, not only for the indirect determination of particle size, but also for the evaluation of corrosion and adsorption data, since chemical reactivity and adsorption are functions of the surface area.

Methods of Testing Packing and Molding Properties

The fitness of a metal powder for molding purposes is determined by testing its performance while filling and settling in the mold, and by testing its behavior under pressure during the formation of a compact.

TESTING OF FLOW RATE

The term "flow rate" is customarily used to designate the ability of a powder to flow under atmospheric conditions through funnels or over chutes into the cavity of a container or mold within a given time interval (see p. 107). It is not to be mistaken for the capacity of the powder—analogous to that of a viscous liquid—to flow under pressure within the die cavity.

For powders which are to be used in commercial molding work where dies must be filled rapidly and uniformly, a determination of the flow rate is essential, since erratic flow will tend to fill the die cavities improperly. The rate of flow can, in practice, be increased by tapping or by the use of vibrators and depends also on the design of the filling device. These factors are not taken into consideration in the conventional test procedures, which consist in the determination of the time required for a given weight or volume of the powder to flow through a standardized funnel-shaped cup with a small orifice at the bottom. A standard apparatus, known as the Hall Tester, has been developed and generally accepted in the industry. The funnel has a 60° angle and is machined to a smooth finish from aluminum or brass to reduce friction at the walls to a minimum. The height and greatest diameter of the cone are about 2 in. The orifice at the bottom is usually 0.1 in. for nonferrous powders and 0.125 in. for ferrous powders; the length of the orifice is $1/8$ in. A brass cup is placed at a fixed distance below the funnel, and the time needed

[54] G. Martin, *Trans. Brit. Ceram. Soc.*, *25*, 51, 63 (1925).
[55] J. Gross and S. R. Zimmerly, *Trans. Am. Inst. Mining Met. Engrs.*, *Milling Methods*, *87*, 7–50 (1930).

to fill the cup is taken as a measure of the flow rate. A sketch of such a flow meter is given in Figure 36.[56] For a 0.1-in. orifice, flow rates of commercial powders used in automatic presses run about 30 seconds per 50 grams.

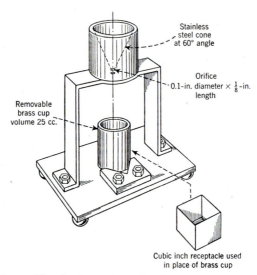

Fig. 36. Sketch of metal powder flow meter (Hardy).[56]

A tentative method of testing the flow rate of metal powders has recently been standardized by the American Society for Testing Materials.[57]

TESTING OF APPARENT DENSITY

The Hall Tester is an equally handy tool for the measuring of the apparent density of the powder—a property that not only determines the setting and movements of the molding tools, but also predetermines largely the density and properties of the compacts after molding. By standardizing the size of the metal cup and the distance of the discharge end of the funnel from the base of the cup, a convenient and reproducible method of testing the "loading weight" of the cup is established. Before the weight of the cup's content is measured, care must be taken that the powder is leveled off flush with the top of the cup. The result is generally

---

[56] C. Hardy, in *Symposium on Powder Metallurgy*, ASTM, Philadelphia, Pa., 1943, p. 1.
[57] ASTM Standard B213–46T.

taken as weight per unit volume or as relative density (expressed in grams per cc.). However, it is very important for the consistency of the results that the cup be filled by an absolutely uniform method, and that the apparatus be shielded from any mechanical disturbances and vibrations which tend to give denser packings, thus obscuring the results.

Another instrument is the well-known Scott volumeter used in many powder metallurgy laboratories. It consists essentially of a 1-in.^{3} receptacle and a series of glass baffles for controlling the height of the fall of the particles, thus eliminating human errors. A similar method provides for filling of a 1-in.^{3} receptacle by letting the powder slide on an incline, or by letting it fall from one incline to another, or to a third incline, thus reducing the velocity of the falling powder.

A somewhat different method of determining the apparent density provides for the filling of a weighed amount of powder into a calibrated glass receptacle on a rocking arm. The measure is rocked at a frequency of 2 cycles per second. The height of the powder column after 5 minutes of rocking is noted, and the density computed. Because of the packing which occurs, the value of the apparent density obtained by this method is considerably higher than that obtained by other methods; the method is useful, however, for comparison tests on the same type of powder.

For the purpose of rapid testing, simple graduates are often used into which the powder is poured through a glass funnel. A procedure frequently used for tungsten and similar powders employs settling of a given weight, e.g., 10 grams, by tapping the graduate gently with a glass rod and measuring the volume (in cc.) occupied by the powder. The value thus obtained—referred to as "tap density," does not necessarily bear a direct relationship to the apparent density. Various other methods, based on the same principle, but applying more systematic and reproducible shocks to the testing device, are in use for certain powders. One severe restriction of the vibratory methods, however, lies in their tendency to allow segregation of various particle sizes, as well as segregation of particles of metals of different specific gravity in the case of powder mixtures, thus introducing serious errors by upsetting the uniformity of the packing.

A tentative method of testing the apparent density of metal powders has recently been standardized by the American Society for Testing Materials.[58]

TESTING OF COMPRESSIBILITY AND COMPACTIBILITY

Numerous attempts have been made to standardize procedures of testing the briquettability of metal powders, as knowledge and control

[58] ASTM Standard B212–46T.

of this property are most desirable from the standpoint of industrial molding of briquettes. It must be realized, however, that the terms "compactibility" and "compressibility" cannot designate definite, clearly defined properties, but must rather refer to the performance of a powder under certain closely controlled conditions.

The compressibility of a powder, in its true sense, denotes the ability of the particles to group themselves into a closer packing by force of outside pressure. Hence, the compressibility is best given in terms of volume

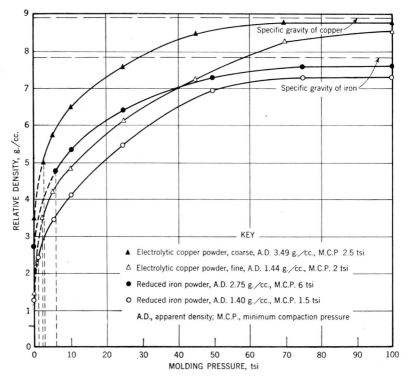

Fig. 37. Compressibility curves and minimum compaction pressures for different copper and iron powders.

or density in relationship to pressure. Simple shapes, such as cubes or cylinders, serve best as test samples. Curves showing the relative density as a function of pressure are easily obtained and are used for research and industrial control work. As shown in Figure 37, these curves originate at the value of the apparent density of the particular powder for zero pressure. The curves are dotted to the point where sufficient pressure is applied to form a compact which does not collapse during removal from the die or upon manual handling. Additional points for the curves are obtained for higher pressures, with the relative density values approaching the specific gravity of the particular metal at high pressures.

Since this method, though accurate, is rather tedious, several simplifications have been introduced. One provides for standardizing the die cavity—usually chosen as a 1-in. diameter—and simply correlates the height of the compact with different pressures. Further simplification is achieved if only one point of the curve is established, e.g., a certain minimum height of a cylindrical compact of 1-in. diameter, compressed from two opposite directions at a fixed pressure of 10 or 20 tsi. This method, however, is adequate only for comparison testing of various lots of the same type of powder to be used for a definite molding procedure, since errors may originate due to the fact that some powders compress well only at low pressures, while others form satisfactory compacts only at high pressures, depending on the properties and condition of the powder.

Another simplification is merely concerned with the slope of the pressure–density curve and requires only the determination of the apparent density and one relative density value for the specific pressure usually employed in production work. According to Smith,[58a] a compressibility factor can be computed by means of the simple formula:

$$C_f = (\rho_r - \rho_a)/\sqrt[3]{P}$$

where C_f is the compressibility factor, ρ the relative density of the compact for the specific pressure P used in the test, and ρ_a the apparent density of the powder; P is to be dimensioned in tsi. The use of this empirical formula has been found helpful in the establishment of quick comparison tests of successive batches of the same type of powder.

The point at which the dotted part of the curve becomes solid is of particular interest as it gives a minimum value for the pressure at which coherent compacts become practically possible. A clear definition as to the degree of cohesiveness of the compact formed at this pressure is of course necessary. For a number of applications involving iron, copper, bronze, etc., it has been found helpful to use as the criterion for the "minimum briquetting pressure" a sufficiently high green strength (see p. 154) to resist any attempts to crush the compact between the fingers, regardless of the position of the compact. Using this method as a base, minimum briquet-

[58a] G. B. Smith, Proc. Fourth Annual Spring Meeting of Metal Powder Association, Chicago, April 15–16, 1948, p. 29.

TABLE 13

Minimum Briquetting Pressure^a of Different Metal Powders

Powder	Pressure, tsi
Copper	
Electrolytic, coarse	2.5
Electrolytic, fine	2
Reduced, coarse	2
Reduced, fine	1.5
Iron	
Electrolytic, hard	25
Electrolytic, annealed	6
Hydrogen-reduced, fine	20
Hydrogen-reduced, sponge	1
Carbonyl, hard	25
Brass, annealed	3.5
Bronze, annealed	5
Aluminum	0.5
Lead	0.5
Tin	0.5
Cast iron, crushed	50
Steel, crushed	Not compressible

^a This term designates the lowest pressure at which the powder is compactible into a cylinder whose length equals its diameter, pressure being applied simultaneously from both ends, and which resists hand pressure.

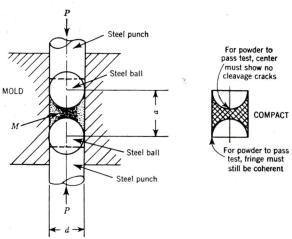

Fig. 38. Compactibility test for curved shapes: P, pressure applied; d, diameter of die; a, not to exceed $^5/_4$ of d; M, powder mass; either P or a to be kept constant.

ting pressures have been established for different commercial powders, and the results are reproduced in Table 13.

If a powder is to be molded into complicated shapes, simple compaction tests are often inadequate, as the complicated shapes must be represented in the test pieces. This refers, for example, to the molding of parts having large curved faces, where compactibility tests on simple cylinders would give no satisfactory answer to the important question of uniform density. By simply compressing the powder between two hardened steel balls within the cylindrical testing die, actual conditions can be closely approached and conclusions can be drawn as to the fitness of a certain powder for this particular application. Figure 38 shows a schematic view of the testing device.

It has recently been suggested by Schwarzkopf[59] that the terms *compressibility* and *compactibility* be standardized, the first indicating the extent to which the powder density is increased by application of a given pressure, and the second designating the minimum pressure required to produce a given green strength.

TESTING OF PLASTICITY AND SOFTNESS

Closely connected with the compactibility is the capacity of a powder to retain a uniform packing and to flow into corners, recesses in the punches, or curved sections, when under compression. This property is much affected by the coefficient of friction, but is also influenced by the plasticity of the particles, which may deform under pressure rather than slip past their neighbors.

If odd-shaped or disproportionate compacts are to be molded from the particular powder under investigation, the flow under pressure can be tested either microscopically or by sectioning the compact and measuring the relative density of the different sections. For very small compacts the microscopic method is preferable, but for larger compacts the density test is practical, especially when high cylinders are the object, for which the ratio of height to diameter exceeds 2 : 1.

Testing of the basic hardness of the individual particles of a powder may give certain indications of their plasticity and strain-hardening capacity, and may even permit offhand conclusions as to the powder's performance under pressure and the possible effects of deformation and distortion of the particles into planes of preferred direction. In addition to conventional microscopic examination, the microhardness tester should become a most useful tool for this kind of test, and Steinitz[60] has demonstrated that this method can be applied to a variety of powders. The microhardness can be determined both for solid and for porous particles, and even very spongy particles, such as hydrogen-reduced sponge iron, can

[59] P. Schwarzkopf, *Powder Metallurgy*. Macmillan, New York, 1947, pp. 19, 20.
[60] R. Steinitz, *Metals & Alloys, 17*, 1183 (1943).

be subjected to this test (see photomicrographs of Figure 26, p. 110). Disregarding slight errors introduced by varying the load, microhardness values of these iron particles have been found to range from 47 Brinell for sponge iron to 111 Brinell for electrolytic iron prior to annealing. This novel method of testing offers great possibilities for research, especially in the direction of testing the individual particles after they have been subjected to compaction. One of the objectives would be the possibility of appraisal of a powder without actually subjecting it to a series of compacting tests.

TESTING OF GREEN STRENGTH

High compressibility values alone cannot be the deciding factor in the selection of a suitable powder, but should always be considered in conjunction with a satisfactory "green strength" of the briquette, under which term such properties as compactness of the shape, sharp corners or extensions, smooth faces free from fissures, laminations or other defects, and sufficient rigidity to withstand rough handling, are generally understood. The testing of these properties is usually a matter of convention, and no general procedures apply. Rather each application demands its specific technological testing method. One widely used method is known as the "break test," applying pressure to a bushing of convenient size (e.g., 1-in. diameter, 1-in. length) and noting the break point as compressive strength. There are, however, two other testing procedures which are of sufficient interest to be reported here, one employing static compression or transverse breaking on a compact, and the other a dynamic impact.

The compression test has been developed by Bal'shin[61] and was employed in his previously described investigations. By subjecting cubic compacts to compression tests, both in the direction of the original axis of applied pressure and perpendicular to it, different values are usually obtained—the compacts being stronger when tested in the direction of the pressure axis. The difference in compressive strength is a function of the anisotropy of the compact; Bal'shin was able to correlate for different powders some of their other properties with their coefficient of anisotropy.

The cross-beaking test is an effective and simple means of testing the green strength of successive batches of a powder, or of comparing this property in like powders from different sources. The test is well-established for brittle materials in general and the only modification lies in applying very small loads with sufficient accuracy. The size of the test bars is kept small and the load applied usually ranges from 5 to 50 pounds.[61a]

The impact test has developed from simple drop tests. Cylindrical or

[61] M. Yu Bal'shin, *Vestnik Metalloprom.*, *16*, No. 18, 82 (1936).
[61a] R. P. Koehring, *private communication*.

rectangular compacts are kept stationary in a socket and are split by a pendulum. The energy consumed is extremely small in comparison with standard impact values, so that absolute values are of little importance. For purposes of comparison, however, this method has the advantages of the simplicity of the apparatus (it can be built from common laboratory ware) and of offering the possibility of making rapid checks in considerable quantities.

An interesting method for testing the green strength of metal powders has been described by Kelton.[62] Although purely arbitrary, the "rattler test" is based on the experience that in many fabricating methods small compacts are ejected from the die and mechanically pushed out onto the bed of the press, from which they drop into a container or slide onto a tray. Then they are racked or otherwise positioned into a boat either automatically or by hand before passing through the sintering furnace. This handling invariably involves some dropping and abrasion. With green strength a serious factor in many powder fabricating operations, the rattler test constitutes an attempt to simulate this handling by a mechanical test in such a manner that good comparisons can be obtained among various types of powders. As specimens for the determination of the green strength, cylindrical compacts 0.50-in. diameter by 0.250-in. thickness, compacted at the desired pressure (30 to 50 tsi) are to be recommended. The thickness of these specimens must be kept to rather close tolerances, not exceeding 0.003 in. either way. Five of these specimens at a time are then tumbled for a certain period in a 0.75-in. diameter 14-mesh bronze screen cylinder 4 $1/8$ in. long and containing one 0.50 in. high by 0.25 in. thick radial baffle fastened lengthwise along the inside surface. A 0.50-in. diameter axle runs through the axis of the cylinder. The cylinder is then rotated horizontally on its axis for 1500 revolutions at a speed of 83 r.p.m. The percentage weight loss on the five specimens is taken as the inverse measure of the green strength.

In the case of operations for which there are requirements other than resistance to dropping or abrasion, different methods of testing the green strength have been found useful. A simple transverse test has, for instance, answered the purpose in the case of pressed thin clutch plates, sliding from the press bed onto flat trays.

Testing of Chemical Properties

Chemical analyses of metal powders are usually made by standard methods,[63] and need little elaboration here. Together with microscopic

[62] E. H. Kelton, lecture on *Powder Metallurgy,* before Am. Soc. Metals, New York, Oct. 30, 1944. Also *private communication.*

[63] See, for example, ASTM Standards C18–41 and C46–43.

examinations, spectrographic and chemical analyses serve to establish evidence of alloying ingredients or impurities. While one must depend on the microscope for determining the nature and form of these ingredients in the powder, analytical methods offer quantitative results and are the accepted basis for most powder specifications.

CHEMICAL DETERMINATION OF OXYGEN

Probably the most controversial topic in setting up chemical specifications for powders such as iron or copper is the determination of the oxygen content, especially in the case of relatively pure powders for which the oxygen content is small.

At present the generally accepted method provides the determination of weight loss of the sample during a reduction treatment in hydrogen (e.g., at 1050°C. ± 15°C., 1920°F. ± 30°F., for 60 min. for Fe; 875°C. ± 15°C., 1785°F. ± 30°F., for 30 min. for Cu[64]). Advantages of the "hydrogen loss" method lie chiefly in the fact that it is possible for tests to be made in the same manner by both supplier and user, and that test conditions by agreement can be made to reproduce actual plant conditions. It must be borne in mind, however, that in practice this method gives results short of the true quantitative amount of oxygen present in the powder. In order to approach the true oxygen content the test would have to be carried out at very high temperatures, and for a considerable length of time, to permit complete diffusion of the hydrogen into the cores of the particles in which the last traces of oxide are usually lodged. The tendency of most powders to sinter into a solid body further inhibits the rate of final reduction, and complete removal of all oxygen may possibly take days. A different approach to the problem would be the determination of the oxygen content by chemical analysis. In the case of iron, for example, the oxygen content can be determined by the difference between the analyses of metallic iron and total iron. Inaccuracies, however, are introduced by the relatively large figures obtained for both analyses.

CHEMICAL DETERMINATION OF INSOLUBLES AND GREASE

In investigating powder samples produced from ore, it is frequently sufficient to determine the over-all content of the gangue residue. Instead of analyzing for SiO_2, Al_2O_3, MgO, CaO, etc., a rapid test can be made by determining the percentage of matter insoluble in dilute nitric acid. The content of grease, usually smeared around the particles during comminution, and sometimes present in rather impressive quantities, is conveniently determined by the Soxhlet extraction method.

[64] Metal Powder Assoc. Standard 2–48T, 1948.

Summary

Standard methods of sampling have recently been adopted in the field of powder metallurgy, and all procedures based on individual specifications agreed upon by powder producers and users and modeled to apply to specific applications are bound to be superseded in due course.

Testing procedures to determine particle characteristics, such as size and shape, include four principal methods:

(*1*) Microscopic examination—for the investigation of structure, shape, purity, and surface conditions of particles, as well as for the determination of particle size distribution by means of a count of particles of various sizes, or for the purpose of calibrating other methods.

(*2*) Sieve analysis—for control purposes of most industrial powders of coarse and medium-fine grades, as well as for control of powdered raw materials from which metal powders are produced.

(*3*) Sedimentation methods—based on the application of Stokes' law, and including fractionation and nonfractionation methods.

(*4*) Testing methods based on surface area—determinations by such methods as adsorption and permeability.

Of these only the sieve analysis has so far been standardized for metal powders.

Testing procedures to determine packing and molding properties include such technological methods as flow and apparent density measurements and various methods to test the compactibility of the powder and cohesive strength after compaction. Only the first two methods have been standardized to date.

Chemical examination of metal powders generally follows standard procedures of quantitative analysis, with the determination of the oxygen content presenting a special problem.

CHAPTER VI

Commercial Powders

Since many details regarding production and properties of metallic powders are considered trade secrets, the known facts remain scant, being restricted to information publicized in the literature or given out by industrial concerns in the form of advertisements, catalogues, etc. Hence, any accumulation of data concerning commercial or experimental metal powders must of necessity remain incomplete and can, serve only as a basis for general information. For additional facts industrial manufacturers or laboratories must be consulted.

In the following survey commercial powders are classified into five groups according to chemical considerations: (*1*) metal powders, (*2*) alloy powders, (*3*) compound powders, (*4*) composite powders, and (*5*) nonmetallic and semimetallic powders. This arrangement disregards the position of the individual materials on the market with reference to price and quantity production. A 1948 price list of some common metal powders is, however, given in Table 14. The prices are generally based on the (then current) market prices of ingots plus a fixed charge, and refer to ton lots. Table 15 lists some commercial metal powders arranged according to their approximate annual production. These data were obtained from various sources and must be considered as being only relative. It can be observed, however, that only iron and copper powders are produced in quantity for powder metallurgy applications, while the large quantities of aluminum powder are produced almost exclusively for the pigment and pyrotechnics industry. It might be interesting to note that the metal powder production has tremendously increased in volume during the war years and that for 1945 a peak production output of 200 million pounds of all types of metal powder has been reported.[1]

METAL POWDERS

Metal powders, arranged in an order corresponding approximately to their temperatures of fusion, can be classified into seven groups: (*1*) re-

---

[1] H. E. Hall, *Communications at the Second Annual Spring Meeting* of the Metal Powder Association, New York, June 13, 1946.

159

TABLE 14. Prices of Metal Powders[a]

Material	Price, dollars[b]
Brass, minus 100 mesh	0.24 to 0.285
Copper, electrolytic 100 and 325 mesh	0.30625 to 0.34625
Copper, reduced, 150 and 200 mesh	0.305 to 0.32
Iron, commercial, 100, 200, 325 mesh, 96+% Fe	0.10 to 0.17
Swedish sponge iron, 100 mesh, c.i.f. New York, carlots, ocean bags	0.074 to 0.085
Domestic sponge iron, minus 48 mesh	0.10
Iron, crushed, 200 mesh and finer, 90+% Fe, carload lots	0.05
Iron, hydrogen reduced, 300 mesh and finer, 98+% Fe, drum lots	0.63 to 0.80
Iron, electrolytic, unannealed, 325 mesh and coarser, 99+% Fe	0.44
Iron, electrolytic, annealed, minus 100 mesh, 99+% Fe	0.395
Iron, carbonyl, 300 mesh and finer, 99–99.8+% Fe	0.90 to 1.75
Aluminum, 100, 200 mesh, carlots	0.23 to 0.29
Antimony, 100 mesh	0.44
Cadmium, 100 mesh	2.00
Chromium, 100 mesh and finer	1.025
Lead, 100, 200, and 300 mesh	0.205 to 0.255
Manganese, minus 325 mesh and coarser	0.59
Nickel, 150 mesh	0.515
Silicon, minus 325 mesh and coarser	0.29
Solder powder, 100 mesh	0.085 plus metal
Stainless steel, type 302, minus 100 mesh	0.75
Tin, 100 mesh	0.90
Tungsten metal powder, 98–99%, any quantity, per lb.	3.05
Molybdenum powder, 99%, in 100-lb. kegs, f.o.b. York, Pa., per lb.	2.65
Under 100 lb.	2.90

[a] *Iron Age, 161*, No. 1, 271 (1948).
[b] Per lb., in ton lots, f.o.b. shipping point.

TABLE 15. Approximate Annual Powder Production for Powder Metallurgy Products in U.S.A., in Tons[a]

Copper
  Reduced, commercial grades ... 3000
  Electrolytic, commercial grades ... 3000
  Electrolytic, light grades ... 400
  Atomized and other ... 200
    Total ... 6600
Iron
  Reduced ... 1200
  Electrolytic ... 300
  Carbonyl ... 300
  Decarburized and other ... 300
    Total ... 2100
Lead ... 600
Tin ... 600
Tungsten ... 600
Molybdenum ... 450
Tantalum ... 20
Cobalt ... 15
Carbides ... 10

[a] Figures based on prewar production. Present production is estimated to be 50 to 100% higher. For nickel and chromium, production figures not available; powders used only for specialized products. For zinc, aluminum, and magnesium, production figures not available; large quantities produced for nonmetallurgical purposes.

fractory metals, (2) precious metals, (3) iron, (4) nickel, cobalt, chromium, and manganese, (5) copper, (6) light metals, and (7) low-melting metals.

Refractory Metal Powders

Of the refractory metal powders, three principal metals are used extensively in industry: tungsten, molybdenum, and tantalum. Other refractory metal powders, such as columbium, thorium, titanium, vanadium, zirconium, and uranium, have only limited production and application as metals, but some are commonly used as alloying elements.

Tungsten Powder. Two different ores are commonly used for the production of tungsten powder, namely iron and manganese tungstates, $(Fe, Mn)WO_4$, known as wolframite, and calcium tungstate, $CaWO_4$, known as scheelite. When manganese predominates, wolframite is known as hübnerite whereas if the ore is substantially free of manganese it is called ferberite, $FeWO_4$. Wolframites are found primarily in China and Australia, and are imported into the United States in the form of concentrates containing about 70% tungsten trioxide, WO_3; ferberite and tungstenite, WS_2, are also recovered in this country. Scheelite is found in large deposits in the United States, and its mining has been emphasized recently.

Pure tungsten trioxide is prepared from the rich ores or concentrates by a variety of mechanical and chemical treatments and is then reduced to tungsten powder. All methods provide for pulverization by grinding the ore or concentrate to fine size. Chemical treatment includes hot digestion with caustic alkali, purification of the alkali tungstate by repeated crystallizations, and final precipitation of tungstic acid, H_2WO_4, with hydrochloric acid. For wolframite, purification of tungstic acid is achieved by dissolution in ammonia followed by precipitation with hydrochloric acid, while scheelite can be precipitated with calcium chloride. Tungstic acid may also be converted to ammonium paratungstate which can either be heated to form tungsten trioxide or decomposed by hydrochloric acid to form tungstic acid which is converted by drying to tungsten trioxide.[1a]. The oxide obtained by heating is denser and less soluble in ammonia than the other type. These properties are usually preferred for the manufacture of filaments. Effective control of grain size and purification can also be achieved by using a chlorination process as the last step, wherein the tungsten trioxide is heated in a stream of chlorine or a chlorine compound, and the

[1a] P. E. Wretblad, in J. Wulff, *Powder Metallurgy*, Am. Soc. Metals, Cleveland. 1942, p. 422.

tungsten chlorides thus formed are decomposed by water.

Filtration, drying, and crushing of the oxide are followed by reduction in hydrogen. Particle size distribution, of utmost importance in incandescent lamp wires, can be controlled by varying the type and particle size of the oxide and the reduction conditions, and by the action of addition agents, impurities, or moisture content. The effect of moisture content during reduction is particularly important, because the small particles, when oxidized by water vapor, sublime as oxide and decompose on the surface of larger grains. Therefore, fine particle sizes can be obtained only if the oxide is dry, if the hydrogen is well desiccated and depleted of all oxygen, and if the hydrogen flow is large enough to remove rapidly all water vapor formed as a reaction product.

A low reduction temperature favors fine particle size. Very pure tungsten trioxide requires a temperature of only 650° to 700°C. (1200° to 1290°F.) for complete reduction within a reasonable time (i.e., 2 hours). This temperature may even be lowered to 550°C. (1020°F.) if alkaline salts are used as catalytic addition agents. In practice, however, reduction temperatures are higher, and may even reach 1000°C. (1830°F.). Rapid heating and long periods at high reduction temperatures must be avoided since they promote large particle size. The reduction process is customarily carried out in tubular furnaces heated by gas or electricity and operated by automatic temperature control. Vessels containing charges of tungsten trioxide are continuously passed through the tubes, and the hydrogen flows through the tubes in a counterstream. Some manufacturers have successfully adopted a two- or multiple-step reduction cycle: they produce lower oxides in one or more operations, sometimes using carbon as the initial reducing agent; these oxides are then broken down, sieved, or mixed with fresh tungsten trioxide before being finally reduced to pure tungsten powder.

The particle sizes range from 0.5 to 3 microns for powders destined for incandescent lamp filaments, to 8 to 20 microns for powders to be used in x-ray targets and electrical contacts. The sintering properties of these powders may be improved by mixing coarse and fine sizes, or by strain-hardening the particles in a mortar or ball mill. Tungsten powders are also available in other grades, such as hydrogen-reduced powders of 150 and 200 mesh (containing 99%+ tungsten), and carbon-reduced powders of 80, 100, and 200 mesh (98%+tungsten). By crushing and pulverizing scrap tungsten from sintered and swaged bars, it is also possible to obtain powder in the form of sintered granules of various mesh sizes. This powder has proved rather valuable in the manufacture of composite metals where resistance against abrasion is the object, and its considerably reduced price (about one-half of the regular market price for tungsten powder) makes its use here particularly attractive. Since commercial tungsten powders are of the highest purity, ranging from 99.0 to 99.99% tungsten content, it should be added that protection from moisture is very important.

For a detailed description of the production of tungsten powders, the books by Smithells[2] and by Li and Wang[3] should be consulted.

Molybdenum. The manufacturing methods for molybdenum powder are very similar to those described for tungsten. Practically unlimited ore supplies are found in the United States. Molybdenum sulfide is recovered from the high-grade ore concentrates and, to a smaller extent, as a by-product of copper flotation manipulations. Only a small amount of the metal is used for the production of molybdenum powder or molybdenum compounds, the bulk being used in the steel industry. Pure commercial compounds, notably ammonium molybdate, are also used as raw materials in the production of fine powders for molybdenum products.

Like tungsten, molybdenum powders are marketed in various forms and purities. For the manufacture of wire and sheet, powders with a 99.9%+ molybdenum content are generally preferred, but powders containing 99.5% of the metal are also on the market. A typical analysis of such a powder shows:

Component	Analysis, %	Component	Analysis, %
Metallic molybdenum	99.00	Carbon	0.25
Molybdenum in molybdenum		Iron	0.25
trioxide	0.30	Sulfur	0.03
Oxygen	0.15	Phosphorus	0.01

Mesh sizes ranging from 80 to 325 mesh are commercially available, but wire manufacturers generally prefer very fine grades. Molybdenum, like tungsten, is also available in the form of sintered granules obtained by comminution of sintered and swaged metal.

Tantalum. Of the many methods for producing metallic powders, the one used for the manufacture of tantalum is considered to be one of

[2] C. J. Smithells, *Tungsten.* 2nd Ed., Chapman and Hall, London, 1936.
[3] K. C. Li and C. Y. Wang, *Tungsten.* 2nd Ed., Reinhold, New York, 1947.

the most difficult. Tantalum ores of economically suitable grades are scarce, principal deposits being found in Australia and Africa, although medium-grade deposits are also worked in this country—in South Dakota. The ores, known as tantalite or columbite, contain iron and manganese tantalates and columbates, columbium always being closely associated with tantalum.

The production of tantalum and columbium powders[4] from the respective minerals is based essentially on fusion of the ores with sodium hydroxide, followed by purification of the tantalates and columbates by digestion with hot hydrochloric acid, which converts them to tantalic and columbic acids. The mixed oxides are then treated with potassium fluoride, forming double fluorides of potassium and tantalum and potassium and columbium. The separation of tantalum from columbium is achieved by crystallizing the less soluble tantalum salt, K_2TaF_7, from the more soluble columbium salt, K_2CbOF_5. Tantalum fluoride is filtered off and crystallized from an aqueous solution containing a small quantity of hydrofluoric acid to prevent hydrolysis, resulting in the formation of oxyfluorides.

Tantalum metal powder is obtained by electrolyzing the fused potassium tantalum double fluoride. Balke[5] has given a detailed description of the process by which the electrolysis is carried out in an apparatus consisting of a pot-shaped cathode and a vertically movable carbon anode. Contamination by iron is negligible, since the salt freezes near the surface of the pot. A fine network of metallic tantalum crystals is built up in which the salt gradually solidifies and which carries the current. In order to prevent electrolysis from being stopped by the formation of gaseous films around the graphite anode when fluorine is liberated, an oxygen-carrying compound that liberates oxygen and causes gradual burning of the graphite anode is usually added to the bath. After completion of the process the anode is withdrawn and the mass is allowed to cool. Removal of the cake is facilitated by shrinkage. The product then consists of the solid salt interwoven with a fine network of metallic tantalum crystals. The metal particles are obtained by crushing, pulverizing, and air separation, followed by thorough washing to dissolve all the salt. A typical analysis of the tantalum powder shows 0.05 to 0.2% carbon, 0.02% iron, with traces of residual fluoride salt and hydrogen as the only significant impurities; others, such as silicon dioxide and aluminum oxide, are negli-

[4] C. W. Balke, *Ind. Eng. Chem., Ind. Ed., 27,* 1166 (1935).
[5] C. C. Balke, *Iron Age, 147,* No. 16, 23 (1941).

gible. Tantalum powders are usually coarser than the fine tungsten and molybdenum powders, and it has been reported that mixed sizes result in a more uniform grain size in the final products.[5a]

Columbium. Because of the greater solubility of its fluoride and because impurities concentrate with it, columbium is more difficult to purify than tantalum. Columbium fluoride is recrystallized from an aqueous solution and columbium metal powder can be produced by a method similar to that used for tantalum. Columbium, however, is used almost exclusively in the form of ferrocolumbium for steel manufacture.

Titanium. Although the metal is present in abundant quantities in ferrous ores, titanium powder is produced in only very small quantities for the chemical and cemented carbide industries, and for the manufacture of getter elements in the high vacuum industry.[5b] Like columbium, titanium is used principally at present as the ferroalloy for scavenging in steel manufacture, and as the oxide in the pigment industry. The manufacture of titanium powder is based on the reduction of titanium oxide with metallic calcium, or on the reduction of titanium tetrachloride in a pressure vessel with alkali metals or magnesium.[5c] For the combination of the metal with other refractory metals into multiple carbides, hydrogen reduction of powdered titanium oxide is frequently used. Reduced titanium powder with a purity varying between 92% and 98% titanium is being marketed in 200-, 250-, and 325-mesh sizes.

Zirconium. This metal is also available principally as a ferroalloy; it is employed in this form both for scavenging and alloying purposes. Zirconium powder is manufactured by processes resembling those employed for the production of titanium and thorium powders. It is used for the manufacture of the pure metal which is very resistant to acids. Reduced zirconium powder of 250-mesh size and finer is available, with a metal content of about 98%.

Vanadium. Although vanadium powder is obtainable in small quantities for chemical and research purposes, no commercial market has yet been found for the pure metal. Instead, this metal is used exclusively as an alloying agent in steel manufacture in the form of ferrovanadium.

[5a] R. Kieffer and W. Hotop, *Pulvermetallurgie und Sinterwerkstoffe.* Springer, Berlin, 1943, p. 258.
[5b] R. Kieffer and W. Hotop, *loc. cit.*, p. 175.
[5c] W. Kroll, *Z. anorg. allgem. Chem.*, *234*, 42 (1937); *Z. Metallkunde*, *29*, 189 (1937).

Thorium.   A procedure resembling that employed for the manufacture of tantalum and columbium is used to produce thorium by electrolysis of the potassium double fluorides in fused salt baths. Reduction of the oxide by calcium or other alkali metals has also been carried out successfully, but the product is somewhat less pure than the powder produced by electrolysis. Like titanium, thorium has found applications in the manufacture of elements for the high vacuum industry.

Uranium.   Because it is unstable in air, uranium powder has found little use except for research purposes. Its manufacture closely resembles that of thorium; the purest metal is obtained by electrolysis of the fused salt. An apparatus[6] for the fusion electrolysis of uranium is shown in Figure 15, Chapter III.

Precious Metal Powders

Three powders in the precious metal group chiefly used in the field of powder metallurgy are platinum, gold, and silver. Other precious metal powders, such as palladium, rhodium, osmium, and irridium can also be produced in powder form, but have found no important uses. Their production is based either on decomposition of a compound, as in the case of osmium powder used in early filament manufacture,[7] or on precipitation from a salt, as in the case of palladium powder. Wollaston, as early as 1805,[8] produced rhodium powder by reducing sodium rhodium chloride with hydrogen.

Platinum.   The Wollaston process for producing platinum is still used today (in its essentials)—chiefly in England.[9,10]   With improved chemical methods of precipitating the ammonium platinum chloride from dilute solutions, and decomposing the precipitate into sponge powder (platinum black) by ignition and heating, a powder can be obtained whose purity is 99.9% platinum and better. This type of powder is suitable for dry briquetting at low pressures; ingots heated to a sufficiently high temperature may surpass the cast metal in physical and metallurgical properties.

Gold.   A method for producing very fine gold powder has been described by Trzebiatowski.[11]   Precipitation of dilute chloroauric acid ($HAuCl_4$) with dilute alkaline hydrogen peroxide, followed by reduction

[6] F. Skaupy, *Metallkeramik*. Verlag Chemie, Berlin, 1943, p. 23.
[7] U. S. Pat. 976,526.
[8] A. W. Deller, in J. Wulff, *Powder Metallurgy. Am. Soc. Metals,* Cleveland. 1942, p. 581.
[9] D. McDonald, *Chemistry & Industry, 9,* 1031 (1931).
[10] R. H. Atkinson and A. R. Raper, *J. Inst. Metals, 59,* 207 (1936).
[11] W. Trzebiatowski, *Z, physik. Chem., A169,* 91 (1934).

with formaldehyde, results in a very pure powder of 999-plus fine gold, with a particle diameter of about 1 micron. Gold powder can also be obtained by other methods, one of them being the electrolysis of cyanide leaching solutions employed in leaching ores.[12] Rapid movement of the cathode relative to the electrolyte, and high current densities, yield fine-grade (200 and 325 mesh) powders. Another method[13] involves the granulation of a gold alloy, with gold present only as a minor constituent. After the alloy is parted in dilute nitric acid, the gold is obtained in the form of coarse particles, while the base metal goes into solution. The gold granules are cleaned with fresh nitric acid and an agent capable of dissolving the nitrate of the base metal.

Silver. Silver powder is produced in quantity by precipitation with copper or iron from a silver nitrate solution and, to a lesser extent, by electrolysis. A variety of mesh sizes can be obtained. The coarser (electrolytic) grades, varying from 100 to 325 mesh, are sometimes referred to as "crystalline" in the trade, while the highly dispersed (precipitated) grades are referred to as "amorphous." The purity of all these grades is 999-plus fine silver.

Iron Powders

The many varieties of powdered iron are receiving ever-increasing attention from metallurgical, mechanical, and electrical engineers, as well as from those associated with powder metallurgy. In view of the important role which iron powder plays in powder metallurgy, the topic will be treated in somewhat greater detail in the following survey of the various commercial grades and their methods of production.

In our discussion of materials suitable for the production of sintered ferrous metals, a distinction must be made between powdered iron and sponge iron. Sponge iron is reduced directly from ore and may be in coarse, granular form or in large chunks, with varying degrees of purity, but generally not in excess of 96% iron content. It has a variety of uses, chiefly as melting stock, but also as a cement waterproofing material. It is suitable for molding powdered metal products only when pulverized to fine mesh sizes and purified further.

Many processes are known for direct reduction of iron ore to produce sponge iron, including those based on: (*1*) horizontal or inclined cylindrical rotary kilns, (*2*) a horizontal, cylindrical steel muffle with internal transport screws, and (*3*) rotary hearth muffle furnaces—the hearth is either circular or annular and is heated with fuel combusted in, or above,

[12] U. S. Pat. 1,251,302.
[13] U. S. Pat. 1,155,652.

silicon carbide tubes or arches. Other processes utilize the by-product coke oven type of furnace or electric resistance furnaces of various designs. In most of these processes either solid or gaseous carbonaceous reducing agents are employed, but other reducing gases, such as hydrogen or dissociated ammonia, have also been tried.

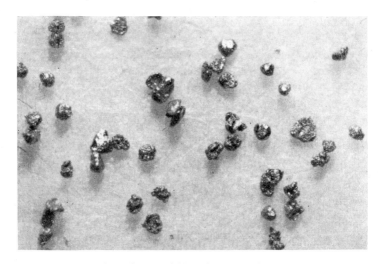

Fig. 39. Shape and surface characteristics of reduced iron powder particles (×50) (courtesy of Metals Refining Co.).

One of the chief reasons for the astonishing development of ferrous powder metallurgy in recent years has been the acknowledgment that only high-purity iron powders give satisfactory results. Therefore, since powders with a minimum iron content of 98% are employed for most present applications, the methods used for the production of sponge iron are obviously inadequate. In fact, it has been recognized that for many products which demand very pure and plastic powders, suitable iron powders cannot be prepared by reduction of ores, but must be prepared by reduction of rolling or drawing scale, oxidized scrap ("synthetic mill scale"), or other ferrous wastes.

The most important methods for producing sponge iron or iron powder now used in the United States are the following: (1) production of sponge iron and powdered iron by reduction in horizontal furnaces; (2) production of sponge iron by reduction in vertical furnaces; (3) production of sponge iron by low-temperature reduction in a pulsating atmosphere; (4) preparation of iron powder by the carbonyl process; (5) preparation of iron powder by electrolysis; (6) production of iron powder by decarburizing low-carbon steel scrap; and (7) production of powdered iron by granulation or shotting of pig iron or scrap.

REDUCED POWDERS

Horizontal furnace reduction is widely practiced in this country, and at present more powder is produced daily by this method than by all other processes combined. Figure 39 shows the shape and surface characteristics, and Figure 40 shows microsections of different types of reduced iron powder. Several plants are known to operate with this process and are producing iron powders by reduction of scale or iron ore concentrates in quantities exceeding five tons per day. A number of variations of this method have been worked out by different producers, especially with reference to the heating cycle and to details of furnace construction, but all employ cracked hydrocarbons, dissociated ammonia, or hydrogen as the chief reducing atmosphere—usually at temperatures ranging from 750° to 1050°C. (1380° to 1920°F.). Several manufacturers are known to employ a duplex reduction cycle which yields very pure and plastic iron powder. A low-temperature cycle (700° to 800°C.; 1290° to 1470°F.) uses a continuous horizontal or rotary kiln furnace and carbonaceous reducing agents, e.g., charcoal. After this treatment the powder is approximately 70% reduced and has an average iron content of 94%. Final reduction is obtained by means of a high-temperature cycle (900° to 1100°C.; 1650° to 2010°F.) in a continuous horizontal furnace with the aid of pure hydrogen or dissociated ammonia. Sintering of the iron particles during the high-temperature cycle—a very serious problem in continuous furnace practice—is held within tolerable limits by special treatment of the walls of vessels, conveyors, etc., by the use of proper addition agents, and by careful selection of the raw material.

There are several methods for producing sponge iron in vertical reduction furnaces.[14] Some provide for externally heated vertical retorts through which preheated iron ore is showered, to be reduced during its fall through the vessel and collected at the base in a suitable chamber. One very recent method provides for reduction of the ore at low temperatures (around 700°C.; 1290°F.) by hydrogen at higher than atmospheric pressures.[15] Another method produces sponge iron in a vertical reduction tube, heated by electrical induction, into which roasted iron ore is fed. As the ore descends the tube it is reduced by means of a reducing gas (e.g., natural gas) which is forced upward through the descending ore.

[14] A. H. Allen, Steel, 109, No. 14, 58 (1941); 104, No. 15, 43 (1939).
[15] U. S. Pats. 2,277,067; 2,287,663; 2,290,734; 2,296,498; 2,316,664; and 2,316,665.

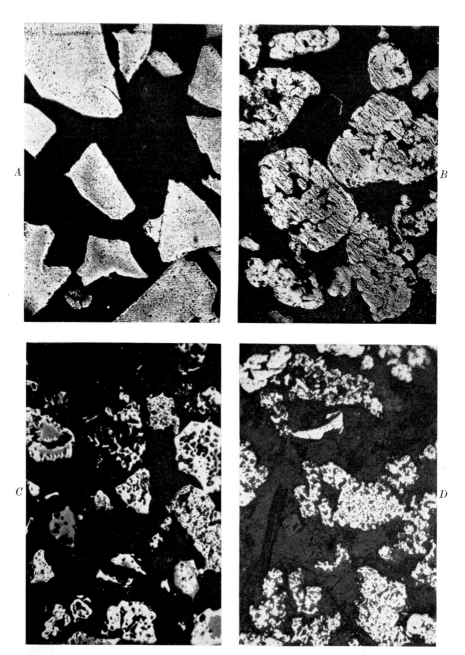

Fig. 40. Microsections through various types of iron powder particles (×200) reduced at: A, low temperature and higher than atmospheric gas pressure from iron ore concentrate; B, medium temperature and higher than atmospheric gas pressure from iron ore concentrate; C, high temperature and atmospheric gas pressure from iron ore concentrate; D, high temperature and atmospheric gas pressure from rolling mill scale.

When the resulting sponge iron is pulverized, a powder is produced which has been found suitable for certain molding applications.

Sponge iron has also been produced in pressure vessels by low-temperature reduction of preheated ore in natural gas subjected to pulsating surges. Reducing temperatures employed in this process are, according to Allen,[14] as low as 60°C. (140°F.), while the gas pressure builds up to a peak of 50 psi. Other methods concerned with low-temperature reduction of iron oxide are mostly of European origin.[16-19] The effects of high pressures on the rates of reduction are discussed in papers by Tenenbaum and Joseph,[20] and by Diepschlag.[21]

Users of high-purity iron powder have by now generally recognized that the best powder for their particular use is one prepared from mill scale or from synthetic oxides. The close association between silicon gangue and iron ores (easily broken down in the blast furnace) is not readily severed by magnetic separation and low-temperature reduction. Furthermore, it appears that the substantially more solid ore particles are less susceptible to low-temperature reduction than the fluffier scales which yield softer and more plastic metal particles. So far, these two factors have been the chief obstacles to the large-scale use in the metal powder industry of iron powder reduced directly from the ore. There is evidence, however, that great effort will be made in the near future to open the vast resources of directly reduced iron ores to powder metallurgy by improving and controlling the purity and plasticity of the products. The price factor —of great importance in the postwar period—will probably be the main incentive for this trend.

At the present time, emphasis is placed on iron oxides which are free from silicon dioxide, and have a uniform iron content, because they simplify the reduction considerably. In addition to the use of oxidized scrap and scales mentioned above, several other methods have been developed that yield pure iron oxides. Allen[22] mentions 2 interesting methods. One employs waste pickling liquors as the starting material, from which ferrous sulfate is crystallized. The ferrous sulfate is then dissociated into pure iron oxide and sulfurous gases—suitable for the manufacture of sulfuric acid. The other method produces iron oxide as a by-product of the

[16] Brit. Pat. 507,277.
[17] German Pat. 716,025.
[18] Brit. Pats. 507,494; and 507,581.
[19] U. S. Pat. 2,266,816.
[20] M. Tenenbaum and T. L. Joseph, *Trans. Am. Inst. Mining Met. Engrs., 135,* 59 (1939); *140,* 106 (1940).
[21] E. Diepschlag, *Arch. Eisenhüttenw., 10,* 179 (1936).
[22] A. H. Allen, *Steel, 109,* No. 14, 58 (1941); *104,* No. 15, 43 (1939).

chlorine method for manufacturing sulfuric acid from pyrites. Gaseous ferric chloride and sulfurous gases are evolved, and after the ferric chloride is condensed, it is broken down into iron oxide and free chlorine, the latter being used over again. The production of chlorine yields hydrogen, which can be utilized effectively for the reduction of the oxide. The abundance of iron pyrites in the United States and Canada and the economy of the method favor this process.

Another method, utilizing Minnesota's carbonate slate, has been suggested by Firth.[22a] Although the ore is highly siliceous, it appears to be amenable to chemical processing. The ore utilized in this process has no value at present, and in certain places it constitutes a portion of the material which must be removed to permit access to the good iron ore underneath. After being crushed and ground to —48-mesh size, the ore is digested with dilute sulfuric acid. More than half the iron content is put into solution and the acid is neutralized. The residue is filtered, and the sulfuric acid can be regenerated. The strong liquor effluent from the digestors is evaporated to remove excess water and the iron sulfate (copperas) is then crystallized. The final purity of the product depends mainly on subsequent dehydration and calcination; iron sulfate srystals are contaminated during digestion by sulfates of aluminum, manganese, calcium, and magnesium. Calcination is achieved under controlled conditions when ferrous sulfate is decomposed and the iron oxidized to ferric oxide, leaving the contaminated sulfates unaltered; soluble impurities are then removed by leaching the calcined product; washing is followed by briquetting the oxide and heating to 1050°C. (1920°F.) under oxidizing conditions. Thus, any sulfur retained because of incomplete washing is removed. The oxide is made uniform prior to reduction by a process which makes use of a vertical furnace employing blue water gas (hydrogen + carbon monoxide) at a temperature above 950°C. (1740°F.).

CARBONYL POWDER

Carbonyl powders are obtained by a process involving passage of carbon monoxide over iron particles at elevated temperatures; liquid iron pentacarbonyl is thus formed. This compound is subsequently decomposed under carefully controlled temperature and pressure conditions. yielding the carbonyl iron powder and carbon monoxide gas, which can be recirculated. Figure 41 shows a microsection through a typical powder

[22a] C. V. Firth, *Information Circ. No. 3,* Univ. Minnesota Mines Experiment Station, Minneapolis, Minn., May, 1943.

grade. A thorough study of the nature, properties, and applications of carbonyl iron powder has recently been published by Pfeil in England.[22b]

During the war, more than 100 tons per month of carbonyl iron powder were produced in Germany by I. G. Farbenindustrie at Oppau[23] for the production of magnets, high-frequency cores, and accumulator plates. The details of the process follow. Spent iron pyrite from an adjoining

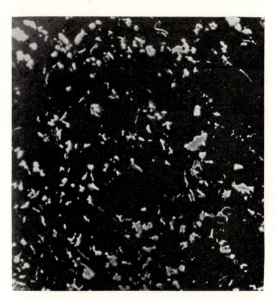

Fig. 41.   Microsection through carbonyl iron powder particles (×200).

sulfuric acid plant was melted with steel scrap in a rotating drum furnace, cast into ingots containing about 8% sulfur, balance iron. These ingots were crushed to pebble size, introduced in charges of about 5 tons into four large reaction chambers (20 in. in diameter by 20 ft. long), and kept at 220°C. (430°F.). Preheated carbon monoxide was allowed to pass over the charge and, after five days, about three-quarters of the iron was converted to iron pentacarbonyl gas. Cooling of the gas to about 50°C. (120°F.) caused liquefaction of the carbonyl gas, which was then tapped off and stored at over atmospheric pressure (5 atmospheres). The carbonyl compound was decomposed in special retorts (3 ft. in diameter by 10 ft.

[22b] L. B. Pfeil, *Symposium on Powder Metallurgy*. The Iron and Steel Institute, Special Report No. 38, London, 1947. p. 47.

[23] T. P. Colclough, "I. G. Farbenindustrie—Oppau Works, Ludwigshafen", *Report to the Combined Intelligence Objectives Subcommittee*; see also *Metal Progress*, *50*, No. 2, 232 (1946).

high), by being heated to its boiling temperature of 102°C. (215°F.) with the aid of steam coils. The carbonyl vapor passed to the top of the retort, held at 240°C. (465°F.) and normal pressure, and there decomposed into fine iron powder that settled to the bottom of collector hoppers. This crude powder, usually containing about 1% carbon as the major impurity, was annealed in moist hydrogen at 400°C. (750°F.) to reduce

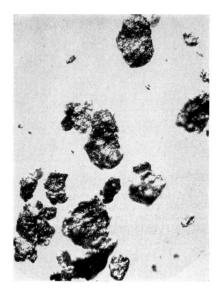

Fig. 42. Shape and surface characteristics
of electrolytic iron powder particles (×60).

the carbon content. Breakdown of the cake, ball milling, and screening completed the process. The apparent density of the powder was about 3.0 g./cc. and the particles were nearly spherical in shape and very small. A limited variation in average particle size could be obtained by changing the conditions in the decomposition chamber—higher temperatures and shorter falls yielding finer particles.

ELECTROLYTIC POWDER

The electrolytic method offers a more economical process for producing iron powders of similar high purity. By this process, ferrous wastes, *e.g.*, scrap, mill scale, etc., as well as low-grade ores can be used as starting materials, with ferrous sulfate or (occasionally) ferrous chloride as the electrolyte. The sulfate bath is usually credited with the ability to produce a more highly refined product, while chloride baths, when operated

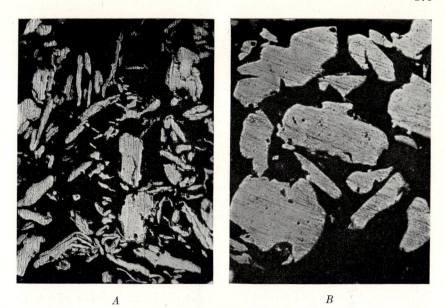

A   B

Fig. 43.   Microsections through electrolytic iron powder particles (×170): A, plate powder; B, angular powder.

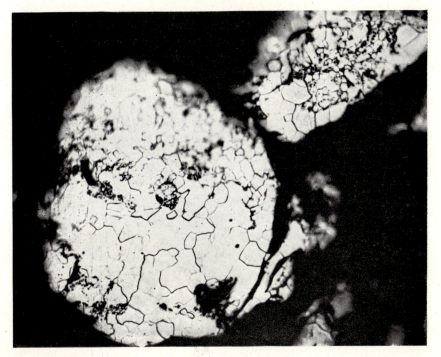

Fig. 44.   Microstructure of electrolytic iron powder particles (×500) (courtesy of Plastics Metals Div., National Radiator Co.).

at higher temperatures, result in greater rates of deposition.[24] These baths are usually operated with soluble anodes, although insoluble anodes can also be used. The purity of the product is dependent upon the purity of the anode material. Steel, cast iron, briquetted reduced sponge iron, and chemically pure iron (Armco[24a]) yield high-grade products whose uniformity is insured by close control of anodes, electrolyte, and bath temperature. The cells are acid resistant and the cathode sheets are usually of stainless steel. Iron deposits on these sheets as a spongy or solid mass and must be stripped. The deposits must be milled to desired screen sizes, and then annealed to dispel work-hardening resulting from milling and to eliminate any hydrogen remaining from the electrolysis of the iron. Figure 42 shows the shape and surface of a typical electrolytic iron powder, while Figure 43 exhibits microsections, and Figure 44 the microstructure of typical particles.

Interesting details of the production of electrolytic iron powder on a 400 pound-per-day pilot plant scale have recently been published by Trask.[25] In this process an electrolyte with an iron chloride base is used in preference to a sulfate solution because of the undesirable effects of sulfur compounds on the resultant electrodeposit. Insoluble contaminants, such as carbon or silica, originating from the anode, or ferric hydroxide, formed by slight oxidation of the bath, are removed from the electrolyte by periodically forcing it through a filter press. Plates of comparatively impure ingot iron have been found satisfactory for use as anode material. The electrolytic deposit is formed as a smooth plate upon thin cathode sheets of stainless steel. Current densities of about 25 amp. per ft.^{2} are employed, and the potential drop from anode to cathode is approximately 1.4 volts. The anodes and cathodes each have an effective area of about 18 ft.^{2} The power consumption is about 0.75 kw.-hr. per pound of iron deposit produced and, with a current efficiency of nearly 100%, a four-cell unit is fed by a 4000-amp., 9-volt, d.c. generator driven by a 60-horsepower synchronous motor. When about ⅛ in. of metal has been deposited the plate is stripped by flexing the stainless steel sheet. Pulverizing of the dried brittle deposit into fine particles of angular structure is accomplished by ball milling in a Hardinge conical mill (3 ft. by 8 in.) operated in an airtight system with a Hardinge-type loop classifier. Cyclone separation feeds the oversize back to the ball mill and collects the fines for further treatment in a continuous annealing furnace. This annealing process has the

[24] Anonymous, *Iron Age, 153*, No. 22, 55 (1944).
[24a] Product of the American Rolling Mill Co., Middleton, Ohio.
[25] H. V. Trask, *Metal Progress, 50*, No. 2, 279 (1946).

twofold purpose of reducing surface oxide films and softening the particles to render them compactible. Close control of conditions prevailing during the annealing operation is essential for maintenance of powder quality. The cycle consists of treatment of the powder at 870°C. (1600°F.) for 2 hours, preferably with dissociated ammonia as the reducing atmosphere. The caked powder is finally reground in a high-speed, hammer-type pulverizer which imparts a minimum of work-hardening. A vibrating 100-mesh screen is connected directly to the pulverizer in an hermetically sealed system. Homogenizing of the product is accomplished by blending the individual batches in a 400-pound, double-cone blender. On the basis of the limited pilot plant production, the cost of the powder was estimated to lie between 15 and 17 cents per pound for full-scale production.

The production of electrolytic iron powder with the daily output running into several tons in 1948 has lately been described by Granberg.[25a] The process, developed at Tacoma, Wash., differs from that previously described mainly in that insoluble anodes and an expandable electrolyte are used. This method demands an inexpensive and abundant source of electric power, available supplies of hydrogen and hydrochloric acid, and a ready source of scrap steel borings and turnings (at a production level of 10 tons per day of powder, approximately 8000 tons annually of scrap are required, for recovery of iron from scrap amounts, roughly, only to 50%). The scrap is first leached with ferric chloride whereby the latter is converted into ferrous chloride which is used as the electrolyte. After filtration, a second leaching operation serves to remove all traces of ferric chloride and any excess hydrochloric acid. The electrolytic process takes place in 40 rectangular cells which contain graphitized carbon anodes and cathodes of stainless steel sheets. The maintenance of high hydrogen overvoltage facilitates the proper deposition of the iron particles at the cathodes. The ferrous chloride is introduced at the bottom of the cells, and the depleted electrolyte leaves by overflowing the cells. The power requirements for the electrolysis are 365 volts and 5000 amp. The current density employed depends upon the conditions desired, and varies from 40 to 60 amp. per ft.[2] An automatic stripping device removes the deposited powder from the cathode, and is believed mainly responsible for the economy of the process.

Another recent account of the production of iron powder by electrodeposition has been made by Gardam in England.[26] A dendritic iron

[25a] W. J. Granberg, *Iron Age, 160,* No. 26, 70 (1947).

[26] G. E. Gardam, *Symposium on Powder Metallurgy.* The Iron and Steel Institute, Special Report No. 38, London, 1947, p. 3.

powder with about 1% oxygen was made in a pilot plant at a rate of one-quarter lb. per hr. by electrolysis of 10% ferrous ammonium sulfate solution. It had to be annealed, preferably in hydrogen, at 700°C. (1290°F.) for 1 hour. Compacts made from it had a tensile strength of about 21,000 psi and 7% elongation. With an output of about 1 ton per week, the cost of the powder was estimated to be about 20 to 27 cents per pound.

DECARBURIZED POWDERS

Pulverizing low-carbon steel scrap results in distinctly equiaxed and practically solid particles. After decarburization in a controlled atmosphere, a practically pure, fairly soft, and compressible iron powder is obtained,[27] which is particularly suitable for certain impregnation work. When it is mixed with sponge iron in definite proportions, porous products are obtained which can be readily impregnated with lead and other low-melting metals for certain types of bearings or parts requiring a high specific gravity.

GRANULATED POWDER

A powder with similar particle characteristics can also be obtained by granulation or shotting. After the pig iron or scrap is fused in a cupola, the liquid metal is forced into a Bessemer converter to reduce the carbon to the 0.10–0.50 range. After removal of the slag, the converter is tipped and the metal is blown against a rotating steel wheel in a shotting tank. The resulting porous, disk-shaped granules are crushed and pulverized in impact mills, and are finally annealed in a reducing atmosphere. During the war, a substantial part of the iron powder produced in Germany (amounting to as much as 50 tons per day[27a]) was produced by the DPG rotating disk process (Chapter III, p. 43). Further reference to this process and the resulting products will be found in Chapter XXV, Volume II.

QUALITY-COST RELATION

A great handicap to the development of iron powder metallurgy has been the problem of obtaining powder of high purity at reasonable prices. The price question has been discussed extensively by Allen,[28] Fellows, [29] Hardy,[30] and others, and may be summarized as follows.

[27] Anonymous, *Iron Age, 149*, No. 8, 49 (1942).
[27a] G. J. Comstock, *Communication at Powder Metallurgy Colloquium*, New York University, April 26, 1946.
[28] A. H. Allen, *Steel, 109*, No. 14, 58 (1941); *104*, No. 15, 43 (1939).
[29] A. T. Fellows, *Metals & Alloys, 12*, No. 3, 288 (1940).
[30] C. Hardy, *Metal Progress, 43*, No. 1, 62 (1943).

(1) Iron in the lowest price range must be classified as sponge iron—not suitable for the powder process. As early as 1927, Williams, Barrett, and Larsen of the U. S. Bureau of Mines[31] published figures of production test runs with cost estimates of $12.80 per ton for a 100-ton capacity and $16.00 per ton for a 20-ton capacity. In a later publication from the same source,[32] a cost estimate of $16.53 per ton was given for a 10-ton pilot plant.

(2) Sponge iron of a purity between 90 and 96% iron may be used for special applications in powder metallurgy when there are no strict requirements as to prop-

TABLE 16
Properties of Reduced (Sponge) Iron Powders

Type[a]	1	2	3	4	5
Particle structure....	Very spongy	Very spongy	Moderately spongy	Spongy	Moderately spongy
Screen analysis, fractions in %					
+100 mesh...	0–1	0.5–1.5	0	0–2.5	0
−100 +150	12–20	11–15	21.5	25–30	20.8
−150 +200	13–22	18–32	16.5	18–22	19.2
−200 +250	9–12	2–5	12.5	10–14	18.5
−250 −325	9–15	16–26	11.5	9–12	14.5
−325 +400	12–20	8–14	14.8	9–12	7.7
−400	30–38	20–25	23.2	15–20	19.3
Subsieve size range −30 μ fraction, %..	8–12	6–10	5	1.5–3	5
Apparent density, g./cc..............	1.9–2.2	2.0–2.3	2.2–2.5	1.9–2.1	2.0–2.7
Flow rate, sec./100 g. (0.1-in. orifice).....	40–50	37–44	30–35	33–37	47
Compressibility	Good	Good	Fair	Good	Fair
Microhardness, Brinell units.......	47	49	60	50	
Chemical composition, %					
Iron...............	98.0 min.	97.0 min.	97.5 min.	98.0 min.	96–98
Oxygen (hydrogen loss).............	1.3–1.5	1.5–2.5	1.0–2.0	1.0–1.5	1.0–3.0
Total carbon......	0.10 max.	0.10 max.	0.4–0.6	0.3–1.5	0.4–0.45
Combined carbon..	—	—	0.05–0.1	0.05–0.1	0.02–0.05
Graphitic carbon...	—	—	0.3–0.55	0.2–0.45	0.35–0.43
Sulfur	0.05 max.	0.05 max.	0.05 max.	0.05 max.	0.08
Phosphorus	0.05 max.	0.05 max.	0.05 max.	0.05 max.	0.04
Manganese	0.5	0.5	0.5	0.4	0.15–0.20
Nitric acid, insolubles	0.08	0.10	0.12	0.10	—
Silicon.............	—	—	—	—	0.15–0.20
Others.............	Traces	Traces	Traces	Traces	Traces

[a] (1) Product of Metals Disintegrating Co., Inc., Elizabeth, N. J., Type MD-111. (2) Product of Metals Disintegrating Co., Inc., Elizabeth, N. J., Type MD-131. (3) Product of Metals Refining Co., Hammond, Ind., Type I-284. (4) Product of Metals Refining Co., Hammond, Ind., Type I-297 (soft annealed). (5) Product of Plastic Metals, Inc., Johnstown, Pa., Type "Plast-Sponge" No. 2290.

[31] U. S. Bur. Mines Bull., No. 270 (1927).
[32] U. S. Bur. Mines Bull., No. 396 (1936).

erties or dimensions. Although the demand for this material is still limited, it has been speculated that a demand up to 500 tons per day could be fostered, and that powder of this type would sell for 2 to 4 cents per pound. Manufacturers of wear-resistant parts, brake linings, clutch facings, etc., have been suggested as potential users of this material, but the author fails to see the market for such tremendous quantities of low-quality powder.

(3) Reduced iron powder of a purity between 96 and 98% iron is suitable for numerous applications in powder metallurgy, particularly when the quantity is large and dimensional accuracy and physical property demands are moderate. This applies to porous bearings as well as to dense parts for general machinery. The price range of this powder has been stabilized at between 7 and 10 cents per pound.

(4) Iron powders of a purity between 98 and 99% iron, usually of the reduced type, are suitable for most applications for which high accuracy and quality of the finished products are required. They have found extensive use for bearings and elec-trical and machine parts. The price of domestic powders has been kept within the range of 11 and 16 cents per pound, while the imported Swedish grades sell in the United States for 7 to 10 cents per pound, depending on specifications.

(5) Iron powders of a purity exceeding 99% iron, such as carbonyl, electrolytic, and special types of hydrogen-reduced powders, are used for bearings, electrical, and general machine parts, and for special applications, e.g., magnetic coils, for which highest qualities but comparatively moderate tonnages are required. Prices of these powders range all the way from 15 cents to $1.00 per pound.

TABLE 17

Properties of Hydrogen-Reduced Iron Powder[a]

Particle structure	Fine, irregular, compact
Screen analysis, fractions in per cent	
+200 mesh	0
−200 +250	0.3
−250 +325	0.2
−325 +400	44.0
−400	55.5
Subsieve size ranges, fractions in per cent	
50–40 μ	11.3
40–30 μ	20.0
30–20 μ	29.3
20–10 μ	19.2
10–0 μ	20.2
Apparent density, g./cc.	2.3
Flow rate, sec./100 g. (0.1-in. orifice)	Nil
Compressibility	Poor
Chemical composition, per cent	
Iron	99.00 min.
Oxygen (hydrogen loss)	0.05
Carbon	0.04
Sulfur	0.012
Phosphorus	0.025
Copper	0.145
Manganese	0.20
Nickel	0.14
Zinc	0.21
Silicon	0.10

[a] Product of Charles Hardy, Inc., New York, N. Y., Type "Mephan."

	1	2	3	4	5	6	7	8	9	10
Type[a]										
Particle structure	Irregular, dendritic	Dendritic, globular	Irregular, flaky, nodular	Irregular, flaky, nodular	Irregular, globular	Dendritic, globular	Dendritic, globular	Dendritic, globular	Dendritic, globular	Dendritic, globular
Screen analysis, fractions in %										
+100 mesh	—	0	—	—	0	Trace	1.0 max.	1.0 max.	—	—
−100 +150	—	0.1	0	—	18.2	—	—	—	—	—
−100 +200	0	—	—	—	—	15–25	40–50	40–50	1.0 max.	1.0 max.
−150 +200	16	2.6	0.5	—	28.6	—	—	—	—	—
−200 +250	—	0.7	2.0	0	14.4	15–25	25–35	30–40	15–20	15–20
−200 +325	18	—	—	—	—	—	—	—	—	—
−250 −325	—	16.6	19.3	0.5	12.2	55–65	20–30	20–30	80–85	85–80
−325	24	—	—	—	—	—	—	—	—	—
−325 +400	42	23.7	29.0	33.5	9.3	—	—	—	—	—
−400	—	56.3	49.2	66	17.3	—	—	—	—	—
Subsieve size range, −30μ fraction, %	17	14	23	27	0.5	—	—	—	—	—
Apparent density, g./cc.	2.5	2.5	2.3	2.3	2.7	2.60–2.70	2.35–2.45	2.65–2.75	2.70–2.80	2.70–2.80
Flow rate, sec./100 g. (0.1 in. orifice)	—	33	37	49	29	33–31	—	—	—	—
Compressibility	Fair	Good	Fair	Fair	Good	Good	Good	Good	Good	Good
Microhardness, Brinell units	—	—	59	—	52	—	—	—	—	—
Chemical composition, %										
Iron	99.0 min.	99.0 min.	99.0 min.	99.0 min.	99.5 min.	99.6+	99.6+	99.5+	99.6+	98.5+
Oxygen (hydrogen loss)	0.5	0.5	0.3–0.5	0.3–0.5	0.2	0.3 max.	0.3 max.	0.4 max.	0.3 max.	
Carbon	0.10	0.005	0.01	0.01	0.02		0.008c			1.0b
Sulfur	0.07	0.004	0.004	0.004	0.002		0.004c			
Phosphorus	0.01	0.002	0.001	0.001	0.001		0.002c			
Manganese	0.04	0.027	0.01	0.01	0.15		0.019c			
Nickel	Trace	Trace	0.01	0.01	0.01		—			
Silicon	0.01	0.01	0.015	0.015	0.01		0.006c			
Copper	0.05	0.01	0.01	0.01	0.01		0.006c			
Zinc	0.14	0.09	—	—	0.01		—			

[a] (1) Product of Charles Hardy, Inc., New York, N. Y., type "Frel," annealed. (2) Product of Charles Hardy, Inc., New York, N. Y., type "Harshaw," annealed. (3) Product of Plastic Metals, Inc., Johnstown, Pa., type "Plast-Iron" No. 1008, annealed. (4) Product of Plastic Metals, Inc., Johnstown, Pa., type "Plast-Iron" No. 1007, annealed. (5) Product of Titanium Div., National Lead Co., New York, N. Y., type "Lectrofer" No. 2P-3J9-A, annealed. (6) Product of Buel Metals Co., Painesville, Ohio, type RA. (7) Product of Buel Metals Co., Painesville, Ohio, type CA. (8) Product of Buel Metals Co., Painesville, Ohio, type DCA. (9) Product of Buel Metals Co., Painesville, Ohio, type RB. (10) Product of Buel Metals Co., Painesville, Ohio, type RB "Graphiron." [b] Added as graphite. [c] Typical chemical analysis.

It must be understood that purity, price, and quantity are by no means the only deciding factors in evaluating commercial iron powders for particular powder metallurgy applications. In Chapter IV the properties of metal powders in general were discussed, and all observations made there apply to iron powders as well. Thus, particle structure, shape and size distribution, as well as density, flow, plasticity, and compressibility, must be controlled uniformly in every batch and must meet requirements called for by the specific application. The decision to use either powdered iron or regular steel as the raw material frequently depends upon these properties. Typical figures for the properties of various types of commercial iron powders are given in Tables 16–19 on pages 179–182.

TABLE 19

Properties of Carbonyl Iron Powders

Type[a]	1		2		3		4		5	
Particle structure..	Spherical, very fine		Spherical, very fine		Spherical, very fine		Spherical, very fine		Spherical, very fine	
Screen analysis, fractions in %										
+325 mesh	0		0		0		0		0	
−325 +400	1		1		0.5		0.5		0.5	
−400	99		99		99.5		99.5		99.5	
Particle size range, μ, (80% min.)...	10–20		4–10		2–7		1–5		1–3	
Average particle diameter, μ.....	20		10		8		5		3	
Apparent density, g./cc............	1.8–3.0		2.5–3.0		2.5–3.5		2.5–3.5		2.5–3.5	
Tap density,[b] g./cc.	3.5–4.0		4.4–4.7		4.4–4.7		4.4–4.7		4.7–4.8	
Compressibility ...	Good		Good		Poor		Poor		Poor	
Chemical composition, %										
Iron............	99.5 min.		99.4 min.		98.0 min.		98.0 min.		98.0 min.	
Oxygen.........	0.1–0.2		0.1–0.3		0.45–0.6		0.5–0.7		0.7–0.8	
Carbon.........	0.005–0.03		0.03–0.12		0.65–0.8		0.5–0.6		0.5–0.6	
Nitrogen	0.005–0.05		0.01–0.1		0.6–0.7		0.5–0.6		0.5–0.6	
Effective permeability and "Q" values	E.P.	Q	E.P.	Q	E.P.	Q	E.P.	Q	E.P.	Q
At 1 kc...........	4.16	3.25	3.65	3.97	3.09	3.86	2.97	7.81	2.17	1.69
10 kc.........	100	78.0	94	102.5	81	101.3	81	213.0	62	67.6
150 kc........	96	74.9	100	109.0	94	117.5	93	244.2	71	88.8
200 kc........	90	70.2	98	106.8	100	125.0	98	257.8	78	126.4
1 mc.........	43	33.5	72	78.4	97	121.2	100	263.0	84	136.0
100 mc.......	1	0.78	3	3.27	30	3.75	54	142.0	100	162.0

[a] Products of General Aniline Works, New York, N. Y. (1) Type "L"; (2) Type "C"; suitable for powder metallurgical and high frequency applications. (3) Type "E"; (4) Type "Th"; (5) Type "SF"; suitable only for high frequency applications.
[b] Tap density is obtained by filling a 100-cc. graduate cylinder to the top, closing it with a stopper and tapping on a wooden surface 100 times from a height of 8 in.:

T.D. = weight of sample, g./volume after tapping, cc.

Nickel, Cobalt, Chromium, and Manganese Powders

Nickel and cobalt powders are used in powder metallurgy to a certain extent, and are essential for certain applications, e.g., sintered magnets and cemented carbides. Chromium and manganese, however, have found little use in powder metallurgy and are employed chiefly as ferroalloys in the steel industry.

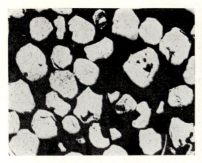
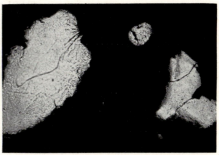

Fig. 45. Microsection through pulverized nickel powder particles (×200).

Fig. 46. Microstructure of electrolytic
nickel powder particles (×650).

Nickel. Nickel powders can be classified into three groups: (1) pulverized electrolytic nickel, (2) carbonyl nickel powder, and (3) reduced nickel powder. In the United States, at the present time, only the first type—used in magnetic materials and sintered nickel parts—is produced in commercial quantities, with Canadian ores serving as the chief source of supply. The amount of carbonyl nickel powder available in this country is insignificant, but it was produced extensively in Germany. Spongy nickel powder has been produced in small quantities from both oxides and oxalates by low-temperature reduction with carbon, hydrogen, or cracked hydrocarbons.[33] Production of reduced nickel on a larger scale in this country will depend on future demands; although it can be produced in uniform size, the material cannot compete in purity with the other types (reduced powder rarely exceeds 99.0% nickel, whereas electrolytic and carbonyl powders range from 99.0 to 99.9% nickel). The properties of commercial grades of pulverized (Fig. 45) and annealed electrolytic nickel (Fig. 46) are summarized in Table 20. In Table 21, similar data are given for representative lots of carbonyl nickel and reduced nickel powders.

[33] R. Kieffer and W. Hotop, *Pulvermetallurgie und Sinterwerkstoffe.* Springer, Berlin, 1943, p. 208.

Cobalt. Although cobalt powder can be produced electrolytically, the reduction process is the sole method by which it is produced in commercial quantities for the hard metal and sintered magnet industries. Ore deposits in the Belgian Congo and, to a lesser degree, in Canada serve as the raw material. After concentration, the oxide is reduced at relatively low temperatures, generally below 800°C. (1470°F.), to obtain the necessary degree of fineness—a prerequisite for the manufacture of cemented

TABLE 20

Properties of Pulverized Electrolytic Nickel Powder[a]

Type[b]	1	2	3	4	5
Particle structure	Solid, globular, very coarse	Solid, globular, coarse	Solid, globular, medium fine	Solid, globular, fine	Solid, globular, fine
Screen analysis, fractions in %					
+ 60 mesh	0	—	—	—	—
− 60 +100	26–30	0–0.5	0	—	—
−100 +150	35–38	9–13	0–0.5	—	—
−150 +200	12–15	27–30	3–8	0	—
−200 +250	8–10	20–23	5–10	2–6	—
−250 +325	4–6	12–15	8–15	8–12	0
−325 +400	4–6	6–9	25–35	30–40	40–50
−400	2–4	14–16	30–40	40–50	50–60
Apparent density, g./cc.	4.35	4.1	3.2–3.6	3.2–3.6	3.2–3.6
Flow rate, sec./100 g. (0.1-in. orifice)	35	32	—	—	—
Compressibility	Fair	Good	Good	Good	Good
Chemical composition, %					
Nickel	99.0 min.	99.0 min.	99.0 min.	99.0 min.	99.0 min.
Oxygen (hydrogen loss)	0.3	0.3	0.3	0.3	0.3
Carbon	0.02	0.03	0.02	0.02	0.03
Sulfur	0.0015	0.012	0.015	0.18	0.12
Iron	0.2	0.2	0.3	0.25	0.25
Copper	0.18	0.15	0.15	0.2	0.2
Nitric acid, insolubles	0.12	0.12	0.08	0.08	0.10

[a] After annealing at 700°C. (1290°F.) for 1 hr. in hydrogen.
[b] Products of Metals Disintegrating Co., Inc. (1) Type MD 60. (2) Type MD 130. (3) Type MD 151. (4) Type MD 201. (5) Type MD 301.

carbides. The temperature, however, must be high enough to permit complete reduction, since imperfectly reduced cobalt causes laminated fractures and other failures in the finished materials. Although carbon is the most commonly used reducing agent, complete reduction can also be obtained with hydrogen. Cobalt powder reduced from the oxalate is pyrophoric and not practical for commercial applications.

Grades preferred for cemented carbides pass through 325 mesh, while coarser grades, e.g., 150- and 200-mesh powders are suitable for sintered magnets. A minimum cobalt content of 98% is required in all cases—with purer powders preferred. Chief impurities are nickel and iron, each amounting to about 0.5%, with carbon, calcium, and manganese present to the extent of 0.1%.

Chromium. Although they are basically brittle, chromium powders are fairly compressible if they are thoroughly annealed in dry hydrogen or in a vacuum. Two types are produced commercially, namely electrolytic and pulverized reduced shot. (The reduction method with calcium in a barium chloride–calcium chloride melt, as described by Kroll,[34] is hardly suitable for commerical application.) The electrolytic type is of high purity, containing approximately 99.5% chromium, with 0.25% iron as the

TABLE 21

Properties of Carbonyl and Reduced Nickel Powders

Type	Carbonyl nickel[a]	Reduced nickel[b]
Particle structure	Spherical, very fine	Spongy, coarse
Screen analysis, fractions in %		
+100 mesh	—	0
−100 +150	—	29
−150 +200	—	28
−200 +250	0	23
−250 +325	0.5	23
−325 +400	17.5	20
−400	82.0	—
Apparent density, g./cc	3.0	2.65
Flow rate, sec./50 g. (0.1-in. orifice)	No flow	32 sec.
Compressibility	Fair	Fair
Chemical composition, %		
Nickel	99.0 min.	99.4 min.
Oxygen (hydrogen loss)	0.5	0.1
Carbon	0.1	—
Sulfur	0.005	—
Iron	0.1	—
Copper	0.05	—
Nitric acid, insolubles	—	Nil

[a] Product of Charles Hardy, Inc., New York, N. Y.
[b] Product of Superior Metal Powders Corp., Toledo, Ohio.

chief impurity. The powder is fine, all passing through 250 or 325 mesh. The reduced powder is coarser and less pure, with an approximate chromium content of 98%, and approximately 0.5% each of iron, carbon, and lithium as its chief impurities (besides oxygen). A commercial type, all passing through 150 mesh, contains about 35% of +200-, 35% of +325-, and 30% of −325-mesh powder of an average apparent density of 3.2 g./cc.

Manganese. This material can also be produced by electrolysis or

[34] W. Kroll, *Z. anorg. allgem. Chem.*, *226*, 23 (1935).

by disintegrating solid or reduced metal. The latter method, however, is used exclusively for the powders generally available. The products are of medium size, all passing through 150 mesh, with a manganese content of approximately 98%. Iron and lithium (the reducing agent) are the chief impurities, each to the extent of about 0.7%. Oxygen and carbon are present in smaller quantities, normally not exceeding 0.3%.

Copper Powders

Of all nonferrous powders, copper powders are by far the most important. They constitute the raw material for nearly three-quarters of all powder in the United States alone is estimated to exceed 500 tons per month. Only aluminum powder is produced in larger quantities, but so far only an infinitesimal fraction of this material is used in powder metallurgy, its chief uses being in the pigment and pyrotechnical industries.

The production and properties of commercial copper powders have recently been discussed by Cordiano.[35] The three important methods in use today are: (1) electrolytic deposition, (2) gaseous reduction of copper oxide, and (3) atomization of molten copper. The production of copper powder by chemical precipitation, e.g., from sulfate solutions or from mine water flowing over scrap iron, has not yet found application in powder metallurgy. Its potential place in the industry, however, is indicated by its substantially lower cost picture, and present efforts in the direction of producing a low-cost, high-purity (99% + copper) powder of closely controlled commercial grades may well influence the entire copper powder market in the not distant future.

ELECTROLYTIC POWDER

This type of powder is produced in quantity in lead-lined tanks using electrolytes consisting of copper sulfate and sulfuric acid, copper anodes, and finely divided antimonial lead cathodes. Copper is removed periodically by scraping or brushing. In some cases, the cathodes are in the form of revolving disks from which the spongy deposits are automatically scraped; in others, stationary cathodes are brushed off manually at regular intervals. The copper is collected at the bottom of the tanks and washed free of the electrolyte in a centrifuge. The wet mud is dried and annealed by being passed through a continuous furnace under a reducing atmosphere.

In the electrodeposition of copper powder, the anode material consists

---

[35] J. J. Cordiano, *Trans. Electrochem. Soc.*, 85, 97 (1944).

of commercial grade copper, generally of an electrorefined type. In this process, further refining of the copper takes place to separate that group of impurities which goes into solution and is not codeposited with the copper. Some impurities are therefore present in lesser quantities than in the anode material, while others are present in equal quantities. In the final product, however, some impurities are present in greater amounts than in the anode material because of mechanical addition through further processing. For instance, the oxygen content of electrolytic copper powder is high compared with other copper products because of the tremendous increase of surface area per unit weight. This oxygen content in electrolytic copper powder manifests itself as a thin surface film of cupric oxide (CuO), the particles sometimes having a dark-brown or even black color, with an oxygen content of less than 0.5%. Occasionally, the oxide film will be either cuprous oxide (Cu_2O) or a mixture of cupric and cuprous oxides; also, though rarely, particles with a surface of Cu_4O have been found.[36] The microstructure of particles of various grades of electrolytic copper powders is shown in the photomicrographs of Figure 47.

Regulation of temperatures, current density, bath circulation and conductivity, impurities, and suitable addition agents permits close control of the particle characteristics. Compared with electrorefining, the copper content of the electrolyte is considerably lower, and the electrode current density much higher. While more detailed information on commercial production is not available at present, reference is made to several patents describing various methods for producing electrolytic copper powders.

Koehler[37] describes a method for producing copper powder, using the following data:

Content of electrolyte..........................0.5–3.5% by wt.
Free acid content of electrolyte................0.5–10% by wt.
Bath temperature.......................24–38°C. (75–100°F.)
Electrode spacing..3 in.
Voltage..2–7 v.
Cathode current density..........................70 amp./ft.²

Drouilly[38] produces copper powder by the electrolysis of a copper sulfate solution containing the carbonaceous reaction product of glucose and sulfuric acid (glucose sulfate)—substantially in colloidal dispersion. Sugar carbon is another helpful addition agent. Fitzpatrick and his collaborators[39] describe a method which employs lead-lined wooden tanks,

[36] Anonymous, *Iron Age, 153,* No. 22, 57 (1944).
[37] U. S. Pat. 1,777,371.
[38] U. S. Pat. 1,799,157.
[39] U. S. Pat. 1,804,924.

188

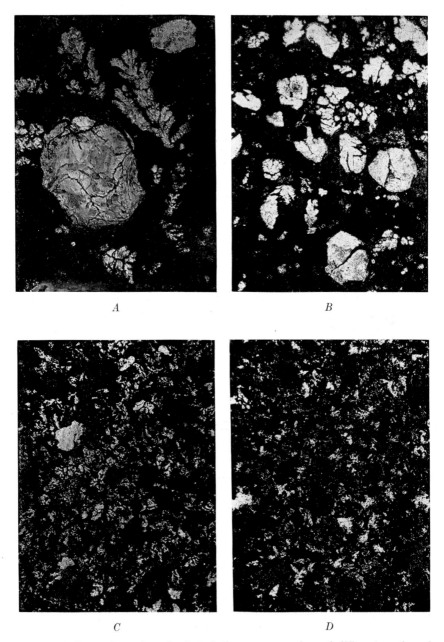

A   B

C   D

Fig. 47. Microsections through electrolytic copper powders of different grades of
fineness (×130): A, course; B, medium; C, fine; D, ultrafine.

a copper sulfate bath, six rectangular copper anodes—providing a surface of approximately twelve square feet—and twelve cylindrical copper cathodes—providing a surface of approximately 2.5 square feet. Circulation of the bath is obtained by gassing at the electrodes. Other data are as follows:

```
Average bath temperature.....................54°C. (130°F.)
Average voltage...........................................1.5 v.
Average current.......................................894 amp.
Average anode current density..................74.5 amp./ft.²
Average cathode density.......................380 amp./ft.²
```

After the powder is washed, it is dried in a cold vacuum drier in which the water is evaporated at 52°C. (125°F.). The resultant powder passes through 160 mesh and has an average copper content of 99.56%.

Fisher[40] describes a more efficient process than was previously attainable. With less current and acid consumption, the method provides for a reversal of the current every 5 to 25 minutes and removal of the deposits by sweepers which traverse the space between the electrodes every 4 minutes. A pump provides for circulation of the electrolyte. The electrodes are vibrated and consist of 21 copper plates, 24 by 20 by 0.5 in. in size, spaced approximately $5/16$ in. apart. Other data are:

```
Copper content of electrolyte..............................1%
Copper sulfate content.....................................4%
Sodium sulfate content.....................................8%
Water...............................................Remainder

Direct current—applied to end electrodes..............460 amp.
Voltage—applied to end electrode.....................15.5 v.
Voltage between plates..............................0.775 v.
Cathode current density.......................138 amp./ft.²
```

The resultant product is very pure and extremely light and fine; apparent densities are less than 1 g./cc., and the powders all pass through a 325-mesh sieve.

REDUCED POWDER

Various raw materials are used for the production of copper powder by gaseous reduction (Fig. 48). They are chiefly copper oxides, obtained by the oxidation of chemical precipitates, or finely divided electrolytic copper. Oxides in the form of oxidized scrap or of scales from wire drawing or sheet-rolling operations are not considered sufficiently pure for most powder metallurgy applications.

[40] U. S. Pat. 2,216,167.

The oxides are comminuted and reduced to the metal by treatment in continuous furnaces in a reducing atmosphere. This heat treatment is carried out under closely controlled conditions to insure complete reduction without excessive sintering of the particles; otherwise subsequent pulverization would be necessary, causing distortion and possible work-hardening of the particles. Commercial reduction temperatures range from 400° to 600°C. (750° to 1100°F.) The reducing atmosphere is

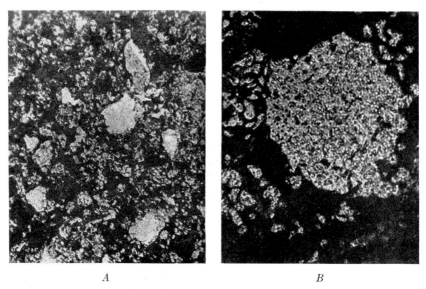

A   B

Fig. 48. Microsections through reduced copper powder particles: A, ×150; B, ×760.

usually a controlled, partially combusted gas, containing chiefly carbon monoxide, hydrogen, and nitrogen. Hydrocarbons and carbon dioxide, as well as oxygen, and water vapor are kept at a minimum, since they impede the reduction cycle. A typical gas analysis is given by Hall:[41]

> Hydrogen.......................................33%
> Carbon monoxide...............................32%
> Methane.......................................12%
> Other hydrocarbons............................ 8%
> Balance—nitrogen, carbon dioxide, etc...........15%

This atmosphere is used in a continuous process for reducing powdered copper oxides which contain 60% to 90% cuprous oxide and the balance cupric oxide. The charge is passed through a 35-ft. heating section and an

[41] U. S. Pat. 2,252,714.

18-ft. cooling section on an endless belt at an average speed of 8 ft. per hour. The heating chamber is divided into four zones in which the temperature is stepped up from 400° to 550°C. (750° to 1020°F.) and back to 400°C. (750°F.). The charge travels countercurrent to the gas flow. Scraping and breaking down of the caked mass of copper is facilitated by furrows in the transparent belt; the lumps are pulverized in saw-tooth crushers.

ATOMIZED POWDER

Commercial production of atomized copper powder from molten ingot copper is based on the flow of the liquid in a thin stream through an orifice, followed by dispersion with high-velocity jets of air. The particles are quenched at the moment of contact with the air, and oxidation takes

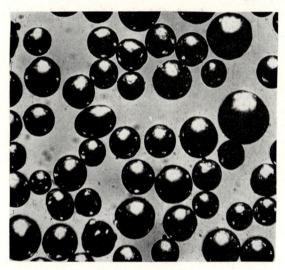

Fig. 49.   Atomized spherical copper powder (×100)
(courtesy of New Jersey Zinc Co.).

place only on the particle surfaces. The over-all oxygen content is not excessive, since the particles are substantially solid and spherical or tear-drop-shaped (Fig. 49), and the films are generally so thin that they are transparent.

A recent process for atomizing copper is given by Best.[42]   The liquid metal, kept at about 50° to 100°C. (90° to 180°F.) above its melting point, is directed vertically downward through a discharge tube. Congealing and variation in the flow of the metal are prevented by heating the

[42] U. S. Pat. 2,308,584.

TABLE 22. Electrolytic Copper Powders
Effect of Particle Size on Apparent Density and Flow Rate

Type[a]	1	2	3	4	5	6
Screen analysis, fractions in %						
+ 20 mesh	0–5	—	—	—	—	—
− 20 + 60	20–70	—	—	—	—	—
− 60 +100	25–60	0–0.5	0–0.5	0	0	0–0.3
−100 +150	5–15	20–30	3–8	0–3	0	0.5–1
−150 +200	0	30–45	15–25	7–12	0–0.5	0.5–1
−200 +250		10–20	2–5	1–4	0.2–0.5	0.5–1
−250 +325		5–10	25–35	18–25	2–5	3–5
−325		15–25	43–47	63–67	94–97	92–96
Apparent density, g./cc.	3.5–4.5	2.5–2.8	2.5–2.6	2.4–2.6	1.8–2.5	1–1.5
Flow rate	Nil	50 g. in 0.50 min.	50 g. in 0.48/0.55 min.	Nil	Nil	Nil
Chemical composition, %						
Copper	95–99	99.5 min.	99.5 min.	99.5	99.2	99.0 min.
Oxygen (hydrogen loss)	0.49	0.23	0.111	0.25	0.26	0.30
Iron	0.35	0.021	0.004	0.13	0.12	0.14
Silicon	0.15	0.015	0.015	0.05	0.05	0.05
Lead	0.18	0.04	0.02	0.01	0.01	0.01
Antimony	0.02	0.04	0.02	0.01	0.01	0.01
Tin	0.15	0.05	0.01	0.01	0.01	0.01
Grease	0.046	0.047	0.049	0.04	0.04	0.045
Nitric acid, insolubles	0.54	0.045	0.07	0.22	0.22	0.25

[a] Products of Charles Hardy, Inc., New York, N. Y. (1) Type "O." (2) Type "A." (3) Type "B." (4) Type "M." (5) Type "C." (6) Type "LC." ,

TABLE 23. Reduced Copper Powders
Effect of Particle Size on Apparent Density

Type[a]	1	2	3	4	5	6
Screen analysis, fractions in %						
+ 20 mesh	0–15	0	—	—	—	—
− 20 + 40	40–44	0.5–1	—	—	—	—
− 40 +100	42–49	55–60	0–0.1	0–0.1	—	—
−100 +150	3–6	10–15	2–5	3–5	—	0–1
−150 +200		5–8	20–22	22–25	—	1–2
−200 +325		5–8	25–30	25–30	4–5	5–8
−325		10–15	48–55	40–45	95–96	90–92
Apparent density, g./cc.	4.5–5	2.8–3.2	2.75–2.85	2.5–2.6	2.2–2.5	1.4–1.8
Chemical composition, %						
Copper	—	—	98.64	98.56	98.48	—
Oxygen (hydrogen loss)	—	—	0.26	0.29	0.44	—
Iron	—	—	0.26	0.35	0.12	—
Lead	—	—	0.18	0.18	0.15	—
Antimony	—	—	0.02	0.02	0.02	—
Tin	—	—	0.16	0.15	0.05	—
Zinc	—	—	0.12	0.15	0.05	—
Silicon	—	—	0.15	0.15	0.10	—
Nitric acid, insolubles	—	—	0.40	0.34	0.28	—
Grease	—	—	0.05	0.045	0.04	—

[a] Products of Metals Disintegrating Co., Inc., Elizabeth, N. J. (1) Type MD No. 45. (2) Type MD No. 41. (3) Type MD No. 152. (4) Type MD No. 154. (5) Type MD No. 301. (6) Type MD No. 181.

lower end of the tube with an oxyacetylene flame. The thin stream of metal is directed slightly off-center into a high-velocity, U-shaped air trough in which it is dispersed and congealed. The air trough travels at right angles to the flow of metal, enclosing it completely. The air stream is formed by compressing cold air at 75 to 180 psi and forcing it through orifices arranged to give the U-form. The atomized powder is collected in a settling chamber and passed through a cyclone and bag room.

PROPERTIES

The properties of various powders of the electrolytic type are summarized in Table 22; Table 23 provides similar data for several reduced powder types. The close connection between structure, shape, size distribution, and apparent density has already been discussed. The particle shape of electrolytic powders varies from equiaxed, coarse particles of high apparent density to irregularly shaped, fine particles of low apparent density. The purity of all commercial grades is greater than 99% copper. The apparent density of the reduced powders is controlled mainly by particle size distribution, since shape and structure do not change with size. The apparent density range is similar to that for electrolytic powders. The purity is somewhat lower—with all commercial grades falling between 98% and 99.5% copper.

Atomized powders are not generally used in molding work, because the particle size is coarser and the particles are not very plastic—due to their solid structures and the presence of oxide films. The apparent density is much higher than that of either electrolytic or reduced copper powder and may be as high as 5.5 g./cc. Because of the spherical shape of the particles, the flow rate is also very high.

Light Metal Powders

Aluminum. The production of aluminum powder experienced a tremendous boom during the war—the 1945 production capacity has been reported to be close to 20,000 tons. This material, which is either coarse granular or of the flake type (Fig. 50), is used chiefly in the pigment and pyrotechnic industries, and is therefore beyond the scope of this book. The reader interested in the flake type of powder is referred to papers by Edwards and co-workers[43-45] and to the ASTM specifications[46] covering

[43] J. D. Edwards, in J. Wulff, *Powder Metallurgy*. Am. Soc. Metals, Cleveland, 1942, p. 124.

[44] J. D. Edwards, *Aluminum Paint and Powder*. Reinhold, New York, 1936.

[45] E. L. McMahan, R. L. Wray, and J. D. Edwards, *Ind. Eng. Chem., Ind. Ed.*, *31*, 729 (1939).

[46] ASTM Standard D266-39, Part II, 1939, p. 596.

the same subject. Two basic patents for producing flake aluminum are held by Hall.[47]

The powder metallurgy industry consumes but a small fraction of the total granular aluminum powder output (Fig. 51). Aluminum powder of flaky form is poorly adapted to molding purposes because the size and shape of the particles tend to yield bodies with laminary air inclusions. Futhermore, films of the oxide and the lubricant on the flakes easily inter-

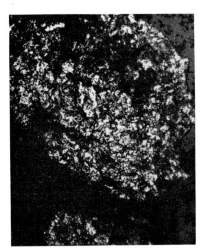 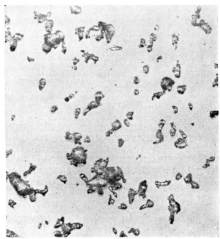

Fig. 50. Microsection through flake aluminum powder particles (×75).

Fig. 51. Shape and surface characteristics of 200-mesh atomized aluminum powder (×100).

fere with pressing and sintering operations. Recently, however, the author's attention has been called to some excellent compacts formed from flake aluminum without the use of a vehicle. Friction against the die walls was overcome by the application of aluminum bronze paint as a die lubricant. Comminuted forms of aluminum, other than flake, include granulated, grained, and atomized material, of which only the last type has been found practical for powder metallurgy. The first two forms result in coarse and impure products. The process of atomization as it applies to aluminum was discussed in Chapter III. It involves a method which permits the production of large quantities of powder at a relatively small cost as compared with other powdered metals. This has undoubtedly been an incentive for developing new applications for atomized aluminum, and it is

[47] U. S. Pats. 1,569,484 and 2,002,891.

believed that in the future several of these applications will require a considerable share of the annual aluminum powder production. The use of a metal of high aluminum content for atomization has also contributed greatly in facilitating the molding, ejection, and heat treatment of compacts from the resulting powders.

Atomized aluminum has a purity of 99.0 to 99.5%. Its apparent density ranges from 0.5 to 1.0 g./cc. Densities and screen sizes of various commercial grades are given in Table 24.

TABLE 24

Properties of Atomized Aluminum Powders

Type[a]	1	2	3	4	5	6
Screen analysis, fractions in %						
—10 +100 mesh	75	—	—	—	—	—
—20 +100	—	75	—	—	—	—
—30 +100	—	—	75	—	—	—
—40 +100	—	—	—	75	—	—
—50 +100	—	—	—	—	75	—
—60 +100	—	—	—	—	—	75
—100 +150	25	25	25	25	25	25
—150	0	0	0	0	0	0
Apparent density, g./cc.	0.75–0.90	0.80–0.95	0.80–0.95	0.80–0.95	0.80–0.95	0.80–0.95

Type[a]	7	8	9	10	11	12
Screen analysis, fractions in %						
—40 +80 mesh	85	—	—	—	—	—
—80 +100	10	Trace	0	—	—	—
—100 +150	5	4	2	—	—	—
—150 +200	0	6	6	0	0	—
—200 +325	—	15	7	1	15	0
—325	—	75	85	99	85	100
Apparent density, g./cc.	0.75–0.90	0.85–1.05	0.85–1.05	0.50–0.70	0.80–1.00	0.75–0.95

[a] Products of Metals Disintegrating Co., Inc., Elizabeth, N. J. (1) Type No. 11. (2) Type No. 21. (3) Type No. 31. (4) Type No. 41. (5) Type No. 51. (6) Type No. 61. (7) Type No. 48. (8) Type No. 101. (9) Type No. 102. (10) Type No. 105. (11) Type No. 201. (12) Type No. 301.

Magnesium. Powdered magnesium, too, gained importance during the war; the quantities used for pyrotechnics and military explosives were tremendous. For certain purposes, especially bombardment flares, magnesium was used in preference to aluminum,[48] and milling facilities had to be expanded greatly to meet the demand.

[48] L. Pasternak, *Metal Powders in Pyrotechnics and Military Explosives*. Address given at the First Annual Spring Meeting of the Metal Powder Association, New York, May 5, 1944.

Magnesium powder is available commercially only as a milled or machined material.[49] To cope with the fire hazard, the particles must be kept rather coarse, and most types do not pass through a 100-mesh sieve. Limited quantities have been available in 100- and 200-mesh sizes, but all are larger than 325 mesh. The powder generally has a purity of between 97% and 99%, but the tenacious oxide films around each flake-like particle make the use of the milled powder for powder metallurgy practically impossible. Although the molding of compacts may be accomplished, friction against the die walls and galling during ejection would require split dies rather than commercial ejector-type molds—a very uneconomical procedure. Other processes have been developed which result in products more suitable for our purposes. The Nichols process[50] consists essentially of atomizing liquid magnesium in a closed system which contains nitrogen under pressure; the nitrogen is recovered, recompressed, and reused; the product contains spherical particles. Another method involves electrodeposition of magnesium powder from fused magnesium chloride. However, considerable technical difficulties have been encountered in washing the powder free of electrolyte and an uneconomical product has resulted. Other methods involve the production of magnesium in powdered form by alkali-reduction of the oxide or vaporization of the metal from the melt in an electric arc furnace, followed by condensation of the magnesium vapor into droplet-shaped powder particles.[51]

Although sintering of magnesium compacts has met with some success in the laboratory, difficulties have generally been experienced because of the large surface oxide areas, especially if milled powder is used. For this reason the powder metallurgy applications of magnesium have remained relatively unimportant to date.

Beryllium. This light metal in particular has a strong affinity for oxygen. However, in spite of the fact that it is very difficult to obtain oxide-free beryllium powder, it *has* been produced for powder metallurgy applications. Because of the brittleness of the metal, mechanical comminution offers no difficulties, and milling in a protective atmosphere has been suggested. Another process for producing beryllium powder involves heating beryllium chloride and potassium in a platinum crucible, followed by cooling and washing out of the beryllium in the form of a fine powder. Molding of the powder is facilitated by annealing, for which a vacuum treatment at 500° to 600°C. (930° to 1110°F.) has been suggested. Hot-

[49] W. W. Moss, Jr., *Light Metal Age, 2,* 17 (March, 1944).
[50] E. J. Groom, *Light Metals, 1,* 33 (Feb., 1938).
[51] U. S. Pats. 1,884,993; 2,074,726; and 2,088,165.

ρressing in steatite molds with graphite-coated steel punches has also been suggested,[52] while another method prefers sintering in a vacuum at about 1000°C. (1830°F.).[53] Since the molding of beryllium objects has not yet gone beyond laboratory scale, the quantities of beryllium powder available are insignificant.

Calcium. Calcium has been produced in the form of a relatively coarse powder by mechanical pulverization of the bulk metal—obtained by fusion of electrodeposits of anhydrous, fused calcium chloride. Pulverization is carried out in a reducing atmosphere. Calcium powder can also be obtained by direct electrodeposition. When briquetted with another, more plastic metal, this powder has found limited use as a reducing agent. It has also been used in briquettes with lead to facilitate alloying for sheathing electric cables.[53a]

Low-Melting Metal Powders

Zinc, tin, and lead are the only metals of this group which have found limited applications in the field. Although powdered zinc, either as a by-product of the metallurgy of zinc (zinc black, zinc dust), or as a refined powder, could also be produced on a large tonnage basis, its use in powder metallurgy has been restricted so far to alloying additions for bearings and for certain types of brass parts. Tin is produced on a larger scale as an alloying ingredient for porous bronze bearings and for some solders. The production of lead powder for powder metallurgy is quite small; it is used chiefly as an alloying or addition agent for bearings or parts whose specific gravity must be high. During the war, however, large quantities of lead powder were used as ingredients in shot for target practice.

Zinc. In the form of dust, zinc is produced in large tonnages, and is used in pigment industries and to a lesser extent as reducing agent and for Sherardizing and spraying operations. Pure zinc powder for powder metallurgy has been used only on a laboratory scale to date. As described in Chapter III, a distillation process is used to obtain the powdered metal. The so-called "blue" powder is a normal by-product of the regular distillation process used for refining zinc. The powder has a high oxide content and is therefore not suitable for molding or for sintering. However, this type of zinc can be refined further by a second condensation with careful control of the carbon dioxide which is in contact with the condensing vapor. Powders having a zinc content of 97% and better can be obtained by this method.

[52] Brit. Pat. 385,629.
[53] Swiss Pat. 100,240.
[53a] C. Hardy, *private communication.*

Several processes for refining zinc have been published in recent years especially in the German literature, e.g., the Thede method[54] and Solutier process.[55] Powdered zinc can also be produced by other methods, notably by atomization (Fig. 52) and by electrolysis.[56,57] The resulting powders are very pure and fine, but considerably higher in price. Zinc powder of a purity reaching 99% is available commercially in various mesh sizes ranging from 100 to 300 mesh. Table 25 gives screen sizes and apparent densities of several grades.

Fig. 52. Atomized zinc powder (×100) (courtesy of New Jersey Zinc Co.).

Tin. In contrast to zinc, tin powders are used in large quantities in powder metallurgy work, principally in the production of porous bronze bearings, as coatings of decorative papers, and as a constituent in soldering and tinning pastes and powders. A good account of the recent development of the powder metallurgy of tin—including an extensive literature and patent survey—is given by Watkins.[58]

[54] I. Thede, *Metall. u. Erz., 35,* 317 (1938).
[55] H. Masukowitz, *Elektrowärme, 8,* No. 9, 233 (1938).
[56] U. S. Pat. 2,313,338.
[57] Brit. Pat. 506,590.
[58] H. C. Watkins, *Metals & Alloys, 15,* No. 5, 751 (1942).

Most tin powder used commercially is produced by one of the following methods: shotting, graining, atomization, chemical precipitation, or electrodeposition. Only the last three methods, however, yield products fine enough to be suitable for blending and molding work. Tin shot and grain tin are used principally for solders or for various chemical applications.

TABLE 25

Properties of Commercial High-Purity Zinc Powders[a]

Type[b].................................	1	2	3	4
Screen analysis, fractions in %				
—24 +35 mesh.....................	26	—	—	—
—35 +60.........................	58	—	—	—
—60 +100........................	16	0	0	—
—100 +200.......................	—	10.0	2.0	0
—200 +325.......................	—	20.0	3.0	3.0
—325	—	70.0	95.0	97.0
Apparent density, g./cc...............2.5–2.75		2.8–3.0	2.8–3.0	2.8–3.0

[a] 98% Zn min.
[b] Products of Metals Disintegrating Co., Inc., Elizabeth, N. J. (1) Type No. 241. (2) Type No. 102. (3) Type No. 101. (4) Type No. 201.

Shotting results in relatively coarse and spherical powder particles which solidify in air after the molten metal is poured into water through screens or small orifices. Irregularly shaped particles are obtained when the molten tin is poured into the water, while comparatively fine powder is obtained when a stream of compressed air or steam is directed against the molten metal stream before the latter reaches the water.

Tin produced by graining is coarse as well as impure, with lead as the chief impurity. Pulverization is accomplished by hammering the metal at a temperature slightly below the melting point of the pure metal, at which point the fusible constituent formed by the impurities causes the tin to disintegrate at the grain boundaries.

Atomization results in a fine product whose particle-size distribution can be regulated over a wide range. Furthermore, the product is of surprisingly high purity; excessive oxidation, as encountered with other atomized products, does not occur because of the rapid chilling effect of the expanding gases released through the nozzle. Thus, the oxygen content of atomized tin is normally below 0.2%; the thin film of oxide produced during atomization with steam or air is sufficient to inhibit further oxidation of the particles. A coarse powder of high purity (with a minimum of fine particles) can be produced by the same method, except that the orifices for the molten metal are kept at right angles to the compressed air or steam jets. This cross-jet system causes slower cooling and somewhat larger particles. The properties of several commercial grades of atomized tin powders are given in Table 26.

TABLE 26
Properties of Commercial Tin Powders

Type[a]	1	2	3	4	5	6	7	8
Screen analysis, fractions in %								
−40 +50 mesh.....	5.0 {							
−50 +100.........	85.0 {	1.0	—	—	—	—	—	—
−100 +150........	8.0)				0	—	—	—
−150 +200........	2.0)	39.0	2.0	2.0	1.0	0	—	—
−200 +325........	0	25.0	4.0	3.0	34.0	3.0	0	0
−325	—	35.0	94.0	95.0	65.0	97.0	100.0	100.0
Apparent density, g./cc..............	3.0–3.5	3.7–4.0	2.9–3.2	2.6–3.0	3.6–3.8	2.5–2.9	2.3–2.5	1.5–1.75

[a] Products of Metals Disintegrating Co., Inc., Elizabeth, N. J. (1) Type No. 51. (2) Type No. 104. (3) Type No. 103 ("Superfines" removed). (4) Type No. 101. (5) Type No. 204. (6) Type No. 201. (7) Type No. 302. (8) Type No. 105 ("Superfine," lightweight).

Extremely fine tin powder, particularly suitable for coating paper and for molding work, is produced from a stannous chloride solution by precipitation with scrap zinc. This process is used in the recovery of tin from tin-plated scrap iron. After the latter is heated with chlorine, the resulting stannic chloride is distilled off and then reduced to stannous chloride. The precipitate is a gray powder with an average tin content of 80%; the impurities include stannous chloride, zinc chloride, ferrous chloride, and water. The purity is improved by careful washing and drying.

Few details are known about the commercial production of electrolytic tin powders. Patent literature suggests[58] the use of an acid solution of tin chloride, or of an electrolyte consisting of approximately four ounces per gallon of tin perchlorate with a slight excess of perchloric acid. The rules for the electrodeposition of other metal powders, e.g., copper and iron, also apply to tin. Acid electrolytes used for production of powders have a relatively low metal concentration and a high acid concentration compared with electrolytes used for electroplating. A loosely adherent spongy product is favored by high cathode current densities, by good circulation of the electrolyte near the cathode, and by the use of carbonaceous addition agents in the bath. Except where it is to be used in wet products, as in solder pastes, the powder must be dried carefully after all the electrolyte has been removed (by washing in a centrifuge). The process permits the production of very fine powders (325 mesh and finer), as well as powders of any size, to conform to specified screen analyses and apparent densities. The purity (99.8% and better) of the particle and uniformity of characteristics make this type of powder very suitable for use in powder metallurgy.

Lead. This is the third low-melting metal powder produced in large quantities. From the point of view of usefulness in powder metallurgy it may be placed between zinc and tin. While the bulk of comminuted lead is used in the pigment and chemical industries, a substantial part is employed as an alloying ingredient for nonferrous and ferrous porous bearings and for sintered parts in which the specific gravity must be kept high.

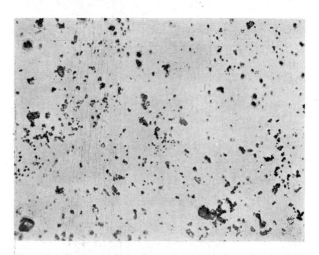

Fig. 53. Shape and surface characteristics of 100-mesh antimony powder (×100).

Shotting, graining, and atomization are the chief methods of production, but only the last method yields powders of sufficient fineness and purity. In Table 27, screen sizes, apparent densities, and, in one case, chemical analysis are given for several commercial powders.

Antimony. This powder is produced by atomization or by electrolysis. Atomized material is available in two grades: one of 100-mesh size with 70% passing through 325 mesh (Fig. 53), and the other grade with 100% passing through 325 mesh. The purity of this material is about 99% antimony. Electrolytic antimony is even purer, with metallic anti-

TABLE 27
Properties of Commercial Lead Powders

Type[a]	1	2	3	4	5	6
Screen analysis, fractions in %						
−40 +100 mesh	1.0	1.0	0	Trace	Trace	—
−100 +200	64.0	32.0	0.5	5.0	2.0	—
−200 +325	15.0	22.0	39.5	8.0	4.0	0
−325	20.0	45.0	60.0	87.0	94.0	100
Apparent density, g./cc.	5.6–6.0	5.8–6.1	5.2–5.8	4.2–4.6	4.1–4.5	4.0–4.4

Type[b,c]	7
Screen analysis, fractions in %	
−100 +150 mesh	0
−150 +200	1.5
−200 +250	1.5
−250 +325	14.0
−325	83.0

[a] Product of Metals Disintegrating Co., Inc., Elizabeth, N. J. (1) Type 104. (2) Type 103. (3) Type 204. (4) Type 102. (5) Type 101. (6) Type 105.
[b] Product of Charles Hardy, Inc., New York, N. Y.
[c] Chemical composition, %: lead, 99.93; bismuth, 0.05; silver, 0.0006; iron, 0.0005; antimony, 0.0003; and zinc, 0.0001.

mony analyzing as high as 99.9%. This type of powder, however, is available only in very small quantities and is generally superseded by the atomized material. Antimony powder is used almost exclusively at present for research and development work on bearing materials.

Cadmium. This metal, available in the form of atomized powder, has also been used for development work in the bearing field; no appreciable amount, however, has been used in commercial production. The grade available on the market contains a minimum of 99% cadmium and is of 100-mesh size, with 90% of it passing through a 325-mesh sieve.

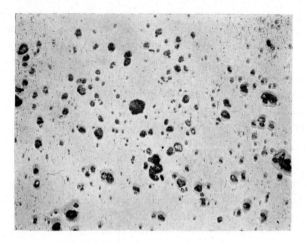

Fig. 54. Shape and surface characteristics of 200-mesh bismuth powder (×100).

Bismuth. Like cadmium and antimony, bismuth is produced as a powder, but only in small quantities (for experimental and development work). Two grades, a 200- and a 300-mesh atomized powder, of a purity as high as 99.9%, are available on the market. Figure 54 shows the surface appearance of the 200-mesh grade.

ALLOY POWDERS

Alloy powders may be classified in four groups: (*1*) refractory alloys, (*2*) ferroalloys, (*3*) ferrous alloys, and (*4*) nonferrous alloys.

Refractory Alloy Powders

Alloy powders in this group are rare and not available in appreciable quantities. Except for various tungsten-molybdenum alloy powders, which are used for special applications in the lamp industry, other refractory, metal-base alloy powders have been produced only for laboratory and development work. Titanium–copper and zirconium–copper master alloys are obtained by briquetting and sintering powder mixtures, and not by using alloy powders.

Tungsten-Molybdenum. Since tungsten and molybdenum are isomorphous and crystallize within the same system with the formation of a solid solution, it is possible to obtain every ratio between the two metals in powder form. This may be achieved by simultaneous reduction of the mixed oxides, the alloy powder being formed by sintering at the high-temperature stage of the reduction treatment. To assure complete inter-diffusion, the high-temperature reduction cycle must often be extended beyond the time normally used in reducing the individual metal powders. The products thus obtained are usually slightly coarser than the pure metal powders, but can be readily worked into wire form for such parts as springs, hooks, supports, heaters in radio tubes, and resistance wires for high-temperature electric furnaces. The alloy powders are produced with tungsten–molybdenum compositions of 90–10, 80–20, 72.5–27.5, 50–50, and 20–80.

Ferroalloy Powders

Ferroalloys include a variety of refractory metal–iron combinations. and are used in very large quantities for alloying or for scavenging oxygen and nitrogen in steel manufacture. In fact, this application consumes approximately 95% of all refractory metal production. The ferroalloy group includes alloys between iron and tungsten, molybdenum, columbium, titanium, zirconium, vanadium, chromium, manganese, and silicon.[59]

Most ferroalloys are produced either in an electric furnace or by the aluminothermic process, because the high melting temperatures of the refractory metal components exclude other methods. Since the products are usually coarse, granular, and of mediocre purity, only selected materials of the finer grades are suitable for alloying in powder metallurgy.

Ferrotungsten. This material is used principally for the manufacture of high-speed steels which contain up to 20% tungsten—generally in the crushed form with granules up to one inch in size. Fine grades, passing through 100 mesh, are suitable for the powder metallurgy of alloy steels. Ferrotungsten has the following standard composition[59]: 70–80% tungsten with a maximum of 0.60% carbon. Other grades commercially available are:

[59] G. K. Herzog, in *Metals Handbook*. Am. Soc. Metals, Cleveland, 1948, p. 337.

Constituent	Grade, %		
	1	2	3
Tungsten	65–72	78–82	78–82
Carbon	2.50, max.	0.25, max.	1–2, max.

Ferromolybdenum. This material is also used principally in high-speed steels in which the molybdenum content may be as high as 10%, replacing part of the tungsten. The size of the crushed granules is up to one inch, but fine powder, all passing through 100 mesh, is available for powder metallurgy work. Ferromolybdenum is available in two grades which are distinguishable from each other by their carbon contents. The following is an approximate analysis:[59]

Constituent	Grade, %		
	1	2	3
Molybdenum	55–75	55–75	55–75
Iron	22–42	22–42	20–40
Carbon	0.10, max.	0.50, max.	2.00, max.
Silicon	1.50, max.	1.50, max.	1.50, max.
Sulfur	0.25, max.	0.25, max.	0.25, max.

Ferrocolumbium. Columbium is added to chromium and chromium-nickel stainless steels as a stabilizer because of its ability to form a stable columbium carbide constituent and so inhibit intergranular corrosion. The amounts of columbium added to steel are very small, varying from about five to ten times the carbon content. For this purpose ferrocolumbium is used in crushed granular form; the finer grades may be used for special alloying work in powder metallurgy. Ferrocolumbium analyzes at 50–60% columbium, the balance is iron—with carbon 0.4%, silicon not in excess of 8%, and manganese of 5%.

Ferrotitanium. Several grades of titanium alloys, generally in coarse, granular form, are used in steel making. The fines of these grades, if they pass through a 100-mesh sieve, are suitable for development work on sintered alloy steels. The different grades of titanium alloys vary in composition and history. One type, high in carbon (grade 1), is known as ferro–carbon–titanium and is used as a deoxidizer, scavenger, and inhibitor of segregation. Other grades are known as low-carbon ferrotitanium and are made either by aluminothermic process (grade 2) or in an electric furnace. Some typical grades have the following approximate compositions:[59]

Constituent	Grade, %			
	1	2	3	4
Titanium	15–16	17–20	20–25	40–45
Iron	71–76	70–77	68–80	45–60
Carbon	6–8	3–5	0.1, max.	0.1, max.
Silicon	2–3	2–3	4, max.	4, max.
Aluminum	1–2	1–2	3, max.	7, max.

Ferrozirconium. Zirconium, in amounts not exceeding 0.1%, is used as a deoxidizer and scavenger for steel. It is commonly alloyed with iron in the presence of a large proportion of silicon, thus constituting a ternary ferro–silicon–zirconium alloy. While generally available as crushed granules for steel making, it may be used as a 100-mesh powder for alloying sintered steels. Two typical grades analyze approximately as follows:[59]

Constituent	Grade, %	
	1	2
Zirconium....................	12–15	35–40
Silicon.......................	39–43	47–52
Iron.........................	40–45	8–12

Ferrovanadium. Vanadium is generally added to engineering steels in amounts 0.10–0.25%, to tool steels up to 1%, and to high-speed steels up to 5%. For this purpose ferrovanadium is used either in lumps weighing several pounds or in the form of crushed granules up to one inch in diameter. It may be ground to a fine powder to accommodate powder metallurgy work. The approximate analysis range of ferrovanadium grades is as follows:[59]

```
Vanadium....................................35–55%
Iron........................................30–60%
Silicon.....................................1.50–12%
Carbon......................................0.20–3.5%
```

Ferrochromium. Chromium is added to steels in the form of ferrochromium, which is available in two grades differing according to carbon content. For steels containing medium carbon and chromium, a high-carbon ferrochromium (grade 1) is used. For low-carbon steels, however, this grade is unsuitable, and low-carbon ferrochromium of different carbon content is used (grade 2). Furthermore, ferrochromium of a high nitrogen content (up to 1.0%) is available in both the high-carbon and low-carbon grades. The alloys are generally furnished in lumps of 75 pounds and up, but also in the form of crushed granules, and as a 100-mesh powder suitable for sintering alloy and stainless steels. Approximate analyses of the two grades of ferrochromium are as follows:[59]

Constituent	Grades, %			
	1	2	3	4
Chromium	65–70	60–65	67–72	62–66
Iron	18–29	22–28	25–33	20–33
Carbon	4–9	4–6	0.03–2.00	1.25, max.
Silicon	2–3	4–6	1.00, max.	4–6
Manganese		4–6		4–6

Ferromanganese. Manganese is generally introduced into molten steel in the form of ferromanganese, in various grades depending on the carbon content of the alloy. Four principal types are used, namely, standard ferromanganese (1), spiegeleisen (2), medium-carbon ferromanganese (3), and low-carbon ferromanganese (4). Their approximate analyses are as follows:[59]

Constituent	Grade, %			
	1	2	3	4
Manganese	78–82	16–28	80–90	80–85
Iron	12–16	65–80	6–18	12–19
Carbon	6–8	6.5, max.	0.07–0.75	1.50, max.
Silicon	1.00, max.	1–3, max.	1–7, max.	1.50, max.
Phosphorus	0.30, max.	0.15, max.	0.2, max.	
Sulfur	0.05, max.	0.05, max.		

Because of the large quantities required in steel making, ferromanganese is generally used in the form of heavy lumps weighing up to 75 pounds. Although it is difficult to disintegrate, especially in the case of the standard grade and spiegeleisen, it is available in crushed form, the granules being between one and two inches in diameter. Pulverization may produce a 100-mesh powder suitable for alloying in ferrous powder metallurgy.

Ferrosilicon. Ferrosilicon is available in a number of grades which may be distinguished by silicon content. Six typical grades—15, 25, 50, 75, 85, and 90% ferrosilicon—are used in steel alloying practices. The first type is used only in the form of ingots weighing up to 100 pounds. Alloys high in silicon are available in lumps or as crushed granules; the fines may be used for powder metallurgy purposes. The different grades of ferrosilicon analyze as follows:[59]

Constituent	Grade, %					
	1	2	3	4	5	6
Silicon	14–18	25–30	46–52	74–79	80–90	90–95
Iron	81–85	69–74	47–53	20–25	9–19	4–9
Carbon	1.00, max.					
Phosphorus	0.05, max.					
Sulfur	0.04, max.					

Iron-Base Alloy Powders

The alloy powders surveyed here include: iron–carbon compositions, carbon-containing alloy steels, austenitic stainless steels, iron–nickel alloys, and iron–aluminum alloys.

IRON–CARBON COMPOSITIONS

Hypoeutectoid Steel. Plain carbon steel powders with different carbon contents (up to 0.8%) have been produced by several processes. The most important ones involve decarburization of pulverized high-carbon alloys or carburization of sponge iron or pure iron powders. The decarburization method yields powders of distinctly equiaxed particle shape, which are not readily compressible because of the solid and rigid structure of the individual particles. Hence, powders produced by this method must be mixed with other, more plastic powders. They are used in large quantities for parts involving infiltration of a low-melting metal such as lead, since the substantially spherical particle shape has proved advantageous for this application. Carbon steel powders obtained by carburization have also been produced successfully, although, to date, mostly on an experimental basis. Carburization of the powder has been obtained either by gas carburizing or by diffusion treatment with powdered solid carbon. Steel powders obtained in this way have been found to be more readily compressible than the decarburized types—even carburized electrolytic iron powders are not exceptions. On the other hand, the spongy or dendritic structure of these steel particles has interfered in many instances with complete diffusion of the carbon during sintering and resulted in nonuniform structures and erratic physical properties.

Hypereutectoid Steel. The compressibility of carbon steel powders naturally decreases with increasing carbon content, and hypereutectoid compositions are no longer suitable for molding work. Powdered steel of approximately eutectoid composition is available, however, in the form of crushed powder (Diamond Crushed Steel) in various mesh sizes[60] (see Table 28). During its manufacture the material is made martensitic by a special quenching process, thus rendering it especially suitable for sand-

---

[60] Products of Pittsburgh Crushed Steel Co., Pittsburgh, Pa.

TABLE 28

Properties of Crushed Iron and Steel Powders

Type	Steel 1[a]	Steel 2[a]	Steel 3[a]	Iron 1[b]	Iron 2[b]	Iron 3[b]	Iron 4[c]	Iron 5[c]	Iron 6[c]	Iron 7[c]
Screen analysis, fractions in %										
+60 mesh	—	—	—	—	—	—	22.0	2.2	0	0
−60 +80	—	—	—	—	—	—	20.0	23.3	0	0
−80 +100	0	0	0	0	0	0	40.8	26.0	0	0
−100 +150	14.5	1.9	0.5	7.3	4.4	6.2	6.1	19.1	1.5	0.5
−150 +200	74.0	18.7	8.9	62.7	4.9	9.8	6.0	15.0	19.0	6.6
−200 +250	4.6	9.7	7.7	11.5	57.7	24.4	0.5	2.2	1.5	2.0
−250 +325	4.8	34.5	38.4	17.1	33.0	59.6	3.0	14.0	33.3	22.2
−325	2.1	35.2	44.5	1.4			1.6	7.2	44.7	68.7
Flow, sec./50 g.	20	19	19	28	28	28	—	—	—	—
Apparent density, g./cc.	3.9	3.4	3.3	3.1	3.3	3.2	—	—	—	—
Microhardness (Brinell units), as is	950	—	—	—	—	—	750	—	—	—
annealed	269	—	—	—	—	—	168	—	—	—

[a] Product of Pittsburgh Crushed Steel Co, Pittsburgh, Pa.: (1) Diamond Crushed Steel No. 120. (2) Diamond Crushed Steel No. 150. (3) Diamond Crushed Steel No. 180.

[b] Products of Pittsburgh Crushed Steel Co, Pittsburgh, Pa.: (1) Crushed Iron No. 120. (2) Crushed Iron No. 170. (3) Crushed Iron No. 200.

[c] Products of Harrison Abrasive Corp, Manchester, N. H.:(4) Shot No. 90. (5) Shot No. 120. (6) Grit No. 180. (7) Grit No. X.

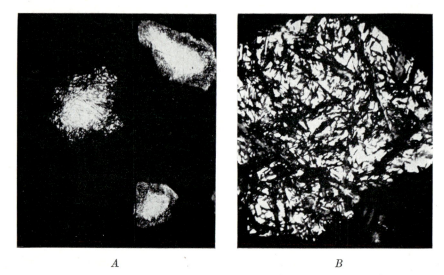

A

B

Fig. 55. Microstructure of particles of crushed steel after quenching: A, incomplete transformation to martensitic structure in center, martensite case (×225); B, complete transformation of martensitic structure (×450).

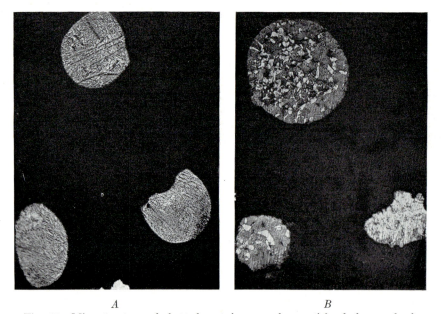

A

B

Fig. 56. Microstructure of shotted cast iron powder particles before and after hydrogen annealing (×150): A, after shotting; B, after shotting and annealing at 1000° C. (1830° F.) for 1 hour.

blasting. But even in the annealed condition the powder has been found to be too hard and too impure for powder metallurgy applications. The microstructure of quenched powder particles is shown in Figure 55.

Hypoeutectic Iron-Carbon Alloys. Iron–carbon alloys of high carbon content are commercially available in comminuted form, commonly known as "crushed irons" or "crushed cast irons."[60,61] Although less brittle than the Diamond Crushed Steels, these materials are also used in large quantities for sandblasting purposes. Usually the alloys analyze between 2.5 and 3.5% carbon, and come in various mesh sizes (Table 28). The powders are not directly suitable for molding work, but can be decarburized by heat treatment in steam or by admixture with a certain proportion of iron oxide. The resulting iron powders are plastic enough for certain molding applications, and they are of particular interest because of their low price. Figure 56 shows photomicrographs of shotted cast iron before and after annealing treatment.

ALLOY STEELS

Recently, powdered alloy steel compositions have been made available as commercial products.[62] They comprise tungsten and molybdenum high-speed steel compositions as well as manganese, chromium, and nickel steel powders. Since all these materials are manufactured by the chemical disintegration method, the individual particles are single crystals of considerable rigidity. Poor compressibility due to resistance to deformation restricts the use of these powders to applications in which binders can be used or high molding pressures are in order.

STAINLESS STEELS

Austenitic corrosion-resistant steel powders of the 18–8 and 35–15 types are produced commercially by the same chemical disintegration method.[62,62a] Unfortunately, because of the type of process involved, these have so far found only limited use. To a large extent their utilization for the manufacture of molded corrosion-resistant parts will have to wait until the plasticity of these materials can be improved sufficiently to permit the use of the customary high-output presses of relatively low pressure capacity.

Austenitic steel powders have also been produced by the sintering method. The plasticity of these materials has been found to be vastly

[61] Products of Harrison Abrasive Corp., Manchester, N. H.
[62] Products of Unexcelled Manufacturing Co., New York, N. Y.
[62a] Products of Charles Hardy, Inc., New York, N. Y.

improved, thus allowing the molding of more intricate shapes at ordinary working pressures. On the other hand, powders made by this method lack uniformity of composition. The sintering method imposes definite temperature limits beyond which it is impossible to pulverize the powder cake without severe strain-hardening of the particles. In the case of the stainless steels this temperature limit is at about 1050°C. (1920°F.) for practical heat-treating periods. However, even treatment at this temperature for prolonged periods of time (up to 4 hours) has resulted in only partial diffusion of iron and nickel, with chromium remaining practically free. Improved diffusion has been observed when iron and chromium were used as an alloy in the form of ferrochrome powder, or when nickel and chromium were used as an alloy in the form of pulverized Nichrome. In these cases, however, better homogeneity of the alloy powder was secured only at the expense of impaired compressibility, because of increased pro-

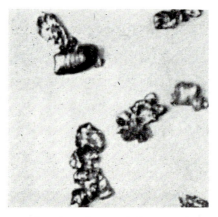

Fig. 57. Shape and surface characteristics of pulverized Permalloy (80–20 Ni–Fe) powder (×100) (courtesy of W. D. Jones).

portion of hard and brittle chromium-rich particles. Hence, none of these methods has yet led to a satisfactory commercial product that does justice to both important demands—good compressibility and homogeneity of the alloy—so essential from the standpoint of corrosion resistance.

OTHER IRON ALLOYS

Iron–Nickel. Production methods for iron–nickel powders include pulverization, carbonyl decomposition, electrolysis, and sintering. Powders of high magnetic permeability (Permalloy) are manufactured by the first method. These powders which have characteristic particle shape—angular contours and smooth edges (as shown in Figure 57)—are not well suited for sintering purposes, but are widely used for magnetic core materials. Pulverization of iron–nickel alloys is made possible by embrittlement of the alloy, either by omission of the deoxidizer or by the addition of ferrous oxide or ferrous sulfide to the molten alloy. Since the particle size of the powder coincides with the grain size of the alloy, it must be decreased by hot-rolling after casting. During subsequent rolling at lower temperatures, the alloy is fragmented and finally ball-milled to fine sizes.[63]

A joint decomposition of iron and nickel carbonyl results in nickel–iron alloy powders of extreme particle fineness and a wide range of compositions.[64] Alloy powders containing between 50% and 80% nickel have been found to be quite suitable for the sintering of high-permeability alloys, while powders containing between 30% and 50% nickel have been successfully sintered into alloys of low and consistent coefficients of thermal expansion. Nickel–iron alloy powders have also been produced by sintering. This method has been found to be advantageous when incomplete interdiffusion of the two metals is desirable because of the need for constant, although low, permeabilities.[65]

Iron–Aluminum. 50–50 iron–aluminum alloy powders are used in the manufacture of sintered permanent magnets of the "Alnico" type. The use of the alloy powders greatly facilitates the diffusion of the aluminum, since the tendency of the latter to form tenacious oxide films prevents its use in virgin form. The extremely brittle 50–50 master alloy can be readily disintegrated into a sufficiently fine powder by ball milling. The hardness of the alloy particles does not markedly impede molding of the magnets, since the iron-aluminum alloy constitutes only about one-quarter of the total composition.

Nonferrous Alloys

NICKEL- AND COBALT-BASE ALLOY POWDERS

Nickel–Chromium. Binary alloys of chromium and nickel with a chromium content analogous to that of alloys made by the conventional fusion process, are produced by joint reduction of the corresponding oxide with the aid of the hydride process.[66] Alloys produced in this fashion

[63] J. C. Chaston, *Metal Treatment, 1*, No. 1, 3 (1935).
[64] G. Hamprecht and L. Schlecht, *Metallwirtschaft, 12*, No. 1, 281 (1933).
[65] Brit. Pat. 383,691.
[66] P. P. Alexander, in J. Wulff, *Powder Metallurgy*. Am. Soc. Metals, Cleveland, 1942, p. 154.

come in the form of fine powders of 100 to 400 mesh. They are well suited for molding, since they are not produced by a method which induces work-hardening, nor is sintering of the alloy impeded by oxidation of free chromium.

Cobalt–Chromium. Alloy powders of chromium and cobalt are also produced via the hydride process by joint reduction of their respective oxides. Alloy powders with different chromium contents are available, and have found use in the development of sintered articles of high strength and hardness.[66] Alloys of the Vitallium and Stellite compositions have recently been produced in powder form on a small scale by the inter-granular corrosion method.[66a]

Nickel–Molybdenum. Nickel-base alloy powders of the Hastelloy types, containing molybdenum and iron as major alloying constituents, have also lately been produced on a limited scale by the chemical disin-tegration of contaminated bulk metal.[66a] Like the cobalt-base alloy powders, they have found application in development work on sintered, high-strength, high heat-resistant components for power engines and fur-nace parts.

COPPER-BASE ALLOY POWDERS

Copper-base alloys comprise the bulk of the nonferrous alloy powders commercially available. Binary alloy powders of copper with zinc, tin, and nickel are produced in different grades and compositions. Other copper alloy powders are so far available only in experimental quantities.

Copper–Zinc. Brass powders are produced either by atomization or by sintering. Atomization results in very uniform powders with particles displaying solid, frequently spherical shapes and smooth contours. There is no variation in composition between the individual particles. The pow-ders are available in various sizes from −100 to −325 mesh. Several standard compositions are on the market in particular copper–zinc ratios of 90–10, 70–30, 65–35, and 60–40.[67] Powders of this type are readily compressible and excellent for sintering. Considerable quantities of atom-ized brass powders are used in electronics and machine parts industries. Figure 58 shows the microstructure of particles of the 90–10 and 70–30 alloy powders.

Similar compositions can be obtained by sintering. Copper and zinc powder mixes of desired proportions are subjected to a diffusion treatment under protective atmospheric conditions. Temperatures as low as 400°C.

[66a] Products of Haynes Stellite Co., New York, N. Y.
[67] Products of Chas. Hardy, Inc., New York, N. Y.

(750°F.), *i.e.*, below the melting point of zinc, have been employed successfully, though the best treating temperature usually ranges between 450° and 550°C. (840° and 1020°F.). The powders obtained by this method are usually coarser than their atomized counterparts (—100 and —150 mesh), as pulverization of the sintered cake must be kept at a mini-

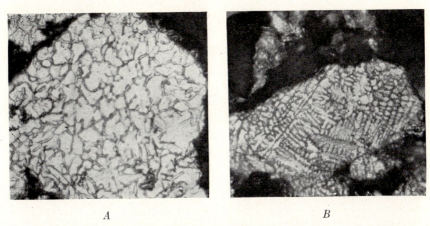

| *A* | *B* |

Fig. 58.   Microstructure of atomized brass alloy powder particles (×500):
A, 90–10 Cu–Zn; B, 70–30 Cu–Zn. (Courtesy of New Jersey Zinc Co.)

mum in order to retain the excellent plasticity of the spongy brass particles. Variations in composition among the individual particles are usually quite evident and make the powders less suitable for the sintering of those parts for which high physical properties are required. On the other hand, this type of powder is ideally suited for the manufacture of porous parts.

The expansion of brass powder metallurgy has been restricted chiefly by the high price of the powder, which is far out of proportion to the price of the bulk metal.

Copper–Tin.   Among the various methods for producing bronze powder, atomization, electrolysis, and sintering are the most important, although no method has led to large-scale production of the alloy powder. In the production of porous bearings it is the general practice to employ virgin powders, which are more easily compressible, and to obtain an alloy by interdiffusion.

Atomized spherical bronze powders of various closely controlled grades have recently appeared on the market for use in the manufacture of porous metal filters. These powders are relatively hard to compress, even when the tin content is below 10%. Figure 59 shows the surface ap-

216

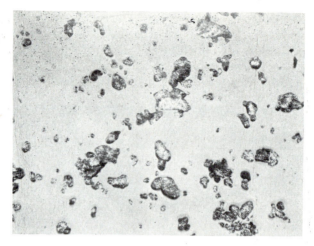

Fig. 59. Shape and surface characteristics of atomized
bronze powder particles (×100).

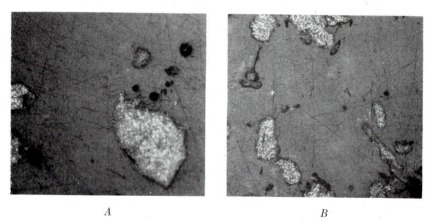

A   B

Fig. 60. Microsections through atomized bronze alloy powder particles. (×100):
A, 95–5 Cu–Sn; B, 90–5–5 Cu–Sn–Zn. (Courtesy of New Jersey Zinc Co.)

pearance of such powder, while in Figure 60 microsections are shown for
different alloy compositions. Similar press difficulties can be experienced
with electrolytic alloy powders obtained by simultaneous electrodeposition. In this case, however, the powders can be made more plastic by extended annealing after the customary drying treatment for electrolytic
powders. Alloy powders made by interdiffusion of copper and tin at a
temperature well above the melting point of tin result in the most plastic

type of bronze powders, and compositions containing as much as 15% tin have been found readily compressible.[68] As in the case of brass powders, this type of material is coarser (—100 mesh) and the particle-to-particle composition is less uniform than powders made by the other two methods. The most commonly used composition of bronze powders contains 10% tin.

Copper–Nickel. Any copper–nickel composition, such as one containing 40% or 67% (Monel) nickel, can be produced by the diffusion method. The spongy particles are very plastic and the materials are easily compressible. Sintering, however, is quite dependent on the degree of uniformity of the powder with regard to composition. Since, however, excellent physical properties of regular Monel metal are not easily attained by ordinary powder metallurgy manipulations, the use of copper–nickel powders remains limited. Although atomized powders result in more homogeneous alloys, they have as yet been made only on an experimental basis. The high melting point of the nickel-rich composition (above 1300°C.; 2370°F.) excludes the type of atomizing equipment used for the commercial production of lower melting metals.

OTHER NONFERROUS ALLOY POWDERS

In addition to copper-base alloy powders, only a few special compositions are of general interest to the powder metallurgist. A few examples are sintered or atomized silver–gold and copper–gold alloy powders, used for jewelry and for ornamental work; electrolytic and atomized lead-tin and lead-antimony alloy powders, used for solders, bearings, anodes, etc.; and atomized aluminum–zinc, aluminum–silicon, and aluminum–magnesium alloy powders, used in the manufacture of molded light metal parts.

COMPOUND POWDERS

Depending upon their final use, powdered metal compounds may be classified as refractory oxides, compounds for hard metals, and compounds for special applications. Inasmuch as refractory oxide powders are basic materials in the field of ceramics, it is felt that a detailed discussion of the subject would be beyond the scope of this text. The reader who is interested in this topic is advised to study the abundant literature on ceramic materials and processes. Only a few oxide powders have found direct uses in powder metallurgy. Aluminum oxide and silica powders are used

---

[68] C. G. Goetzel, *Trans. Am. Inst. Mining Met. Engrs.*, **161**, 580 (1945).

in small amounts (from 2 to approximately 7%) as nonmetallic friction-producing agents in metallic friction materials such as brake linings, clutch facings, etc.; iron oxide in the form of finely ground powders is used for certain types of magnetic cores; magnesium, aluminum, and zirconium oxide powders are used together with powdered diamonds in certain types of grinding wheels.

Metal phosphides and metal hydrides are compounds which have special applications. Copper phosphide powders are added to some bronze compositions as hardeners and have also been used as brazing compounds. Powdered metal hydrides have found use in the previously described hydride process for the reduction of the more stable metal oxides (Chapter III, p. 58).

Those compounds which serve as a basis for the manufacture of hard metals are of great metallurgical importance. In this group, metal carbides take first place, with other hard compounds, such as borides, nitrides, and silicides playing minor roles. Although it is general practice to form these hard tool alloys during the presintering and high-sintering stages of manufacture, the carbide, boride, and silicide components are first produced in powdered form. The same applies to the various grades of silicon carbide powders used in grinding wheels and friction materials. In spite of excessive particle hardness, all of these compound powders can be formed into simple shapes with the aid of special metallic and nonmetallic binders. (For further information on the subject, consult Volume II, Chapter XXII.)

COMPOSITE POWDERS

The term "composite" usually designates a combination of materials which form neither alloys nor compounds. This includes both metal–metal and metal–nonmetal combinations. In our discussion of powders, however, the term will refer only to combinations of two metallic phases within each particle. Powders whose particles are partly metal and partly nonmetal are extremely rare and do not belong in a discussion of commercial powders.

Composite powders usually consist of so-called "coated" powder particles. Each individual particle consists of a core of one particular metal surrounded by a shell of another metal. The two metals may or may not combine to form an alloy or a compound. There are several methods by which the shell can be applied; the best known are electrodeposition and chemical replacement, but condensation and mechanical coating processes are also in use. Although the first two methods permit a wide range of proportions between coat and core, the last two methods are practical only when very thin coats are the objective.

Probably the best known example for the replacement method is the copper-coated lead powder described in Chapter III (p. 61). Powders containing from 30% to 50% lead can be made by this method, with each particle substantially spherical in shape and analogous to the initial atomized lead. The high proportion of copper in each particle permits complete closure of each shell. Other examples include copper-coated iron and steel powders, with a copper content of 5–25%. The electrolytic method can be used to produce a variety of combinations.[69,70] For example, tin can be plated on copper particles by employing tin anodes and a tin plating bath or, conversely, copper can be plated on tin powder by using a copper anode and a copper plating bath. Similarly, nickel may be deposited on copper, or copper on nickel. Metals employed for base particles include tungsten, molybdenum, iron, nickel, manganese, copper, tin, and antimony; metals which may be used for coating the particles are tin, lead, cadmium, zinc, copper, nickel, cobalt, chromium, silver, gold, and platinum. More than one metal coating may be applied by the same method, and various complex combinations may be obtained, as, for example, iron–nickel–chromium, copper–zinc–tin, or copper–tin–lead.

We may cite the following examples: tin-coated copper—resulting in a combination of 88–92% copper and 12–8% tin; nickel-coated copper—resulting in a combination of 30–40% copper and 70–60% nickel; and tin–lead–coated copper—resulting in combinations of 80–90% copper, 10–8% tin, and 2% lead. Other interesting combinations reported by Hardy[71] include cobalt–coated tungsten with 4% cobalt, copper-coated carbon with 5% copper, copper-coated silver with 30–35% copper, silver-coated molybdenum with 40–45% silver, silver-coated nickel with 12–40% silver, cadmium-coated nickel with 5% cadmium, lead-coated nickel with 1% lead, nickel-coated iron with 6–25% nickel, nickel-coated chromium with 58% nickel, tin-coated antimony with 10% tin. Most of these powders pass through 200 mesh, with more than 80% passing through a 325-mesh sieve. For example, the apparent density of the silver-coated nickel powder is 2.65, that of the silver-coated molybdenum, 2.75 g./cc. Some experimental data on the electrolysis of coated powders are reproduced in Table 29.

The production of these coated powders must still be considered to be in its initial state, with controls for larger scale production not yet worked out. However, it appears that this kind of composite powder, particularly

[69] U. S. Pat. 1,986,197.
[70] U. S. Pats. 2,027,532; 2,033,240; and 2,182,567.
[71] C. Hardy, *Contribution to discussion*, Massachusetts Institute of Technology Conference, 1941; also *private communication*.

TABLE 29

Electrolysis of Coated Powders[a]

Powder	Cobalt-coated tungsten	Cobalt-coated copper	Copper-coated graphite	Copper-coated silver	Silver-coated molybdenum	Silver-coated nickel
Analysis of coated powder........	4% Co 96% W	31% Co 69% Cu	(1) 2% Cu 98% C (2) 5% Cu 95% C	(1) 3% Cu 97% Ag (2) 34% Cu 66% Ag	1% Ag 99% Mo	12% Ag 88% Ni
Treatment........	Single	Single	Double	Double	Single	Single
Electrolyte, g./l..	$CoSO_4$, 120 NH_4Cl, 15 H_3BO_3, 15	$CoSO_4$, 120 NH_4Cl, 15 H_3BO_3, 15	(1) $Cu(CN)_2$, 22.5 NaCN, 33.0 Na_2CO_3, 15.5 (2) $CuSO_4 \cdot 5H_2O$, 135 H_2SO_4, 46.5	(1) $Cu(CN)_2$, 22.5 NaCN, 33.0 Na_2CO_3, 15.5 (2) $CuSO_4 \cdot 5H_2O$, 135 H_2SO_4, 46.5	AgCN, 30 NaCN, 23 K_2CO_3, 30	AgCN, 30 NaCN, 23 K_2CO_3, 30
Type of powder to be coated......	−200 mesh tungsten	−100 mesh electrol. copper	(1) −80 mesh graphite (2) 98/2 Cu-coated C	(1) Crystalline silver (2) 97/3 Cu-coated Ag	−200 mesh molybdenum	−150 mesh electrol. nickel
Anode material..	Cast lead	Cobalt powder pressed at 30 tsi	Cast copper	Sheet copper	Cast lead	Cast lead
Anode current density, amp./ ft.²...........	48	144	(1) 24 (2) 110	38	16	32

220

TABLE 29 (Continued)

Powder	Cobalt-coated tungsten	Cobalt-coated copper	Copper-coated graphite	Copper-coated silver	Silver-coated molybdenum	Silver-coated nickel
Cathode material	1000-cc. copper beaker	#18 gage copper wire	1000-cc. copper beaker	#18 gage copper wire	1000-cc. copper beaker	1000-cc. copper beaker
Cathode current density, amp./ft.^{2}	5				2	4
Total current, amp.	1.5	10	(1) 2 (2) 10	4	0.5	1
Total voltage, v.	4	9	(1) 0.5 (2) 5	4.5	2	2
Period of electrolysis, hrs.	7$^1/_2$	9$^1/_2$	(1) 16 (2) 4	(1) $^1/_4$ (2) 3$^3/_4$	3	3
Current efficiency, %		97		62		42
Kw.-hr. per lb. of coat produced						1.5
Approx. percentage of powder coated	16		(1) 20 (2) 75	(1) 50 (2) 100		100

(Continued on page 222)

TABLE 29 (Continued)

Powder	Cadmium-coated nickel	Lead-coated nickel	Nickel-coated iron	Nickel-coated chromium	Tin-coated copper	Tin-coated antimony
Analysis of coated powder......	5% Cd 95% Ni	1% Pb 99% Ni	(1) 6% Ni 94% Fe (2) 25% Ni 75% Fe	58% Ni 42% Cr	9% Sn 91% Cu	10% Sn 90% Sb
Treatment......	Single	Single	Double	Single	Single	Single
Electrolyte, g./l.	CdO, 22.5 NaCN, 97.5	$PbSiF_6$, 120	(1) $NiSO_4 \cdot 6\,H_2O$, 240 $NiCl_2 \cdot 6\,H_2O$, 30 H_3BO_3, 60 (2) $NiSO_4 \cdot 6\,H_2O$, 240 $NiCl_2 \cdot 6\,H_2O$, 15 H_3BO_3, 30	$NiSO_4$, 180 NH_4Cl, 22 H_3BO_3, 22	Na_2SnO_3, 117 NaOH, 10 $CH_3CHOHCOOH$, 0.7	Na_2SnO_3, 117 NaOH, 10 $CH_3CHOHCOOH$, 0.7
Type of powder to be coated......	−150 mesh electrol. nickel	−150 mesh electrol. nickel	−100 mesh electrol. iron	−150 mesh pulverized chromium etched with 10% aqua regia	−100 mesh electrol. copper	−100 mesh atomized antimony
Anode material..	Cast cadmium	Cast lead	Cast nickel	Cast nickel	Tin powder pressed at 20 tsi	Tin powder pressed at 20 tsi
Anode current density, amp./ ft.²	108	48	(1) 64 (2) 80	128	46	46

TABLE 29 (Concluded)

Powder	Cadium-coated nickel	Lead-coated nickel	Nickel-coated iron	Nickel-coated chromium	Tin-coated copper	Tin-coated antimony
Cathode material	1000-cc. copper beaker	1000-cc. copper beaker	(1) copper beaker. (2) #4 gage copper wire	Steel ball	1000-cc. copper beaker	1000-cc. copper beaker
Cathode current density, amp./ft.^{2}	6	12	(1) 32		37	37
Total current, amp.	1.5	3	(1) 8 (2) 10	18	9.5	9.5
Total voltage, v.	3.5	2	(1) 4 (2) 7	12	5.5	5.5
Period of electrolysis, hrs.	3.5	5	(1) 14 (2) 10$^1/_2$	5	4$^1/_2$	4
Current efficiency, %	23		49.5		20	25
Kw.-hr. per lb. of coat produced	3.3		8.6		0.85	4.75
Approx. percentage of powder coated	75	50	(1) 75 (2) 100	10	100	100

ª Courtesy of Chas. Hardy, Inc., New York. Data abstracted from laboratory reports.

in connection with ferrous metal powders, offers some definitely interesting possibilities.[72,73]

The condensation method is suitable for the production of zinc-coated powders, but has also been suggested for other low-melting metals. The method is similar to the well-known Sherardizing process. After zinc dust is intimately mixed with the metal powder to be coated (e.g., iron), the mixture is heated in a closed chamber—under controlled atmospheric conditions—to a temperature above the melting point of the zinc. This is followed by rapid cooling, whereupon the zinc condensate forms solid films around the iron particles. Another method provides for passing the metal powder to be coated through a furnace chamber saturated with zinc vapor. Upon cooling, a thin film of zinc is condensed on the surface of the particles. The ratio between coat and core is usually about 1:10.

Mechanical coating is commonly used in the manufacture of cemented carbides and steels. Cobalt coatings are applied to refractory metal particles and graphite films to iron particles by means of ball milling. This is often carried out with the aid of a liquid phase, e.g., carbon tetrachloride—in some cases, at elevated temperatures under controlled atmospheric conditions. The method is also practiced for other combinations, such as copper-coated tungsten, silver-coated molybdenum, etc. The proportion by weight between shell and core does not usually exceed 1:5.

Certain composite powders are produced by chemical coprecipitation of the metallic ingredients, and each particle of the resulting powder consists of an intimate mixture of the two metals. Typical examples of this type of material are tungsten–copper, tungsten–silver, molybdenum–copper, and molybdenum–silver powders having an average particle size of 0.1–0.2 μ, which apparently are especially suitable for the manufacture of contact metals.[74] In the first instance, the powder is produced by preparing copper tungstate made by treating a 20% sodium tungstate solution at the boiling point with $N/2$ copper sulfate solution in stoichiometric ratio (the pH is maintained by adding sodium hydroxide or ammonia). The copper tungstate is a gelatinous precipitate which, after washing and filtering, is electrolyzed between two diaphragms (the voltage being increased gradually from 20 to 220 volts). The hardy residue is pulverized and reduced in pure hydrogen at 800–850°C. (1470–1560°F.). The other powders are produced from similar starting materials.

[72] U. S. Pat. 2,289,897.
[73] U. S. Pat. 2,358,326.
[74] Brit. Pat. Appl. 32,900/46 (Nov. 29, 1945).

SEMIMETALLIC AND NONMETALLIC POWDERS

Semimetallic Powders

Tellurium. Powdered tellurium is derived from the anode mud from copper and lead refineries[75] or from the flue dust formed in roasting telluride gold ores. After being treated by fusion with sodium nitrate and carbonate, the melt is extracted with water. The resulting hot water solution is acidified with sulfuric acid for the purpose of neutralization, whereupon tellurium dioxide is precipitated as a fine powder when the solution is combined with the flue dust from the sludge wastes. Reduction of the tellurium dioxide is achieved by heating it with powdered coal. Tellurium powder has found some use as a hardener for lead.

Selenium. Selenium powder is obtained from the same sources as tellurium, by continuation of the chemical treatment of the solution.[75] The anode sludge of copper refineries is fused with sodium nitrate and silica, or oxidized with nitric acid, and the water extract is then acidified with hydrochloric and sulfuric acid. When sulfur dioxide is passed over the solution, free selenium is precipitated in the form of a very fine powder. This powder has found a commercial application in photoelectric cells, since its electrical conductivity is changed by variations in light intensities.

Nonmetallic Powders

Silicon. In spite of the fact that silicon, occurring as silicon dioxide in various forms, is the most abundant element in the solid crust of the earth, its use as an engineering material in connection with metals has been rather limited because of the brittle, metalloid nature of the element. Accordingly, the use of the element in comminuted form has also been restricted to a few minor applications, such as alloying ingredients for soft magnets or certain copper alloys, addition agents for certain types of deoxidizers, and for general chemical and research work. It is also used for siliconizing steels. The metalloid in its comminuted state may be obtained either by direct reduction to an amorphous powder, or by mechanical disintegration of crystalline bulk silicon. Reduction is achieved by a complicated high-temperature process with aluminum or magnesium, or in an electric arc furnace with carbon as the reducing agent.

Powdered silicon is available as a 100- and 200-mesh powder, with approximately 90% passing through a 325-mesh sieve. The apparent den-

[75] D. O. Noel, in J. Wulff, *Powder Metallurgy*. Am. Soc. Metals, Cleveland, 1942, p. 121.

sity varies between 0.5 and 0.7 g./cc.; the purity is 97% and better—the chief impurities originating from the reduction and disintegration processes. A typical analysis of a commercial silicon powder shows:

Silicon	97.50%	Calcium	0.32%
Iron	0.75%	Manganese	0.05%
Aluminum	0.55%	Carbon	0.04%

Carbon. Powdered carbon is used in powder metallurgy in a variety of forms, ranging from diamond dust to coarse graphite machinings and to colloidal carbon collected from soot deposits. Commercial lampblacks include such well-known products as Thermatomic Carbon Black, Thermax, Wyex, and Micronex, and commercial types of graphite powders, such as those used in the manufacture of bearings and brushes, include natural and synthetic types—crystalline, amorphous, or flaky—as produced by a number of industrial firms.[76,77] Powdered graphite is obtained either by pulverizing natural graphite (from Ceylon deposits) or by heating anthracite coal to the temperature of the electric furnace. Other varieties of carbon are prepared by heating certain organic materials, e.g., powdered coconut charcoal for gas adsorption, powdered bone charcoal and activated carbon for decolorizing solutions of organic substances, powdered gas carbon by coking the residue in the distillation of petroleum, and powdered coke and wood charcoal by the destructive distillation of coal and wood, respectively. Lampblack or soot in various degrees of dispersion, as used for deoxidizing purposes in the refractory metal and cemented carbide fields as well as in the pigment industry (in large quantities), is deposited by incomplete combustion of petroleum, natural gas, or hydrocarbons. Diamond powders of industrial grades, used for grinding wheels and special drills, consist of grinding dust and small diamond fragments, the particles being considerably smaller than 1/10 of a carat.

Because of the great variety of these forms of powdered carbon, a detailed list of the commercial products would be beyond the scope of this book. The reader especially interested in this information is advised to consult the literature of carbon and graphite manufacturers.

Summary

Metal powders are commercially available in a great variety of types and grades. Monometallic powders, especially iron, copper, aluminum, lead, tin, and tungsten, comprise the bulk of present powder production.

[76] Products of the Acheson Graphite Div., National Carbon Co., Inc., New York, N. Y.
[77] Products of the American Graphite Co. and Joseph Dixon Crucible Co., Jersey City, N. J.

TABLE 30

Metal Powders and Their Suppliers (Schumacher and Souden[78])

Material	How produced	Purity or composition	Available meshes	Suppliers of specific powders[a]
Aluminum	Atomized	99.0+%	−10 to −325	Hardy, MD, Reynolds
	Flake-milled	—	−325	Alcoa, MD
Aluminum alloy	Atomized	96–98%	−100	Hardy, Reynolds
	Hot-milled	Dural	−60 to −300	Unexcelled
Antimony	Milled	99%	−100 to −325	Hardy, MD
Beryllium	Reduced	97+%	−200	Hardy
Beryllium alloys	Milled	2.5% Be–Cu	−60 to −300	Unexcelled
	Hydride	10% Be–Ni	−100	Hydrides
Bismuth	Milled	99.9%	−200	Hardy, MD
Brass	Atomized	60/40 to 90/10	−100 to −325	Hardy, MD, N J Zinc
	Atomized	60/40	−30	N J Zinc
	Atomized	70/30	−100 (Spher)	N J Zinc
Bronze	Flake-milled	Cu–Zn–Al	Various, to −325	Am Bronze
	Reduced	—	Various	MD
	Hot-milled	77% Cu, 8% Sn, 15% Pb	−60 to −300	Unexcelled
Cadmium	Milled	99.5%	−100, −300	Hardy
	Atomized	—	−325	MD
Chromium	Milled	98+%	−150 to −325	Hardy, MD
Cobalt	Reduced	97.5–99%	−100 to −300	Hardy
	Reduced	99.9%	−325	R & R
Columbium	Reduced	95.0+%	−50	Fansteel
Copper	Electrolytic	99.5+%	−100 to −325	A M Co, Gen Met, Hardy, MR
	Reduced	99.5+%	−40 to −325	MD, PM & A
	Atomized	—	−30 to −200	N J Zinc
	Flake-milled	—	−325	Am Bronze
Copper–nickel	Hot-milled	70% Cu, 30% Ni	−60 to −300	Unexcelled
Graphite	—	97–98% C	−200, −325	Dixon
	—	95+% C	−325	Dixon
	—	90–92% C	−200	Dixon
	—	95, 97% C	5 micron	Dixon
	—	—	−200 and finer	National Carbon

(Continued on page 228)

227

TABLE 30 (Continued)

Material	How produced	Purity or composition	Available meshes	Suppliers of specific powders[a]
Iron...........	Reduced	99+ %	Many, to −325	MD, PM & A, Hardy
	Reduced	97–98.2%	−8 to −325	MR, Plast Met
	Electrolytic	99.5%	−8 to −325	Plast Met, E & T
	Carbonyl	98–99.9%	−180, −400	GAW
Lead	Atomized	99.5–99.9%	−100 to −325	A M Co, Hardy, MD, MR
Magnesium...	Milled	96–99.9%	−15 to −325	Apex, Magna, NS, Nat Mag, NE Mag
Manganese ...	Milled	99.9%	−100 to −325	Hardy, MD
	Electrolytic	99.75%	−20 to −325	Plast Met
Molybdenum ...	Reduced	99.90%	−150, −200, −300	Callite, Hardy
	H-reduced	99.7+ %	−80	Fansteel, NAP
	Reduced	99+ %	−80	Hardy
Nickel........	Milled	Ni+Co 99%	−150 to −325	Hardy, MD
	Flake-milled	—	−325	Am Bronze
	Reduced	99.5%	Various	P M & A
Nickel–copper..	Hot-milled	70% Ni, 30% Cu	−60 to −300	Unexcelled
"Nickel Silver".	Hot-milled	Cu–Ni–Zn	−60 to −300	Unexcelled
Silicon.........	Milled	96, 97+ %	−100 to −325	MD, Plast Met, Hardy
Silver.........	Electrolytic	99.90%	−200 to 325	A M Co, H & H, Hardy, MD
Solder........	Flake-milled	—	−325	Am Bronze
Stainless Steel..	Atomized	50/50; 40/60	−40 to −325	MD
	Corrosion	18/8	−60 to −300	Unexcelled
	Milled and reduced	35/15	−60 to −300	Unexcelled
Steel	Reduced	97–99% Fe	Various	P M & A
Tantalum......	Electrolytic	99.8+ %	80% −400	Fansteel
Tantalum carbide......		Ta C	−200	Fansteel
Thorium........	Reduced	95+ %	−100	Hardy

228

TABLE 30 (*Concluded*)

Material	How produced	Purity or composition	Available meshes	Suppliers of specific powders[a]
Tin............	Atomized	99.5+%	−40 to −325	MD, MR, Hardy
	Flake-milled	—		Am Bronze
Titanium	Reduced	99.5%	−325	Hardy
	Hydride	98–99.5%	−250	Hydrides
	Hydride	98–99.5%	−100	Hydrides
Titanium hydride...	Reduced	99.9+%	−325	Callite, Hardy, R & R, Fansteel, NAP
Tungsten	Reduced		−150, 200, 300	Callite, Hardy
Tungsten carbide...	Reduced	99.0 WC	−80 to −300	Carboloy, Fansteel
Vanadium	Milled	90%	Various	Hardy
Zinc	Atomized	99+%	−80 and finer	M D, Hardy, N J Zinc
Zirconium	Reduced	99.5%	−24 to 325	Hardy
Zirconium hydride...	Hydride	99–99.9%	−250, −200	Hydrides

a General list of powder suppliers: (Alcoa) Aluminum Co. of America, Pittsburgh. (Am Bronze) American Bronze Div., Metals Disintegrating Co., Verona, N. J. (AM Co) American Metal Co., Ltd., New York. (Apex) Apex Smelting Co., Chicago. Baer Bros., New York. Belmont Smelting & Refining Co., Brooklyn, N. Y. (Callite) Callite Tungsten Corp., Union City, N. J. (Carboloy) Carboloy Corp., Detroit. (Dixon) Jos. Dixon Crucible Co., Jersey City, N. J. (E&T) Ekstrand & Tholand, Inc., New York. (Fansteel) Fansteel Metallurgical Corp., No. Chicago, Ill. (Gen Met) General Metals Powder Co., Akron, O. (GAW) General Aniline Works Div., New York. Goldfield Consolidated Mines Co., San Francisco. (H&H) Handy & Harman, New York. (Hardy) Chas. Hardy, Inc., New York. Harshaw Chemical Co., Cleveland. O. Hommel Co., Pittsburgh. (Magna) Magna Mfg. Co., Inc., New York. McAleer Mfg. Co., Rochester, Mich. (Hydrides) Metal Hydrides, Inc., Beverly, Mass. (MD) Metals Disintegrating Co., Elizabeth, N. J. Metals, Incorporated, San Francisco. (MR) Metals Refining Co. of Penna., Palmerton, Pa. (NAP) North American Phillips Co., Cleveland. (Nat Mag) National Magnesium Corp., New York. (NS) National Smelting Co., New York. (NE Mag) New England Magnesium Co., Inc. Malden, Mass. (NJ Zinc) New Jersey Zinc Co. of Penna., Palmerton, Pa. (NAP) North American Phillips Co., Inc., New York. (Plast Met) Plastic Metals, Inc., Johnstown, Pa. (PM&A) Powder Metals & Alloys, Inc., New York. Pyron Corp.; Niagara Falls, N. Y. (R&R) Reduction & Refining Co., Newark, N. J. (Reynolds) Reynolds Metals Co., Louisville, Ky. (Unexcelled) Unexcelled Mfg Co., New York, N. Y. U. S. Metal Powders Co., Closter, N. J. U. S. Magnesium Co., Pleasant Valley, N. Y.

Alloy powders, usually less adaptable for industrial molding operations, have not yet been developed on a major scale, except for certain steel and brass powders. Compound powders, such as carbides or oxides, are used only in comparatively small quantities. Composite powders consisting of coated particles have interesting possibilities. Carbon, in a variety of forms, is by far the most important nonmetallic powder used in powder metallurgy. A tabulation (Table 30) showing the types and sources of supply for the most important domestic products taken from Schumacher and Souden's manual on powder metallurgy products,[78] should be of interest to the reader.

[78] E. E. Schumacher and A. G. Souden, *Metals & Alloys* (Materials & Methods Manual), *20*, No. 5, 1340 (1944).

CHAPTER VII

Powder Conditioning and Function of Addition Agents

Conditioning and preparation of metal powders by mechanical or thermal treatment, as well as the functions of any required addition agents (nonmetallic alloying ingredients, binders, lubricants, etc.), have received little attention in general accounts of powder metallurgy processing. With the exception of literature and patent descriptions relating to the manufacture of. cemented carbides, there are few references to the problems connected with the protection of powders from oxidation prior to their use, blending of various materials to obtain uniform and reproducible mixes without affecting important powder characteristics, and uniform mixtures with nonmetallic ingredients for the purpose of lubrication or volatilization. And yet, it is probably no exaggeration to say that conditioning of the metal powder for molding and sintering is as important as powder manufacture, pressing, and sintering operations. One possible reason for the paucity of printed accounts of these steps in powder processing may be the fact that the preparation of powders for use has remained at the level of an art, based largely on the skill and inventive spirit of the individual powder user.

In discussing this subject a distinction should be made between work done by the powder producer and that left to the user. For example, the user expects to receive dry material which is (*1*) free from impurities incidental to the mode of manufacture, (*2*) suitable for molding (removal of stresses incurred during manufacture), and (*3*) suitable for sintering (surface oxides removed by a high-temperature reduction treatment). Above all, the user rightfully expects to be supplied in subsequent shipments with uniform lots and identical grades of powder in accordance with specifications agreed upon by the manufacturer and the user.

These stipulations have not always been fulfilled, and minor differences in the raw material have led to great difficulties in processing. This has frequently been the case even though consecutive powder lots were of the same type and from the same manufacturer. The necessity for frequent changes in the pressure, or in the pressing or heating cycles has been

231

the sad experience of all powder users. The situation is made worse when substitute powders (possibly from another supplier) are used, which may even require the redesign of dies and tools. It is easy to see that a few such drastic changes may easily mean the difference between profit and loss on a given contract. Patch,[1] in a recent address, voiced the opinion of most powder users when he stated that up to about 50% of all industrial molding and sintering equipment is withdrawn at all times from the production line for the sole reason that permanent readjustments and redesigns are necessitated by batch-to-batch fluctuations or by differences in powders from different sources (due to different interpretations of specifications).

It is obvious from the foregoing that closest cooperation between the powder producer and the user is of paramount importance. It is no accident that, during recent years, those industries which control both powder production and powder use have expanded and advanced most rapidly. The cemented carbide industry is a striking example, giving ample evidence of the significance of powder control throughout the entire complicated process.

In the majority of cases, however, the manufacturer of powder metallurgy products does not produce his own raw material, and must therefore select his powders with great care. Before experimental work begins, the delivery of a sufficient quantity of the identical grade of commercial powder to carry through the initial order must be assured. A contract for additional deliveries of identical grade for future orders should be governed by rigid specifications drawn up on the basis of the first powder supply. The same precautions apply if lack of capital or storage facilities prohibits the purchase of all the powder needed for the first order. In any event, to avoid serious production difficulties, the powder used in the initial work successfully should remain standard throughout the entire order.

TREATMENTS IMMEDIATELY AFTER MANUFACTURE

Washing the Powder

The chief object of the manufacturer is to produce the metal in powder form. To make the powders suitable for specific applications, however, it is necessary that the product fulfill certain requirments. As has already been amply explained, the physical characteristics of powders depend to a great extent upon the raw materials and process by which they are

---

[1] E. S. Patch, *Technical Requirements of Metal Powders,* Address given at the First Annual Spring Meeting of the Metal Powder Association, New York, May 5, 1944.

prepared. Although certain methods of powder manufacture invest in the powder the desired characteristics (*e.g.*, the reduction method), in many cases the properties of the freshly made powder are not satisfactory and further treatment is necessary to meet the user's requirements. Powders made by electrolysis or precipitation are typical examples. Although these processes are flexible enough to control certain physical characteristics, such as particle size distribution, apparent density, and flow, these properties may be affected by chemical or atmospheric conditions to such an extent that the powder becomes unsuitable for use. Precipitated powders, for example, are generally of such fine particle size that they have a tendency to oxidize strongly during washing, filtration, or drying. In some cases, this leads to particle agglomeration, and in other cases, to pyrophoric reactions. Electrolytic powders, moreover, are wet when they leave the cell, and must be washed and dried—the latter operation also resulting in particle agglomeration. Since these agglomerations generally exceed the permissible upper limits of particle size, mechanical comminution becomes necessary to avoid large percentages of oversize material.

To free the powder particles from adherent traces of electrolytic or other nonmetallic impurities incidental to the particular process, the powders are washed in a centrifuge after leaving the electrolytic or chemical tank. Before the wet powder can be removed and dried, sufficient time must elapse in order to eliminate any acidity in the wash water.

Drying the Powder

The drying operation is usually coupled with a reduction treatment to remove the slight surface oxide films caused by wet handling.

Batch-type furnaces as well as continuous furnaces are used for this purpose, and various atmospheres—especially partially combusted hydrocarbons, dissociated ammonia, and hydrogen—are used as reducing agents. The wet powder is placed in pans or sometimes briquetted and then placed directly on a mesh-belt conveyor. Neither the method of drying nor the heating cycle has yet been standardized, and it is known that the drying temperature for the same metal powder varies by as much as 250°C. (450°F.) among different plants.

In connection with the drying cycle the question of restricting grain growth during sintering has aroused great interest. The idea has been advanced that it may be possible to predetermine the dimensions of a sintered part more closely if the initial powders have been previously treated at sufficiently high temperatures to allow substantial grain growth to take place. Although experiments of this nature are very interesting and undoubtedly have been carried out by many powder manufacturers, little has been made public about their results. Even Drapeau[2,3] in his extensive studies of dimensional changes during sintering of copper and bronzes fails to refer to possible effects of high-temperature heat treatment prior to compression.

It must be realized, of course, that the effects of such heat treatments are largely dependent on the process of manufacture and the structure of the particles. Thus, considerable structural changes may be effected in precipitated or in pulverized and mechanically work-hardened powders while these effects may be less marked in the case of electrolytic powders. Very little effect can be expected in the case of reduced, spongy powders previously treated at high temperatures, but several points remain to be clarified experimentally, especially with regard to the intraparticle changes caused by preliminary heat treatment. It has been found that even in unworked powders recrystallization within the particles becomes apparent after normal recrystallization temperatures have been reached. In many cases growth of the crystallites has been observed, which in certain instances, e.g., in nodular electrolytic particles, has led to single-crystal particles when the powder was treated at sufficiently high temperatures. In a recent study of the sintering of powders without pressure, Delisle[4] made some very interesting observations concerning external particle grain growth, For carbonyl iron, growth, and consequently sintering, began at 300°C. (570°F.) whereas for electrolytic iron, particle growth was not uniformly accomplished even at 1300°C. (2370°F.). Bal'shin,[5] in his extensive experiments on annealed copper and iron powders, found that particle growth and change in the packing column could be observed at a temperature of about 200°C. (390°F.). Whether the powders are work-hardened or not, this temperature lies very close to the point of "incipient" sintering, which, in 1923, was established experimentally by Tammann and Masuri[6] as the temperature at which a stirrer stops rotating in a mass of powder. Finally, the close relationship between grain growth and sintering imposes a practical limit upon the temperature and extent of the preheating cycle. Particle growth is accompanied by agglomeration and eventually results in a stable porous cake which can be comminuted only by mechanical means. Thus strains are imposed on the particles that may have adverse effects on the behavior of the powder during pressing or sintering.

[2] J. E. Drapeau, Jr., in J. Wulff, *Powder Metallurgy*. Am. Soc. Metals, Cleveland, 1942, p. 323.
[3] J. E. Drapeau, Jr., in J. Wulff. *loc. cit.*, p. 332.
[4] L. Delisle, *Trans. Electrochem. Soc.*, 85, 135 (1944).
[5] M. Yu Bal'shin, *Vestnik Metalloprom.*, 16, No. 18, 82 (1936).
[6] G. Tammann and Q. A. Mansuri, *Z. anorg. allgem. Chem.*, 126, 126 (1923).

Other Thermal Treatments

In addition to drying, heat treatment of powders may serve a variety of purposes: (1) purification by reduction, (2) softening through annealing, (3) softening by changing the composition, and (4) alloying by diffusion treatment. The temperatures required for purification are relatively low—reduction of oxide films on electrolytic copper particles is accomplished at about 300°C. (570°F.), that of iron or nickel at about 700°C. (1290°F.). Generally, temperatures required for reduction do not exceed 900°C. (1650°F.); even for complete elimination of carbon, oxygen, or sulfur in carbonyl iron and nickel, higher temperatures were not necessary.[7] The annealing operation is frequently performed after mechanical treatment in order to soften the work-hardened powder. The temperatures and time periods most suitable for annealing or alloying are largely dependent upon the sintering tendency of the particular powder. Sintering into a hard mass would cause renewed work-hardening during pulverizing. Generally speaking, temperatures and times can be increased with an increase in particle size.

Heat treatment of steel powders to insure decarburization, decomposition of the iron carbide, or modification of the iron carbide structure may result in a substantial softening of these ferrous powders. The heat-treating cycles depend on the nature of the individual process.[8,9]

Pulverization

Under normal conditions, using moderate temperatures for thermal treatment, pulverization of the agglomerated masses causes no noticeable strain-hardening of the particles. Comminution is effected by preliminary breakdown in a jaw crusher or disk grinder, as shown in Figure 61. This operation is usually followed by grinding in ball or tube mills. Final pulverization is obtained in hammer or impact mills, such as a micropulverizer which is also shown in Figure 61. This type of equipment tends only to sever the interparticle bonds without deforming the actual particles.

In producing some electrolytic powders, e.g., certain iron, nickel, and chromium powders, the metals are not immediately deposited in powder form—the electrodeposited metal adheres to the cathodes in the form of flakes, plates or coherent sponges. These deposits are stripped at definite intervals from the cathodes, which are generally constructed and prepared

[7] W. Eilender and R. Schwalbe, Arch. Eisenhüttenw., 13, No. 6, 267 (1939).
[8] M. Yu Bal'shin and N. G. Korolenko, Vestnik Metalloprom., 19, No. 3, 34 (1939).
[9] U. S. Pats. 2,175,850; 2,289,570; and 2,301,805.

to facilitate periodic scraping. Prior to the drying operation, the deposits of basically ductile materials, such as iron or nickel, are pulverized immediately after washing. In this manner, advantage is taken of the hydrogen embrittlement of the metal during electrodeposition. Drying and reduction at elevated temperatures, which have an annealing effect, tend to make the metal so ductile that grinding can be carried out only with difficulty, resulting in flattened or otherwise severely deformed and strained particles.

Fig. 61. Pulverizing, grading, and tumbling of metal powders. Left to right: 200-lb. oblique tumbler; stabilized gyrating screener; micropulverizer; disk grinder. (Courtesy of American Electro Metal Corp.)

Whether pulverization is carried out on dry or wet material, great care must be taken in the choice of grinding equipment.[10] Although in the case of ball or tube mills customary types of mills may be employed, the

[10] Sources of grinding equipment suitable for powder metallurgy uses: Abbe Engineering Co., New York, N. Y.; American Pulverizer & Crusher Co., St. Louis, Mo., Franklin-McAllister Corp., Chicago, Ill.; Hardinge Co., Inc., York, Pa.; Jeffrey Mfg. Co., Columbus, Ohio; Patterson Foundry & Machine Co., East Liverpool, Ohio; H. K. Porter Co., Inc., Pittsburgh, Pa.; Pulverizing Machinery Co., Summit, N. J.; Raymond Pulverizer Div., Combustion Engineering Co., Inc., Chicago, Ill.; Stedman's Foundry & Machine Works, Aurora, Ind.; Sturtevant Mill Co., Boston, Mass.

linings deserve special attention. Thus, refractory metal powders may be ground suitably in tungsten mills with tungsten or tungsten carbide balls, whereas lower melting metals, such as chromium, nickel, or copper, require alloy steel mills or steel mills whose linings are hard-chromium plated. These precautions are essential in order to avoid contamination with iron. In one case in which crushed tungsten was ground in a regular steel mill for one hour, its purity decreased from 99.5% to almost 98%, the difference being iron introduced by the milling equipment. Correct fill of the mill is necessary to minimize the contamination of the powder with iron. Stain-

Fig. 62. The Tyler Ro-Tap automatic sieve shaker with nest of screens. (Courtesy of W. S. Tyler Co.)

less steel mills have also been reported to reduce contamination in cemented carbide powders.

Grading the Powder

Following pulverization or mechanical breakdown of agglomerates, dry powders are graded by being passed through large sieves. Many different types answer the purpose, such as pulsating, vibrating, or eccentrically moving screens of the so-called "stabilized gyrating" type,[11] also shown in Figure 61. Rotating sieves, so arranged that the prescribed mesh sizes and subdivisions can be withdrawn, are particularly favored by many powder producers. Where small batches or samples are involved, the "Ro-Tap" screen analyzer (Fig. 62, see also p. 132, Chapter V) or the "Combs" auto-

Fig. 63. Two-ton, double-cone blender. (Courtesy of American Electro Metal Corp.).

matic vibratory sieve shaker[11a] (Fig. 65, p. 248) can also be used. The oversize material is usually passed once more through the mill. Subdivisions are obtained by employing standard mesh screens, usually of the 100, 150, 200, 250, and 325 denominations. The fractions of the powders that pass the respective sieves are collected in separate receptacles. Different proportions are then taken from these collecting pans to be composed into a single lot which has a particle-size distribution that meets the consum-

[11] Product of the J. H. Day Co., Cincinnati, Ohio.
[11a] Combs Gyratory Sifting Machine, product of Great Western Mfg. Co., Leavenworth, Kan.

er's specifications. The individual size proportions are thoroughly blended in mixers, of which various types may be used. Cone blenders of the type shown in Figure 63 are preferred for this operation by many powder manufacturers.

The limit of commercially available large-size screens is 325 mesh, but many powders of refractory metals, as well as of certain lower melting metals (cobalt, nickel, copper), either pass through this sieve entirely or to a large extent. Other methods of classifications must be resorted to in this case. Separation of these fine sizes is conveniently performed by some system of elutriation, preferably employing air or gas. Cyclone separators are widely used in classifying or eliminating certain subsieve size ranges or fine metal dust.

Densification

Very porous or voluminous aggregates of particles may result in powders with poor flow characteristics and low apparent densities, which may cause unsatisfactory performance during pressing or sintering. The apparent density of such powders can be increased by mechanical working, and agglomerating the fine particles into coarser granules. In the case of a somewhat aggregated carbonyl iron powder, for example, a prolongation of the ball milling period from 12 to 96 hours can effect an increase of 20% in the apparent density.[11b] Mechanical compression followed by comminution in an impact mill constitutes another method of powder densification.[11c] Briquetting of the ultrafine powder into large ingots which are afterwards mechanically comminuted into a coarse granular powder is practiced by some cemented carbide manufacturers in order to facilitate the molding of hard metal shapes in automatic pelleting machines. In general, work-hardening of the more plastic metals cannot be completely avoided in such procedures and subsequent softening of the particles by annealing becomes necessary.

Packing, Shipment, and Storage

Powders blended to the desired screen analysis are ready for packing, storage, or shipment. To avoid contamination by oxygen or moisture, transfer from the mixer to shipping containers should be made as quickly as possible. Moreover, hermetic closure of the containers is imperative. Although usually shipped in steel containers, certain types of commercial iron and copper powders have been shipped without undue contamination by oxygen or moisture in wooden barrels lined with oil-impregnated,

[11b] E. K. Offermann, *Mitt. Kohle- u. Eisenforsch. G.m.b.H.*, 1, No. 5, 85 (1936).
[11c] U. S. Pat. 2,306,665.

watertight paper bags. The covers of shipping containers should be firmly locked and sealed by means of rubber gaskets. It is good practice to fill the containers to the very top in order to keep the air column at a minimum and to avoid shifting and possible strain-hardening of the powder during transport. If high humidity is encountered during packing, balling-up of the powder may be increased by excessive shifting.

Another important precaution is to make certain that packing containers are kept absolutely dry before they are used. Washing with alcohol is general practice, but may not always be sufficient. Certain powders (especially sensitive to oxidation) are stored and shipped in desiccating containers. Silica gel, calcium chloride, activated alumina, etc., are used to condition the atmosphere of the container; they are placed either in false bottoms, or in bags on top of the powder.

In a number of cases rigid precautions against oxidation are necessary. Very fine powders, like carbonyl iron (average particle size of one micron), as well as fine lead powders, may be pyrophoric, and consequently require an inert atmosphere in the packing container. Carbon dioxide or nitrogen is used for this purpose. Other fine powders, such as titanium, are best handled in the presence of occluded or combined hydrogen. The powders are either packed under hydrogen or, if the hydrogen is removed during annealing, they are packed "wet" in a proper vehicle.[12]   They may also be packed and sealed under hydrogen or *in vacuo*.

In general, metal powders used in powder metallurgy are not of fine enough mesh size to create a pyrophoric condition. It is, however, advisable that the powder producer and the user take proper precautions against fire hazards. Special caution is to be observed in the manufacture and use of fine aluminum and magnesium powders and of superfine powders of other metals prepared by precipitation or reduction. Dust hazards are usually confined to the final stage of manufacture—pulverizing, screening, and packing—or to the initial stages of processing, such as sampling upon reception, preparation for use, and feeding the presses. The danger of explosion and fire is substantially decreased as soon as the exposed surface of the powder is reduced during compression. Only in rare cases (*e.g.*, uranium) is the frictional heat engendered during compression great enough to be a hazard.[12]

There has been a tendency recently to produce, store, condition, and press metal powders in air-conditioned premises. Manufacturers in the tungsten and tungsten carbide fields were the first to use air conditioning in their plants. They have found this very useful not only for continuous

[12] E. Pletsch and T. Edwardsen, in J. Wulff, *Powder Metallurgy*. Am. Soc. Metals. Cleveland, 1942, p. 548.

removal of dust particles from the air, but also for stabilizing the moisture content of the air and, indirectly, the flow conditions of the powder. In the United States where climatic conditions are subject to such drastic changes, especially during the summer months when the humidity sometimes exceeds 90%, it is frequently found that changing powder characteristics and consequently erratic properties in the finished products make operations impossible. Agglomeration and interparticle adhesion inhibit flow and uniform filling of molds. Adsorption of moisture causes undue oxidation or growth of the compacts in the sintering furnace because of excessive degasification. The installation of air conditioning, however, assures uniform temperature and humidity throughout the year, permitting smooth plant operation under standard conditions. The cost of installation can be absorbed quickly by the increased productive output of uninterrupted work, as well as by a substantial reduction in defective products. With the steady improvement in quality and increase in popularity of such installations during the last decade, air conditioning has been generally adopted by many plants engaged in various branches of powder metallurgy.

MIXING OF METAL POWDERS

Monometallic Mixtures

While most molding applications involve polymetallic structures which are generally produced from mixed powders, there are several industrially important monometallic powder applications in which mixing is employed for close control of the process. The reasons for such a procedure are (1) to assure consistently uniform powder batches, and (2) to synthesize batches of specific particle-size distribution or other physical or chemical characteristics.

Even with the most rigid enforcement of specifications, variations in the properties of individual shipments of the same powder will always occur—their limits in many cases, exceeding the permissible tolerances; this is particularly true for size distribution. Stricter control of shipments by the powder manufacturer is not always successful and, as a result, powder shipments must be received as manufactured to avoid serious losses or delays. In such cases powder consumers have tried to remedy the situation by several expedients, such as retreatment of the powder, exchange arrangements with other concerns, etc. The best solution of the problem probably consists in analysis and storage of several shipments which are then, fully or in part, made up into large lots with uniform characteristics. Since the physical and chemical properties of consecutive lots from the same supplier approach statistical averages, these averages can be taken as the basis of workable production specifications, which, incidentally, may often fall short of the ideal originally set by the user. On the other hand, this is offset by the advantage of having a continuous supply of the same raw material—for which equipment and operating cycles have to be worked out only once.

Example: a commercial hydrogen-reduced iron powder used in manufacture of a soft magnetic part of complicated contours varied in weekly 2-ton shipments over the period of several months within the following ranges:

Apparent density..1.88–2.23 g./cc.
Flow..0–30 sec./50 g.
—325 mesh fraction...12–48%
Sub 30 μ fraction..4–14%
Compressibility....................0.48–0.62 in. ($^3/_4$ in.φ, 15 g., 5 tsi)
Hydrogen loss...........................1.9–3.3% (1000°C., 1 hour)
Carbon 0.02–0.12%
Silica 0.06–0.14%

The great fluctuations in some of these properties, especially hydrogen loss, compressibility, and apparent density, required continuous changes in the molding tools, sintering cycle, and flow of the reducing atmosphere during sintering. In order to use all shipments of this powder, four different sets of press tools and sintering schedules had to be employed. Needless to say, this situation made the molding process uneconomical. This was changed, however, by stabilizing conditions through the use of large mixtures composed of parts of at least three individual shipments. In this way the above properties were confined to the following ranges:

Apparent density..1.97–2.14 g./cc.
Flow..33–35 sec./50 g.
—325 mesh fraction...23–33%
Sub 30 μ fraction..7–10%
Compressibility....................0.53–0.57 in. ($^3/_4$ in.φ, 15 g., 5 tsi)
Hydrogen loss...........................2.0–2.5% (1000°C., 1 hour)
Carbon 0.04–0.11%
Silica 0.07–0.11%

The substantially narrowed ranges of the different properties made it possible to carry on with only two sets of tools and one sintering cycle, thus converting the order into a profitable contract.

Production of soft magnetic iron parts of high permeability requires a relatively dense structure, controlled grain size, a minimum of impurities, and no strains in the crystal lattice. Annealing in hydrogen at high temperatures is a common expedient for improvement of permeability. Iron powders of close size range are not suitable for this purpose. Comparatively coarse powders, *e.g.*, passing through 100 mesh, but not 200 mesh, would be satisfactory from the standpoint of grain size and impurities (which appear mostly at the particle interstices and grain boundaries), but not with reference to density. As previously mentioned (Chapter IV), the porosity is enhanced in the case of large particles within one close size range. Furthermore, coarse powders generally result in compacts of only mediocre mechanical properties. Iron powders of any other close size range have all the disadvantages of the finer size, such as an increased amount of impurities (due to increased particle surface area), in addition to possible impaired pressing properties (due to internal stresses), while retaining the large porosity. Also, fine powders cause gross shrinkage in many cases, interfering with the aim of holding to the closest possible tolerances to assure optimum magnetic performances. An appreciable variation in density throughout the finished product may also be caused by the comparatively high molding pressures required by fine powders. However, by resorting to mixing, a powder of definite particle size distribution can be obtained. This will assure the very closest packing into a solid, dense structure, during molding and sintering. Filling of the interstices can be attempted by mixing different particle size ranges of the same type of powder, and even more effectively mixing two or more powders of different size and particle shape. To illustrate this point, it was reported that magnetic permeability values have been almost doubled when mixing 95% electrolytic iron powder (dendritic particles which all pass through 100 mesh, 30% through 325 mesh), with 5% carbonyl iron (substantially spherical particles smaller than 10 μ in diameter).[12a] A similar improvement in flux density has also been obtained in mixing 70% coarse electrolytic powder (nodular particles, all passing through 100, but none through 200 mesh) with 20% hydrogen-reduced iron (spongy particles, all passing through 200 mesh), and 10% of the same carbonyl powder.

Polymetallic Mixtures

Most polymetallic powder compacts are produced from powder mixtures. The sole major exceptions of industrial importance to date are parts molded from brass powder. In this case a simple method for producing the alloy powder by atomization favors the use of alloy powders as starting materials. This is due to the high plasticity of the alloy particles and difficulties inherent in the sintering of mixtures (containing free zinc).

[12a] G. J. Comstock, *private communication.*

Today all other important branches of powder metallurgy require mixing of different metallic, and in certain cases nonmetallic, powders prior to being molded into various shapes and being sintered into coherent homogeneous or composite structures. Powder mixing constitutes an integral process in the manufacture of cemented carbides, duplex contact metal structures, self-lubricating bearings, friction materials, or alloy steel parts, to name some of the most important products. In these cases at least two finely divided metals are joined intimately. Thorough and uniform mixture is always essential regardless of whether these metals are to form an alloy of certain desired new properties, or to remain in the compact as independent constituents while retaining their individual characteristics. The mechanism of mixing metal powders is influenced greatly by several factors, including (1) specific gravity of components, (2) particle size of components, (3) particle shape and structure of components, (4) particle surface conditions of components, (5) external conditions during mixing, and (6) type of mixing equipment.

Segregation is favored by a difference in specific gravity between the individual components. If, for example, powdered aluminum and lead are blended in a porcelain bowl, a large concentration of the heavy lead particles will be noticed at the bottom, while the upper part of the mix will contain a large concentration of aluminum. If powders of this kind are mixed dry, it is almost impossible to prevent some segregation, which is favored by the impact caused by the starting or stopping of the operation, by vibrations, turbulence, or eddy currents during the operation, etc. Nevertheless, it has become possible to overcome most of these difficulties and to obtain intimate mixes of materials of unequal densities (i.e., aluminum and lead or tungsten and copper) by taking certain necessary precautions. If dry mixing must be employed, segregation can be prevented successfully by static fixing of the relative positions of the particles. Thus, very uniform iron-lead mixtures have been obtained on an experimental basis by keeping the iron powder in a magnetic field during mixing. This is accomplished by attaching permanent magnets to the tumbler, and results in the formation of closely knit particle aggregates whose relative densities resemble that of the individual lead particles. Alternate magnetization and demagnetization have also been effective in mixes with one ferromagnetic component. In order to counteract the effects of gravity, experiments have been conducted in a magnetic field, but the results were not very encouraging.

Another widely used expedient is to employ organic substances with adhesive properties, generally for the double purpose of impeding segregation during mixing and filling of the mold, and of facilitating the bonding of the particles during pressing. Such additions usually amount to only a fraction of one per cent; they are added either dissolved in a commercial solvent, or after being heated to a low-viscosity liquid state. Satisfactory coverage of the particle surfaces of one of the components—in most cases the heavier one—is necessary to prevent segregation. Lorol[13] (lauryl alcohol) has proved satisfactory in many cases in preventing segregation, the required amount ranging between 0.5 and 1%, depending on the surface area of the powder mixture. Another example is camphor, which is used by several manufacturers of refractory alloys and cemented carbides. If handling the powder in a moistened condition is allowable, water as well as certain commercial solvents (particularly alcohols of high molecular weight) is effective in preventing or impeding segregation. However, care must be taken that the homogeneous mix does not separate while it is being dried or fed into the molds. In certain cases, this may even necessitate molding the mixture while it is still moist.

Classification (separation according to particle size) during mixing or handling prior to pressing deserves similar attention. The problem is, generally speaking, not as serious as that of segregation, since most commercial powders have a wide range of intermediate particle sizes between the extremely coarse and fine sizes; these intermediate sizes tend to impede classification caused by gravity, turbulence, or centrifugal forces which act during the mixing operation. If, however, powders of unfavorable size distributions must be blended, it is frequently necessary to resort to expedients similar to those used to prevent segregation. Thus, for example, a 100-mesh molybdenum powder of which only 20% passes a 325-mesh sieve cannot be satisfactorily blended with a very fine "amorphous" silver powder unless camphor or another organic binder is added, or the mixture is moistened into a mud with the aid of a commercial solvent; this is necessary in spite of the fact that the specific gravities of silver and molybdenum are almost identical.

Mixing of powders with different particle shapes and structures may also cause difficulties. Although perfectly uniform mixtures can be obtained in blending electrolytic and reduced powders, a tendency toward dissociation during mixing can frequently be observed in the case of combinations of atomized and electrolytic or reduced powders. In practical applications, such as in making mixtures for porous bearings, this phenomenon is usually overcome by the addition of nonmetallic substances, especially graphite. The graphite seems to coat the spherical tin or lead particles, at the same time filling the surface depressions and discontinuities of the dentritic, electrolytic, or spongy, reduced powder particles. Its

---

[13] Product of E. I. du Pont de Nemours & Co., Inc., Wilmington, Del.

action tends, therefore, to equalize the different shapes and thus to homogenize the mixture. On the other hand, large areas of concentration of spherical and rugged particles have been observed in mixes of carbonyl iron and finely reduced cobalt powder, or of pulverized steel and electrolytic copper powders. In this case, too, wet mixing, or the addition of organic substances, has been found most helpful in securing uniform mixtures.

Two additional factors concerning the particle surface affect the homogeneity of the mixture in a similar manner. From the foregoing it is apparent that surface conditions, *i.e.*, the actual physical structure of the particle surface as well as of the surrounding medium play important roles in producing homogeneous mixtures. Depending upon whether this medium is a dry gas or a liquid (perhaps with considerable wetting properties) neighboring particles may either slip past each other during mixing, or tend to form agglomerates or even lumps. Thus, it is common to see powders in a moist atmosphere form lumps and "balls," which often cling together with considerable tenacity. In many conventionally designed mixers such agglomerations naturally impede thorough blending. On the other hand, mixing in controlled atmospheres of nitrogen or carbon dioxide, or even *in vacuo*, has gone far to assure uniform mixing—especially when very fine powders with hygroscopic or pyrophoric tendencies are involved. Furthermore, the degree of uniformity of a mixture depends to a certain extent upon whether the particle surfaces are smooth or rough, clean or contaminated with natural or artificial impurities. Accordingly, the details of the operation must be worked out from case to case, and few general rules can be established. Whereas in one case a short time of 10 minutes, for example, is sufficient to assure complete homogenization of the mix, in other cases, 24 hours and more may be required. However, it is impossible to establish a long mixing period as general practice, since unduly prolonged mixing may cause work-hardening or comminution through attrition. Of course, the time necessary to obtain uniform mixing depends a great deal upon the type of equipment used, and also upon external conditions, such as the type of atmosphere or liquid surrounding the particles.

Various types of mills and blenders are used for mixing.[14] Both ball and rod mills are used extensively for blending powders effectively, and especially for those powders of hard metals such as carbides, refractory

[14] Products of: Abbe Engineering Co., New York, N. Y.; Baker Perkins, Inc., Saginaw, Mich.; Franklin-McAllister Corp., Chicago, Ill.; Hardinge Co., Inc., York, Pa.; Lancaster Iron Works, Inc., Brick Machinery Div., Lancaster, Pa.; National Engineering Co., Chicago, Ill.; Patterson Foundry & Machine Co., East Liverpool, Ohio; H. K. Porter Co., Inc., Pittsburgh, Pa.; Pulverizing Machinery Co., Summit, N. J.; Read Machinery Co., Inc., York, Pa.; Sturtevant Mill Co., Boston, Mass.

metals, steel, etc., where size changes must be controlled closely. In Figure 64 a battery of such mills is shown for the production of powders. These mills, however, are less suitable for softer metal powders in which the danger of particle deformation and work-hardening is greater. For such powders various types of blenders have given better service, with the

Fig. 64. Batteries of ball mills and vibratory screens for the production of cemented carbide powders (courtesy of Carboloy Co.).

double cone-type blender (Figure 62)—available in various sizes from 500-lb. to 5-ton capacity—probably the most widely used. These blenders can be fitted with baffles to intensify the tumbling motion of the powder. One particular type of construction resembles that used in the cement industry; other types of mixers promote tumbling of the powder by rocking, eccentric, gyratory, or otherwise complex motions, by predetermined separations with the aid of vertical wings which divide the cone sections into compartments[14a] (Figure 65), or by intensive mixing with the aid of baffles, mullers, etc. (Fig. 66).

[14a] MacLellan Batch Mixer, product of Anglo American Mill Corporation, Owensboro, Ky.

As an illustration of the importance of the type of mixing equipment used, one manufacturer claims that the use of his cone blender has reduced the mixing time for cemented carbide powder batches from 2 days (neces-

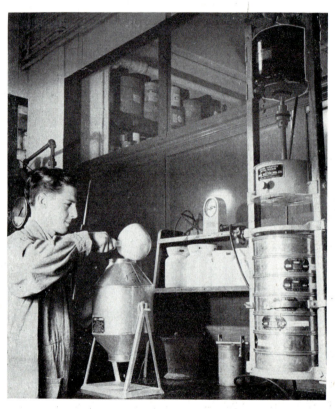

Fig. 65. Loading of sample batch mixer containing internal vertical baffles and compartments[14a] (left), and testing screen size distribution in automatic vibratory sieve shaker[14b] (right). (Courtesy of Sintercast Corporation of America.)

sary with conventional ball mills) to ½ to 1 hour, depending on batch size and composition.

COATING OF METAL POWDERS

Coating of metal powder particles with metallic or nonmetallic films is employed for a variety of purposes. The metallic films and the particles to be coated may have the same or different compositions. Similarity of

[14b] Combs Gyratory Sifting Machine, products of Great Western Mfg. Co., Leavenworth, Kan.

composition aids in preventing classification when a very wide size distribution range is used. Separation due to marked differences in particle size can be overcome by causing the fine particles to adhere to the surfaces of the coarser particles with the aid of some kind of adhesive (*e.g.*, gum arabic, glycerine, water, etc.). This method also applies to mixtures of several metals. A metallic coating having a composition different from that of the base powder is frequently used to modify the chemical properties or the susceptibility to plastic deformation of the powder particles. Nonmetallic coatings are used to insulate the individual metal particles (*e.g.*, in magnetic materials) to aid the pressing operation (lubricants and binders), or to accomplish alloying effects during sintering (graphite with iron).

Fig. 66. Milling, screening, and mixing of metal powders. Left to right: scale for weighing screen fractions; 200-lb. ball mill; Ro-Tap sieve shaker; Intensive Mixer with mullers and baffles for wet mixing (courtesy of American Electro Metal Corp.).

Coatings on metal particles can be achieved by four methods: (*1*) ball milling, (*2*) electroplating, (*3*) precipitation, and (*4*) spraying.

The ball-milling procedure involves milling relatively compact particles of a hard metal with a small amount of a finely powdered soft material. It is the procedure generally used for the production of cemented carbide powders.[15] The adherence of the fine, soft particles to the surfaces of the coarser particles can be facilitated by moistening the fine particles, as suggested for the coating of copper particles by tin.[16] The methods involving electrolysis (the particles to be plated form the cathode) and precipitation or displacement (the coat is precipitated from a salt solution upon the surface of the less noble base particles) were mentioned in Chapters III and VI, especially in connection with the coating of lead particles with copper. Such powders can be obtained either by electrodeposition (using mixed aqueous solutions of lead acetate, copper acetate, and acetic acid[17]), or by stirring lead powder into a copper acetate solution at approximately 60°C. (140°F.). If the lead particles are completely coated with copper, the compacted powder can be sintered at temperatures considerably higher than the melting point of lead. Another example is the protection of copper particles against oxidation by coating them with up to 1% of a more electropositive metal having a melting point below 500°C. (930°F.) (e.g., tin, bismuth, cadmium, lead, or zinc).[18]

The coating principle was employed in many recent efforts to make steel particles more susceptible to plastic deformation during molding. Production of hard iron and steel powders with particles coated with ductile metal films, either of iron[19] or of lower melting soft metals, such as copper[20] or lead,[21] has resulted in reduced compacting pressures, lower sintering temperatures, higher densities, and better mechanical properties.

In the manufacture of magnetic cores, iron, nickel, and special alloy powders are coated with thin films of nonmetallic materials of high electric resistance. These insulating coatings reduce eddy current losses in the coils. Although ceramic-type materials appear to be most effective for this purpose,[22] coatings of iron and zinc salts,[23] or of phosphates precipitated from ammoniacal solutions[24] have also been suggested. Oxide films obtained by controlled surface oxidation can accomplish the same purpose.[25]

[15] A. McKenzie, *Wire and Wire Products, 17*, 574 (1942).
[16] U. S. Pat. 2,273,832.
[17] U. S. Pat. 2,182,567.
[18] U. S. Pat. 2,286,237.
[19] U. S. Pats. 2,289,897; 2,342,799; and 2,352,316.
[20] R. G. Olson, *Communication during discussion* at the First Annual Spring Meeting of the Metal Powder Association, New York, May 5, 1944.
[21] U. S. Pat. 2,327,805.
[22] E. E. Schumacher, in J. Wulff, *Powder Metallurgy*. Am. Soc. Metals, Cleveland, 1942, p. 166. Also German Pat. 712,575.
[23] U. S. Pat. 2,261,425.
[24] German Pat. 703,669.
[25] German Pat. 717,111.

MIXING WITH NONMETALLIC SUBSTANCES

Nonmetallic additions to metal powders may be classified as (a) transitory additions which disappear during later operations, or (b) permanent additions which are incorporated in the final product. They may fulfill one or more of the following purposes: (1) form an alloy composition; (2) act as free ingredients; (3) promote controlled porosity; (4) act as binders; and (5) act as lubricants.

Mixing of these nonmetallic additions with the metal powders is usually not a simple matter, because the organic particles tend to agglomerate before making satisfactory contact with the metal surfaces. For this reason dry mixing is not always possible, and some binders are dissolved in a commercial solvent prior to mixing. If the particles are substantially solid, as is usual with the pulverized hard metal powders which are most in need of a binder, perfect coating of the individual particles is readily attained by the wet-mixing method. Rotating drum or "intensive-type" mixer[26] (see Figure 66) have been found suitable for this method.

Nonmetallic Alloy Ingredients

Carbon and silicon are two important representatives of nonmetals or semimetals which form alloys. Probably the best-known example is that of cemented carbides. Carbon (approximately 6%) in the form of lampblack is mixed with tungsten powder and heated for a few hours at 1400° to 1500°C. (2550° to 2730°F.).[27,28] Mixing is carried out either in cone blenders or in ball mills, which are preferably lined with tungsten, Stellite, or stainless steel (see Fig. 64); the balls are either tungsten carbide or stellite. The same type of mill is used later for grinding the carbide powder to the desired grain size for processing to a given structure, and for mixing with the softer binder metal. The object of the latter operation (which may last from a few hours to a few days[29]) is to coat each particle with a continuous layer of the cementing metal.

In several processes, carbon in the form of graphite is mixed with iron powders for the production of sintered steel parts[30]; the amount of graphite added may vary between 0.1 and 2%. Mixing of this kind is gen-

[26] Simpson Intensive Mixers, products of National Engineering Co., Chicago, Ill.
[27] K. Becker, *Hochschmelzende Hartstoffe und ihre technische Anwendung*. Verlag Chemie, Berlin, 1935, p. 25.
[28] F. Skaupy, *Metallkeramik*. Verlag Chemie, Berlin, 1943, p. 176.
[29] E. W. Engle, in J. Wulff, *Powder Metallurgy*. Am. Soc. Metals, Cleveland, 1942, p. 439.
[30] F. V. Lenel, in J. Wulff, *loc. cit.*, p. 503.

erally carried out in tumblers, blenders, or rotating drums, since ball-milling tends to impair compressibility by strain-hardening the iron particles. Similarly, silicon powder may be added to iron or steel powders for the production of soft magnetic articles. Tumblers and rotating drums are again preferred to ball or rod mills for the reason mentioned above. Other examples of lesser importance include mixtures of copper and bronze powders with 7–8% phosphorus for brazing purposes,[31] cadmium and lead powders with approximately 1% sulfur for special bearings,[32] tungsten and other refractory metal powders with 1–5% boron for certain hard alloys,[33-35] and copper and gold powders with up to 1% radium salts for certain special radioactive compositions.[36] Ordinary rotating drum mixers or cone blenders are used for all these mixtures with the exception of those containing boron, which are most efficiently prepared in ball mills.

Free Nonmetallic Ingredients

Combination of metal powders with nonmetallic materials which retain their individual properties and structures forms the basis of several important powder metallurgical applications. Antifriction and friction elements are perhaps the best-known examples; in both cases, graphite (up to 6%) is the chief nonmetallic constituent. In bearings, the graphite should remain substantially free throughout the forming and heat-treating operations, so that its lubricating qualities are present in the final products. In friction elements, graphite additions counteract abrasion during molding and impart lubricating properties to the product, e.g., to facilitate smooth engagement of the brake or clutch lining; at the same time discontinuance of the metallic matrix enhances frictional properties. Friction elements usually contain at least one other nonmetallic ingredient which acts as an abrasive.[37] Powdered asbestos, silica, or emery is suitable for this purpose, and is added in amounts of 2–7% by weight.

High-temperature, wear-resistant materials (e.g., valve seats) contain various amounts of abrasive powders such as silicon carbide, silica, and other refractory metal oxides. In the case of composite hard metals for

[31] H. E. Hall, in J. Wulff, loc. cit., p. 30. Also Canadian Pat. 357,942 and French Pat. 759,367.
[32] C. Hardy, private communication.
[33] U. S. Pat. 2,059,041.
[34] Brit. Pat. 379,681 and Canadian Pat. 346,719.
[35] French Pat. 757,419.
[36] U. S. Pat. 2,326,631.
[37] J. E. Kuzmick, Symposium on Powder Metallurgy. ASTM, Philadelphia, March, 1943, p. 46.

tools, such as grinding wheels, high-speed drills, etc., diamond dust constitutes the superhard component.

Mixing of the metallic and nonmetallic ingredients is usually carried out in one of the powder mixers or ball mills mentioned previously. In the case of copper, tin, or lead mixtures for bearings, the graphite readily forms a protective film around the particles, facilitating interparticle lubrication and slippage during tumbling, and greatly reduces the possibility of strain-hardening. In all other cases, however, precautions must be taken to avoid undue work-hardening of the metallic components during mixing. This is especially important for very hard compositions containing diamond or silicon carbide powders. In such cases the constituents of the softer metal tend to form films around the abrasive particles, which results in severe plastic deformation of the metal—making it almost impossible to compress such mixtures into coherent compacts. To avoid this possibility, it has been found practical to eliminate the use of ball mills, reduce the time of mixing to a minimum, and prepare the mixtures in a commercial solvent which cushions the impact. Very voluminous, and preferably spongy, metal powders are believed to be most suitable for molding such mixtures, since they tend to engulf the fine, solid, abrasive particles in their pores.

Transitory Nonmetallic Ingredients

The manufacture of articles of controlled porosity is facilitated by the use of transitory ingredients which volatilize during sintering, leaving behind pores and channels of a definite size and structure. Although porosity can be controlled by other methods inherent in the pressing or sintering operations, the employment of auxiliary materials has been favored in those industries in which very close control of the interconnected pores is essential. Thus, in mixtures for porous bearings and filters, small quantities of such organic substances as salicylic acid,[38] camphor,[39] or certain metal stearates are added for the sole purpose of facilitating disruption of the structure of the compressed mixture prior to sintering. The transitory addition agent escapes from the compact in the form of a vapor at a temperature so low that only superficial bonding of the metal particles takes place. Entire particle aggregates, however, must sometimes be shifted to allow spontaneous escape of the vapor in channels whose direction may even be controlled by certain manipulations. In many cases the pore-creating agent also serves as a medium for facilitating molding.

[38] Brit. Pat. 284,532.
[39] C. Hardy, *private communication.*

NONMETALLIC BINDERS

Those branches of the field dealing with hard and brittle powder particles have produced a successful technique for the use of plastic auxiliary materials as cements or binders. Refractory metal and carbide powders, as well as steel or permanent magnet powders, are examples of such hard materials which normally require very high pressures for forming. Even if such pressure could be attained commercially, the resulting compacts would require very careful handling upon removal from the die, thereby eliminating the use of the conventional and economical automatic upward ejection method. Furthermore, compacts formed in this manner would have imperfect forms, impaired by transverse laminated fractures or by cone-shaped fractures at 45-degree angles; broken corners, deteriorating edges, or in the case of extremely hard powders, pulverization upon pressure release, also rule out this procedure. In the case of cemented carbides, matters are somewhat improved because of the addition of the soft metallic cementing materials, cobalt or nickel, but in other cases in which the final composition is present in the powder, e.g., in crushed steels or in aluminum–nickel–iron powders, addition of a soft metal powder is not permissible, making the addition of nonmetallic binders the only solution.

In investigating the many organic compounds which can be used for cementing metal particles, it is natural to concentrate on those which would have the least effect on the final product. In order to keep impurities in the final product as low as possible, removal of the nonmetallic component during sintering is the most important requirement. This restricts the amount of binder to a few per cent to avoid excessive porosity in the product, which in turn might seriously impede the sintering of the compact into a solid body. In addition, the temperature of evaporation of the binder must be low enough to make its removal possible before the consolidating shrinkage forces (at higher sintering temperatures) tend to close the escape corridors within the compact; otherwise, explosive disruption of the coherent shapes may result. In other words, only those organic compounds can be used as binders which are solid and extremely plastic at room temperature and evaporate quickly at slightly elevated temperatures. In order to be effective in very small additions, e.g., 1% by weight or less, the binder must have good adhesive ("wetting") properties with reference to the metal surfaces, and also have enough lubricating qualities to permit uniform distribution during mixing.

Most commercial plastics of the formaldehyde or cellulose types do not fulfill these requirements. Their binding properties become effective only when pressure is applied in the presence of heat; this may lead to undesirable surface oxidation of the metal particles before they are cemented together. Even then, a relatively large proportion of these plastics is necessary to create an uninterrupted network of binder material. This consideration, and their rather high boiling points, limit the use of these resins to straight molding work, such as in the manufacture of (plastic-bound) powdered iron high-frequency coils, (plastic-embedded) granular Alnico speedometer magnets, etc.

Among materials found suitable for use as binders in powder metallurgy, the following are probably best known: (1) camphor, (2) paraffin, (3) ammonium chloride, (4) mineral oil, (5) starch, (6) talc,[40] (7) magnesium carbonate,[41] (8) Pyroxylin, (9) Glyptal resins, (10) synthetic thermoplastic resins e.g., vinyl resins (Vinylite A and V,[42] Alvar, Forenvar,[43] vinyl acetate), and (11) various inorganic and organic colloids.[44] Most of these materials have been found useful because of their affinities for metallic particle surfaces, and because of the other advantages they offer, such as promotion of porosity, and lubricating qualities. Vinyl resins added in amounts up to 5% may also serve as carriers to permit lateral flow into projections.[45]

Since most organic particles tend to agglomerate before making satisfactory contact with the metal surfaces, dry mixing of the nonmetallic binder substances with the metal powders is generally not as effective as wet mixing with the aid of a commercial solvent in which the binder material is previously dissolved. Perfect coatings of the individual particles are attainable in this way, and the wet mixing method is universally employed in cemented carbide powder preparation work. Rotating drums, tumbling barrels, cone blenders and intensive-type mixers are well adaptable for wet mixing purposes.

NONMETALLIC LUBRICANTS

Lubrication is perhaps the most important purpose of nonmetallic additions in powder metallurgy. Friction problems play a decisive role in the molding of metal powders and have serious effects on uniform density of the compacts, on complete filling of all projections in complicated shapes, on the useful life of the mold and tools and, finally, on the smoothness of the surface finish of the compacts. Friction between the metal

[40] U. S. Pat. 1,642,348.
[41] U. S. Pat. 1,642,349.
[42] Products of Carbide and Carbon Chemicals Corp., New York, N. Y.
[43] Products of Shawinigan Products Corp., New York, N. Y.
[44] German Pats. 707,897; 707,898; and 711,650.
[45] H. L. Strauss, Jr., Am. Machinist, 89, No. 20, 113 (1945).

particles and the faces of dies and punches is as harmful as friction between the metal particles themselves. Interparticle lubrication has been a great aid in reducing the coefficient of friction in both cases.

In addition to powdered graphite, many organic and inorganic compounds have been employed successfully as lubricants. They include boric acid,[46] stearic acid, benzoic acid, paraffin,[47] beeswax, Acrawax,[48] tallow, talc, salicylic acid,[49] inorganic and organic colloids,[50] ammonium chloride, magnesium carbonate, zinc stearate, calcium stearate, lithium stearate, aluminum stearate, Sterotex,[51] vegetable oils of the drying type (*e.g.*, linseed oil), and lubricating and petroleum oils. Except for the last item, these lubricants can either be added dry in the form of powders, or dissolved in a suitable solvent, such as ether, acetone, alcohol, benzine, carbon tetrachloride, carbon disulfide,[52] or even plain water.

Obviously, the lubricant is more effective in reducing friction and increasing uniformity of pressure distribution if it is applied while wet. Mixing of this kind is best carried out in tumbling barrels, revolving drums, cone blenders, or in intensive-type mixers; but it is not always practical to mix lubricants in solution. In the case of very spongy or rugged powder particles, the large surface would require an excessive amount of lubricant to produce a continuous film; in these cases dry mixing must be used. Care must be taken to have a maximum metal surface come into instantaneous contact with the nonmetallic powder in order to avoid "balling-up" of the particles. One effective method of dry mixing provides for a preliminary mix with the aid of screens. The lubricant is forced through a 20-mesh sieve with the aid of steel balls. The sieve is located on top of a large-size 100-mesh screen through which the metal powder passes. The two sieving operations are synchronized by adjusting the size and amount of steel balls. A coarse screen is necessary for the lubricant in order to avoid clogging of the mesh. After this preliminary "contact blending" the substantially uniform mixture may be further homogenized in tumblers or blenders of conventional design.

Additions of nonmetallic substances for lubricating purposes are usually very small. The exact amount must of course depend on various factors, such as powder type and particle size distribution, apparent density and plasticity of the metal powder, amount of surface of dies and

[46] U. S. Pat. 2,178,529.
[47] W. P. Sykes, *Metal Progress, 25*, No. 3, 24 (1934).
[48] Product of Glyco Products Co., Inc., Brooklyn, N. Y.
[49] Brit. Pat. 364,546.
[50] German Pats. 707,897; 707,898; and 711,650.
[51] A soybean product of The Capital City Products Co., Columbus, Ohio.
[52] Brit. Pat. 441,177.

punches exposed to powder and compact, and type of lubricant. In general, the amount to be added for purely lubricating purposes does not exceed 1% by weight. It is desirable to keep the amount of lubricant addition to a minimum in order to reduce operating difficulties during sintering. Reaction products from the lubricant (*e.g.*, carbon or zinc) have a tendency to contaminate the furnace chambers—especially in the cooler sections where the reaction products condense or precipitate. However, these lubricating materials may have to be added in excess of this amount if they serve other purposes, such as promoting large and controlled porosity, or acting as bonding agents during molding.

Summary

The conditioning of powders for molding involves a number of different practices. Cleaning, drying, and storage under thoroughly controlled conditions are as much the responsibility of the powder manufacturer as of the user. Contamination of the powder during shipment and handling can be avoided effectively by a number of precautionary measures.

Conventional mixing techniques serve a variety of purposes. They facilitate homogenizing of large powder lots, and enable the powder user to blend various grades of the same powder, different powders of the same metal, or different metals. At the same time, mixing of metal powders with powdered nonmetallic substances may serve a variety of purposes, essentially to affect the properties or porosity of the finished product, or to facilitate molding of the powder.

CHAPTER VIII

Behavior of Powders under Pressure

MECHANISM OF INTERPARTICLE BONDING

In order to understand the many complications inherent in the molding of metal powders, we must first consider what actually happens when the powder is subjected to compression. There has been a great deal of speculative discussion concerning the mechanism of powder compression, and several widely divergent theories have been advanced in attempts to explain the complex phenomena occurring during compaction. All of these theories, however, agree that the primary purpose of powder compression is to bring a sufficient number of particles close enough to each other to produce interparticle adherence—often referred to as "bonding." The forces of this interparticle adhesion, in turn, result in the cohesive strength of the powder compact. To understand the following material, it appears to be sufficient to give only a brief account of these various concepts.

Among the many causes advanced for the bonding of powders, those to be discussed here are: (1) interatomic forces (surface adhesion, cold-welding, and surface tension, (2) liquid surface cements, and (3) mechanical interlocking. Interatomic forces will be discussed more thoroughly than the other theories because it appears to the author to be the most substantial advanced so far.

Interatomic Bonding Forces

NATURE OF FORCES

Atomic forces play a fundamental role in the powder compaction process. Atoms grouped in the regular space lattice of the particular metal are the building blocks of the metallic crystals. Bonding between two contacting crystals implies a rearrangement of the atoms in the surface layers of the crystals involved, and, therefore, a certain mobility of the atoms. This mobility and tendency to change places are related to the temperature level at which the particles—or, more exactly, their surface areas—are subjected to particular pressure conditions. In addition, the rearrangement of atoms may transgress a particle boundary under certain circumstances, *i.e.*, may involve two different surfaces, forming a direct and intimate contact. In general, rearrangement of atoms in the surface layers of two adjacent particles will produce atomic links between these particles, provided that the contact between the two surface areas is so close (*i.e.*, of a distance in the order of the diameter of the atom) that atomic field forces can come into play. The greater the cumulative forces of these atomic links between two adjacent surfaces (and, in turn, the coherence of the entire powder mass), the more points of intimate contact exist, and the larger is the total area of contact. Therefore, it is desirable to have the largest possible surface area available (*i.e.*, a powder of greatest possible dispersion) in order to produce the greatest number of contact points.

Theoretically speaking, the metal particles could be mathematically perfect cubes or parallelepipeds, which could be piled up as in brick laying. If particle after particle were brought into intimate contact, a solid piece of metal would result, even without pressure or heat. Rigidity of such a "brick work" would be due entirely to the atomic bonds which are effective at the particle surfaces. Possible bridging effects could be overcome by jarring the pile until proper orientation of each particle was obtained. If we further assume that all atomic lattices of each particle are oriented in one preferred direction, and that there are no internal stresses, recrystallization would not take place on heating, and the formation of a single crystal of solid metal could take place.

However, this relatively simple picture does not exist in practice; it is complicated in several ways. We have already learned that close packing with a maximum possible number of contact points depends on having a variety of particle sizes, following definite laws of distribution. The same is true for the factor of particle shape, *i.e.*, it is conceivable that mixed shapes may lead to closer chance packing than single shape types. The denser these chance packings become during feeding of the die cavity, the greater is the cumulative area of contact points at the outset of compression, which, in turn, will further increase the total contact area formed during the compression cycle.

In connection with the discussion of particle size effect, an interesting view is advanced by Bal'shin,[1] who takes issue with the concept that the density, and accordingly the number of interparticle contact points of a compact increases with decreasing particle size, since the effect of atomic forces becomes stronger. This concept completely overlooks the fact that,

[1] M. Yu. Bal'shin, *Vestnik Metalloprom.*, *18*, No. 2, 124 (1938).

during compression, energy is expended to overcome these forces and that, in the overwhelming majority of cases, density and cohesive strength of a compact decrease with increasing dispersion of the powder.

There are other phenomena which come into play during the course of compression. The coefficient of friction between particles usually eliminates further contraction beyond chance packing obtained when the mold is filled. But additional densification of powder mass—as a basis for further increase in interparticle atomic links—can be affected by: (1) tendency of the particles to shift and reorient themselves into certain strata; (2) the presence of lubricating films whose thickness exceeds that of the maximum distance allowed for atomic attraction forces; and (3) plastic deformation of the particles, enhanced either by pressure (directly) or by heat of friction (caused by pressure).

EFFECT OF PLASTICITY

Susceptibility of a metal to plastic deformation depends largely upon its crystal structure. Most metals with a face-centered cubic lattice, such as nickel, copper, and especially silver, gold, aluminum, and lead, are easily deformable. The same is true for many solid solutions having a face-centered structure, such as α-brass. Some of these metals, especially gold, aluminum, and lead, display a very low resistance to deformation, since they are not subject to appreciable work-hardening at room temperatures. Accordingly, only a low specific pressure is required to form coherent compacts from their powders. Metals with body-centered, cubic crystal lattices, such as tungsten, molybdenum, tantalum, columbium, vanadium, and ferritic iron, are less easily deformable, and their powders require higher compacting pressures. Generally, these metals have strong work-hardening tendencies and combine easily with metalloids, decreasing still further their plastic deformability.

There is no doubt that the obstructions caused by irregular particle shapes can be overcome to some extent by plastic deformability of the metal, which leads to greater areas of intimate contact where the atomic forces can come into play. However, in powders consisting of very hard metal particles plastic deformation will produce few contacts close enough to permit bonding even if exceedingly high pressures are applied. In these cases, contacts are limited to certain projecting points, or are obtained only at locations where superficial abrasion of the neighboring surfaces has occurred. Particles of basically soft and plastic metals will deform at the main areas of friction, i.e., at the contact areas between large particles. The degree of deformation is dependent on several factors, especially upon the nature of the metal, the nature and history of the powder, and the magnitude of specific pressure applied. These factors are chiefly responsible for the frictional conditions prevailing during molding at the die walls, as well as between the individual particles. The resulting heat of friction in conjunction with the specific pressure applied will, in many cases, increase particle plasticity and foster new interparticle contacts. It is generally known that pure metals become increasingly deformable at higher temperatures. Therefore, a rise in temperature caused by frictional heat may offset, at least in part, work-hardening of the particles caused by excessive deformation when very high pressures are applied.

Comstock,[2] in discussing plasticity, attributes the formation of bonds more to recrystallization phenomena than to plasticity—a term to which he gives too generalized a meaning. However, it may be said that the initial formation of bonds between particles is promoted by ready deformability of particle surfaces, for which increased atomic mobility is required. According to Comstock, plasticity becomes effective in the formation of new particle bonds only if the temperature necessary to promote it reaches (at least locally) levels required for recrystallization under the particular pressure conditions. No bonding of this type (i.e., in the absence of a liquid phase) takes place at temperatures below that limit.

The fact that rising temperatures (and, therefore, increased plasticity) promote greatly extended areas of contact where atomic interparticle links can be established, leads us ultimately to hot-pressing, which is known to result in structures of firmly consolidated particles. Under stress of the compacting pressure, particles deform until resistance to plastic flow is built up to the point where sufficiently strong interparticle contacts result in strain-hardening of the particle agglomerate.

The physical properties of the metal also influence the capacity of the powder to form large areas of contact. For example, powders of soft metals, such as tin, lead, or silver, are most susceptible to plastic deformation, even under moderate pressures, with a large number of contact areas to allow action of atomic forces. Consequently, the particles are pulled together into a strong, coherent agglomerate which yields a strong and stable compact. Basically hard and brittle metals, such as chromium, steel, refractory metals, or carbide powders, when compacted at similar pressures, lack sufficient contact areas to yield coherent compacts. In fact, the particles may be so hard that formation of contact points by plastic deformation is impossible. In such cases, chance contacts obtained in filling the mold are too insignificant to matter, and the powder may remain incompactible even at the highest possible pressures. Metal powders

² G. J. Comstock, *Metal Progress, 35,* No. 6, 576 (1939).

having an intermediate position between these two extremes, such as copper, nickel, or iron, act in an intermediate fashion when subjected to pressure.

Degree of dispersion of a powder and particle distribution also influence plastic behavior under pressures. As compression progresses, intraparticle forces begin to play an important role, and compaction is increasingly aided by filling of the voids within the particles through their deformation (without breaking the bonds already established between those particles which have common contact areas). When comparing powders of different degrees of dispersion, it often becomes apparent that the finer particles undergo deformation more readily; because of better flow conditions, than in the case of a coarse powder, filling of the voids is also rendered easier.

EFFECT OF PARTICLE SURFACE CONDITIONS AND DISSOLVED GASES

An obstruction to the formation of interparticle atomic bonds may arise from conditions prevailing at the particle surface, especially from the presence of filmlike foreign matter wedged between the metallic surfaces. Interference may result from foreign phases regardless of whether they are in a solid, liquid, or gaseous state, or whether they are free or combined in the form of adsorbed or chemically bound films. Thus, for example, such impurities may be chemical compounds (particularly oxides) but may also consist of common atmospheric gases condensed on the particle surface or adsorbed by the surface layers. They may also be dispersed throughout the entire metal particle, particularly in powders prepared in a gaseous atmosphere, *e.g.*, in the reduction of metal oxides, or by electrolysis. Often, this causes greater hardness of the metal particles, which become less apt to deform under pressure because effective contacts with neighboring particles cannot be made during powder compression. Balke[3] cites tantalum an an interesting example of this phenomenon. Normally, tantalum powder may dissolve 150 volumes of hydrogen and as many as 700 volumes under extreme conditions. Unless this gas is removed the metal remains too brittle to be compressible into the desired shapes.

Other obstructions that adversely affect surface conditions are impurities, work-hardening, and basic hardness of the particles. As an example of the last case, tungsten carbide particles will not form effective contacts unless completely surrounded by a thin film of a ductile metal (cobalt).

[a] C. C. Balke, *Iron Age, 147,* No. 16, 23 (1941).

Compression of the powder then becomes possible, even though the particle interiors remain hard and brittle. Powders prepared by certain mechanical methods often contain a majority of particles whose surfaces are burnished or strained from work-hardening. Accordingly, plasticity of the particles is greatly impaired, and the formation of extensive contact areas during compression is prevented. Sometimes, such burnished surfaces may even cause flow disturbances and packing of the powder in the mold prior to compression. The same conditions apply to surface films of entirely foreign substances. If, for example, a metal powder is ball-milled for an extensive period, in a porcelain mill, with flint pebbles, nonmetallic coats may form around the particles, and the brittleness of these coats may seriously hamper the formation of contacts between the particles.

Probably the most common interference at particle surfaces is caused by the presence of oxides. These oxide films are usually formed through reaction of the metal with the atmosphere, either during production of the powder, or during storage, handling, or processing, prior to pressing. Oxidation of particle surfaces may also result from frictional effects during pressing. Dies[4] studied this frictional oxidation for various pairs of metallic materials that were rubbed together under different loads with and without lubrication. Although no frictional oxidation was noticed with lubrication, it was found that without lubrication, the amount of wear depended upon the hardness and adherence of the abrasion product, as well as upon the hardness of the base materials. In the case of aluminum, Dies proved by x-ray and chemical analysis that excessive wear led to the production of corundum. Serious die wear, generally encountered when pressing aluminum powders, may thus be considered to be a combination of at least three factors: (1) the high cohesive affinity between aluminum and iron surfaces, causing excessive cold-welding; (2) the high rate at which freshly cleaved or abraded aluminum particles oxidize; and (3) the very abrasive character of the resultant aluminum oxide surface layers.

When it is recalled that the atomic forces of a metal can become effective only when the distance between the two metallic surfaces is small, the seriousness of such interference of solid foreign matter can be appreciated. In this case, only a bond between the oxide layers could be formed, which —being non-metallic—would be weak. This can be seen in the case of iron. A ferric oxide layer, which can be distinguished by its pale yellow color, is formed at the relatively low temperature of about 300°C. (570°F.). It has a thickness of approximately 40 Angström units, i.e.,

4 K. Dies, Arch. Eisenhüttenw., 16, 399 (1943).

about ten times the height of the unit prism of ferric oxide. If two iron particles, with nonmetallic surface films 40 Angström units deep, are brought into very close contact, it is obvious that no bonds due to attraction of the metallic atoms can be established, since the width of the interfering phase exceeds by far the effective field of atomic attraction force (for iron). Even a film of only a few Angström units might effectively prevent metallic bonding. Contact between the metal faces can be established only if these interfering films are punctured locally or otherwise destroyed, e.g., by shear forces acting during compression.

The basic necessity of clean surface conditions to permit the action of atomic forces cannot be better demonstrated than by comparing powders of precious metals with those of baser ones. For example, platinum or gold powders are very readily consolidated under pressure, with obstruction by oxide films practically nonexistent. On the other hand, other powders which are equally plastic readily acquire a thin film of oxide on the surface of the particles and, consequently, require higher pressures for the disruption of the films during consolidation.

Even if produced and stored in the absence of oxygen, and exposed to the atmosphere only for the shortest time necessary for transfer from an airtight container into a mold, powders of such metals as lead, aluminum, or magnesium can never be compressed under truly oxygen-free conditions. In some cases (e.g., Swedish sponge iron powder or oxidized copper powder), during compression, the oxide films surrounding the particles are pierced by their projections or ruptured, since the plasticity of these particular metals permits considerable deformation. But if the films are allowed to become exceedingly heavy, even such plastic deformation does not disrupt the continuity of the oxide films. Figure 67 shows the microstructure of a heavily pressure-deformed magnesium powder whose oxide surface films were too heavy to rupture. Instead, the oxide is deformed along with the metal particles, forming an almost continuous magnesium oxide network with a preferred orientation imposed by the pressure.

Other Concepts of Bonding Forces

SURFACE ADHESION AND COLD-WELDING

As we briefly survey some of the other theories on bonding of metal powder particles, it should be kept in mind that all of these concepts are based on the previously described mechanics of the attractive force field of the metal atoms in the surface layers. Regardless of the name assigned to these forces or the peculiarities of the mechanism referred to, atomic forces and movements in contact regions between clean surfaces are fundamental in creating the desired bonds.

When discussing the attractive forces between individual particles (especially during compression), it has become customary to designate them simply as adhesive forces. Sauerwald[5] in his early studies on synthetic metal bodies uses the concept that adhesive forces between particles

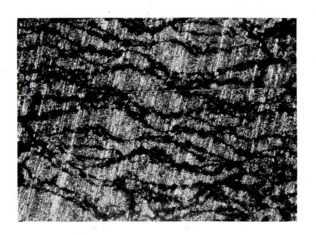

Fig. 67.   Magnesium powder compressed at 80 tsi showing heavily deformed metal particles and oxide films (×200).

come into effect if the particles are brought close enough together. These forces are still dominant in the early stages of subsequent heat-treatment and are replaced by major atomic movements only when the temperature of recrystallization is reached. As these adhesive forces are considered to be strictly surface phenomena, Sauerwald concludes that when particles are finer, more contact areas are created and the sum total of these forces is greater, i.e., the compact is more stable.

In practice, however, it has been proved by many examples that these forces can also be observed when adjacent surfaces are not of this small magnitude. Although it is true that many phenomena in metal powder particles appear to be different in larger metallic bodies, there are certain analogous phenomena in massive metals which are worthy of brief consideration. Jones[6] has given a detailed account of what he describes as "cold-

[5] F. Sauerwald, Z. anorg. allgem. Chem., 122, 277 (1922).
[6] W. D. Jones, Principles of Powder Metallurgy. Arnold, London, 1937, p. 2.

welding." If two freshly parted halves of a material are brought into immediate and close contact, they adhere to each other with sufficient strength to necessitate considerable force to reseparate them. The same applies to optically flat glass plates, which require a force of as much as 650 psi for separation.[7]   In his review, Jones reports many interesting studies concerning the nature of the force that makes one surface adhere to another so tenaciously. Dokos,[8] investigating the variations occurring in the coefficient of friction (between clean metal surfaces moving under high loads) over a wide range of sliding velocities, found the results dependent not only on the bulk properties of the contacting metals, but also on surface effect; it was estimated that the pressures on microscopic areas could reach 30 tsi, although the external loads might not exceed 200 pounds. The phenomena of glass beads[9] or metal filings[10] drawn over clean glass plates, as well as the strong adherence of two highly polished and optically flat gage blocks, have shown that these adhesion forces may reach considerable magnitudes. Thus, for example, it has been found that gold leaf adheres to glass with a force ranging between 1800 to 46,000 psi.[10]

There are other common occurrences in which these phenomena are equally apparent. Consider, for example, freshly sheared-off surfaces of such metals as gold, lead, tin, cadmium, etc., which are pressed into intimate contact by hand immediately after shearing. These will adhere to each other to such an extent that considerable force is required to separate them. Another example of a similar effect is the case of glass drawn out to a fine thread. If this thread is broken, after cooling, and the two sections are placed parallel to each other and not far apart, it can be observed that the attractive forces are sufficient to make the threads draw together and adhere to each other with a force great enough to prevent separation (without breaking in cross section[11]). Or, if immediately after being drawn, one thread touches the other at one end, the attractive force between them can be observed from the deflection of the two threads. If one thread is fairly stiff, the forces may even suffice to support the entire weight of the other (finer) thread.[9]

It is obvious from these experiments that only part of these attractive forces can be attributed to atmospheric pressure; thus, a residual force of considerable strength is left, which for simplicity's sake may be called adhesion or cold-welding.

[7] Lord Rayleigh, *Proc. Roy. Soc. London, A156,* 326 (1936).

[8] S. J. Dokos, *J. Applied Mechanics, 13,* 148 (1946).

[9] G. A. Tomlinson, *Proc. Roy. Soc. London, A115,* 472 (1927). .

[10] G. T. Beilby, *Aggregation and Flow of Solids.* Macmillan, London, 1921.

[11] W. J. Baëza, *A Course in Powder Metallurgy.* Reinhold, New York, 1943, pp. 56 and 57.

SURFACE TENSION

Several well-known authors dealing with the subject have expressed the belief that surface tension may be chiefly responsible for the attractive forces acting to consolidate powder into a compact.[11,11a] Whether these surface tension forces are caused or facilitated by differences in surface potentials of the individual particles,[11] remains an open question as long as the possible difference (*i.e.*, either positive or negative, and of equal magnitude) in potential of the minute surfaces cannot be measured. Since surface tension in solids is generally believed to be mainly a manifestation of interatomic forces, interpreting the bonding mechanism in terms of surface tension appears to be only another form of describing bonding as due to the attractive forces of atoms.

LIQUID SURFACE CEMENTS

The process of forming bonds between individual particles has been described by the term "cold-welding" when it is compared with adhesion phenomena in massive materials; the term "atomic welding" has also been occasionally employed to designate the same phenomenon. It is the author's opinion that use of the term "welding" in this connection is misleading, since welding always implies liquefaction of the metal; this has not been proved to be the case in compression of single metal powders. There is ample evidence that some increase in temperature occurs because of a change of the kinetic energy evolved during compression into heat energy, promoted by friction between the particles and between border particles and the die walls. This rise in temperature may well lead to greater plasticity, increased atomic mobility, and, therefore, to further strengthening of interparticle bonds. But it is difficult to conceive that, at a relatively moderate over-all compact temperature (usually far below the melting point of the metal), the particle surface fuses while the interior of the particle remains solid. Even in the case of gold particles which convert from angular shapes into spheres at 900°C. (1650°F.),[11a] it is difficult to conceive that local fusion at the contact points takes place several hundred degrees below the melting point.

The liquid cement theory has been mentioned on several occasions,[12,13] but little evidence has been brought forward to substantiate it. Except for very low-melting metals pressed in high-speed presses at very high pres-

[11a] C. C. Balke, *Iron Age, 147*, No. 16, 23 (1941).
[12] W. J. Baëza, *A Course in Powder Metallurgy*. Reinhold, New York, 1943, p. 64.
[13] C. Hardy, *Metal Progress, 35*, 171 (1939).

sures, local temperature increases in most metals of practical importance are not sufficient to overcome heat conductivity from the contact areas to the inner regions of the individual particles.

A simple experiment will substantiate this point. When a commercial metal powder (e.g., electrolytic iron) is pressed in a tableting press to a convenient pellet size, preferably with a grooved top, a definite equilibrium temperature can be obtained after a number of pressings—one or two hundred. This temperature is, of course, different when a harder or a softer metal is used, when a lubricant is added, or when a certain percentage of an abrasive such as silica or alumina is added to the metal. Immediately after ejection, the over-all temperature can be checked by inserting a thermometer or other temperature indicator into the groove between two pellets. If finely atomized Wood's metal (melting point, 68°C.; 155°F.) is added to a portion of the powder mixture, preferably by coating during ball milling, fusion of these low-melting metal films would give definite indication of an increased temperature at the contact points. The results of such an experiment are listed in Table 31. From these data it is apparent that the temperatures involved in compression of common metal powders are very low indeed. They are insufficient to cause even local fusion, since evidence of melting of the Wood's metal phase could be observed only for abrasive-containing mixtures pressed at high pressures and high speeds. Indeed, it can safely be assumed that the temperatures do not even become sufficiently high to cause appreciable particle surface oxidation during compression of the more common metal powders. Since measurements of the over-all temperatures and the microscopic picture of the structure of low-melting metal films or particles are in such close agreement, it appears improbable that any marked difference exists between the temperature at the particle surface contacts and that of the particle interiors at any time during compression. The experimental results point toward quick dissipation of heat throughout the compact with no apparent high-temperature peaks at contact points, thus disproving any theory of local fusion at these points.

MECHANICAL INTERLOCKING

The theory which explains particle bonds by mechanical interlocking of particles with specially suitable contours is the only concept which cannot be traced to atomic forces and to transgression of atoms of the top surface layers beyond particle boundaries.

Several facts must be considered when discussing the possibilities of interlocking between particles. The compressibility of acicular particles

TABLE 31

Increase in Temperature from Friction during Compression
of Metal Powders

Material	Approx. pressure, tsi	Press speed, pressings per min.	Surface temp. °C.	Surface temp. °F.	Microstructure of compact[a]
Electrolytic iron powder (−100 mesh) plus ½% stearic acid as lubricant	10	1 5 50	21 21 22	69.8 69.8 71.6	Very loosely agglomerated iron particles; few contact points discernible; particle structure consists of plate-like crystallites, originating from electrolysis.
	20	1 5 50	21 22 23	69.8 71.6 73.4	Particles are more densely agglomerated; number of contact points appears to have increased slightly, but no distortion of particles or crystallite plates detectable.
	50	1 5 50	22 23 33	71.6 73.4 91.4	Dense agglomerate of particles, pores, and interstices greatly diminished; original intraparticle structure not changed, though some evidence of distortion at contact points.
Electrolytic iron powder (−100 mesh) plus 5% quartz powder (−100 mesh); no lubricant	15	1 5 50	32 37 44	89.6 98.6 111.2	Very loosely agglomerated iron particles; few contact points; free silica inclusions in abundance; no inclusions wedged between metal particles.
	25	1 5 50	44 49 60	111.2 120.2 140.0	More densely agglomerated particles; contact points increased; no deformation of metal particles at contact points, but some silica particles wedged between particles.
	50	1 5 50	77 88 97	170.6 190.4 206.6	Dense agglomerate of particles; silica inclusions often wedged between metal particles or lying freely in pores; some deformation of metal at contact points.

[a] By observation at 500, 1000, and 1500 diameters magnification.

(Table continued)

TABLE 31 (*Concluded*)

Material	Approx. pressure, tsi	Press speed, pressings per min.	Surface temp. °C.	Surface temp. °F.	Microstructure of compact[a]
Electrolytic iron powder (−100 mesh) plus 5% quartz powder (−100 mesh) plus 5% atomized Wood's metal (−100 mesh), tumbled	10	1 5 50	29 34 39	84.2 93.2 102.2	Very loose agglomerate of iron particles; few contact points; silica present in form of free inclusion; low-melting metal particles apparently undistorted.
	20	1 5 50	41 46 50	105.8 114.8 122.0	More densely agglomerated particles; silica mostly free, some wedged between particles at contact points; Wood's metal particles mostly wedged and flattened out, but no evidence of their fusion.
	50	1 5 50	69 79 89	156.2 171.2 192.2	Dense agglomerate of particles; most silica inclusions wedged between metal particles; Wood's metal particles found less frequently, but larger in size and frequently forming coatings of iron particle facings, having obviously passed through liquid state.
Electrolytic iron powder (−100 mesh) plus 5% quartz powder (−100 mesh) plus 5% atomized Wood's metal (−100 mesh), ball milled	15	1 5 50	31 33 40	87.8 91.4 104.0	Very loosely agglomerated iron particles; few contact points; silica present as free inclusions; Wood's metal particles flattened out occasionally as free particles but mostly adhering to iron particle surfaces.
	25	1 5 50	39 47 55	102.2 116.6 131.0	More densely agglomerated particles; silica free or wedged between particles, Wood's metal mostly on surface of particles, grossly deformed at some contact points, but no evidence of fusion.
	50	1 5 50	66 75 87	150.8 167.0 188.6	Dense agglomerate of particles with silica inclusions as before; Wood's metal agglomerated into a few larger areas that adhere to surface of iron particles; evidence that Wood's metal had liquefied.

is definitely affected by the capacity of the material for plastic deformation. Hard electrolytic copper and iron powder particles, although substantially dendritic in shape, show considerable resilience in compaction; in certain cases they are even inferior in relation to compressibility to spherical carbonyl iron. Only thorough annealing treatment of the powders makes the particles sufficiently plastic to allow deformation and increased contact areas—needed for stable bonding of the particles into a coherent mass. Prolonged annealing in a reducing atmosphere also tends to reduce the last traces of oxides, leaving intraparticle cavities which effect an increase in specific surface.

It must be understood that these feathery or spongy particles are not always solid units, but may vary from single particles (with rigidly anchored branches) to agglomerations or clusters of tiny particles which

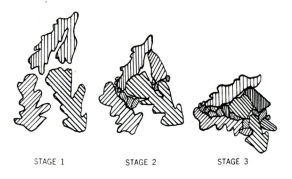

STAGE 1   STAGE 2   STAGE 3

Fig. 68.   Packing scheme of dendritic particles during compaction.

cling loosely to the trunk or main body. It is not easy to study these conditions in a loose powder with the microscope. Any polished section through such a particle represents a cut in only one plane; small particles often found next to the main body and apparently not connected to it, may actually be distant branches, with the trunk in a plane at an angle to the cut face, and covered by the mounting mass at the roots. In spite of these difficulties, such microscopic investigations have been accurate enough to permit general conclusions concerning the contours of these particles. By observing powder structure after compaction to increasing relative densities, it has been found that in a majority of particles these projected branches tend to shear off and shift into larger interstices at the slightest compression. The sketches in Figure 68 illustrate the by-passing and fragmentation of three dendritic particles during compression.

The fact that particles tend to move in the direction of applied pressure (along the path of least resistance) involves relative movements of some particles toward or past others ("slippage"). Depending on shape and location, some particles may move further or faster than others, the driving force being external pressure as well as particle concentration gradients. This slippage is, of course, affected considerably by interparticle friction, which, in turn, is a function of the surfaces and contours of the particles. Consequently, spherical particles will slip more easily than rugged ones, and only the latter are capable of mechanical interlocking (or keying). It is reasonable to assume, however, that while rugged surfaces offer resistance to slippage, projections must be exceptionally strong to prevent shear caused by the driving force. Generally, commercial metal powders do not have particles with such strong projections, most particles having only thin or weak branches projecting outward, as in the case of spongy or dendritic particles. Such projections are readily deformed or sheared off by the very large friction forces which come into play during compression.

Experimental evidence indicates that in some special cases interlocking plays a part in compaction. The superior compacting properties of spongy copper powder as compared to spherical copper particles serves as an example of such cases. However, for the reasons outlined previously, we cannot generalize to the effect that interlocking plays the predominant part in compaction.

EFFECTS OF APPLIED PRESSURE

In the foregoing, we have attempted to analyze the behavior at the boundaries of two or more minute metal particles brought into intimate contact. The various theories dealing with the extraordinary phenomenon of interparticle bonding agree that closest proximity is indispensable for the establishment of vital links. Since loose filling or even comparatively close packing of the (generally) irregular particles does not fulfill this prerequisite satisfactorily, application of pressure becomes a fundamental requirement.

Theoretical Aspects Based on Experimental Evidence

In seeking a more precise answer to the complex problem of the behavior of a metal powder when subjected to external pressure, it is advisable to discuss in some detail the theories advanced by Bal'shin,[14] Unckel,[15] Seelig and Wulff,[16] and Kamm, Steinberg, and Wulff.[17] /

[14] M. Yu Bal'shin, *Vestnik Metalloprom.*, *18*, No. 2, 124 (1938).
[15] H. Unckel, *Arch. Eisenhüttenw.* *18*, 161 (1945).
[16] R. P. Seelig and J. Wulff, *Trans. Am. Inst. Mining Met. Engrs.*, *166*, 492 (1946).
[17] R Kamm, M. Steinberg, and J. Wulff, *Trans. Am. Inst. Mining Met. Engrs.*, *171*, 439 (1947)

A review and summary of the literature have recently been presented by Seelig[17a] during a seminar on the pressing of metal powders. The reader is advised to refer to this summary, and in particular, to study the subsequent discussion which (in part) refers to the above-mentioned theories and contains much basic and highly revealing information. Unfortunately, lack of space and the late date of publication of the seminar proceedings prohibit a detailed discussion of it here. Further reference, however, is made in Volume II, Chapter XXXV, in connection with a summary of the theoretical aspects of the bonding and sintering of metal powders.

WORK OF BAL'SHIN

In compacting a powder, the pressure applied to the punch of the die performs a certain amount of work. This expenditure of energy is, according to Bal'shin, composed of energy expended on: (*1*) overcoming the adhesive forces between the powder particles as they change their relative positions; (*2*) deforming the particles; (*3*) transforming the individual particles into an aggregate of fragments; and (*4*) overcoming elastic and residual stresses in pressing these aggregates. As the density of the compact is increased, the magnitude of the pressure required for further compaction must also increase.

The action of adhesive forces intensifies with an increase in the number and size of contact areas, as well as with a decrease in interparticle interstices. Work-hardening of the metal occurring with advancing compaction counteracts deformation of the particles. Increased pressure may also cause so extensive a fragmentation of some particles that adhesive forces are overcome. The internal stresses in a compact grow with increasing compaction. From the (theoretical) relationship between the decrease in volume of a powder and the necessary increase in pressure (analogous to the elastic deformation of a metal body under load), Bal'shin arrives at the conclusion that if the height and relative volume of the compact decrease in an arithmetical progression, the pressure must increase geometrically. This relationship is expressed by the equation:

$$\log p = LV_r + C$$

where p is the pressure, V_r the relative volume, and L and C are constants depending on the nature of the powder and conditions of compacting. In analogy to the modulus of elasticity, L is designated as modulus of pressing. The straight-line pressing curve, $\log p = f(V_r)$, shown in Figure 69, is characterized by two parameters, two points, or one point and the tangent of the angle of the slope toward the abscissa. This tangent is the modulus of pressing, L, and, as the second parameter, it is possible to use the logarithm of the hypothetical maximum pressure, $\log p_{max}$, corresponding to $V_r = 1$ (*i.e.*, the theoretical volume of the particular metal). Actually,

[17a] R. P. Seelig, *Trans. Am. Inst. Mining Met. Engrs.*, *171*, 506 (1947).

the pressing curves, $\log p = f(V_r)$, deviate somewhat from a straight-line course, as shown by a number of examples in the diagrams of Figure 70. Bal'shin attributes this deviation to changes in the initial properties of the powder during compression.

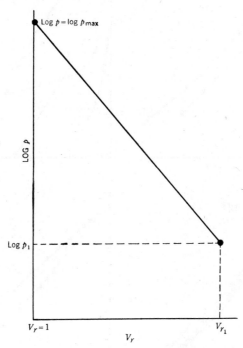

Fig. 69. Theoretical pressing curve (according to Bal'shin[18]);
V_r is the relative volume $\left(\dfrac{\text{volume of compact}}{\text{volume of bulk metal}}\right)$ and $\log p$
the Briggsian logarithm of the specific pressure.

The behavior of a powder during compression—expressed by this pressing curve—is dependent on certain conditions inherent both in the power and in the pressing operation. The chief factors affecting this are degree of dispersion, particle shape, and composition.

The action of adhesive forces between individual particles counteracts the work expended during cold-pressing. It intensifies with increasing dispersion of the powder. In the initial stages of the pressing operation, pressure is absorbed primarily by shifting of particles when bonds between them are broken, with deformation making only a secondary contribution. Therefore, for equal apparent density, the initial (minimum briquetting)

[18] M. Yu Bal'shin, *Vestnik Metalloprom.*, *18*, No. 2, 124 (1938).

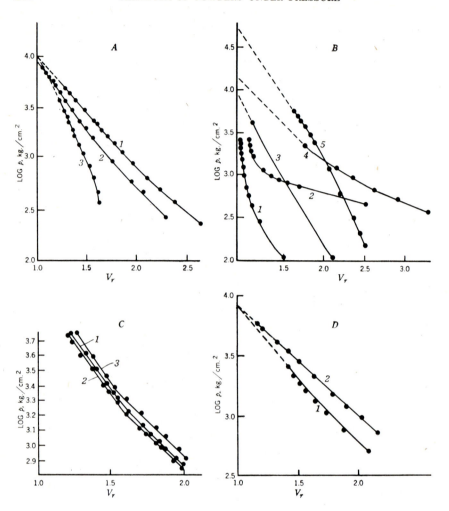

Fig. 70. Actual pressing curves (according to Bal'shin[18]). *A,* for copper powders of different grades; curve *1*: fine electrolytic copper, a.d., 0.97 g./cc.; curve *2*: medium-size electrolytic copper, a.d., 1.50 g./cc.; *3*: coarse granulated copper with spherical particle shape, a.d., 4.50 g./cc. *B,* for metal powders of different apparent density; curve *1*: coarse powdered lead, 3.98 g./cc.; curve *2*: fine electrolytic tin, 1.10 g./cc.; curve *3*: coarse electrolytic iron, 2.70 g./cc.; curve *4*: fine reduced iron 0.57 g./cc.; curve *5*: reduced tungsten, 4.18 g./cc.; *C,* for electrolytic copper powder compacted to different heights; electrolytic copper, a.d., 1.42; curve *1*: $h_k = 0.6$ mm. (0.024 inch); curve *2*: $h_k = 4$ mm. (0.157 inch); curve *3*: $h_k = 8$ mm. (0.315 inch). *D,* for electrolytic copper powder compacted into different cross sections; electrolytic copper, a.d., 1.50 g./cc.; curve *1*: die diameter, 15 mm. (0.590 inch); curve *2*: die diameter, 10 mm. (0.394 inch).

pressures are always greater for very fine powders. Intraparticle forces begin to play an important role only in subsequent stages of compression, and densification of the compact is effected to a large degree by smoothing of crevices and filling of voids of the particles through deformation without breaking the contact bonds. As a rule, the finer particles undergo deformation more readily, since improved flow causes better filling of the voids. The capacity of a powder to undergo plastic deformation is expressed by the coefficient of deformability $(1/L)$, which increases with increased deformation, or by the modulus of pressing (L) which, of course, acts in reverse. This relationship applies to particles as individuals as well as to agglomerates. Hence, the modulus, *i.e.*, the angle of slope of the pressing curve, is greater for coarse powders, as can be seen from the graph in Figure 70*A*. Consequently, in spite of the fact that initial pressures are lower for such a powder, final pressures for very high ultimate densities may be higher than in the case of fine powders (of the same apparent density).

Considering now the shape of the particles, it has been shown that smooth particles lend themselves to packing more readily than rough-faced ones. Therefore, initial compaction pressures for smooth particles are lower—other conditions being equal. Deformability of smooth particles, *e.g.*, spheres, is inferior, however—leading to a higher modulus of pressing.

The influence of composition and nature of the powder on deformability and compressibility have already been discussed. With constant relative density of the compact, hard metals require higher pressures (both initial and final) than soft metals having the same particle structure. By the same token, the harder the metal, the higher the value for the modulus of pressure. The effect of the basic ductility and hardness of a metal on the pressing curves can be seen in Figure 70*B* (reproducing Bal'shin's results for various metal powders).

Compressibility may also be affected by other factors, *e.g.*, large structural subdivision of the particles, local cold-working effects, or surface tension effects caused by the enveloping gas phase. Although little is known about the extent of these specific effects, it may generally be assumed that they all tend to lower deformability, thus increasing the numerical value of the modulus of pressing.

In addition to the numerical value of the pressure and the method and directions for pressing which will be discussed later, there are a number of factors inherent in the pressing operation that influence the behavior of a powder during compression. They include: (*1*) height of the compact,

(*2*) cross section of the compact, (*3*) frictional conditions at the die walls, and (*4*) speed of pressing.

In discussing the height or thickness of a compact from a purely theoretical standpoint, Bal'shin[18] finds a maximum density for a predetermined height of compact. With constant pressing speed, the relative rate of deformation (and work hardening) of the compact is known to *decrease* with increased height of the desired object which, in some cases, lowers the modulus of pressing and the pressure of compaction; on the other hand, pressure losses in the body also increase with height which, in turn, *increases* the modulus of pressing and the pressure necessary for the desired compaction. The action of these factors working in opposite directions causes a maximum density with a predetermined height of the compact, usually in the order of 0.04 to 0.08 inch (1 to 2 mm.). If this optimum height is increased or decreased, the density of the compact decreases, as may be seen from the diagram of Figure 70*C*, showing press-

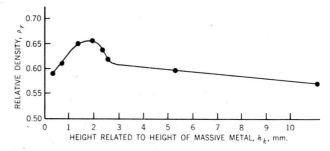

Fig. 71. Relative density of a tungsten powder compact as a function of height for constant pressure (according to Bal'shin[18]); $\rho_r = \dfrac{\text{density of pressed metal}}{\text{density of massive metal}}$; h_k = height related to height of massive metal; p is constant.

ing curves for electrolytic copper compacts of different heights. In Figure 71 relative density is shown as a function of compact height related to the height of the solid metal for tungsten powder compacted at constant pressure. Elastic overworking contributes to the decrease in density for thinner compacts. Thus, for instance, a 0.040-inch (1 mm.) thick compact consisting of particles of 40 microns diameter has only 25 layers of particles in a vertical plane. On the other hand, with greater height, planes of equal pressure become curved instead of straight (as for small heights). Curvature of the plane tends to impede spreading of elastic overworking. Thus a practical lower limit in the height of a compact appears to be approximately $^{1}/_{32}$ inch.

The effects of the cross-sectional area of a compact on the course of the pressing operation are twofold. If the diameter of a compact is increased while the specific pressure is held constant, the effect of friction at the wall of the die decreases. As this decrease becomes greater in proportion to the increase in diameter, the necessary specific pressure will decrease with increasing cross sections. On the other hand, an increase in diameter of the compact exerts an opposite influence on the quantity of particles being deformed per unit of die wall surface and of time. Inasmuch as this quantity increases in proportion to the square of the diameter, while the wall surface increases linearly, the speed of deformation at the wall increases in proportion to the diameter, leading, in some cases, to an increase in the modulus of pressing and a decrease in the density of the compact.[18] In such cases, a reduction in pressing speed is necessary with increased compact diameter. Figure 70D shows Bal'shin's pressing curves for two sizes of compact pressed from the same kind of powder. In practice, the size of a compact is limited both at the lower and upper ends. While the specific effect of friction at the die walls imposes a minimum diameter size of about ⅛ inch, the maximum is usually set by the capacity of the press. The largest cross-sectional area that can (at present) be pressed satisfactorily is about 50 square inches. (An iron or bronze compact of this size, which is only 80% dense, requires at least a 2000-ton press.)

Frictional conditions at the die wall are governed by the finish and by possible lubrication. Roughness of the walls entails an increase in compacting pressure—especially for small compacts. Die wall friction and die wear are, however, reduced appreciably by the addition of lubricants, either mixed with the powder or applied directly to the walls (p. 287). According to Bal'shin, effectiveness of lubricants is limited, and the course of the pressing curves is only slightly influenced by such additions. These poor results can be accounted for by the fact that a liquid lubricant is forced out at high pressures. A decrease in the interstices due to their being filled with lubricant has no bearing on the density of the compact and little bearing on adhesive forces. Although, in certain cases, small amounts (e.g., 1% by weight) of a lubricant reduce friction at the die walls to some degree, most lubricants promote fluidity of the powder only when they are added in considerably larger proportions. Unless pressure sintering follows, such large quantities of volatile foreign matter result in excessive porosity after sintering. Therefore, if the lubricant is added for any other purpose than to promote increased porosity, its choice generally involves balancing the advantages of reduced friction and increased uniformity in particle distribution against the disadvantage of ultimately increased porosity.

Bal'shin's theory has been discussed in detail because it represents the first effort to interpret the phenomena taking place during compaction. The assumptions upon which his theoretical deductions are based, particularly with regard to the analogy between the modulus of elasticity and the modulus of pressing, are, however, questionable, and appear to overlook the fact that deformation taking place during compaction is essentially plastic. It is thus not surprising that the theoretical curves deviate from the experimental ones. Nevertheless, Bal'shin's experimental work is superb and solid, and has been corroborated, in its essential parts, by later investigations.

WORK OF UNCKEL

The first attempts to gain further insight into the mechanism of compaction of metal powders were made by Unckel,[19] who investigated the pressure distribution in solid and cored cylindrical compacts on the basis of the distribution of density and hardness throughout the cross section. This work deserves particular credit, since it takes into account such important factors as die wall friction and displacement of powder particles during compaction, and also includes a theoretical treatment of the pressing process.

For the experiments dealing with density distribution measurements, Unckel used electrolytic copper powders and Swedish sponge iron powders, which he compacted to cylinders of $7\!/\!8$-inch and $2\,7\!/\!8$-inch diameters, respectively. The smaller cylinders were compressed at about 35 tsi, the larger at about 43 tsi. Determination of density distribution was then made by sectioning the small cylinders parallel to the pressing axis, and the large cylinders by first cutting a $3\!/\!8$-inch thick slice through the center of the compacts (parallel to the pressing axis), and then subdividing the slice into small $3\!/\!8$-inch cubes. The results of these tests are reproduced in Figure 72, sections A to C indicating the density and hardness distribution for a mixed grade of copper, a fine grade of copper, and a fine grade of sponge iron, respectively. In Figure 72D the density and hardness distribution of cylindrical compacts from the mixed grade of copper are compared with those of compacts having a 1 : 50 taper. Figure 72E shows the distribution of hardness and density at different heights along the radius, as well as the change in hardness along the surface of the center slice.

In his determination of friction forces acting on the die wall, Unckel used the smaller die size for his work. The die was placed on top of a soft iron plate, which in turn was set on three half-sunk, hardened steel balls. The friction forces acting on the die wall produced an impression

[19] H. Unckel, *Arch. Eisenhüttenw., 18,* 161 (1945).

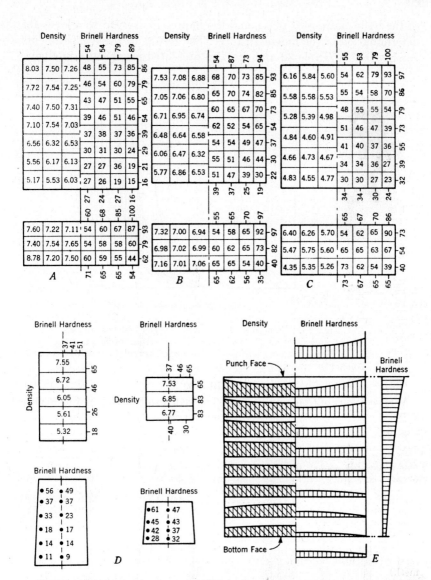

Fig. 72. Distribution of density and Brinell hardness in cylindrical compacts (according to Unckel[19]: *A*, copper powder of mixed grade, 3- and 1-kg. weights of compact; 43-tsi pressure, center slice; *B*, fine copper powder, 3- and 1-kg. weights of compact, 43-tsi pressure, center slice; *C*, iron powder, 3- and 1-kg. weights of compact, 43-tsi pressure, center slice; *D*, copper powder of mixed grade, ⅞-in. compacts of cylindrical and conical shape; and *E*, radial distribution of hardness and density at different heights, and hardness changes along the surface of a longitudinal section.

of the balls on the soft plate, and the impression served as a basis for the determination of the indenting force. The results of these experiments are summarized in Table 32 and are also presented graphically in Figure 73.

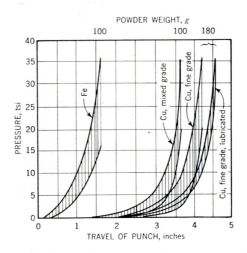

Fig. 73. Compaction pressure (upper curves of each band) and fraction thereof exerted at bottom of ⅞-in. compacts (lower curves) as function of punch travel (according to Unckel[19]).

TABLE 32

Values of Part of Compaction Force Translated to Die Wall and of Required Ejection Pressure (Unckel[19])

⅞-in. die	Compaction force of punch, kg.					Ejection force, kg.
	2000	4000	8000	12000	18000	
Copper, mixed grade,[a] 100-g. specimen.....	1600	2700	4300	—	6580	5800
Copper, fine grade,[b] 120-g. specimen.......	1550	3170	5600	7600	8500	7350
Copper, fine grade, 120-g. specimen, lubricated	900	1550	2150	2250	2370	2300
Iron, 100-g. specimen.....................	1400	2480	—	6920	9670	7760
Iron, 50-g. specimen.....................	—	—	—	2310	2970	2700
Iron with 3% graphite, 100-g. specimen.....	—	2610	—	6140	7300	4675

[a] Mixed grade of copper contained: 25%, 0.15–0.3 mm. particles (−50 +100 mesh); 37%, 0.04–0.15 mm. (−100 +325 mesh); 38%, smaller than 0.04 mm. (−325 mesh).

[b] Fine grade of copper contained: 100%, smaller than 0.04 mm. particles (−325 mesh).

The upper curves for each type of metal used represent the pressure acting on the punch; the lower curves represent the indentation pressure on the bottom plate; the distance between two complementary curves (measured parallel to the ordinate) constitutes the loss in pressure through

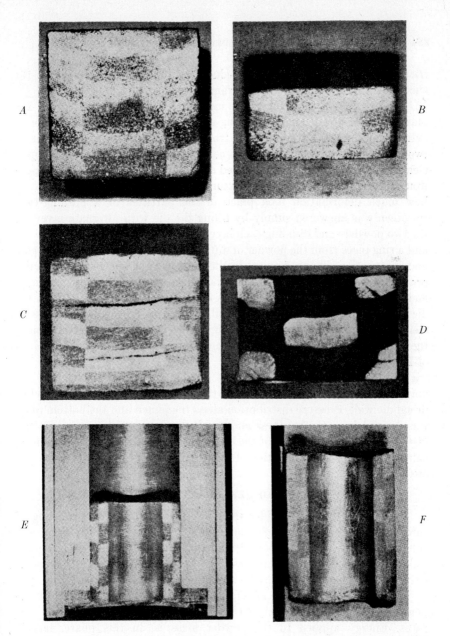

Fig. 74. Deformation of specimens composed of ring-shaped layers (according to Unckel[19]): A, cylinder pressed dry from copper powder of mixed grade—ratio of length to diameter about 1; B, same—ratio of length to diameter about 0.3; C, cylinder pressed while lubricated from copper powder of mixed grade—ratio of length to diameter about 1; D, cylinder pressed dry from iron powder—ratio of length to diameter about 0.3; E, cored bushing pressed from copper powder of mixed grade; and F, cored bushing pressed from iron powder.

friction. This loss is considerably reduced by lubrication, as shown in the curves at the right side of the diagram.

Unckel used a very ingenious method for the investigation of the plastic flow of powder masses during compaction. Solid, as well as cored, cylindrical compacts were formed from alternate layers and sections of two types of powder that differed in color. In the experiments with copper, clean and oxidized powders were used alternately; in the case of iron, pure iron and a 75–25, iron–copper mixture were employed. The question of how much deformation took place in the particles of a bushing-type specimen was answered simply by filling the die with alternate layers of the two powders and dividing each layer into a core piece from one powder and a ring piece from the powder of different color. In adjacent layers, core and ring sections changed color. In the case of cored specimens, a similar procedure was followed except that the layers were subdivided into inner and outer rings. This subdivision was easily obtained by placing into the die cavity a thin cylinder of tin foil, which was raised (in successive steps) as the alternate layers were poured into the die. The deformation of the multiple-layered solid and cored compacts is shown in the photographs of Figure 74.

The diagram of Figure 75 shows graphically the distribution of the normal and shear stresses exerted on the punch, bottom plate, and cylindrical die wall. Pressure distribution along the punch and the bottom plate was derived from the hardness and density measurements, while pressure distribution along the walls of the die cavity was determined by extensive calculations. For the calculation of shear stresses, Unckel used a coefficient of friction of 0.2.

WORK OF SEELIG AND WULFF

Seelig and Wulff[20] consider the compaction process as consisting of three steps: (1) packing, (2) elastic and plastic deformation, and (3) cold-working (with or without fragmentation). The three steps do not follow in sequence, but usually overlap in practice; under many conditions at least one of them may be absent.

Packing (first step) is affected by particle-size distribution of the powder and by the chemical and mechanical nature of the particle surfaces. Energy applied to the powder mass is, in this phase, largely absorbed by interparticle friction. The second step calls for a certain amount of plasticity in the powder, and is therefore practically absent in very hard powders, such as tungsten carbide. In this phase, the effect of die wall friction is predominant and absorbs a considerable part of the

[20] R. P. Seelig and J. Wulff, *Trans. Am. Inst. Mining Met. Engrs.*, *166*, 492 (1946).

applied energy. The third step calls for a powder which is susceptible to cold-working or fragmentation, and can, in practice, be neglected for soft metal powders, such as lead and tin. Cold-working produces high residual stresses which may lead to dimensional changes and even failures after pressing, but may also be beneficial for the subsequent sintering operation.

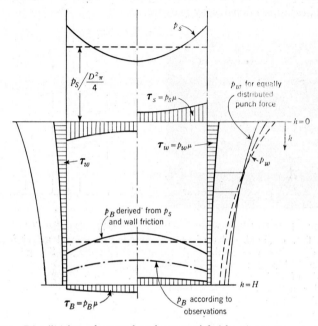

Fig. 75. Distribution of normal and tangential (shear) stresses at punch, bottom plane, and cylinder wall (according to Unckel[19]); where p_S = punch pressure; p_W = wall pressure; p_B = bottom pressure; τ_S = shear stresses at punch; τ_W = sheer stresses at wall; τ_B = shear stresses at bottom; h = distance between two cross sections; and H = maximum distance.

The first two steps are illustrated by the photoelastic picture of Fig. 76, obtained by Vose,[20a] who pressed small celluloid disks of three different diameters in a thin die having glass walls, and photographed the disks under pressure in polarized light. The picture shows that many voids are left between the particles, and also that some particles are not even elastically stressed. It is obvious that, in the case of undeformable powders, the addition of a certain percentage of extremely fine powders would permit an increased powder feed and thus a more effective utilization of the die space, as well as an increase in the areas of contact between par-

[20a] R. W. Vose, Massachusetts Institute of Technology.

ticles. Although with undeformable powder particles a further increase in packing can be achieved only by fragmentation, powders susceptible to plastic deformation can be further densified by using higher pressures, thus increasing the deformation. Without considering the effects on the sintering performance, it can be stated that for the softer powders (*e.g.*, iron and copper), addition of extremely fine powders is not imperative in order to obtain high density by pressing.

Fig. 76. Polarized light photograph of small, flat, celluloid disks of three different diameters while subjected to pressure in a thin die; front and back walls made of glass, and bottom, side walls, and punch made of wood (technique of Vose[204]; according to Seelig and Wulff[20]).

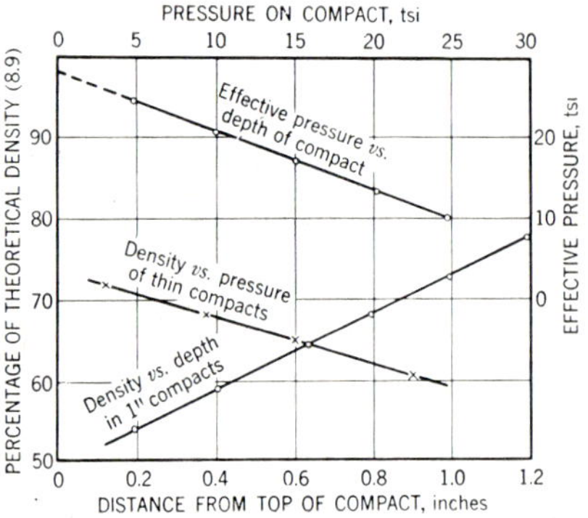

Fig. 77. Density and effective pressure in compressed nickel powder compacts (according to Seelig and Wulff[20]).

Other experimental results of Seelig and Wulff represent a duplication of experiments of Bal'shin, but are of special interest since Bal'shin[21] reported his results without giving details about his experimental procedures. Figure 77, illustrating the second step, was obtained with annealed electrolytic nickel powder. The density *vs.* depth-of-compact curve was obtained by cutting the pressed compact into eight sections of equal thickness and determining the average density of each section. The effective pressure *vs.* depth-of-compact curve was calculated from the density *vs.* pressure curve obtained by pressing thin compacts, approximately $1/8$-inch thick, under various pressures. The results indicate that when nickel powder is compacted at a pressure of 30 tsi, the effective pressure near the bottom of a 1-inch compact amounts to approximately 10 tsi. For iron, pressed at 15–40 tsi, a horizontal pressure gradient—to be attributed to

[21] M. Yu Bal'shin, *Vestnik Metalloprom.*, *18*, No. 2, 124 (1938).

wall friction—was established by carefully cutting sections from the center and sides of compact slides.

Seelig and Wulff demonstrate the practical importance of wall friction by comparing the effect of lubricants added to the powder with that of lubricants applied to the die wall. The results obtained with nickel powder (Table 33) indicate that, under the conditions of these experiments,

TABLE 33

Effect of Lubricants on Green Density of Nickel Powder Compacts
(Seelig and Wulff[20])

Material and treatment	Absolute density, %
Nickel powder	66
Nickel powder +0.5% graphite............................	76
Nickel powder +4% graphite..............................	78
Nickel powder +4.5% silica flour..........................	64
Nickel powder +0.5% stearic acid.........................	75
Nickel powder; die walls lubricated with stearic acid..........	76
Nickel powder; die walls lubricated with colloidal graphite....	74
Nickel powder; die walls lubricated with stearic acid[b].........	76

[a] Pressed at 30 tsi.   [b] This compact $^1/_8$ in. high; all others were 1 in.

effects of the die wall are preponderant and outweigh those of interparticle friction. Here, the authors do not agree with Bal'shin's deductions.

WORK OF KAMM, STEINBERG, AND WULFF

The pressure and density gradients evidenced by the measurements of Bal'shin and of Seelig and Wulff can be directly demonstrated in experiments not involving density measurements. Bal'shin[21] used the method mentioned by Rakovski,[22] placing graphite powder layers between three copper powder sections of a compact; Seelig and Wulff[23] repeated the experiment by placing copper powder layers between iron powder sections. The deformation of these layers during compaction clearly indicated the effect of die wall friction. In a brilliant piece of experimental and theoretical work, Kamm, Steinberg, and Wulff[24,24a] have recently succeeded in demonstrating vertical as well as horizontal pressure gradients.

Kamm, Steinberg, and Wulff sought to obtain quantitative information about the density distribution within cold-pressed powder compacts by inserting a deformable lead grid into the powder mass. After being

[22] V. S. Rakovski, *Fundamental Considerations in the Production of Hard Alloys.* Onti, Moscow-Leningrad, 1935.

[23] R. P. Seelig and J. Wulff, *Trans. Am. Inst. Mining Met. Engrs., 166,* 492 (1946).

[24] R. Kamm, M. Steinberg, and J. Wulff, *Trans. Am. Inst. Mining Met. Engrs., 171.* 439 (1947).

[24a] R. Kamm, M. Steinberg, and J. Wulff, *Metals Technol., 15,* No. 8, T.P. 2487 (1948).

pressed in cylindrical dies, the ejected compact was radiographed, and the pattern of the deformed grid was used for exact measurements of point-to-point deformation as a means of determining the strain distribution and stress trajectories (data from which the coefficient of friction at the die wall could be computed). Lead was used as the grid material because it is softer than the metal powders to be compacted, and could therefore follow the flow of the powder during compaction. In addition, lead has the advantage of exhibiting a higher x-ray density, which permitted radio-

Fig. 78. Radiograph of typical lead grid used in investigation of plastic deformation phenomena in metal powder compacts (according to Kamm, Steinberg, and Wulff[24]).

graphic recording of the deformed grid. In most cases 0.1-inch thick sheets of pure lead were first punched to form a rectangular network with rectangular holes, but in some special cases circular holes were punched into the sheet to facilitate the determination of stress and strain. Careful etching with nitric acid served to remove all asperities from the grids. A representative lead grid used in these experiments is shown in Figure 78.

The grids were supsended vertically in dies 1.125, 0.560, and 0.259 inches in diameter, and the powder was poured around them to the height of the grid. By placing them across the diameter of the die, parallel to the axial direction of the cylindrical die, advantage could be taken of rotational symmetry, permitting simpler mathematical treatment. To maintain accurate grid deflection measurements, it was found essential that the fill density of the powder before compression be uniform throughout the poured powder mass. Care was also employed in locating the grid so that it would not touch the die wall at any point. During the process of powder compaction (in one direction) the lead grid moves downward, following closely the downward movement of adjacent powder particles. Its position after compaction is located by radiography of the compact or, in the case of small thin-walled dies, of the entire die assembly.

Radiographs of a series of 0.560-inch diameter compacts pressed with lead grids are reproduced in Figure 79. All represent carbonyl iron powder compacts with 80% of −325-mesh size. The pressure was varied from 16 to 64 tsi in steps of 16 tsi, and the over-all densities after pressing were determined as 65.6, 76.4, 79.6 and 91.4% of theoretical density. The ratio of height to diameter of the initial powder fill was 1 : 1 for the series in Figure 79 *A* to *D*. The other series, Figure 79 *E* to *H*, represents corresponding radiographs of compacts pressed at the same pressures, but with a ratio of powder fill height to diameter of 1.6 : 1. From the radiographs, it becomes apparent that die wall friction has a retarding influence on the flow of metal powder during compaction. Hence, powder flows faster at the center of the compact than at the sides. Evidently the powder moves only in a vertical direction with no lateral movement, since the vertical streamlines would otherwise not remain equidistant throughout the compact. The density at every point can be determined by measurement of the axial strains throughout the compact, and the distribution of the density can be presented graphically. The graphical diagrams of Figure 80 represent the density distribution found in the specimens shown in Figure 79 *A–D*. The lines numbered 1 to 5 give the density distribution across the centers of the various sets of squares on the radiographs, with density distribution being measured along the curved lines connecting the centers of each horizontal line of squares.

From these illustrations it becomes apparent that a density maximum exists at the top edges of all compacts; that the lowest density in all compacts exists at the bottom edges; that density near the cylindrical surfaces of the compact decreases generally with height (from top to bottom); that the density at the bottom in the center of the compact is greater than the density near the top at the center; and that, with increasing compacting pressure, the average density increases, but the variation in density throughout the compact also increases. Thus, for compacts pressed from one side only, the densest part is at the top outer circumference, and the least dense at the bottom circumference near the stationary punch. For compacts of heights shown in Figure 79 density at the axis is greater

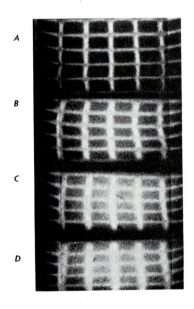
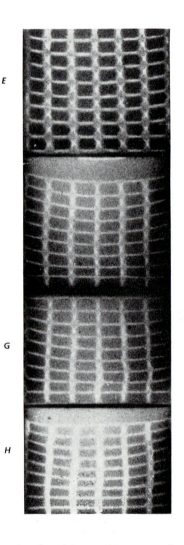

A, ratio of powder-fill height to diameter, 1:1; pressure, 16 tsi.
B, ratio of powder-fill height to diameter, 1:1 pressure 32 tsi.
C, ratio of powder-fill height to diameter, 1:1; pressure 48 tsi.
D, ratio of powder-fill height to diameter, 1:1; pressure, 64 tsi.
E, ratio of powder-fill height to diameter, 1.6:1; pressure, 16 tsi.
F, ratio of powder-fill height to diameter, 1.6:1; pressure, 32 tsi.
G, ratio of powder-fill height to diameter, 1.6:1; pressure, 48 tsi.
H, ratio of powder-fill height to diameter, 1.6:1; pressure, 64 tsi.

Fig. 79. Representative radiographs of compressed carbonyl iron compacts containing lead grids (according to Kamm, Steinberg, and Wulff[24]).

at the bottom than at the top, a fact that must be attributed to the effects of die wall friction.

The effect of compact height on density distribution is apparent from Figures 81 and 82, representing the radiograph of a compact for a 2 : 1

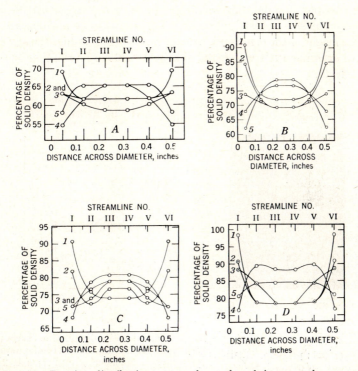

Fig. 80. Density distribution curves for carbonyl iron powder compacts of powder-fill height to diameter ratio 1:1, radiographs Figure 79, A to D (according to Kamm, Steinberg, and Wulff[24]); a.d., 2.64 g./cc.; height before pressing, 0.550 inch; A, 16-tsi pressure; B, 32-tsi pressure; C, 48-tsi pressure; and D, 64-tsi pressure.

ratio of powder-fill height to diameter, and the corresponding density distribution graph. If the height is increased sufficiently, density variations become so great that the density at the center of the base is no longer greater than that at the center of the top face. In experiments with increased compact diameters, it was observed that more uniform density distribution was secured for the same ratio of height to diameter. At the same time, an increase in height caused a less marked increase in density variations than was noticed for smaller diameter compacts. As the ratio of diameter to height of compact was increased, the die wall exerted correspondingly less friction.

Kamm, Steinberg, and Wulff also investigated the effect of lubrication, and showed by means of additional radiographs and density distribution plots, that more uniform density distribution throughout the compact is possible (for the compacting pressures used) only if the die walls are highly finished and well lubricated. This seems to be particularly true when the height of the compact is greater than about one-half the diameter. In these experiments it was found that interparticle lubrication has only minor effects.

Fig. 81. Radiograph of compressed carbonyl iron compact containing lead grid; ratio of powder-fill height to diameter, 2:1; pressure, 43 tsi (according to Kamm, Steinberg, and Wulff[24]).

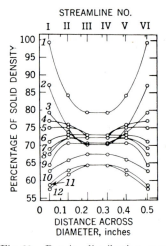

Fig. 82. Density distribution curves for carbonyl iron powder compacts of powder-fill height to diameter ratio, 2:1; a.d., 2.64 g./cc.; height before pressing, 1.10 inches.

The work includes a mathematical determination of the strains and stresses in the cold-pressed powder compact. The radiographs enabled Kamm, Steinberg, and Wulff to measure not only the axial strains necessary for the determination of density variations, but also additional shear displacements near the die walls. Thus, the determination of the principal deformations which take place during powder compaction includes compression and shear deformations. From the directions of the principal strains at every point in the compact, it was possible to draw the stress trajectories, since the direction of principal stress must be in the direction of principal strain. The distribution of principal strains in the two compacts shown in Figures 79A and D, and in Figures 80A and D, respectively, are graphically presented in Figures 83*A* and *B*. It is evident from these diagrams that maximum compressive strain occurs in the streamlines at the outer edges of the lead grid near the top of the compact, whereas minimum compressive strain occurs in the same streamlines near the bottom of the compact. Fewer variations are apparent in the other stream-lines, and the principal tensile strains seem small in magnitude and fairly uniform throughout the compacts. In Figure 84, the directions of the principal strains and therefore of the principal stresses are shown, and from these data the stress trajectories shown in Figure 85 were produced. In these illustrations the trajectories are represented by solid lines, while the

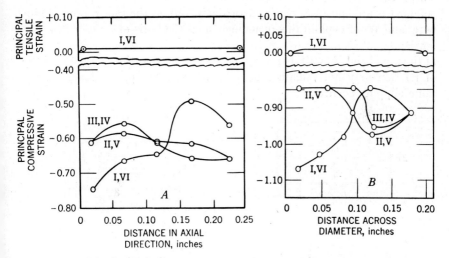

Fig. 83.   Distribution of principal strains in carbonyl iron powder compacts of powder-fill height to diameter ratio 1:1 (according to Kamm, Steinberg, and Wulff[24]): *A*, pressure, 16 tsi (radiograph Figure 79*A*); and *B*, pressure, 64 tsi (radiograph Figure 79*D*).

positions of the lead grids are indicated by the dotted lines. It becomes apparent that the direction of stress is not the same at the die wall as at the center of the compact, and that the change in direction is greatest near the top edges of the compact. The form of the stress trajectories and the strain curves suggests a conical type of fracture during compression.

Finally, Kamm and his co-workers were able to utilize the stress trajectories for the determination of the coefficient of friction between the die wall and the compact. For the two compacts selected from Figure 79, the coefficients of friction at the points of intersection of the horizontal stress trajectories with the edge of the compact were found to vary as follows:

Carbonyl iron compact pressed to 65.6% density at 16 tsi		Carbonyl iron compact pressed to 91.4% density at 64 tsi	
Horizontal stress trajectory	Coefficient of friction	Horizontal stress trajectory	Coefficient of friction
1	0.625	1	0.37
2	0.47	2	0.53
3	0.36	3	0.63
4	0.17	4	0.53
5	0.07	5	0.24
6	0	6	0

These values show that the coefficient of friction at the surface of the compact is not constant, being particularly high at the top. Kamm, Steinberg, and Wulff explain this phenomenon on the basis of a possible work-hardening effect at the top of the powder compact, or a rougher surface at the top of the die wall. It is possible, moreover, that the actual coefficient of friction may not remain constant with greater applied loads on the same slipping surface.

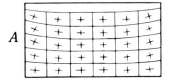

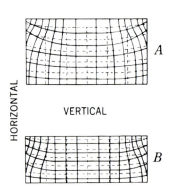

Fig. 84. Directions of Principal stresses in carbonyl iron powder compacts (according to Kamm, Steinberg, and Wulff[24]): A, pressure, 16 tsi (radiograph Figure 79A); and B, pressure, 64 tsi (radiograph Figure 79D).

Fig. 85. Stress trajectories of the principal stresses shown in Figure 84 (according to Kamm, Steinberg, Wulff[24]): A, pressure, 16 tsi (radiograph Figure 79A); B, pressure, 64 tsi (radiograph Figure 79D.)

OTHER WORK

The decrease in density with increased height of compact is a well-known phenomenon, and will be treated further in the discussion of density distribution in Chapter IX. At this point it may suffice to point out that, as a rule, maximum density is approached progressively with greater compact heights as pressing speeds and powder coarseness increase. While there are practical examples of compacts as high as 12 inches, the general rule is to mold considerably smaller compacts. The optimum height is frequently chosen in relation to the cross section according to the equation:

$$h_0 = \sqrt{A}$$

where h_0 is the optimum height of the compact and A is the cross-sectional area. In the case of a cylindrical compact, the optimum thickness can be expressed as a function of the diameter:

$$h_0 = (d/2)\sqrt{\pi} = 0.886d$$

In this simple relationship, however, no consideration is given to the shape of the compact and to those surfaces that are subjected to die wall friction. In a simple body (e.g., a cylinder) the friction effects at the surface of the compact which faces the die wall are less marked in relation to the interparticle friction effects of the compact's interior than in the case of a more complicated shape (e.g., a prism or a gear). Accordingly, the most favorable height of a compact may also be expressed in relation to the ratio of volume to surface:

$$h = F_p (V/S)$$

where h is the height of the compact, V is the volume of the compact, S is the surface causing friction with the die wall (i.e., the curved surface in the case of a cylinder), and F_p is a proportionality factor which, for simplicity's sake, can be set equal to unity. Since both the volume, V, and the surface, S, are functions of the height, the expression can then be re-written as:

$$h = (Ah/Ph) = (A/P)$$

where A is the cross-sectional area and P is the perimeter. In the case of a cylinder, the height then becomes:

$$h = (\pi d^2/4\pi d) = (d/4)$$

and the diameter becomes:

$$d = 4h = 4(A/P)$$

This expression can now be introduced into the original relationship:

$$h_0 = 0.886d$$

to give a modified formula expressing the optimum height as a function of the ratio of cross-sectional area to perimeter:

$$h_0 = 0.886(4A/P) = 3.545(A/P)$$

This formula considers the effect of shape by decreasing the permissible optimum height as compensation for increased perimeter, as may be seen from Table 34, in which for progressively more complicated shapes

TABLE 34

Relationship of a Compact's Optimum Height to Cross Section and to Ratio of Cross Section to Perimeter for Different Prismatic Shapes

Cross section	Area A	Simple formula $h_0 = \sqrt{A}$	Modified formula $h_0 = 3.545\,A/C$	
		Optimum height (h_0)	Ratio of area over perimeter A/P	Optimum height (h_0)
Circle	$\pi d^2/4$	$0.886d$	$\pi d^2/4\pi d$	$0.886d$
Hexagon	$^3/_2\,a\sqrt{3}$	$1.61a$	$3a^2\sqrt{3}/12a$	$1.536a$
Square	a^2	a	$a^2/4a$	$0.886a$
Rectangle ($b = 2a$).	$2a^2$	$1.414a$	$2a^2/6a$	$1.18a$
Rectangle ($b = 4a$).	$4a^2$	$2a$	$4a^2/10a$	$1.42a$
Equilateral triangle..	$a^2\sqrt{3}/4$	$0.658a$	$a^2\sqrt{3}/12a$	$0.512a$

the optimum permissible thickness is expressed by the simple and modified formula.

In excellent agreement with this analysis of the effect of compact shape on density are the recently published findings by Squire.[25] On the basis of an impressive series of experiments involving 32 different sizes of iron powder compacts and 8 different dies, Squire found with remarkable consistency that the density decreases directly with an increase of the ratio of wall area to pressing area, W/P, of the compact. Inversely, it was established that the greater the pressing area with relation to the thickness of the compact the higher is the density. Accordingly, either a reduction in height for a given diameter or an increase in diameter for a given height will effect an increase in density. The ratio of wall area to the pressing area of the compact is introduced by Squire in order to arrive at a relationship that would permit the application of this axiom to any desired shape.

[25] A. Squire, *Trans. Am. Inst. Mining Met. Engrs.*, **171**, 485 (1947); see also *Watertown Arsenal Lab. Rep.*, WAL 671/23 (April 18, 1945) and *Office of Technical Services, Rept. PB. 49079* (April 18, 1945).

In general, the compressibility (*i.e.*, the relative density on the modulus of pressing for a given pressure) is only slightly affected by the speed of pressing. On the basis of Bal'shin's work, we may expect that, theoretically, the modulus of pressing is bound to increase somewhat with increasing pressing speed (at constant pressures), whereas the density should show a slight decrease. Since in practice, however, the process of rearrangement of the particles and of their plastic deformation requires time, the particles will be unable to slip into the most favorable positions for close packing if pressure is applied at greater speeds. Similary, lack of time may favor local absorption of energy in the top layers (nearest the face of the pressure-exerting punch) instead of transmission of the energy into successive powder layers (toward the interior of the compact). There is also the possibility that higher speeds may favor local deformation, especially in multiphase materials, in which more plastic or mobile metal particles, *i.e.*, of a lower melting constituent, are pressed outward, resulting sometimes in nonuniform distribution of the different structural and chemical components. There remain certain secondary effects connected with the speed of pressing, such as removal of entrapped air, etc.; their practical significance will be discussed in Chapter XI, dealing with presses. One method reported to be successful in overcoming some of the friction at the die walls is to accelerate the speed of pressing to such an extent that compression takes place by impact blows (dynamic loading). The process, however, may have to be repeated several times.[26] Apparently the instantaneous realignment of the surface particles during compaction does not permit preferred orientation to take place along the die walls (connected with cold-welding and galling effects).

Practical Effects of Pressure

It now seems advisable to analyze the implications of the applied pressure from a more practical point of view; these effects may be divided into microeffects and macroeffects.

MICROEFFECTS

The microeffects of pressing may be of a dynamic or static nature. The dynamic effects may be resolved into the friction between particles undergoing relative motion, and the change in position of the particles. The static effect of pressing is the gross increase in surface-to-surface contact areas resulting in an increase in adhesive forces. During compression a large part of the pressure is absorbed during deformation or fragmenta-

[26] C. G. Goetzel, discussion by R. P. Seelig and J. Wulff, *Trans. Am. Inst. Mining Met. Engrs., 166,* 505 (1946)

tion of the particles—overcoming elastic or residual stress in the powder agglomerations.

There are two reasons why a heap of loosely packed powder remains unchanged even after a prolonged period of time—namely: (*1*) the total interparticle contact area is small in comparison with the total particle surface area so that atomic forces are rather ineffective, and (*2*) gaseous surface films tend to interfere with the establishment of atomic links if the powder is not packed under vacuum.

As a given mass of particles is contracted into a progressively smaller space by the compressive action of one or more plungers in the die, a considerable degree of attrition must always precede the establishment of these contacts. A grinding action between adjacent particle surfaces will, in many instances, bring about a substantial physical change in microstructure. Aside from a simple case of attrition, involving possible pulverization of protruding projections or loose members of acicular particles, deformation of whole particles or of particle sections will tend to impose a definite pattern of preferred particle orientation. In the photomicrographs in Figures 86 and 87, of compressed copper powder and of a mixture of silver and molybdenum powders, such a pattern of a preferred orientation can readily be recognized.

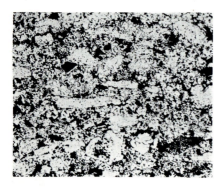

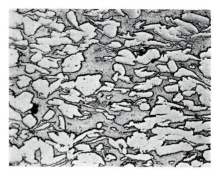

Fig. 86. Compact compressed at 45 tsi from reduced and annealed copper powder, displaying preferred orientation of particles in planes perpendicular to axis of pressure (×100).

Fig. 87. Compact compressed at 125 tsi from mixture of molybdenum and silver powder, displaying preferred orientation of molybdenum particles in planes perpendicular to axis of pressure (×100).

As a result of both attrition and plastic deformation, microscopic—if not molecular—irregularities of the surface regions are obliterated. In the wake of this wear and tear of the boundary areas, the innumerable fresh, clean metal surfaces which are exposed permit the attractive force to operate. This, then, must be considered to be the primary microeffect of the pressure. But there are concurrent effects of equal importance in the establishment of durable bonds. As previously indicated, the surface areas may be subjected to great changes as compression proceeds. Not only will the metallic surfaces be drawn into the zipperlike keying action of the border atoms, but, at the same time, obstructions are likely to be moved, if not entirely eliminated. Thus the displacement of solid surface impurities at contact points and the extraction of adsorbed gases from surface regions near the contact areas, are very valuable secondary microeffects of the pressure.

In the ideal (friction-free) case, complete packing of the particles can be obtained if sufficient pressure and time for plastic flow is allowed. In this case, particle shape and size are disregarded. Actually, this ideal case cannot be approached—even under the most favorable conditions. Even if particle shape and size distribution are controlled carefully, if the metal is extremely plastic, if high pressures are applied, and if the form is geometrically simple, the high coefficient of interparticle friction will prevent complete filling of the space with metal. There will, on the contrary, always be some points at which frictional forces can overcome driving forces (motivated by the exertion of the pressure). This is due largely to the fact that the minute pieces of metal are solids, incapable of transmitting sideways pressure received from above or below. Consequently, the formation of wedges and bridges is a frequent phenomenon in metal powder compaction. As a result, it is nearly impossible, and also impractical, to obtain a relative density of 100% for a compact.

In relation to this, another effect of friction must be considered—its transformation into heat energy. Under the severe stresses implied by the pressure, this heat, in turn, causes a temperature rise which, although small in most cases, is sufficient to increase local atomic mobility, as well as deformability of the metal. In this way, interparticle friction greatly inhibits the establishment of contact areas due to bridge effects, at the same time counteracting this effect by enlarging contact areas (due to increased plastic deformability of the surface regions), as well as by intensifying the action of atomic force fields (once such contact areas have been established).

The chief microeffects of the applied pressure may then be said to be: (1) conjunction of the particles by mutual sliding and rotation into progressively denser agglomerates; (2) breakdown of bridges and reorientation of wedging particles to permit mechanical keying; (3) cleavage and puncture of solid surface impurities; (4) partial extraction of interfering gaseous phases from the contact areas; (*5*) creation of fresh, clean surfaces by shear forces imposed by friction; and (*6*) filling of microscopic asperities on particle surfaces by plastic deformation.

At the same time we should take into consideration certain secondary effects caused by the heat of friction: (*1*) increase in atomic mobility in surface contact areas; (*2*) increased susceptibility to plastic deformation, causing better contact areas; (*3*) extraction of an interfering gas phase in the contact regions; and (*4*) limited crystal growth at the surface, possibly causing the filling of some voids vacated by the gas envelopes.

All these microprocesses result in a considerable increase in total contact area and, therefore, in cohesive strength of the compact. It is possible that pressure may also have a direct influence in increasing the ability of surface atoms to influence each other, thus leading to an increase in intensity of the attractive forces available.[27]

Before completing the discussion of the microeffects of pressure, it may be worth while to consider intraparticle structure for a moment. It has already been demonstrated that as in the slip phenomena within crystals of ordinary metals—where slip occurs more readily in some planes than in others—the random orientation of particles in a powder mass is often replaced during compression by a preferred orientation having a definite relation to the direction of applied pressure. This, however, has not been established for arrangements of crystallites within particles. For example, it has not yet been possible to detect (by x-ray investigation) a preferred direction of the individual crystallites of compressed sponge iron particles in the (111) and (100) planes, lying in the direction of compression, as is possible in the case of compressed polycrystalline bulk iron. Nor has it been possible to detect a preferred direction of the individual crystallites of compressed copper particles in the (110) plane perpendicular to the direction of compression, as evident in compressed bulk copper. Even after applying pressures of 200 tsi to fine copper or gold powders, Trzebiatowski,[28] could find no evidence of a preferred crystallite arrangement, their orientation after compression being substantially the same as in the original powder.

MACROEFFECTS

Under the term "macroeffects" we may group all clearly perceptible phenomena caused by the application of pressure to a metal powder. This refers primarily to the visual formation of a pseudosolid body from a loose agglomerate of individual particles. From a wider viewpoint, we may

[27] W. D. Jones, *Principles of Powder Metallurgy*. Arnold, London, 1937, p. 27.
[28] W. Trzebiatowski, *Z. physik Chem.*, B24, 75 (1934).

include in this term all those interdependent mechanical effects which must be considered to be by-products of the compression mechanism, *i.e.,* such phenomena as cleavage cracks, pressure cones, or laminations in planes perpendicular to the pressure axis, and all results of local excessive strain-hardening. In fact, progressive work-hardening of the metal particles (especially in the surface regions of the compact facing the punches and the walls of the die cavity) with increased pressures, imposes a definite practical limit on the pressure that can be exerted. Beyond this limit, not only are these disturbances encountered within the compact, but severe stresses will also be imposed on all molding tools, resulting in frequent failures.

To clarify the rather complex processes that occur during powder compression, it might be well to review at this point a few basic concepts of the powder molding operation. The most distinctly perceptible effect of the application of pressure is a reduction in relative volume of the powder mass. As the particles reorient themselves to fill voids and gaps, the agglomerate becomes denser. The original relative density, known as apparent density, increases with rising pressures. At any given greater relative density, the powder has been compressed to a certain definite extent. The proportional decrease in relative volume is known as the compression ratio, a practical concept of particular importance (as will be shown later). Depending on the order of magnitude of the apparent density of the powder, the effective compression ratio, *i.e.,* the ratio at which the powder particles are sufficiently agglomerated to yield a coherent compact, may vary considerably. Once the limit of the effective ratio is overstepped, additional densification of the coherent powder compact will proceed disproportionately to the rising pressure, since work-hardening effects will interfere increasingly with the creation of new contact points by reorientation or plastic deformation of the particles. To overcome resistance caused by these strain-hardening effects, increasingly greater pressures become necessary, which in many practical applications involve manipulation beyond the realm of practicability. In fact, it may be said that most commercial powders can be compressed to the theoretical density of the metal only with infinitely high pressures. Naturally, basic plasticity and the condition of the metal powder under consideration are of great importance in this connection. This is clearly expressed by the different curves in Figure 88, which show relative density as a function of pressure for various soft, hard, and work-hardened metal powders. Regardless of material or state, relative density increases with rising pressures, with the curves assuming hyperbolic shapes. The very soft and plastic silver, iron, and copper powders approach theoretical absolute

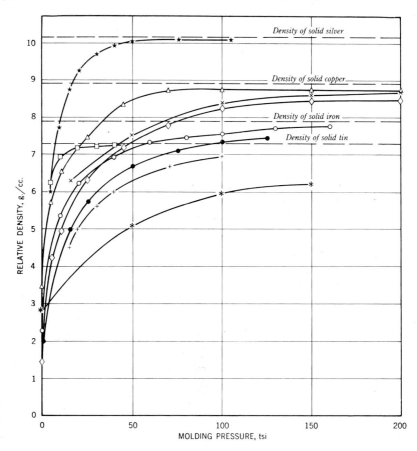

Fig. 88.   Relative density of powder compacts as function of molding pressure. $\square$ atomized tin, -325 mesh (according to J. E. Kuzmick[28a]); $\bigstar$ crystalline silver, -100 mesh; $\triangle$ coarse electrolytic copper, -100 mesh; $\times$ fine precipitated copper, (according to W. Trzebiatowski[23]); $\diamondsuit$ fine electrolytic copper; $\bigcirc$ purified, soft electrolytic iron (according to C. W. Balke[23b]); $\bullet$ hydrogen-reduced iron, -100 mesh; $+$ pure iron (according to P. R. Kalischer[28c]); $*$ annealed crushed steel, -100 mesh.

density values even at moderately high pressures, *i.e.*, between 50 and 100 tsi. On the other hand, hard powders, such as annealed crushed steel, cannot be compressed to a greater density than about three-quarters of the theoretical value regardless of the pressure applied.

From these curves it becomes apparent that pressures above a certain practical limit are not effective. Moreover, very high pressures produce a

[28a] J. E. Kuzmick, *Trans. Am. Inst. Mining Met, Engrs.*, *161*, 612 (1945).
[28b] C. W. Balke, *Symposium on Powder Metallurgy*, ASTM, Philadelphia, Pa., 1943, p. 13.
[28c] P. R. Kalischer, *Iron Age, 149*, No. 6, 41 (1942); No. 7, (1942).

variety of detrimental effects which cannot be overlooked. In a compact of uniform density, excessive strain-hardening of the particles will simply result in great resilience of the agglomerate which will tend to cushion further densification caused by increased pressure. This cushioning action will be supported by the very high compression of air trapped in the many cavities within and between the particles. As a consequence, phenomena such as volumetric expansion and lateral disjunctions appear upon release of pressure. The appearance of shear fractures—with a tendency toward a 45° angle—constitutes further proof of excessive embrittlement by overstraining. The trend of shear fractures can be recognized from Figure 89, in which a photograph is shown of the cross section of a highly compressed cylindrical iron compact in a split, segment-type die. Laminary fractures (resulting from overpressure) in a bar-shaped copper alloy compact are shown in Figure 90.

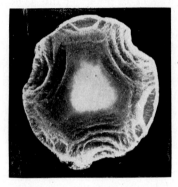 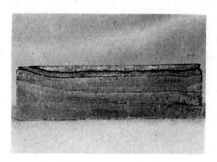

Fig. 89. H o r i z o n t a l section through cylindrical iron powder compact compressed at 125 tsi in collapsible, segment-type die, displaying typical shear failures.

Fig. 90. Laminary fractures of bar-shaped copper–nickel alloy compact compressed at 100 tsi.

The different forms of failure in metal powder compacts have been discussed by Wulff[29] and Bailey.[29a] According to Wulff, three types of failure seem to exist. The first is the 45° shear failure which usually occurs in *slow* pressing of hard powders at moderate or high pressures, and of soft powders at very high pressures (when excessive strain-hardening is produced). This type of failure can usually be eliminated by the use of a lubricant. The second type of failure also begins as a 45° shear failure at

[29] J. Wulff, Communication on "Principles of Pressing," *Powder Metallurgy Colloquium*, New York University, June 14, 1946.
[29a] L. H. Bailey, discussion during *Powder Metallurgy Colloquium*, New York University, June 14, 1946.

the edge of the compact, but proceeds inward in the form of a lamination; it occurs primarily in very *rapid* pressing, and is usually overcome by slowing down the pressing speed. The third type of failure is caused by uneven die fill which creates shear stresses in addition to straight compression stresses; it also results in laminar fractures. Bailey cites two major causes for lamination in the compacts: one, mentioned earlier, is due to expansion of trapped air when pressure is released, while the other type of lamination is apparently due to elastic expansion of the compact when ejected from the die. As the compact leaves the die cavity, the top becomes free to expand while the bottom is still retained in the die. The compact is not strong enough to withstand this abrupt change in stress distribution, and failure appears in the form of a lamination in the top surface plane of the die. If ejection of the compact is carried out slowly, with the compact moving out of the die a little at a time, each increment produces another laminar fracture. One way of overcoming this type of failure would be to remove the compact from the die while it is kept under pressure from the two opposing punches.

With such effects evident in relatively simple, geometrically shaped bodies, it becomes obvious that any possible local nonuniformity in the density of the agglomerate will only exaggerate these conditions, regardless of whether this nonuniformity is caused by improper powder fill or by complications in the shape of the desired form. Many molding failures can be traced immediately to such local overstraining due to improper particle distribution within the agglomerate.

The macroeffects of applied pressure have been discussed in considerable detail by Jones.[30] In terms of porosity, the influence of external pressure applied on a powder mass can be expressed as causing a reduction: (*1*) by movement of the particles into voids, (*2*) by deformation (promotes keying), and (*3*) on an atomic scale by flattening microscopic or submicroscopic asperities on particle surfaces.

If the total porosity of a compact remains high with moderate pressures, the strength and hardness of the body must be attributed chiefly to the forces causing particle surface attraction. As long as the total contact area is small, strength and hardness must also be expected to be small. At higher pressures, increased areas of contact are chiefly dependent on the deformability of the particles. Consequently, powders from softer and more plastic metals will show a greater increase in strength for any given applied pressure than harder and more brittle materials. As will be shown in more detail later, there are many practical examples to prove this fact.

[30] W. D. Jones, *Principles of Powder Metallurgy*. Arnold, London, 1937, p. 27.

For instance, a specific pressure of 5 tsi is hardly sufficient to mold iron or tungsten into a coherent body, but it is sufficient to compress tin or lead powder into a strong compact.

Still other factors enter the picture at very high pressures. With a decrease in porosity, the number and size of contact areas increase greatly. Deformation of most particles (as effected by changes in their volume or abrasion of their surfaces) must be considered to be a primary cause of work-hardening; wedging or keying of some surface regions undoubtedly contributes further to this condition. The strength and hardness of a powder compact after application of very high pressures are thus due to rather complex conditions, especially when we take into account the relatively small particle size—as compared to grain size—of common cold-worked metals. The basic necessity of particle deformability for the build-up of strength and hardness in highly compressed compacts is probably best emphasized by the fact that pressure alone is not sufficient to obtain these conditions. It is known that hydrostatic pressure acting on a non-porous polycrystalline body has no appreciable influence on hardness. On the other hand, very dense compacts possessing only a small percentage of porosity show a marked increase in hardness and strength after subjection to very high pressures. It is true, however, that the greatest increase in hardness of a compact occurs in the early stages of compression, while consolidation of the powder mass is taking place. With progressively increasing pressures, the hardness increases at a constantly reduced rate. Some small increase in hardness may still be obtained after the compact has been almost entirely consolidated by very high pressures.

As has already been mentioned, the two types of particle deformation that can be distinguished are surface abrasion (caused by friction) and volume deformation (necessitated by the particle adjusting its shape to that of its neighbors). It may now be assumed that in the case of a substantial consolidation the extent of both types of deformation will be fairly constant in amount and independent of the numerical value of the pressure applied. It may further be reasoned that this deformation alone cannot cause such excessive hardness values as have been found for highly compressed powders (e.g., **180** Brinell for copper—found by Trzebiatowski.)[31] Even excessive deformation of a single crystal of copper—as could be obtained by deforming a sphere into a cube with the aid of slowly acting high pressure—cannot be expected to result in even nearly as high a hardness. Therefore, it is safe to conclude that the extreme hardness of compacts after subjection to high pressures is not caused en-

[31]W. Trzebiatowski, Z. physik Chem., B24, **75** (1934).

tirely by work-hardening of the individual particles or crystallites. X-ray investigations conducted by Trzebiatowski[31,32] give ample evidence of a highly cold-worked condition. Besides general plastic deformation, a possible localized distortion (perhaps in the particle surface regions), as well as internal or elastic stresses and superfine particle sizes, may also contribute to the broadening of the Debye-Scherer spectrum and result in the loss of resolution of the Ka doublet. Trzebiatowski[32] himself lends support to the assumption that this hardness is only partly caused by lattice deformation, since he finds that after heating the highly pressed compacts to 350°C. (660°F.), the Ka doublet is once again resolved, although the hardness of the compact is still high.

Jones,[32a] in summarizing the discussion on direct and indirect effects of applied pressure on the unusual physical properties of cold-pressed compacts lists the following contributory factors: (1) deformation and work-hardening of the particles, (2) internal or elastic strains, (3) abrasion of the particle surfaces, (4) action of surface film forces, (5) slip prevention due to surface film forces or to superfine particles, and (6) space lattice conditions of such fine particles.

Magnitude of Pressure

From the foregoing it becomes apparent that the correct choice of applied pressure is of paramount importance in metal powder molding. Obviously, no specific pressure can be singled out, although, in most cases, conditions for an optimum forming pressure exist. We will discuss here the limits of the permissible pressure range in which the optimum pressure can be found.

The lower limit is given by the capacity of the powder to form a coherent compact, i.e., by the capacity of the particle to form sufficient contact areas (permitting the cumulative attractive forces to overcome external repulsive forces which tend to retain the powder in its original loosely heaped state). As shown previously, particle characteristics, as well as the nature and condition of the metal, have a governing influence on this condition. To be more specific, it may be said that the following factors determine the numerical value of the minimum pressure necessary for the forming of a coherent body: (1) basic softness and plasticity of the metal, (2) particle structures favoring the establishment of a maximum number of contact points, (3) particle size distribution favoring closest packing before and during application of pressure, (4) particle surface conditions favoring the by-passing and reorientation of particles by reduction of interparticle friction, (5) selection of a form which does

[23] W. Trzebiatowski, Z, physik. Chem., A169, 91 (1934).
[32a] W. D. Jones, Principles of Powder Metallurgy. Arnold, London, 1937, p. 32.

not obstruct the densification of the powder mass as a whole or in certain sections, (6) selection of materials for the mold and punches that do not interfere with the densification of the powder mass, and (7) selection of proper lubricants to reduce friction between particles and die walls or between the particles themselves.

TABLE 35

Molding Pressures of Various Metal Powders

Material	Minimum briquetting pressure,[a] tsi	Briquetting pressures for $\rho_r = {}^1/_2 \rho_{theor.}$[b] tsi	Briquetting pressure for $BH = BH_n$[c] tsi
Tin, −100 mesh	0.5	3	5
Lead, −100 mesh	0.6	4	5
Aluminum, −100 mesh	1	7.5	10
Silver, amorphous	0.8	3	6
Silver, crystalline	1	5	9
Copper, electrolytic, coarse	1.5	3.5	17
Copper, electrolytic, fine	2.5	7.5	30
Copper, reduced, coarse	1.5	5	20
Brass, 70/30, atomized	3.5	8	30
Iron, hydrogen-reduced	1	5	50
Iron, electrolytic, annealed	2.5	7	50
Iron, electrolytic, hard	10	15	60
Iron, carbonyl	10	12	50
Nickel, electrolytic, annealed	1.5	5	22
Nickel, carbonyl	5	12	40
Chromium, electrolytic, annealed	16	25	75
Steel, crushed, annealed	15	25	75
Molybdenum, reduced	4	10	35
Molybdenum, crushed, annealed	22	30	75
Tungsten, reduced	5	12.5	45
Tungsten, crushed, annealed	25	33	85
Tungsten carbide+6% cobalt[d]	10	14	—

[a] By minimum briquetting pressure is meant the lowest possible pressure at which a coherent compact with sharp edges is formed that cannot be crushed by hand.
[b] ρ_r designates relative density.
[c] BH designates indentation hardness in Brinell units, covering the over-all hardness of the compact with pores overbridged; BH_n is Brinell hardness of bulk metal.
[d] Nonmetallic binder added.

In discussing *practical* lower pressure limits, we must first agree on the required condition of the compact. As explained in Chapters IV and V, many powders require only a very slight pressure to form briquettes which are sufficiently coherent to endure normal handling. This "minimum briquetting" pressure may be about 1 tsi for certain iron powders, 0.5 tsi for certain copper powders, or 0.1 tsi for powders of soft metals such as lead or tin. But this limit is rather arbitrary since there is no clear specification of the limit of force or abuse in handling that the compact must take before excessive deterioration of the edges will take place. Otherwise, the lowest forming pressure can be expressed as a function of the relative density (*e.g.*, at one-half of the theoretical density), but here again we must consider this to be arbitrary, because of the widely varying apparent densities of different powders. Indentation hardness equalling that of the bulk metal may also serve as a reference basis, since it has the advantage of assuring a reasonable stability of the compact during industrial handling. How the numerical low-pressure limit varies for these different conditions is shown in Table 35 for a variety of metal powders under different conditions.

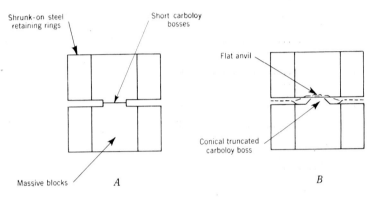

Fig. 91. High-pressure die block arrangements, as used by Bridgman[33]: *A*, for pressures up to 1100 tsi; *B*, for pressures up to 3100 tsi.

Turning now to the upper pressure limit, we must first consider the maximum pressures attainable with present day equipment. Bridgman, the greatest authority in this field, has in recent years made wonderful progress toward reaching ultimate pressures.[33-36] In his attempts to reach the limit of utilizable pressures (using very short cemented carbide punches and high confining pressures), pressures of 1100 tsi (155,000 kg./cm.²) were reached. These pressures, however, could be exceeded by employing the knife-edge principle and by selecting very short carbide punches (*i.e.*, low bosses on massive blocks) which permit a distortion that —beyond a certain pressure—would allow support by the edge (Fig. 91). With a height of the boss of 0.005 inch, support around the edges began at an average pressure on the boss of over 2200 tsi (310,000 kg./cm.²). With a boss height of three times as much (0.015 inch), and with a confining pressure of 190 tsi (27,000 kg./cm.²), a one-sided compression of

[33] P. W. Bridgman, *J. Applied Phys.*, *12*, No. 6, 461 (1941).
[34] P. W. Bridgman, *Phys. Rev.*, *57*, 342 (1940)
[35] P. W. Bridgman, *Proc. Am. Acad. Arts Sci.*, *72*, 157 (1938).
[36] P. W. Bridgman, *Proc. Am. Acad. Arts Sci.*, *71*, 387 (1937).

over 2600 tsi (370,000 kg./cm.^{2}) was obtained, making a total average hydrostatic pressure over the area of contact of 2800 tsi (400,000 kg./cm.^{2}). Still higher stresses were obtained when the geometry was arranged to permit destructive rupture. Thus, for an average one-sided compressive force on the area of contact just before rupture, Bridgman found a pressure of about 2900 tsi (410,000 kg./cm.^{2}). Apparently, the ultimate pressure possible between cemented carbide pieces with confining pressures of about 200 tsi is somewhere between 3100 and 3600 tsi (440,000 to 510,000 kg./cm.^{2}).

It is obvious that the tremendously high pressures made available by Bridgman cannot be used in commercial practice. The useful range must be kept considerably lower, even when taking all possible precautions in designing the dies and punches. Trzebiatowski,[37] in his experimental pressings of copper and gold powders, successfully employed pressures of up to 200 tsi, and the present author later found pressures up to 190 tsi possible, although not practical.[38] On the other hand, specific pressures up to 150 tsi have been repeatedly used with success by Balke[39] in his recent work on iron and steel, and it may be stated that with our present knowledge of die and tool design, pressures up to 100 tsi and, under certain favorable conditions, possibly up to 150 tsi are quite feasible. Constant pressure reversals, as imposed during production operations, entail, of course, critical tool wear and maintenance problems, and it is much safer to lower the practical upper pressure limit for larger scale production work to 75 tsi.

The range of the practical pressures used in powder metallurgy (within the realm of possible application) is shown on the logarithmic scale of Figure 92. At the same time, pressure ranges of related operations are indicated.

Finally, there remains the question of the optimum specific pressure for the molding of a powder. Here, many factors, most of which have already been discussed, must be taken into consideration. The practical lower pressure limit is in most cases controlled by the desired final density of the product. Other requirements, such as satisfactory green strength and rigidity of all contours (especially edges or sharp projections) may demand a higher pressure than is otherwise necessary. The upper pressure limit for most practical applications is limited by certain demands on the

[37] W. Trzebiatowski, Z, physik. Chem., B24, 75 (1934).
[38] C. G. Goetzel, The Influence of Processing Methods on the Structure and Properties of Compressed and Heat-Treated Copper Powders, Dissertation. Columbia University, 1939.
[39] C. W. Balke, Symposium on Powder Metallurgy. ASTM, Philadelphia, Pa., 1943, p. 11.

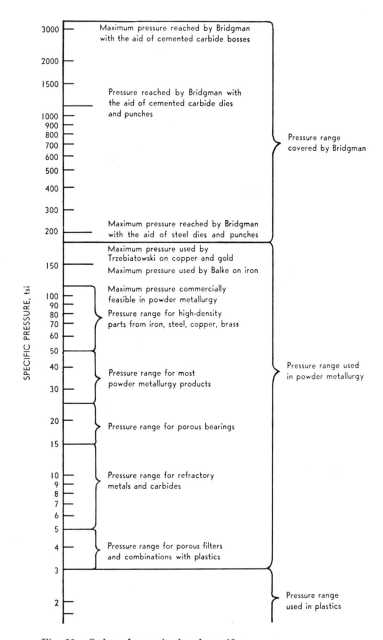

Fig. 92.   Order of magnitude of specific pressures.

finished product. To these belong the final porosity and a desire to retain the molded shape during sintering. Volumetric expansion, as well as mechanical failures, such as cleavage cracks and laminations, are additional objectionable by-products of excessive molding pressures. Finally, wear resistance and the limited life of dies and punches puts a definite upper limit on all practical pressures, even though higher pressure, in some cases, would result in denser, stronger, or otherwise more desirable products. Thus the range for optimum specific pressures in metal powder molding appears to be confined to about 20 to 50 tsi, with slightly lower pressures suitable for compression of porous bearings and refractory metal and carbide compacts with nonmetallic binders; slightly higher pressures are needed for the molding of structural or electrical parts from iron, steel, copper, or brass.

Summary

Any reasonable concept of the mechanism of bonding metal powder particles under pressure must be based on atomic attraction. Atoms located in the surface regions of one particular particle link with atoms located in the surface layer of an adjacent particle, provided that the contact between the two surface areas is so close (*i.e.*, of the order of the diameter of the atom) that atomic field forces can come into play. Obviously, this requirement can be fulfilled in metal powders only with the aid of outside pressure. The various theories advanced in connection with the interparticle bonding can be traced to this basic concept, with the sole exception of the one that explains bonding on the basis of mere mechanical interlocking, a somewhat questionable concept.

The effects of applied pressure on the compressibility of metal powders depend on a number of conditions inherent in the initial powder (*e.g.*, dispersion) as well as in the pressing operation itself. Coarser powders may require low initial pressures, while the final pressures for very high densities may be higher than in the case of fine powders of the same apparent density. Smooth-shaped particles favor closer packing, but their deformability is inferior. The composition of a powder has a logical influence on its hardness and plastic deformability. The magnitude of the pressure itself is of great importance. The lower limit of the practical pressure range is given by the compactibility of the powder, and is usually not below 10 tsi. The upper limit is given by the capacity of the material to withstand overstraining, as well as by the performance of the tools under these pressures. Although considerably higher pressures can be obtained on an experimental basis, 75 to 100 tsi must, at present, be considered to be the maximum commercially feasible pressures.

The size of the compact also affects the compressibility. Thicknesses below $1/32$ inch cause elastic overstraining, while for thicknesses above three inches the pressure losses in the compact cause too great variations in density to satisfy most practical requirements. For parts below ten square inches in cross section, the height of the compact should not be larger than the square root of the cross-sectional area. With regard to the effects of the cross section itself, the lower limit is controlled by the friction at the die walls, and will generally have to exceed $1/8$ inch. The upper limit is, for economic reasons, determined chiefly by the capacity of the press and the necessary forming pressure. Fifty square inches must be considered close to the present practical limit.

The chief microeffects of applied pressure are densification of the agglomerate of particles by mutual sliding, rotation, or reorientation of the particles, creation of clean surfaces by shearing the metal surfaces, or by puncturing interfering solid impurity hulls. These effects may be supported by increased atomic mobility and the greater susceptibility of the metal to plastic deformation because of increased heat of friction. All these microprocesses result in considerable increase in the contact areas and therefore in improved cohesive strength of the compact.

The chief macroeffect of the applied pressure is the formation of a coherent body, a body which becomes increasingly dense and hard at higher pressures. In certain cases the hardness may even exceed that of the severely work-hardened solid metal. Among the defects caused by overpressure are volumetric expansion due to elastic stress release, cleavage fractures due to excessive work-hardening, and laminations perpendicular to the axis of pressure—due to overcompression of trapped air. Lack of sufficient pressure, on the other hand, causes frailness of edges or projections, or complete disintegration of the compact.

CHAPTER IX

The Molding of Powders into Solid Forms

Any study of the behavior of metal powders under pressure must take into consideration the method of applying the pressure. Obviously, the question of whether external pressure is applied from one or more directions is of great importance for the process of densification of the powder mass. On the other hand, the selection of a practical method of pressure application may depend largely on the ability of the powder to follow the laws of hydrodynamics to some extent. It is known that very fine powders —especially when properly aerated—possess many of the properties of a fluid.[1] For example, dry sand flowing through an aperture in the base of a container undergoes a contraction similar to the contracted vein observed in fluids emerging from a narrow orifice. Similar flow characteristics have been demonstrated for differently dyed Bakelite powders when passing through apertures, when passing obstructions of various shapes, or when being compressed slightly with a ram. These illustrations, however, apply only to loose powders in air.

With the application of pressure, especially when a substantial removal of gas is effected, similarity to a fluid is far less apparent, chiefly because of friction forces and increased effectiveness of particle bonding. Especially in the case of very small particles, the applied external force loses its effectiveness at a small distance from the surface of the powder mass, an effect accentuated with increased irregularity in shape. This refers to translation of the pressure in direction of application (e.g., inwardly, from above), as well as in the opposite direction (e.g., inwardly from below), or in a lateral direction (e.g., inwardly, from the walls of the container). However, when pressure is applied in many directions, internal pressure distribution tends to equalize, thus minimizing pressure losses by friction, and resulting in more uniform particle concentration.

---

[1] W. D. Jones, *Principles of Powder Metallurgy*. Arnold, London, 1937, p. 3.

EFFECTS ON DENSITY OF VARIOUS METHODS OF APPLYING PRESSURE

Application of Triaxial Pressure on Prismatic Compacts

On the basis of the foregoing observations, it becomes apparent that the distribution of density in pressed compacts cannot be completely uniform over the height or the cross section of the compact. For this reason the method of applying pressure becomes particularly significant— both from a theoretical and a practical point of view. Let us consider the case of a mass of powder freely suspended in air, where the mass is compressed into a cube or cylinder. As seen from Figure 93, there are three

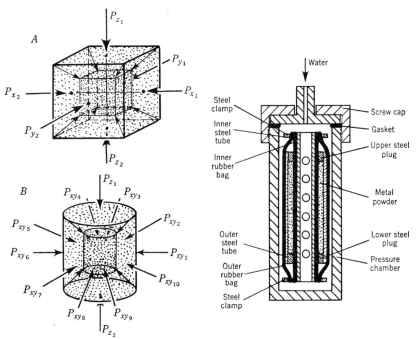

Fig. 93. Three-dimensional compaction (schematic): A, cube; and B, cylinder.

Fig. 94. Hydrostatic pressure apparatus for compacting metal powders (according to Skaupy[2]).

dimensional planes in which pressure may be applied. Compaction into a cube requires pressure in six directions, whereas compression into a cylinder would require pressure from two directions along the axis of the cylinder, and from an infinite number of directions in the plane perpendicular to the axis.

For a cylindrical compact, this theoretical method of a multiple pressure application can be approached by applying pure hydrostatic pressure. Skaupy[2] describes a practical apparatus reproduced in Figure 94 that obtains hydrostatic conditions for molding refractory metal powders. It consists, essentially, of a pressure chamber sealed by a screw cap. Inside the chamber, two polished steel tubes are inserted one within the other—the inner one being perforated. They are protected by rubber bags that extend and close tightly over the ends of the outer, somewhat shorter, steel tube. The metal powder is placed between the inner rubber lining, and the outer steel tube, and is sealed by two end plugs. Uniform compaction from top and bottom, as well as from all around, is accomplished by subjecting the chamber to high water pressure. A similar apparatus[3] consists of steel pressure chambers containing a number of steel tubes plugged at both ends and inserted in rubber hoses. The powders are placed in the tubes and subjected to hydraulic pressure. In this way, it is possible to increase the density considerably (utilizing pressures about ten times higher than would otherwise be possible) while keeping stresses at a minimum. A decided disadvantage in applying this method to practical molding problems, however, is the fact that tolerances, especially of the diameter, can not be held uniformly and closely enough to satisfy most engineering requirements. Furthermore, the surface finish of hydrostatically pressed cylinders is usually poor, requiring a final machining operation where close fits are demanded. To improve these conditions, a number of experiments have been made in which the rubber bags were replaced by smooth steel linings.

Split dies (operating in the manner of an orange peel) have been tried in molding spheres. Gravity forces, of course, prevent the powder particles from packing toward the center of the sphere, but in one case known to the author, the introduction of magnetism overcame this obstacle, and ferromagnetic powders have been pressed (three-dimensionally) into spherical bodies. The method, however, was found unsuitable for molding very dense spheres, since the amount of powder fill was governed by the strength of the magnetic material from which the die segments were constructed. Most of the particles near the die walls retained their preferred orientation after compacting, but this effect was nullified during sintering. The flash on the surface of the spheres, caused by contact of adjacent die

[2] F. Skaupy, *Metallkeramik*. Verlag Chemie, Berlin, 1943, p. 156.
[3] U. S. Pat. 1,081,618; also Brit. Pat. 401,521.

segments, was removed readily when the sintered spheres were tumbled in a ball mill.

Application of Uniaxial Pressure

COMPRESSION OF PRISMATIC COMPACTS

Our present experience in metal powder molding is confined to a simplified method of pressure application. So far, side pressure is limited to the act of clamping dies together in a viselike fashion. The forming pressure is applied along only one axis, either in a single direction or in two opposite directions. Accordingly the cross section of the die cavity remains constant at all stages of compression.

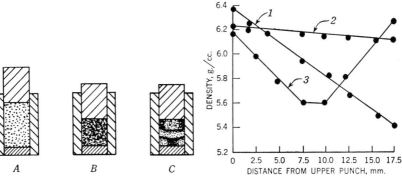

Fig. 95. Compression of a cylindrical compact in a single direction: A, loosely packed powder in die cavity, light areas indicate interstices; B, compressed powder, interstices are considerably reduced in quantity and size; C, curved compression layers formed by graphite powder layers in copper compact.

Fig. 96. Density vs. height of compact (according to Bal'shin[4]): (1) Electrolytic copper, a.d., 1.42 g./cc., without graphite, pressure applied from one side; (2) same, plus 4% graphite; (3) same, without graphite, pressure applied from two sides.

For the sake of simplicity, the application of the pressure in a single direction will be considered first. Figure 95 indicates the vertical section of a cylindrical mold in which a powder is compressed in one direction. That a metal powder does not follow the law of hydrodynamics can be seen from the fact that if, for instance, a copper powder is used, and flat graphite powder layers are placed in between, the layers assume a curved shape upon compression (Figure 95C; see also Figure 74A, p. 283). This curved shape can be explained by the friction effect of the die walls. Accordingly, it may be concluded that a density maximum exists in the upper portion of the cylinder near the die wall, while in the lower portion a zone at maximum density exists in the center. The latter, however, is usually considerably less marked than the density maxima next to the moving top punch. According to Bal'shin,[4] density of the compact decreases with height in a roughly linear manner (as shown in the graph of Figure 96). This decrease is intensified with decreasing diameter of the die. It is less pronounced with coarser or harder powders, or when lubricants are added.

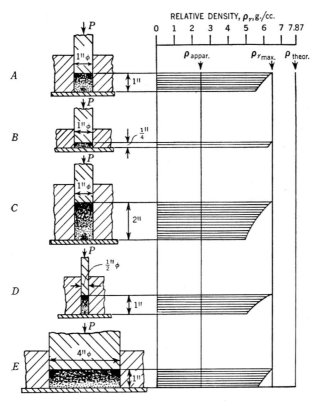

Fig. 97. Effect of compact height and diameter on density of iron powder (schematic). 50-tsi pressure applied in one direction: A, $h = d$; B, $h = \frac{1}{4}d$; C, $h = 2d$; D, $d = \frac{1}{2}h$; E, $d = 4h$.

In Figure 97 the effects of height and diameter on the density are indicated for a compressed iron powder. The minimum density is affected by the height or diameter of the compact, but this is not so for the maximum density in the top layer of each compact. The latter is determined

[4] M. Yu. Bal'shin, *Vestnik Metalloprom.*, *18*, No. 2, 124 (1938).

by the type of metal, the order of magnitude of applied pressure, and the conditions prevailing at the punch face. With a given molding pressure, compacts which are long and slender in the direction of pressing (Figures 97C and D) will be less dense in the bottom portion than compacts that are short and wide (Figures 97B and E). The density throughout the length of the compact will be more uneven as the ratio of length (dimension in the direction of pressing) to width (dimension perpendicular to direction of pressing) increases. In general, this difference in density exceeds 10% if the ratio of length to width is greater than 1 (Figure 97A); it may reach 20% if the ratio is 2 (Figure 97C). Of course, this merely indicates a trend, and each individual powder and condition of pressing implies a certain deviation from this rule.

Recently, the results of Bal'shin were confirmed by Kamm, Steinberg, and Wulff,[5] who used soft metal grids—suspended in the powder mass—as indicators of the pressure distribution (see page 287).

Obviously, the same effects of friction will apply when a compact is pressed from both ends simultaneously, except for a numerical change in density in the various sections, caused by different pressure distribution. This is apparent from Bal'shin's simple experiment shown in Figure 96. The trend of density distribution in iron powder compacts of various dimensions, pressed along one axis from two opposite directions, is shown in Figure 98. Particularly noteworthy is the fact that the total drop in density is considerably reduced by this action, while the minima are shifted to the middle sections. Under otherwise constant conditions, the density gradient in a compact having a ratio of height to width of 1 is generally less than 5% of the value for the solid metal. Thus, pressing prismatic compacts along one axis simultaneously from top and bottom gives a reasonably uniform distribution of density, and therefore substantially facilitates uniform shrinkage during sintering; this will result in uniform properties of the finished product. As a consequence, the entire modern molding technique has been developed along the two-directional pressure application, and will be discussed extensively in the chapters dealing with dies and presses.

COMPRESSION OF MULTIPLE-SECTION COMPACTS

Our studies of the effect of pressure on metal powder have so far been confined to simple prismatic shapes, i.e., compacts having a uniform section in the direction of pressing. There are many applications where such forms are used—especially with center bores—as in the many types of

[5] R. Kamm, M. Steinberg, and J. Wulff, *Trans. Am. Inst. Mining Met. Engrs.*, *171*, 439 (1947).

bushings, sleeves, and rings for the bearing industry. But this class of shapes is small in comparison to the engineering industries' demand for parts with various sectional thicknesses, having flanges, offsets, steps, and the like. These parts, if molded from powders, would have nonuniform sections in the direction of pressing, and it is apparent that the already

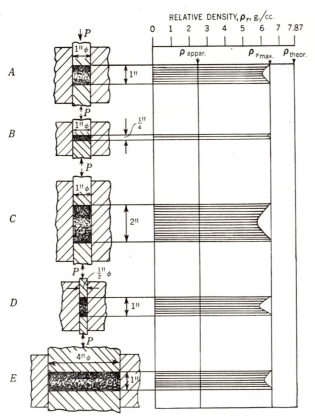

Fig. 98. Effect of compact height and diameter on density of iron powder (schematic). 50-tsi pressure applied in opposite directions: $A, h = d$; $B, h = \frac{1}{4}d$; $C,$ $h = 2d$; $D, d = \frac{1}{2}h$; $E, d = 4h$.

complex conditions prevailing during the molding of prismatic bodies will be still more complicated by such intricacies of contour. In general, it may be said that in such cases a metal powder under compression fails to follow any simple laws, and that such processes as local malcompression due to pressure losses, or side compression due to lateral deflection of particles, may account for mechanical complications involving the design of both the part and the mold.

From the start, it must be borne in mind that the density of such molded bodies will approach uniformity only if the compression ratio remains the same over the entire cross section at any stage during the press operation. This, in turn, is possible only if the powder fill volume corresponds to the compression ratio for each plane parallel to the pressure axis, and if lateral slippage of individual particles during compression of the powder is prevented.

This may perhaps best be illustrated by a simple two-level form having one steplike depression in the center. The nonuniform section is shown in Figures 99A and B. If such a piece is molded in the conventional manner

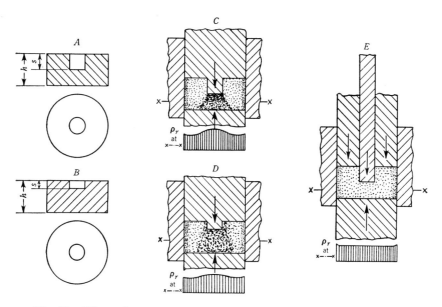

Fig. 99. Effect of compact cross section on density: A, two-level cylindrical specimen, $s = \frac{1}{2}h$; B, same, $s = \frac{1}{4}h$; C, density distribution for single punch compression, $s = \frac{1}{2}h$; D, same, $s = \frac{1}{4}h$; E, density distribution for multiple punch compression, $s = \frac{1}{2}h$.

with single punches the density will be higher in the short middle section than in the long outer section—as shown in Figures 99C and D—because of the difference in compression rates. For applications where this step is comparatively small (Fig. 99B), this difference in density and the corresponding difference in strength, hardness, and ductility may still be permissibly small (Fig. 99D), but in cases where the step exceeds about one-fourth of the total thickness (Fig. 99A), the product becomes too nonuniform for all practical purposes (Fig. 99C). In such cases the rate of compression over the cross section must be equalized by multiple pressure application using several individual punches acting in parallel. The beneficial effect of two telescoping punches on the density distribution is shown in Figure 99E.

Advantages of such a molding procedure are perhaps even better demonstrated if a more exaggerated difference in section is chosen. Figs. 100

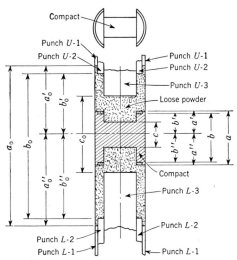

Fig. 100. Idealized powder fill and molding of multiple-level armature body; compression ratio, $a_0:a = b_0:b = c_0:c = 3:1$.

to 102 illustrate the molding of an iron rotor (for a d.c. generator for field telephones) with the aid of multiple punches acting from two opposite directions.[6] Despite the extent of the two steps, the density of the molded compact varies less than 2% throughout the cross section.

Figure 100 illustrates an idealized powder-fill volume, corresponding to the same compression ratio for each plane parallel to the pressure axis, and disregarding any possible lateral slippage of individual particles. The compression ratio chosen for this illustration is 3:1, obtained with an iron powder having an approximate apparent density of 2.5 g./cc. If the relative position of the compact in the die were not to change when applying this compression ratio in every section, a theoretical powder-fill curve of very different levels would be necessary. We have already shown that there exist few, if any, practical means of properly locating the particles, of fixing their positions, or of influencing their paths during compression—

[6] C. G. Goetzel, *Trans. Am. Inst. Mining Met. Engrs.*, *166*, 506 (1946).

especially at its early stages. Hence, this theoretical fill condition is not feasible in actual operation, and the powder must be moved in its relative position in the mold while compression goes on. This expedient enables flush filling of the die cavity and striking off the excess powder from the top face. The level fill is shown in Figure 101A, and the shift of powder toward the center of the die is apparent from Figure 101B. Any possible lateral movements of certain particles—which may result in cave-ins—can be forestalled by initial compression of the sides prior to transfer of the central mass. Molding into a compact of uniform density (throughout the entire cross section) can then be accomplished by applying differential

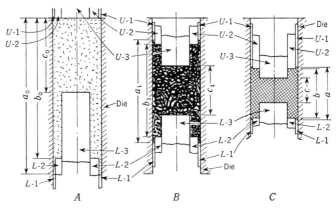

Fig. 101. Flush-level powder fill and molding of armature body (theoretical). A, powder fill; $a_0:a = b_0:b = c_0:c = 3:1$ at $P = 0$. B, transfer of powder in bridge; $a_1:a = b_1:b = c_1:c = 2:1$ at $P = P_1$. C, multiple pressure molding; $P = P_{max}$.

pressures with mulitple upper and lower punches, as shown in Figure 101C. Of course, each of these punches has individual movement and velocity in order to equalize the rate of compression of the powder; for the sake of a uniform density and stress-free structure, each individual punch must exert its own specific pressure. In the case of the armature, shown in Figure 100, the members U_3 and L_3 are solid punches, while U_1 and U_2 as well as L_1 and L_2 represent segments of forklike upper and lower punches. Figure 102 gives the true stages between flush-level powder fill and final compression as actually carried out in a cam-driven mechanical press.

Other methods of overcoming difficulties in the compacting of multiple-level parts have been outlined by Seelig.[7] Flanged designs (Figure

[7] R. P. Seelig, *Powder Metallurgy Bull.*, 1, 54 (1946).

103) are given as examples illustrating these methods with Figure 103*A* showing a part as originally designed for machining operation, and Figures 103*B* and *C* showing alternate designs more suitable for powder metallurgy production. The nominal depth of the hole is the same in all three cases. A punch design suitable for molding parts of this type is shown in Figure 104. The spring-loaded punch penetrates the loose powder prior to compaction by the main punch. Proper adjustment of stroke and load yields satisfactory density distribution. This same general principle can be applied to designs illustrated in Figures 105*A* and *B*. In the first of

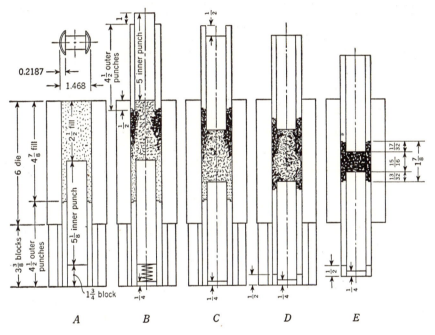

Fig. 102. Flush-level powder fill and molding of armature body (actual): *A*, powder fill; *B*, initial compression; *C*, transfer of powder in bridge; *D*, secondary compression, approximately 3 tsi; and *E*, final compression, approximately 50 tsi.

these examples, the movable, upper punch member is a sleeve, while in the second case it consists of a pair of tongues.

Another design feature requiring special die construction is characterized by projections or bosses as illustrated in Figure 105*C*. A die design suitable for the compaction of this type of shape is shown in Figure 106, where an even density distribution is obtained by using brake shoes to hold back the inner, upper punch. In the compacting cycle, inner, and

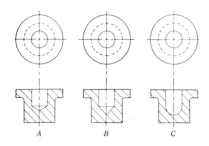

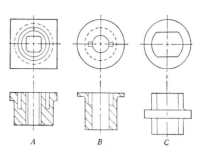

Fig. 103. Flange-type, multiple-level parts (according to Seelig[7]): A, part as originally designed for machining operations; B, alternate design more suitable for molding powders; and C, most suitable design for molding powders.

Fig. 105. Complex flange-type, multiple-level parts (according to Seelig[7]): A, part requiring a sleeve-type, movable upper punch; B, part requiring an upper punch consisting of a pair of tongues; and C, part requiring special die design shown in Figure 106.

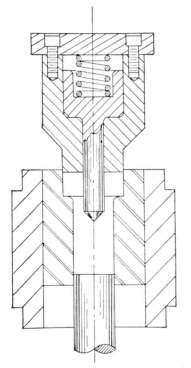

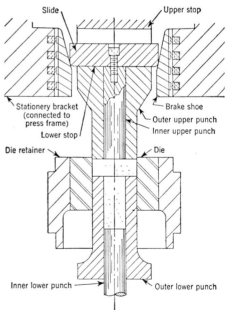

Fig. 104. Tool and die design suitable for molding the flange-type parts of Figure 103 (according to Seelig[7]).

Fig. 106. Tool and die design suitable for molding the flange-type part of Figure 105C (according to Seelig[7]).

outer, upper punches first move downward (remaining flush). As soon as the cavity is sealed, the slide in the ram hits the brake shoes, and the inner punch is held back while the outer punch moves on. Simultaneously, the two lower punches move upward—the inner, lower punch being first. The inner, upper punch does not participate in compaction before the slide hits its upper stop. When the upper ram reverses after completion of the compacting cycle, the outer punch is picked up first, while the inner punch is held by the brake until the two punches are once more flush.

Parts combining a thin flange with a long body require exact timing of the punch movements, in order to prevent the powder from being pushed out of the cavity by the lower punch before the upper punch has a chance to enter the cavity. Certain mechanical presses, however, do not permit a change of the timing. If such presses are used, the loss of powder can be prevented by spring loading the upper punch and thus extending its length prior to pressure application.

COMPRESSION OF CURVED-FACED COMPACTS

The molding of parts having curved faces presents a particularly difficult problem. The loss of pressing force due to internal friction is kept small only if pressing is being done in the direction of the shortest extension of the compact. As long as the curved surfaces are parallel to this shortest axis of pressure, no special difficulties arise. The cross sections perpendicular to the pressure do not vary and the only change during compression is a gradually reduced thickness (Figure 107A). The problem,

Fig. 107. Designs of parts having curved faces: *A,* curved faces parallel to shortest axis; and *B,* curved faces parallel to longest axis.

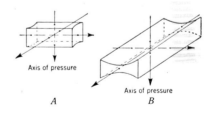

Axis of pressure

Axis of pressure

A   *B*

however, assumes an entirely different aspect if the curved surfaces are perpendicular to the shortest axis of the part (Figure 107B). The density distribution during compression of such a compact is indicated in Figure 108. In accordance with previous observations, we may conclude that in curved shapes: (*1*) the compression ratio of a powder, as given by its apparent density, is the most controlling factor, with the customary trend toward smallest possible compression ratio having a distinctly adverse effect on uniformity; (*2*) the specific pressure used for molding also has adverse effects in the high pressure ranges necessary for complete densification, but is dependent largely on the nature of the powder and on the individual design; and (3) the plasticity of powder particles and the

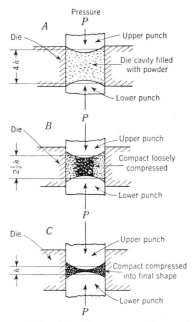

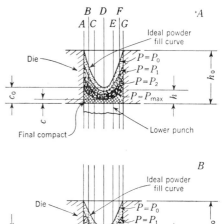

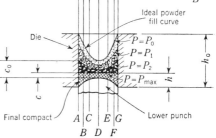

Fig. 108. C o m p r e s s i o n of compact with two curved faces perpendicular to short axis of pressure: A, initial compression ($P = 0$); B, advance compression ($P = \frac{1}{10} P_{max}$); and C, final compression ($P = P_{max}$).

Fig. 109. Idealized molding of concave-shaped body for compression ratio of 4 to 1: A, curved face of final compact formed upwards; and B, curved face of final compact formed downwards.

application of lubricants on die cavity walls and between particles have distinctly beneficial effects on the molding process.

Thus, it becomes apparent that uniformly dense compacts having large concave or convex faces in a plane parallel to the longitudinal axis can be molded only if the compression ratio is large, and if its order of magnitude can be maintained constant in each plane parallel to the axis of pressure throughout the entire compressing operation. This is illustrated in Figure 109 for a concave-shaped body, and in Figure 110 for a part with a convex-shaped face. The illustrations represent idealized cases for a compression ratio of 4 : 1. If this compression ratio is to be maintained over the entire width, the powder of an apparent density of 1.8 cannot be filled flush with the top face of the die. Instead, its top layer of particles must follow an elliptic curve, and the powder fill volume must be strongly concave (Figure 109) or convex (Figure 110). This concavity or convexity of the ideal powder fill curve is only slightly reduced by placing the curved face of the final part downwards (*b*) instead of upwards (*a*).

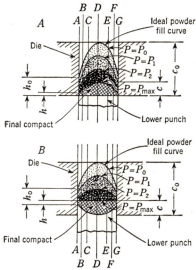

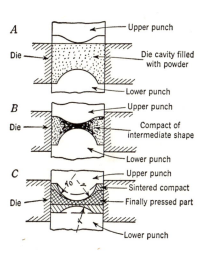

Fig. 110. Idealized molding of convex-shaped body for compression ratio of 4 to 1: *A,* curved face of final compact formed upwards; and *B,* curved face of final compact formed downwards.

Fig. 111. Actual molding of concave-faced pole piece: *A,* flush-level powder fill; *B,* compression of powder into intermediate shape; and *C,* compression of sintered intermediate body into final shape.

This idealized setup must be carried still further. To produce a strictly *isodense* compact, the respective compression ratio must be maintained uniformly throughout the entire cross section at every stage during the actual compression. If, for instance, slight pressure ($P = P_1$) from top to bottom results in one-quarter densification of the powder, the compression ratio must change from the initial 4 : 1 ($P = P_0$) to 3 : 1 throughout the entire cross section. The initial elliptical shape of the top powder particle layer would have to flatten out into another elliptical top face of larger parameter. This, of course, would require two things: *first,* that the top punch change its face to follow the ideal fill curve, flattening out as the pressure and degree of compaction increase; and *second,* that all individual particles move only in planes (A, B, etc.) parallel to the axis of pressure, excluding any lateral movement of particles. The same must be true if the pressure is increased ($P = P_2$) so that the powder is further densified into a compact one-half of the final volume—the compression ratio being reduced to 2 : 1 throughout—or if the pressure is raised to the maximum necessary to obtain the isodense solid body.

In practice, this idealized molding procedure is no more feasible than that previously described for multiple-section compacts, since the shape of the top punch, if made of metal, cannot change during the pressing operation. The ideal elliptic powder fill cannot be obtained by industrial press and molding equipment which, to date, is constructed for level powder fill, flush with the top face of the die. The idealized type of filling would require special loading sheets, complicated feeding devices, auxiliary vibrating or air flow arrangements, and possibly scraping arrangements for removal of excess powder. Finally, there is no way to prevent individual powder particles from deflecting from their straight line course —parallel to the pressure axis—once pressure is applied. The movement of the particles follows complex laws, influenced chiefly by such factors as: (1) interparticle friction, (2) particle shape, (3) plasticity and work-hardening of particles, (4) configuration of die cavity, (5) type of lubricants, (6) speed of pressing, and (7) uniformity of loading, etc. It is known that a powder mass under pressure behaves neither like a liquid nor a solid, and that the properties of a compact, in its capacity to absorb and transmit pressure, approach that of a brittle, solid body only if extremely high pressures are used. It may be said that at low pressures the particles under pressure will follow the way of least resistance, whether in a longitudinal or a lateral direction, or a resultant of the two.

The following practical method of molding such shapes, which may be considered an approach to the idealized case, has given good results in molding certain generator field pole pieces from soft, spongy iron powder.[8] These field pole pieces are flat, rectangular plates approximately 2 in. long, 1 1/2 in. wide, and 3/8 in. thick. One large face, however, is curved, with a radius of approximately 3/4 in. The method of molding this shape involves regular level feeding of the powder, thus making possible the use of industrial molding equipment. The powder to be allotted to each compact is discharged from a feed hopper into the die cavity, and then the excess amount is leveled flush with the top face of the die. As the die cavity and the position of the punches during loading are the same for each press cycle, this volumetric method of controlling the amount of

[8] C. G. Goetzel, *Trans. Am. Inst. Mining Met. Engrs.*, *166*, 506 (1946); see also U. S. Pat. 2,386,604.

powder to be briquetted has been found accurate enough for all practical purposes. However, for extreme accuracy of the final dimensions of the compact, requiring weight tolerances of less than 1%, the batch-weight method has been used successfully. Here, each powder allotment may be weighed with the aid of a photoelectric-cell-controlled, exact-weight scale, and then fed into the die cavity. No levelling of the powder is then necessary, and the lower punch may be set low enough to permit a powder fill level slightly below the die top face, and so prevent losses of powder which would upset the accurate weight method.

To approach theoretically ideal compression ratios, obtainable only with the impractical, variable, curved-face top punch, it was necessary to introduce fractional molding—with intermediate heat treatments—which yielded partly compressed, porous bodies of shapes intermediate between the powder fill shape and the final contours of the dense body. At the same time, favorable material distribution in the intermediate compact was obtained by curving both the upper and lower faces, as shown in Figure 111. Only in the repressing (coining) operation is the one (more shallow) curvature flattened out into a plane face. The number of intermediate steps, and the configuration of the resulting intermediate shapes were found to depend on several factors, namely: (1) order of final density, (2) plasticity of powder particles, (3) their tendency to flow under pressure and to by-pass each other, (4) the use of interparticle lubricants, (5) radius of the curvature in the final compact, (6) conditions prevailing during intermediate sintering, (7) ductility of the metallic body after intermediate sintering, and (8) type of pressing. In the case of the pole piece under consideration, one intermediate shape was already found satisfactory for parts whose density averaged 92% and did not vary more than ± 1 1/2% throughout the cross section.

For this shape, a body was formed whose one radius, r_0, was three-fourths of the radius, r, of the final piece; the opposite curvature had the same radius, r, as the final piece. This design facilitated pressing a sufficiently porous body with considerable material pushed into the heavier side regions to allow coining into a practically isodense part of approximately 5/16-in. thickness at the sides, and 1/16-in. thickness at the center. For the first pressing, a pressure of 5 to 10 tsi was found most suitable, because it yielded sufficiently strong compacts whose thinner center regions, though denser than the sides—were not yet overstrained by excessive local pressure. Moreover, a pressure of this order resulted in bodies of approximately 50% average density, which, after sintering, were sufficiently ductile and compressible to be coined into their final shape at pressures of 50 to 70 tsi.

To obtain these favorable results, the following conditions had to be fulfilled: (*1*) A very plastic iron powder of the reduced sponge iron type had to be used. (*2*) To facilitate interparticle slippage, up to 1% of a powdered metal stearate had to be added to the powder as a lubricant. (*3*) An intermediate porous body, curved on the two opposite sides—the porosity being about 50%—was to be pressed first. (*4*) The compacts were to be pressed from two opposite directions in a single-movement press by means of a floating die setup. (*5*) The porous intermediate bodies were to be heated at a temperature sufficiently high to render the compact ductile.

A similar procedure has recently been suggested by Seelig.[9] Figure 112*A* shows the preform employed for the manufacture of a part having a spherical depression (Figure 112*B*). A punch design producing the preform is shown schematically in Figure 113. Seelig points out that, in practice, individually controlled punch movements would probably be preferable to the spring loading indicated in the drawing.

COMPACTING BY CENTRIFUGING

Although essentially analogous to unidirectional compaction, centrifugal compaction secures a uniform density throughout the depth of the compact. When powders are compacted by subjecting them to centrifugal forces, every individual powder particle is acted on by the pressing force, while in ordinary pressing the pressure must be transmitted from one or two surfaces through the entire powder mass (which does not follow the hydrodynamic laws). It is conceivable that difficulties caused by poor pressure transmission in complicated shapes may be greatly relieved by centrifuging, using the same force as in ordinary pressing. If a liquid lubricant is added to the powder to be compacted by centrifuging, a reduction of interparticle friction should be particularly effective. According to Wulff[10] this method yielded practical results in the manufacture of cemented carbide parts, with glycerine used as a lubricant.

COMPACTING BY EXTRUSION

Compacting of metal powders by extrusion can be achieved at high pressures in appropriate dies, either at room temperature or at elevated temperatures. This method was applied in the historical "paste process"

[9] R. P. Seelig, *Powder Metallurgy Bull.*, *1*, 54 (1946).
[10] J. Wulff, in J. Wulff, editor, *Powder Metallurgy*. Am. Soc. Metals, Cleveland. 1942, p. 255.

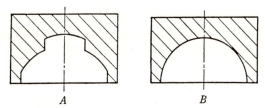

Fig. 112. Design of cup-shaped part with spherical depression (according to Seelig[9]): *A*, intermediate shape having a stepped spherical depression; and *B*, final shape.

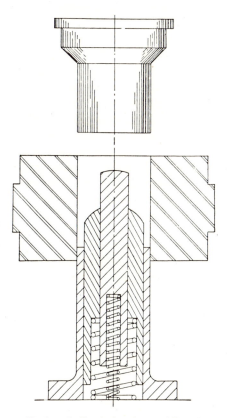

Fig. 113. Tool and die design for molding the part of Figure 112 (according to Seelig[9]).

for the manufacture of lamp filaments from osmium, tungsten, and tantalum. The powder was mixed with organic binders, such as sugar syrup, into a thick paste, which was then forced through diamond dies. The binder was volatilized and the metal powder sintered in a subsequent heat treatment.

In certain instances, the extrusion method may even yield wires directly from the powder without the aid of a binder. These wires may be heat-treated after extrusion. For example, it is possible to extrude bismuth powder into a ductile wire at room temperature. If of sufficiently high purity, the powder can be forced through an orifice as small as $1/32$ of one inch, and the resulting wire will be of sufficient coherence to make it moderately flexible. Prepressing and sintering before extrusion would result in a further improvement in the ductility of the wire.

Although the technique of extruding metal powders has been in the experimental stage ever since the manufacture of lamp filaments from pastes more than 50 years ago, the first serious commercial use of the process was made only very recently. The Carboloy Company, Inc.[11] has made available extruded, cemented carbide rods, spirals, tubes, and nozzles in lengths up to 20 in. Rods can be made in a diameter range of from 0.015 in. to 0.375 in., tubings as small as 0.060 in. outside diameter by 0.030 in. inside diameter. In this process, too, a plastic binder such as starch or gum arabic is mixed with the carbide powder for the extrusion process, the products being subsequently baked, and finally sintered at a high temperature. A similar method is used for extruding powders into special rods for arc welding.[12] In this process, besides the usual binder, a small amount of flux (1% borax) is added to the powder. A typical composition contains 39% copper, 53% phosphor-copper (15% phosphorus), and 8% tin.

A recent publication[13] describes the manufacture of synthetic welding rods by compacting a powder mixture of alloying elements around a bare wire, sintering, and finally applying a flux coating. The use of suitable binders and lubricants permits compaction to take place by extrusion, the procedure being similar to the customary flux-coating method.

For a survey of patents concerned with the direct extrusion of metal powders, an article by Jones[14] should be consulted.

[11] E. W. Engle, in J. Wulff, editor, *Powder Metallurgy*. Am. Soc. Metals, Cleveland, 1942, p. 442.
[12] C. E. Pearson, *The Extrusion of Metals*. Chapman & Hall, London, 1944, p. 201.
[13] F. G. Daveler, *Materials & Methods, 23,* 1917 (1946).
[14] W. D. Jones, *Metal Ind. London, 57,* No. 2, 27, (1940).

COMPACTING BY ROLLING

In certain industrial fabricating processes metal powder is compacted while passing through a rolling mill. This method has been developed especially in connection with the manufacture of certain steel-backed bearings where the steel strip functions as a support during processing and ultimate service of the bearing. The most important applications of this method can be found in the production of copper–nickel–lead[14a] and copper–lead[14b] bearings (see Chapter XXVI, Volume II.)

DESIGN OF INDUSTRIALLY FEASIBLE PARTS

Up to the present time, the applicability of powder metallurgy to a large variety of commercial products has been stressed (in the trade) for promotional purposes. Although this may be helpful in establishing and enlarging markets, it is the firm belief of the author that the industry would best be served by more detailed and critical discussions of basic design rules that limit the use of powder metallurgy parts. Fortunately, the experience of leading powder metallurgists, gained in the last fifteen years with various powdered materials and with many different shapes, has contributed considerably to the scant knowledge of the intricate mechanics of powder compaction. Thus, we are now in a position to give an offhand judgment of many propositions without going into time-consuming and expensive trial-and-error experiments to determine whether a particular part "can be done." In the field today, the design engineer ranks equally in importance with the metallurgist; the mechanical design department has become the clearing house as well as the mechanical brain in any production. Equally gratifying is the fact that in all engineering circles interested in powder metal parts, there can be found an ever-increasing understanding of the particular design problems inherent in the production of these parts.

Basic Rules of Design

Before discussing some basic design rules, we must be aware of the fact that there are often several ways of making, as well as using, a product. Design flexibility is often a controlling factor, not only in the selection of the metal fabricating process, but also—if the powder metallurgy approach is chosen—in the individual molding, coining, or heat-treating

[14a] A. L. Boegehold, in J. Wulff, *Powder Metallurgy.* Am. Soc. Metals, Cleveland, 1942, p. 520.
[14b] E. R. Darby, *Proc. Third Annual Spring Meeting,* Metal Powder Association, New York, May 27, 1947, p. 52.

steps. In studying powder metallurgy fabrication along with drop forging, sand casting, die casting, or screw machine operation, a full understanding of the various design factors and limitations is essential.

Recently, Victor and Sorg[15] discussed the design problems of powder metallurgy parts. Their splendid account is taken as a basis for most of the following discussion, although some aspects have already been covered in the beginning of this chapter.

GENERAL DESIGN FACTORS

The following design factors must be given due attention when considering parts for production on a large scale: (1) size of part, (2) ratio between length and width, (3) lateral projections, (4) undercuts, threads, and re-entrant angles, (5) tapers and bevels, (6) holes in longitudinal and lateral directions, (7) density variations, (8) weak sections and abrupt changes in thickness, (9) insert molding, and (10) tolerances.

Size of Part. This topic has already been discussed in Chapter VIII from the standpoint of compressibility. It may suffice to repeat that sizes of actually produced parts may vary from about $1/8$ in. to 50 in. in cross section and from $1/32$ in. to about 6 in. in height. The limitations are imposed either by wall friction or press capacity and internal friction, that is to say, by factors which could be overcome by increase of the press capacity. It can therefore be said that the limitations are imposed essentially by economic necessity.

Ratio between Length and Width. For reasons mentioned previously, the length of a part in the direction of pressing should be comparable to the cross-sectional area. Since all practical molding is accomplished on a straight-line principle (from top and bottom only), an extremely long piece will have very unequal density, with the middle portion having lowest density and least strength.

Lateral Projections. As a rule, lateral projections constitute a major feature of powder metallurgy parts. A great variety of shapes is possible, such as the ones shown in Figures 114A and B.[16] But parts of columnar longitudinal sections are most economical if they have only one lateral projection or flange (Figs. 114C and D).[17]

Undercuts. If the design requires two lateral projections on opposite ends of the longitudinal section, an undercut results that cannot be molded

[15] M. T. Victor and C. A. Sorg, *Metals & Alloys, 19,* No. 3, 584 (1944).
[16] R. P. Seelig, in J. Wulff, *Powder Metallurgy.* Am. Soc. Metals, Cleveland, 1942, p. 264.
[17] E. H. Kelton, *Machine Design, 16,* 129 (1944).

by regular means and must be machined in a subsequent operation (Fig. 115*A*). The same applies for re-entrant angles and internal or external threads (Figs. 115*B* and *C*).

Tapers and Bevels. Much the same rules apply for tapers and bevels as for undercuts. Slight internal or external tapers, usually not in excess of 5 to 10 minutes, are normally introduced to facilitate ejection, but designs with larger tapers, internal angles, or re-entrant angles must first be molded in full and then machined after sintering (Fig. 116*A*).

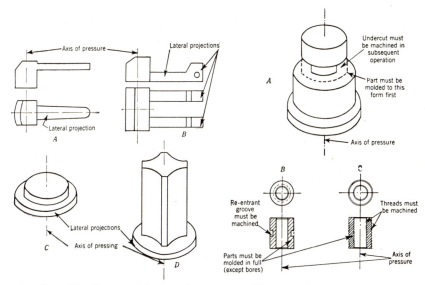

Fig. 114. Forms with lateral projections (*A* and *B* according to Seelig[16]; *C* and *D* according to Kelton[17]).

Fig. 115. Forms with undercuts (*A* according to Kelton[17]; *B* and *C* according to Victor and Sorg[15]).

Bevels must have the smallest possible angle to the transverse plane (Fig. 116*B*). Steep bevels (Fig. 116*C*) require feather-edged punches, which show early failures. Internal angles must have fillers, since with steady wear of the die parts, the powder will also work into the crevices formed by these angles and a fin or flash between them and the molded parts will result.

Holes in Longitudinal and Lateral Direction. There is no difficulty in molding parts with one or even more holes in the direction of pressure (Fig. 117*A*), but those not parallel must be machined after sintering (Fig. 117*B*).

Density Variations. For the sake of high strength, density variations must be kept at a minimum. Since, as has already been explained, metal powders have almost no lateral flow in the die—as a liquid or a plastic would have—they must be placed in the mold with their particles oriented so that the straight-line compression moves them into their final locations. Where there are sections of different length in the direction of

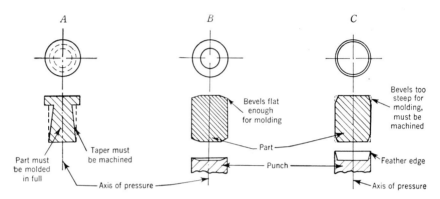

Fig. 116. Forms with tapers and bevels according to Victor and Sorg[15]: A, tapered part; B and C, beveled parts.

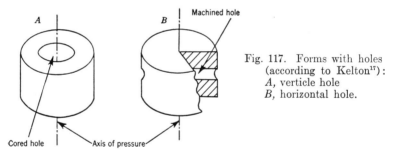

Fig. 117. Forms with holes (according to Kelton[17]): A, verticle hole B, horizontal hole.

compacting, and where the part is round in cross section so that it can be compacted only in the longitudinal direction, the density will vary greatly if single punches are used. Consequently, every level of the finished product requires its own separate die insert or punch segment which moves or presses independently to allow for uniform density through equalized rates of compression in the various sections.

Weak Sections and Abrupt Changes in Thickness. Molded shapes requiring feather edges, very narrow or deep splines, or small pins (Figs. 118A and B) should be avoided, since these parts may easily collapse upon ejection. Dies in powder metallurgy must withstand great pressures, often exceeding 100,000 psi, which expand them elastically, perhaps several thousandths of an inch. After pressure release, the compact must resist the tendency of the die to contract to its original form. But this resistance is constantly diminished upon ejection, until the part still remaining in the die succumbs to the contracting force of the die. Depressions, bosses, and counterbores should be kept as shallow as possible, with sides tapered to permit easy removal from the die. For thesame reason,

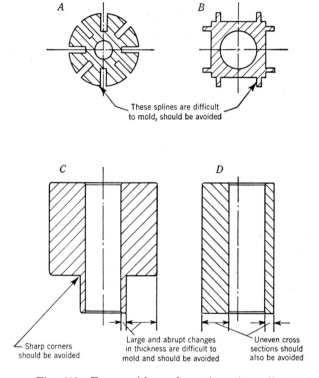

Fig. 118. Forms with weak sections (according to Victor and Sorg[15]): *A* and *B*, forms with splines; *B* and *C*, forms with uneven sections.

it is advisable, on parts having several steps, that each step be at least $1/_{16}$ in. in diameter larger than the body of the preceding step. In addition to these restrictions imposed by the ejection operation, large and abrupt changes in thickness or cross-sectional area (Figs. 118*C* and *D*) cause different rates of volume changes, during sintering, making such parts difficult to size, or causing local fractures or laminations.

Insert Molding. Metal inserts are not usually molded from metal powders, but are pressed into place after molding and sintering. When a powder metallurgy product is to be used as an insert in a plastic molding, vertical sections parallel to the pressure axis can be molded on the outside diameter to serve the purpose of knurling, which would otherwise require an additional operation.[17a]

Tolerances. While from a production standpoint it is always desirable to have tolerances as liberal as possible, powder metallurgy parts can be molded to very close dimensions. Radial tolerances of ± 0.0005 in. and axial tolerances (in the direction of pressing) of ± 0.005 in. can readily be met by the bearing industry and most other branches of the field. Closer tolerances are not impossible, but may require frequent die and tool changes or additional sizing or gaging operations. It is often more advisable to maintain extremely close tolerances by machining or grinding, using the same procedure as on cast metals. Powder metallurgy parts have machining characteristics similar to those of castings, but inherent porosity often necessitates frequent regrinding of tool bits, for which tungsten carbide tools have proved excellent. Of course, no coolant can be used on parts which must later be oil-impregnated.

TYPICAL ROUND DESIGNS

According to Victor and Sorg,[17a] typical round designs in powder metallurgy parts include: (1) cylindrical parts, (2) flanged parts, and (3) profiled parts.

Cylindrical Parts. The vast majority of parts for the bearing industry fall into this category. Of great importance in the design of sleeves and rings is the relationship between the wall thickness and the length and outside diameter. Manufacturers of individual parts have compiled tables of these data based on their particular equipment. It is understood that, for the sake of economy, wall sections should be kept as large as possible. Tolerances readily maintainable at low production costs vary from 0.001 in. per inch on the inside or outside diameter to 0.005 to 0.008 in. per inch in the direction of pressing. Eccentricity is another important factor; a standard tolerance between inside and outside diameter is 0.003 in., while closer tolerances down to 0.001 in. can be maintained at higher costs.

Flanged Parts. In order to maintain the strength of the compact, fillets between the flange and the body are highly desirable, especially since sharp corners are difficult to mold and may require additional operations. A minimum radius of 0.010 in. is advisable. It has also been sug-

[17a] M. T. Victor and C. A. Sorg, *Metals & Alloys, 19,* No. 3, 584 (1944).

gested that the diameter of a flange on a part longer than $3/4$ in. should not be more than 1.5 times the outside diameter of the part itself. In order to facilitate ejection from the die, the length of the flange should be tapered 0.001 in. for each $1/8$-in. flange thickness.

Profiled Parts. Perhaps the best examples of profiled parts ideally suited for fabrication by powder metallurgy methods are oil pump gears used in the automotive industry[18] (Fig. 119A). These gears have true in-

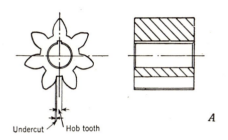

Undercut   Hob tooth

A

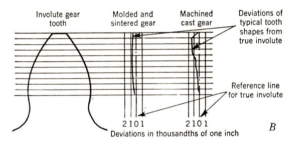

Involute gear tooth   Molded and sintered gear   Machined cast gear   Deviations of typical tooth shapes from true involute

Reference line for true involute

2 1 0 1   2 1 0 1   *B*
Deviations in thousandths of one inch

Fig. 119. Oil pump gear as example of profiled parts (according to Lenel[18]): *A*, gear; *B*, gear tooth.

volute gear-tooth forms which cannot be hobbed due to the profile undercut. Despite the very close tolerances in such parts, only four finishing operations are necessary on the molded gear: burnish broaching of the inside diameter, grinding of the outside diameter, machining of the end length, and chamfering and simultaneous deburring of the teeth. Only a few thousandths of an inch of material is removed in any of these operations, as compared with approximately 64% of material which must be removed when machining the gear from the conventional cast-iron blank. The most important factor, however, is the precision of the tooth shape.

[18] F. V. Lenel, in J. Wulff, *Powder Metallurgy*. Am. Soc. Metals, Cleveland, 1942, p. 502.

If the teeth do not mesh perfectly, oil will leak back from the outside of the pump to the inlet side, and the efficiency of the pump will be reduced. The shape of the teeth should follow the involute curve as nearly as possible (Fig. 119B). The molded and sintered gear shows a deviation of less than 0.001 in. against a deviation of nearly 0.0015 in. for a machined, cast iron gear. The diagram (Fig. 119B) also illustrates that where the teeth mesh, deviations of the involute of a molded gear fall on a straight line, while the machined gear shows its greatest irregularities in the same section.

Parts of Intricate Design Features

In recent years, competition in the manufacture of intricate parts has developed between powder metallurgy and other methods. it is of particular importance that the limitations inherent in the metal powder molding technique be recognized at this point. The chief problem of the pressing operation lies in the fact that it is necessary to fill the mold with powder whose volume is about three times that of the final shape. The fact that metal powders do not act like liquids or plastics imposes some of the general limitations on the powder metallurgy process. On the basis of the preceding discussion, it becomes obvious that the powder cannot be expected to fill intricate projections or to flow around corners. Auxiliary devices, such as vibrators or lubricants, can help to overcome this problem, but cannot be counted on for all cases.

The automatic removal of the compact is essential in all production work, making it imperative to mold a part in such a manner that the largest cross section of the die will be formed on the side toward which the piece will be removed.[19] While, on rare occasions, split dies may be used advantageously to facilitate the removal of intricate shapes (having several lateral projections), customary molding practices rule out this procedure. Therefore, any design of a metal powder part must permit unobstructed compression of the powder, as well as ejection of the compact— both to take place along the axis of pressure. It should be worth while to describe these rules briefly on the basis of a few typical examples. (The cited parts have been molded by Seelig[20,20a] on a commercial scale and his interesting publications should be consulted for further details.)

Figure 120A shows a pin molded from ferrous metal powder. To achieve satisfactory results, special attention must be paid to the ratio of height to thickness. L must not be too great in relation to the area at

[19] R. P. Seelig, *Metals & Alloys, 12,* No. 6, 744 (1940).
[20] R. P. Seelig and E. C. Gordon, *Machine Design, 13,* No. 4, 42 (1941).
[20a] R. P. Seelig, in J. Wulff, *Powder Metallurgy.* Am. Soc. Metals, Cleveland, 1942, p. 264.

the diameter, D. While no definite rule has been set forth, a ratio of 3 : 1 has been found workable for brass and bronze and a ratio of 2 : 1 for ferrous compositions, with higher proportions resulting in unsatisfactory density distribution and mechanical weakness after sintering.

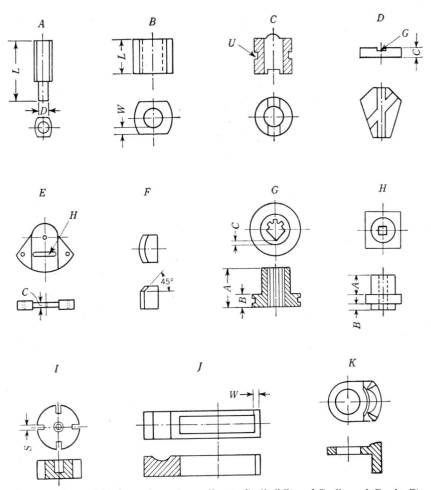

Fig. 120.   Parts of intricate shape (according to Seelig,[19,20a] and Seelig and Gordon[20]).

Another part which could easily be formed is shown in Figure 120B. The center bore could be made either straight or tapered, to facilitate ejection. Again, the ratio between height and thickness is important; it cannot be too great, and the wall thickness (W) must not be too small since, in pressing, the powders would flow poorly at that point, causing nonuniform density. Figure 120C shows a similar part, with an end projection and an undercut (U). Molding the projections is not particularly difficult as long as they are not too large, but machining is required for the undercut.

A shape of the type shown in Figure 120D is very suitable for molding since its size is governed only by the capacity of the press. To allow for a sufficient green strength, its thickness (C) in the grooved sections should, preferably, not be less than $1/8$ in. For the same reason the depth of the grooves (G) should not exceed the thickness of the plate.

The part shown in Figure 120E could readily be pressed from metal powders, if points are kept in mind concerning the depression (C), and the center section with the hole (H). Radical and numerous changes in the depth of C, though feasible, create ejection problems and may result in nonuniform density. An accurate control of the powder fill would require complicated dies and multiple-action presses. The center section including the hole (H) must be thick enough to allow for a die core sufficiently strong to assure accurate centering. It is usually more economical to drill small holes after sintering than to complicate the die with very fine core rods.

In Figure 120F an easily molded bevel part is shown. Caution is necessary, however, in designing such bevels. If the angle is too steep, the male punch would require a very weak feather-edged section which would cause frequent production failures. Usually at least a 60° angle vs. the pressure axis is recommended, but tough die steels permit molding of 45° angles (Figure 120F).

Probably one of the most interesting parts of Seelig's collection is the complicated flange part shown in Figure 120G. Molding this form from metal powders completely eliminates the need for broaching the center hole, thus creating an interesting opportunity for the powder metallurgy approach. When the conventional method of machining is used, this part can only be made of bronze, because the broach breaks regularly in production when steel is used. By powder metallurgy, however, this part can be molded from iron and steel as well as from nonferrous powders. If the wall (C) is to be kept thin, the flange (B) cannot be made too small in relation to the over-all height (A). Molding is facilitated by reducing the difference between (A) and (B), or by increasing the wall thickness. Of course the lateral groove must be machined after sintering.

A part of somewhat less complicated shape is shown in Fig. 120H. If difficulties in fill and ejection are to be avoided, the distances A and B must be quite large. The little wheel in Figure 120I can also be molded if the slots (S) are not too narrow—otherwise the splines in the punches would easily break in production. The center bore may either be cored or drilled afterward. Figure 120*J* shows a part which also can be molded readily if the wall thickness (W) is not too small.

The steel cam shown in Figure 120*K* is interesting in that it must be molded with split punches forming the bottom of the die. During ejection these punches must be locked, to prevent movement of individual sections which might result in shear cracks on the inside of the flange.

Design Factors for Comparison with Competitive Fabricating Methods

No discussion of commercial molding of metal powders into mechanical parts is complete without a proper appraisal of what can be accomplished by the other, long-established methods of fabricating parts. The true competitive possibilities of the metal powder molding technique—disregarding its ability to produce unusual structures and properties—will best be understood after a brief survey of some of the conventional fabricating methods from the standpoint of design engineering. Peters,[21] in a recent critical analysis of this subject, lists eight methods in competition with metal powder molding: (*1*) sand-casting, (*2*) die-casting, (*3*) permanent-mold casting, (*4*) precision-casting, (*5*) machining of bar stock, (*6*) cold-heading, (*7*) drop-forging, and (*8*) stamping and drawing.

Individually, their competitive positions with respect to powder metallurgy are generalized by Peters.

Sand-Casting. Sand-casting of small parts is often an inexpensive method of manufacture, since raw material costs for gray iron are lowest of the various methods, and tools and molds need not be of supreme quality. In design flexibility the method rates about highest among competing processes. The possibility of producing complicated, cored, and undercut parts in one piece puts it in a favorable position over the powder metallurgy method. Multi-impression molds may give higher production rates for small castings than obtained with similarly sized parts pressed from powders. On the other hand, sand castings cannot be produced to great precision, with tolerances usually being as wide as ± $^1/_{16}$ in. per inch. Steel castings are less precise and more costly than iron, with nonferrous castings the most expensive of all. Where close tolerances are required, sand castings must be machined, adding considerably to the general costs of pattern making, mold making, melting, and pouring. Additional cost is entailed by waste caused by gates and risers.

[21] F. P. Peters, *Trans. Am. Inst. Mining Met. Engrs.*, 161, 527 (1945).

Die-Casting. Powder metallurgy has not yet entered into extensive competition with die castings of zinc, aluminum, or magnesium alloys, although both processes have high production rates, minimum waste, and close tolerances in common. But the die-casting method offers certain advantages which are responsible, at least in part, for its superior competitive position. Raw materials are considerably cheaper than those metal powders which would yield comparable physical properties of the product. Tool costs—especially for the extended runs generally practiced in die casting—are lower, and production rates for many small parts extremely high.

Flexibility of design is limited in both cases, but the possibilities in die-casting are greater because of the better flow of the liquid metal. Dies must withstand less pressure than in powder molding, permitting sacrifices in strength for flexibility. Multiple actions, inserts, and slides constitute less of a problem than in the case of powders. Precision in die casting is very high, with tolerances usually about 0.001 in. These tolerances cannot generally be held in powder metallurgy parts (which are simply sintered to compete with the great economy of die castings). Even though coined or sized sintered parts can be made to closer tolerances than 0.001 in.—especially in the radial or lateral direction—economy suffers when this additional operation is added.

Permanent-Mold Casting. This method is, in many respects, somewhere between sand- and die-casting. It is most suitable for medium production runs, where precision requirements are not high enough to warrant die-casting or powder metallurgy methods. Tolerances can usually be held to ± 0.005 in. The intermediate position of this process is also reflected in flexibility, of design, which is less than in sand-casting, but greater than in the more precise methods.

Precision Casting. Many precision casting methods have had their advent in recent years through large-scale use for industrial and war products. The lost wax or similar investment molding processes—in conjunction with centrifugal pressure or vacuum casting—combine the permissible design intricacies of sand-casting with precisions close to that of die castings, with tolerances of ± 0.003 in. being obtainable. Production rates are high, because only one permanent master mold is required for each design. However, economy of this method suffers because of material waste and because of the need for skilled labor. But it is ideally suited for the large-scale production of small parts, too intricate to be molded from powders, whose tolerances are too close for sand castings, and which must be made from metals melting too high to be die-cast.

Machining Bar Stock. Parts machined from bar stock are forms most often regarded to be in closest competition with powder metallurgy parts. Although the comparative factors depend almost entirely on the individual case, the outstanding differences between the methods are summarized by Peters. (*1*) Machining involves considerable waste metal and powder metallurgy almost none. (*2*) Production rates are generally much higher for powder metallurgy than for machining (with the exception of screw machining). (*3*) Machining necessitates a greater tie-up of machines, tools, and skilled labor. (*4*) Mechanical properties of pressed and sintered parts are usually inferior to those of machined bar stock. (*5*) Powders in all cases cost considerably more than bar stock. (*6*) Design flexibility is much greater for machined parts than for parts molded from powders. (*7*) Tool costs are considerably higher in powder metallurgy. To these factors may be added an accounting of the time necessary for tooling up; it is much greater in powder metallurgy than in machining, not only because the latter method has a wealth of accumulated experience at its disposal but primarily because tooling for *any* molding process is more involved than for metal-removing processes. The use of powder molding is indicated where the first three factors predominate over the remainder. The decision may depend on the amount of metal that need be removed if the part were to be machined from bar stock. Tolerances may be held slightly closer by ordinary machining, but the difference, in most cases, is not great enough to make the powder method noncompetitive.

Exceptions, of course, are the automatic screw machine and multiple spindle operations, which normally surpass powder metallurgy, both in speed and in precision. Screw machine applications are therefore beyond competition, except for complex shapes that are moldable from powders and would involve excessive metal removal by the machining operation.

Cold-Heading. Another process for making small parts of moderately intricate design is cold-heading. The process, however, is limited to designs involving upsetting, while axial cavities must be machined. Tolerances of ± 0.002 in. are obtainable; production rates are high and mechanical properties surpass those of singly pressed and sintered parts. Waste during the process is small and surface finishes compete with the best that powder metallurgy can offer. High costs of the heading dies—subject to considerable wear and tear—as well as the need for annealing most parts, make the method less attractive on a cost basis. Similar factors, however, must also be taken into account in high-quality powder metallurgy parts requiring coining and subsequent heat-treatments.

Drop-Forging. The best known advantages in drop forgings are their high strength and toughness, and for this reason, parts processed from metal powders by simply molding and sintering cannot possibly compete. Tolerances, on the other hand, are wider than for powder metallurgy, approximating those obtainable in permanent-mold castings (0.004–0.012 in.). Forgings entail scale, and more waste caused by flashings, than would molding from powders, though less than in sand castings. Tool and die cost are generally high, making the method most attractive for large-scale production.

Stamping and Drawing. This method is a typical mass-production method, involving high tool costs, relatively low raw material costs, some waste, and medium labor charges. It competes favorably with powder metallurgy and die-casting methods where high strength is of importance, and where very thin parts must be formed. Tolerances may be held from 0.001 to as low as 0.0003 in.

Seelig[22] includes in his review of competitive methods the following processes not covered by the survey of Peters: cut extrusion, hot-press forging, and assemblies.

Cut Extrusion. Extrusion and cold-drawing methods frequently compete with powder metallurgy in the production of parts with uniform cross section, that is to say, parts which are particularly suited for powder metallurgy techniques. Cut extrusion can be applied to brass, bronze, aluminum alloys, as well as to steel (including stainless steel). The tolerances are about equivalent to those obtainable by powder metallurgy, and tooling and production costs are very low. On the other hand, one of the main advantages of powder metallurgy, as compared with cut extrusion, is that internal shapes can be produced more readily.

Hot-Press Forging. This method, which has gained importance both for ferrous and nonferrous metals—particularly brass—is subject to similar design limitations as powder metallurgy. High-dimensional accuracy can, however, be achieved only at considerable sacrifice of tool life.

Assemblies. Parts made by joining separately produced components usually exhibit an accuracy inferior to that obtainable by powder metallurgy. Powder metallurgy is employed not only for the production of components to be assembled, but also, in many instances, to replace assembling methods and produce, more economically and in a single operation, parts previously obtained by assembling components.

Summary

Uniformity of density in a compact which controls its final properties depends largely on the method of applying the pressure. While triaxial

[22] R. P. Seelig, *Steel, 117*, No. 21, 116 (1945).

application of pressure is ideal, the closest practical approach to uniformity can be found in a two-directional compression along one axis. The density variation is then considerably less than if the pressure is applied from only one direction.

The primary concern in compacting a powder must be that the compression ratio, *i.e.*, the ratio between the apparent density of the powder and the relative density of the compact at any stage of compression, remains constant throughout the cross section of the form, regardless of the complexity of the contours. In multiple-section compacts, this can be achieved only by keeping the rate of compression constant in each plane parallel to the pressure axis. Multiple punches with individual movements, velocities, and pressures may be needed if the difference in level becomes too great. With curved compacts, other expedients may be helpful, *e.g.*, preforms whose shape may be intermediate between the final form and the powder fill form dictated by practical necessities. Where very long parts, possibly of profiled contours, must be formed with a uniform density throughout the cross section, compacting by extrusion is indicated.

The design of industrially feasible parts that can be molded from metal powders must follow certain basic rules, founded on the necessity compacting a loose· powder into a coherent, solid body of considerably smaller volume, and on the peculiar behavior of the powder when subjected to pressure. The design engineer who considers the powder metallurgy method must be aware of certain general design factors, of which size of part, ratio between length and width, abrupt changes in section, projections, and holes and tolerances are most important. While flanged parts are generally moldable from powders, undercuts must always be machined after sintering. Profiled parts, such as certain types of gears, are especially attractive to the powder metallurgist.

Many intricate shapes are producible from powders and ever-increasing experience has made it possible to reach closer tolerances and higher production rates than considered feasible a few years ago. In molding ordinary machine parts in competition with other fabricating methods, one must concede that the other processes display individually some, and collectively all, of the advantages that the powder method can offer. Perhaps the sole exception to this is material waste, where powder metallurgy appears to be most economical. Fastest production is usually attainable by die-casting, and closest tolerances through machining, including screw machine operations. The relatively high cost of tools makes the molding of ordinary machine parts from metal powders economical only on a mass production basis.

CHAPTER X

Dies and Punches

GENERAL ASPECTS OF CONSTRUCTION

Our survey of the various factors to be considered in the compaction of metallic powders, and also of the questions involved in the design of metal powder parts, has no doubt suggested that many of these parts can be molded by employing only one press of sufficient capacity. The mold, on the contrary, has to be tailor-made for each particular shape and for each different powder to be used. Consequently, since the die problem is of utmost importance in any molding of metal powders, we shall first discuss some general aspects of die construction and then present in detail typical die designs used in powder metallurgy.

Without a mold, no satisfactory pressure distribution can be obtained in a powder mass being consolidated by outside force (perhaps, with exception of rolling thin powder layers onto steel backs); and without a mold no precise form can be retained prior to the application of heat, unless the powder is wet or mixed with a plastic binder. Conversely, any mold to be used for metal powders must fullfill two basic demands: it must give the compact the desired contours, and it must be strong enough to withstand the force necessary to make the form. There remain a number of other important points that must be given close attention. The die must be constructed of a material that will not only be strong enough to withstand, without deflection, the high specific pressures applied, but will also resist the abrasion caused by friction of the fine particles against the side walls of the mold. The construction must permit removal of the compact in such a way that considerably less force is necessary for ejection than for compaction. The construction must permit rapid operation, and yet the tool must be made at an economical cost.

Die requirements include the following essential points. (*1*) The die cavity must be a true negative of the desired shape of the compact. (*2*) The die construction must be of sufficient strength to retain the shape during compaction to closest precision. (*3*) The die construction must facilitate even pressure distribution. (*4*) The die material must satisfactorily overcome friction during compaction and ejection. (*5*) The die material must withstand wear and abrasion. (*6*) The construction must allow for easy refitting or relining of the die cavity, and for replacement of vital sections.

A powder metallurgy die can be designed intelligently only if certain basic information is available concerning the powder and press to be used, and if this is accompanied by a sound appraisal of the elements of die construction.

Prerequisite Information for Die Construction

FACTORS IMPOSED BY THE POWDER

Among the factors involving the powder the following are most important: (*1*) compactibility, (*2*) abrasiveness, and (*3*) the use of lubricants.

The compactibility of the powder governs the specific pressure necessary for the molding of the compact; it thereby effects the maximum unit pressures to which the die is subjected. The die, as well as punches and core rods, must be constructed to withstand the stresses of the applied force. The importance of good powder compactibility has been discussed in Chapter IV, and the order of magnitude of specific molding pressures has been discussed in Chapter VIII (Table 35, Fig. 92). These data serve to evaluate the stresses to which punches and die walls are subjected, with due allowances for complexity in contours or possible misalignments of certain component parts or of the entire die assembly. This subject will be treated more fully in the discussion of die strength. Abrasive particles in the powder usually cause excessive wear on the walls of the die, and the high friction produced by these particles necessitates an increase in the molding pressure. The deleterious effect can be counteracted by the use of lubricants, either incorporated in the powder or applied to the die walls, or by the use of especially wear-resistant materials (cemented carbides) for the die liners. The beneficial effect of using lubricants with the powders, both in decreasing the necessary compacting pressure and in equalizing the density over the length and cross section of the compact, has been discussed extensively in Chapters VII and VIII.

FACTORS IMPOSED BY THE PRESS

The type of press and method of pressing have a deciding influence on the construction of the die. Not only must the necessary pressure be delivered, but it is also essential to decide whether a fast-acting mechanical press or a slow-acting hydraulic press is involved, and whether the press acts in one or two directions. Certain other factors which have a direct bearing on the die design are (*1*) pressing to a uniform density, (*2*) method of filling the die, and (*3*) method of removing the pressed compact. The problem of molding metal powders into uniformly dense parts has been analyzed in detail in Chapters VIII and IX. Although absolute uniformity of density cannot be achieved, certain expedients can aid considerably in approaching this goal. Besides lubricants, multiple punches or spring-loaded punch or die members can transfer sections of the powder into more desirable portions of the die cavity during molding. During each compaction, the die is filled to an accurately predetermined level. A uniform and reproducible fill may be obtained by such methods as evacuating the die cavity (possibly by a sudden withdrawal of the lower punch) or vibrating the die or feed shoe. Ejection of the compact must be simplified in any practical die design. Usually the lower punch ejects the compact upwards, but downwards ejection is also possible (especially with hot-pressing where the lower punch is withdrawn and suction or pressure from the upper punch moves the compact downwards). Where the compacts are very fragile, however, removal may have to be made by hand after the die cavity has been widened by means of a split-die assembly.

Elements of Die Construction

Any die to be used for the compression of metal powders must provide a cavity having the negative contours of the desired shape. The degree of accuracy to which the cavity must be held depends upon several factors, but mostly on the plasticity of the metal powder, which controls the necessary forming pressure, and on ability of the compact to retain its dimensions during the subsequent heat treatment. As a rule the die cavity must be dimensioned below specified part measurements so as to compensate for elastic overexpansion of the die. This becomes more important with higher forming pressures. A good approximation for hardened steel dies is 0.05% undersize for the diameter or width for each 10 tsi applied. It is more difficult to compensate for change of dimensions during sintering, since considerable experience is necessary to control and reproduce these changes in production work. Many powders cause considerable volumetric changes, whose consistency usually depends on many factors, such as particle size distribution, density distribution in the compact, sintering cycle, alloying, or furnace atmosphere conditions. To complicate matters further, dimensional changes during sintering are by no means equal in all directions. Drapeau,[1] for instance, found that copper compacts showed a tendency toward shrinkage in the diameter and growth in the length. These difficulties are aggravated in the case of more intricate shapes, and inability to control the dimensions during sintering has led, in many cases, to a subsequent sizing operation.

[1] J. E. Drapeau, Jr., in J. Wulff, *Powder Metallurgy*. Am. Soc. Metals, Cleveland, 1942, p. 323.

TABLE 36

Die and Punch Materials

Applications[a]				Dies[b]		Punches[b]	
Shape	Powder	Pressure	Inserts and liners	Reenforcement	Facings		Bases
Si	So	L	Hardwood Sn–bronze Cold-rolled steel Low-alloy steel	Experimental	Hardwood Sn–bronze Cold-rolled steel Low-alloy steel		
Si	Ha	Hi	Be–bronze Surface-hardened steel Medium-C steel Tool steel *Hardened to C-45*		Be–bronze Surface-hardened steel Medium-C steel Tool steel *Hardened to C-40*		
C	So	L	Be–bronze Water-hardened steel *Hardened to C-45*	Brass Hot-rolled steel Cold-rolled steel	Be–bronze Surface-hardened steel *Hardened to C-40*		
C	Ha	Hi	Cr–Ni alloy steel Cr–W high-speed steel High Cr–C tool steel *Hardened to C-56*	Machine steel Low-alloy steel	Cr–Ni alloy steel Cr–W high-speed steel Medium Cr–C tool steel *Hardened to C-50*		*Hardened to C-45*

TABLE 36 (*Concluded*)

Production

Applications[a]			Inserts and liners	Dies[b]	Reenforcement	Facings	Punches[b]	Bases
Shape	Powder	Pressure						
Si	So	L	Medium-C steel Low-alloy steel *Hardened to C-58*		Machine steel Low-alloy steel *Hardened and drawn to C-25*	Medium-C steel Surface-hardened steel Low-alloy steel *Hardened to C-50*		Low-C steel *Hardened to C-45*
Si	Ha	Hi	High Cr–C tool steel Cr–W high-speed steel *Hardened to C-62* Hard-chromium plate Cemented carbide		Low-alloy steel High-Mn steel Medium-C steel *Hardened and drawn to C-25*	High Cr–C tool steel Cr–W high-speed steel *Hardened to C-58* Hard-chromium plate Cemented carbide		Alloy steel Tool steel High-speed steel *Hardened to C-50*
C	So	L	High-C steel Cr–C tool steel Cr–Ni alloy steel *Hardened to C-60* Hard-chromium plate Cemented carbide		Machine steel Low-alloy steel *Hardened and drawn to C-25*	High-C steel Cr–C tool steel Cr–Ni alloy steel *Hardened to C-56* Hard-chromium plate Cemented carbide		Low-C steel Tool steel Alloy steel *Hardened to C-50*
C	Ha	Hi	High Cr–C tool steel Cr–W high-speed steel High Cr–Ni alloy steel *Hardened to C-62* Hard-chromium plate Cemented carbide		Alloy steel Medium-C steel *Hardened and drawn to C-25*	High Cr–C tool steel Cr–W high-speed steel Cr–Ni alloy steel *Hardened to C-58* Hard-chromium plate Cemented carbide		Tool steel High-speed steel Alloy steel *Hardened to C-50*

[a] Si = simple; So = soft; L = low; Ha = hard; Hi = high; C = complex.
[b] Hardness values given refer to Rockwell "C" scale.

MEMBERS OF DIE ASSEMBLY

While the die cavity forms the lateral dimensions of the compact, the dimensions in the direction of pressing must be molded by the pressure exerting punches. Depending on how complicated the shape is, the punches may consist of one or more parts moving freely within the die cavity. The punches and the die cavity must be reinforced with sufficient material to withstand the forming pressure. While simple dies are, in most cases, machined out of one piece, dies of several segments must be held together by screws, clamps, or tie rings. In the case of a simple bar, for example, four die segments, usually called inserts, form the cavity; they are held together by a fifth part which may either be a screw clamp or a shrink ring. Complicated shapes may require more than four die inserts. In addition to a tie ring, a large reinforcement frame or ring may be necessary. Compacts with holes require stationary or movable core rods in addition to the moving punches. Finally, additional die accessories, such as spacer blocks, floating bases, etc., may be necessary to aid in compression or ejection.

MATERIALS OF CONSTRUCTION

The two primary physical requirements that must be fulfilled by a die are strength and wear resistance. Accordingly, the choice of the proper material must be made in view of the actual function each part has to perform within the die assembly. Inserts and punches forming the mold cavity must be strong and wear resistant, while reinforcing sections need only be strong. Faces in immediate contact with the powder may have to be very hard, whereas tie rings and the interior or base of punches must be tough and shock-resistant. A large list of materials that fulfill these requirements, ranging all the way from hardwood to cemented carbides, could be compiled. The most important materials for these different purposes are shown in Table 36. For facings lining the powder, high alloy or tool steel compositions, hardened to Rockwell C-58 to 62, are generally employed. Punches usually contain less carbide-forming constituents and are drawn somewhat softer to render them tougher, and to prevent their cutting into the die walls. In each case, however, highest possible finish is essential. When production demands are sufficiently high to warrant the cost, linings of punch sides and die walls facing the powder are made of cemented carbides and brazed onto steels drawn to medium hardness (Rockwell C-50 to 52). Hard chromium layers, electroplated to a thickness of 0.002 to 0.020 inch, are frequently used where cemented carbide liners are not practical. If the plating is highly polished and buffed, its performance often approaches that of cemented carbides.

STRENGTH OF DIE

In designing dies for the compaction or sizing of metal powder parts, the question of die strength is of utmost importance. It must be recognized from the start that conditions are not the same for the two pressing operations. Sizing or coining is carried out on substantially solid bodies which will behave much the same as regular solid metals during compression. Brittle metal cylinders will display the familiar pressure cones when stressed beyond the yield point, while ductile metal cylinders will deform into flatter bodies of a barrel-like shape, the increase in diameter being less marked at the ends than in the center because of the friction at the bearing surfaces.

Any sensible calculation of the dimensions of the die must be based on the stresses acting on the faces in contact with the powder mass under compression. This imposes on the die designer the most difficult task of first analyzing these stresses. The extent of the difficulties has been shown in Chapter VIII in connection with the work of Kamm, Steinberg, and Wulff,[1a] who have analyzed the internal stress distribution in a compact and at its surface and have computed the coefficient of friction acting on the die cavity walls for the most simple case of a cylinder compressed only by action in one direction. The calculations for this case alone have required experimental work of major proportion, and stress analysis work of no less extensive scope. Obviously, this procedure cannot be expected to be adopted by the average die designer who is called upon to produce a safe and workable die construction for production work within the usually very short interval permitted by the production and delivery contract with the customer. Hence, a simpler and more empirical approach to the problem of calculating die dimensions is required.

It is known that, in the case of tension, an axial elongation is always accompanied by lateral contraction of the material, and that, within the elastic limit, the ratio of unit lateral contraction to unit axial elongation is constant for a given material. This constant, known as Poisson's ratio, has been calculated to be 0.25 for isotropic materials, and has been found by experiment to be near 0.3 for structural steel and most other metals. With suitable changes, this phenomenon of lateral contraction during tension can be applied to the case of compression. Longitudinal compression will be accompanied by lateral expansion, and for calculating this expansion for the friction-free case, the same value of Poisson's ratio is used as for tension. The situation is quite different when a powder is being compacted, with only densely compressed compacts under compression in a confined

[1a] R. Kamm, M. Steinberg, and J. Wulff, *Trans. Am. Inst. Mining Met. Engrs.*, *171*, 439 (1947).

space beginning to resemble brittle solids. But it is under these conditions that high pressures are applied and considerations of die strength enter the picture. Since Poisson's ratio of 0.3 refers only to completely dense and isotropic bodies, it cannot apply (without qualification) to powder compacts which are neither completely dense nor completely isotropic in properties. However, empirical figures of a smaller value have been established for certain nonmetallic materials, which, in some respects, resemble powder compacts more closely than solid metals. Poisson's ratio for concrete has been found to be in the neighborhood of 0.1, and that for cork can be assumed to be equal to zero.[2] It may therefore be assumed that Poisson's ratio will increase progressively from $\mu = 0$ for slightly compacted powder, to about $\mu = 0.3$ for highly compressed compacts approaching bulk density.

Since strain is proportional to stress within the elastic limit, Poisson's ratio gives a basis for an attempt to evaluate the stresses acting on the die walls. It must be recalled, however, that stress (pressure) distribution throughout a compact under compression is very complicated and, as discussed in Chapter VIII, a number of factors—notably particle-size distribution, shape of the compact, or mode of applying the pressure—govern the transmission of axial compressive stress in the lateral direction. Accordingly, an exact calculation of the action of stress on the die walls is almost impossible, with only an approximation being possible (with due reference to the particulars of the case).

In this connection, it is interesting to note that, through die expansion measurements during compaction of iron powder and mathematical computations, Goncharova[2a] found that the lateral pressure in the die increased linearly with the compacting pressure up to 23.5 tsi, amounting to 9 tsi (0.38 times). The radial extension of the cylindrical die was evaluated as $U = 2a^2bq/E(b^2 - a^2)$, where U is the outer radial extension, a the internal diameter of the die, b the outer diameter, q the internal pressure, and E the modulus of elasticity.

The simplest approach to this problem can be made by application of the well-known formulas for tubes under hydrostatic pressure. The customary formulas for stresses in thin-walled cylinders are based on the approximate assumption that the circumferential stresses (Fig. 121A), set up in thin-walled cylinders by a uniform internal pressure, are uniform throughout the thickness. Figure 121B represents the cross section of the thin-walled cylinder acted on by internal pressure, while Figure 121C represents the distribution of fluid pressure exerted by the fluid or gas in the upper half upon the semi-cylindrical wall and upon a liquid occupying

[2] S. Timoshenko and G. H. MacCullough, *Elements of Strength of Materials*. Van Nostrand, New York, 1935, p. 69.
[2a] V. N. Goncharova, *Zavodskaya Lab.*, *14*, 575 (1948); see also *Metal Powder Rept.*, *3*, No. 347 (1948).

the lower half of the cylinder. Since the upper half of the cylinder is loaded (Figs. 121B and C) the circumferential tension, P, acting across the element at A must be the same for both cases. If the upper and lower parts of the cylinder in Figure 121C are shown separately, as in Fig. 121D, the force, P, can be calculated from the conditions for static equilibrium of the lower part:

$$2P = p2rl$$

The unit tensile stress in the wall of thickness, t, is accordingly:

$$S = (prl/tl) = (pr/t)$$

or the necessary wall thickness of the tube is:

$$t = (pr/S)$$

Thus, for example, for an internal pressure of 100,000 psi and a tensile strength of the container material of 100,000 psi, the wall thickness becomes equal to the internal radius of the tube.

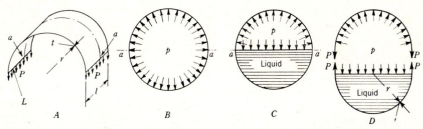

Fig. 121. Stresses in thin-walled cylinders: A, thin-walled half cylinder; B, cross section of cylinder acted upon by internal pressure; C, distribution of pressure exerted on upper semi-cylindrical wall and on liquid in lower cylinder half; D, upper and lower halves of cylinder after separation. l, length of cylinder; r, internal radius of cylinder; t, thickness of cylinder; p, internal pressure, psi; P, circumferential stress; L, longitudinal joint; $a --- a$, cut with longitudinal joint.

Application of this simple formula in calculating the size of powder metallurgy dies, however, is unsatisfactory for the following reasons. (*1*) Dies for metal powder compaction are used for high pressure work; they are not thin-walled, and the circumferential stresses are not uniform throughout the thickness. (*2*) The internal pressure within the compact is not uniformly distributed; hence, its action on the die walls is also not uniform. (*3*) The internal pressure acting over the length, l, of the compact is balanced by the projected area of the die over its total length, L (Fig. 122). (*4*) Friction between the compact and the punch faces materially affects the pressure distribution throughout the compact. (*5*) In order to make ejection of the compact possible, the stresses acting on the die walls cannot be allowed to exceed the elastic limit of the die material. (*6*) Within

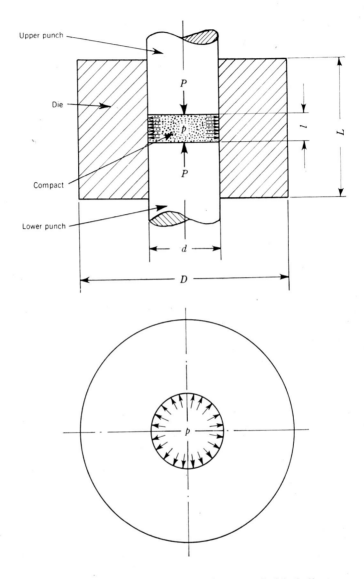

Fig. 122. Internal pressure acting on cylindrical die.

P = specific pressure applied during compaction
p = internal pressure acting on die wall
L = length of die
I = length of compact
D = outside diameter of die
d = diameter of compact

the elastic range, Poisson's ratio is valid and can be applied; it reduces the value of internal stress acting on the die walls to a fraction, the exact value depending on the degree of densification of the compact, *i.e.*, the magnitude of the compacting pressure applied. (*7*) An exact translation of the lateral stresses acting from the compact to the die is not possible; the differences in density involve differences in Poisson's ratio, and for each degree of compaction, different degrees of lateral deformation of the two materials are in equilibrium. (*8*) No consideration is given to the effects of strengthening dies by preloading them with the aid of shrink rings or by using a number of concentric tapered rings.

It is probably impossible to take all these points into consideration collectively when trying to arrive at a more satisfactory way of calculating the die strength, but it appears quite possible to account for some of them individually. The objection of the low pressure range of the simple pipe formula could be met by Barlow's formula:[3]

$$t = (DP/2S)$$

where t is the wall thickness, D the outside diameter of the die, P the internal pressure, and S the allowable tensile stress. Since the inside diameter, I, is given by the size of the compact to be formed, this formula is not practical for use.

Cylinders of heavier wall thickness subjected to high internal hydrostatic pressures are generally calculated with Lamé's formula:[3]

$$t = r \left\{ \sqrt{\frac{S + P}{S - P}} - 1 \right\}$$

or its simplified form:

$$R = r \sqrt{\frac{S + P}{S - P}}$$

where t is the thickness of the cylinder, R the outer radius of the cylinder, r the inner radius of the cylinder, S the maximum allowable fiber stress, and P the pressure within the cylinder. While this pressure is uniform in the hydrostatic case, it should be multiplied with Poisson's ratio in the case of metal powder compression. Using this formula, for example, in the case of a metal powder to be compressed into a 2-inch diameter compact, at 50 tsi, in a hardened steel die having an elastic limit of 40,000 psi, the outer diameter of the die would be:

$$D = d \sqrt{\frac{S + p\mu}{S - p\mu}} = 2 \sqrt{\frac{40,000 + 100,000 \times 0.3}{40,000 - 100,000 \times 0.3}} = 5.3 \text{ inches}$$

[3] *Machinery's Handbook*. Industrial Press, New York, 1942, p. 403.

At this point it may be emphasized again that this formula is only an approximation, since it does not take into account such important factors as the ratio of die length to compact length or the pressure distribution within the compact.

There remains also the important factor of the shape of the compact. Apart from the fact that dies for noncylindrical compacts must in most cases be composed of separate segments to facilitate die making, the internal stress distribution for such shapes becomes still more involved. In such constructions, lateral stresses exerted by the compact are transferred from the individual die inserts to an exterior ring or frame, and corresponding allowances must be made in the stress calculation. There remains, finally, the question of hardened tools. While it is true that static properties, such as tensile strength and yield point, are substantially increased by heat treatment, this is not the case for Young's modulus of elasticity which remains the same. As all stresses applied to the die must lie well within the elastic range, there is no absolute need to harden steel dies from the standpoint of stress absorption. It would be advantageous only for materials having a higher elastic limit and a larger modulus of elasticity than steel, but few materials are in this category. The main reason for the almost universally employed practice of hardening steel dies or die inserts can be seen in the fact that resistance against abrasion is greatly improved.

While the question of safely dimensioning dies to prevent bursting is relatively simple, there remains the problem of elastic overexpansion of the die cavity. Since Young's modulus is not affected by hardening the steel, the elastic spread of the cavity can only be overcome by preloading in the opposite direction. Herein lies the greatest advantage of employing shrink rings. In Table 37 a few practical figures are given for compact growth due to elastic overexpansion of the die cavity. The effect of compact size, pressure, and shrink temperature (controlling the initial counter stresses) are quite apparent. For the sake of smooth ejection, the growth of the compact must be kept to a minimum; obviously this is best achieved in dies whose reinforcement rings are shrunk at temperatures near red heat.

With regard to the strength of punches, ordinary static strength considerations are applicable. Where possible, proper reinforcement of the base of the punch must be provided for, and stress concentrations in sharp corners, weak sections, undercuts, etc., must be avoided. Otherwise, design rules applying to stamping and extrusion tools can be followed.

TABLE 37

Compact Growth Phenomena Caused by Elastic Die Expansion

Material	Compact shape	Size	Type of die construction	Shrink temp.	Forming pressure, tsi	Compact growth (% vs. original dimensions of die cavity)
Copper..	Cylinder	1-in. diam. 1 in. high	S i n g l e die of hardened steel		20 50	0.10 0.35
Iron.....	Cylinder	1-in. diam. 1 in. high	S i n g l e die of hardened steel		50 70	0.30 0.55
Copper..	Square plate	2 in. square $^1/_2$ in. high	S i n g l e die of hardened steel		25 50	0.15 0.35
Iron.....	Square plate	2 in. square $^1/_2$ in. high	S i n g l e die of hardened steel		20 40 70	0.10 0.25 0.50
Copper..	Rectangular plate	$2^1/_4 \times 1^7/_8$ $\times$ $^1/_2$ in.	Hardened die i n s e r t s clamped in soft s t e e l shrink ring 6 in. I.D. (0.020 i n . undersize)	370°C. (700°F.)	25 50 75	0.03 0.10 0.20
Iron.....	Rectangular plate	$2^1/_4 \times 1^7/_8$ $\times$ $^1/_2$ in.	Hardened die i n s e r t s clamped in soft s t e e l shrink ring 6 in. I.D. (0.020 i n . undersize)	370°C. (700°F.)	25 50 75	0.04 0.12 0.20
Iron.....	Rectangular plate	$1^7/_8 \times 1^5/_8$ $\times$ $^1/_2$ in.	Hardened die i n s e r t s clamped in soft s t e e l shrink ring 6 in. I.D. (0.035 i n . undersize)	540°C. (1000°F.)	25 50 75 100	0 0.02 0.10 0.30

WEAR OF DIE AND PUNCHES

The wear resistance of die-cavity walls and punch surfaces facing the powder is closely connected with the static properties, especially the hardness of the molding tools. In order to understand this problem fully, we must recall that when metal powder particles rub against each other and against the confining walls of the die cavity during compression, they create friction. Many nonmetallic additions act as abrasives and aggra-

TABLE 38. Wear and Life Test with Different Die Materials[a]

Powder	Die material	Cavity shape	Punch material	Period of pressing hr.	Period of pressing min.	Approx. No. of pressings	Increase in compact diameter, %
Electrolytic nickel, −100 mesh	SAE 1060 carbon steel	Cylindrical	SAE 1060 carbon steel	0	35	700	0.63
	SAE 1060 carbon steel	Tapered		0	55	1,100	0.70
	High chrome-carbon steel (12% chromium; 2.2% carbon)	Cylindrical	Cemented carbide	1	05	1,300	0.66
		Cylindrical	High chrome-carbon steel	3	25	4,100	0.57
		Tapered	Cemented carbide	4	40	5,600	0.63
	Hard-chromium-plated high chrome carbon steel	Cylindrical	Cemented carbide	5	05	6,000	0.62
		Cylindrical	Hard-chromium-plated high chrome-carbon steel	31	30	38,000	0.55
		Tapered	Cemented carbide	44	10	53,000	0.57
		Tapered	Cemented carbide	47	30	57,000	0.59
	Cemented carbide	Cylindrical	Cemented carbide	485		582,000	0.30
		Tapered	Cemented carbide	664		797,000	0.33
Electrolytic nickel, −100 mesh, + 1/2% benzoic acid	SAE 1060 carbon steel	Cylindrical	SAE 1060 carbon steel	1	15	1,500	0.66
	SAE 1060 carbon steel	Tapered		1	35	1,900	0.69
	High chrome-carbon steel (12% chromium; 2.2% carbon)	Cylindrical	Cemented carbide	1	40	2,000	0.67
		Cylindrical	High chrome-carbon steel	5	30	6,600	0.59
		Tapered	Cemented carbide	7	25	8,900	0.57
	Hard-chromium-plated high chrome-carbon steel	Cylindrical	Cemented carbide	8	45	10,500	0.57
		Cylindrical	Hard-chromium-plated high chrome-carbon steel	42	30	51,000	0.60
		Tapered	Cemented carbide	64	10	77,000	0.62
		Tapered	Cemented carbide	72	30	87,000	0.60
	Cemented carbide	Cylindrical	Cemented carbide	744		893,000	0.29
		Tapered	Cemented carbide	909		1,090,0000	0.30

Powder	Die material	Cavity shape	Punch material	Period of pressing hr.	min.	Approx. No. of pressings	Increase in compact diameter, %
Stainless steel alloy, −100 mesh + 1% stearic acid	SAE 1060 carbon steel	Cylindrical	SAE 1060 carbon steel	0	08	400	0.66
		Cylindrical	Cemented carbide	0	12	600	0.69
		Tapered	Cemented carbide	0	14	700	0.66
	SAE 3250 nickel-chromium steel	Cylindrical	SAE 3250 nickel-chromium steel	0	17	850	0.63
		Cylindrical	Cemented carbide	0	23	1,150	0.66
		Tapered	Cemented carbide	0	29	1,450	0.64
	High-speed steel (18% tungsten, 4% chromium, 1% vanadium)	Cylindrical	High-speed steel	0	16	800	0.65
		Cylindrical	Cemented carbide	0	25	1,250	0.65
		Tapered	Cemented carbide	0	29	1,450	0.62
	High chrome-carbon steel (12% chromium, 2.2% carbon)	Cylindrical	High chrome-carbon steel	0	44	2,200	0.59
		Cylindrical	Cemented carbide	1	07	3,350	0.57
		Tapered	Cemented carbide	1	14	3,700	0.59
Stainless steel grindings, −100 mesh containing 2% abrasives	SAE 1060 carbon steel	Cylindrical	SAE 1060 carbon steel	0	03	150	0.87
		Cylindrical	Cemented carbide	0	05	250	0.89
		Tapered	Cemented carbide	0	06	300	0.88
	SAE 3250 nickel-chromium steel	Cylindrical	SAE 3250 nickel-chromium steel	0	06	300	0.79
		Cylindrical	Cemented carbide	0	11	550	0.77
		Tapered	Cemented carbide	0	14	700	0.79
	High-speed steel (18% tungsten, 4% chromium, 1% vanadium)	Cylindrical	High-speed steel	0	07	350	0.77
		Cylindrical	Cemented carbide	0	12	600	0.75
		Tapered	Cemented carbide	0	14	700	0.75
	High chrome-carbon steel (12% chromium, 2.2% carbon)	Cylindrical	High chrome-carbon steel	0	19	950	0.69
		Cylindrical	Cemented carbide	0	31	1,550	0.70
		Tapered	Cemented carbide	0	35	1,750	0.66

a Courtesy of American Electro Metal Corp., Yonkers, N. Y.

vate this friction. The effects of this abrasion on the molding tools are significant from two points of view: dimensional tolerances are gradually lost, and harmful shear stresses during ejection act upon the compact to an ever-increasing extent, eventually causing mechanical failure.

There are three distinctly different wear phenomena that can be observed when metal powders are compacted in continuous operation: (1) abrasion of die cavity walls during compaction, (2) abrasion of punch facings during compaction, and (3) abrasion of die cavity walls during ejection. Although, to a certain extent, these phenomena are interrelated and form a single problem, it is helpful to segregate them for the purpose of study. The wear effects caused by ejection can be eliminated by employing excessively tapered cavities or partly split dies which open slightly in orange-peel fashion. The joint action of wear on die and punches can be separated by using alternately die and punch linings from mirrorlike polished cemented carbides.

By following essentially such a procedure, a series of tests were carried out, the most important results of which are presented in Table 38. All experiments were carried out in a 10-ton tableting press, producing 20 to 50 pellets per minute. The pellets were approximately $1/4$ inch and $1/2$ inch in diameter, depending on the dies available; the applied pressure was kept constant at 50 tsi. Wear effects during ejection were eliminated by substituting a conical die for the cylindrical die; wear effects on the punches were eliminated by using highly polished carbide-tipped punches. Various die materials were tried on different metal powders and mixtures with lubricants and abrasives.

The rather revealing conclusions based on the results obtained by this elimination method may be summarized as: (1) Over-all tool abrasion is decreased by the addition of lubricants to the powder; it is increased by abrasives mixed with the powder. (2) High finish and great hardness generally decrease wear effects. (3) Wear effects are more pronounced on the die cavity wall than on the punch facings. (4) Generous tapers tend to decrease or completely eliminate wear effects during ejection. (5) Close clearance between punches and dies tend to diminish die wear. The results further appear to indicate that wear in ejection is much less pronounced than in compaction. This conclusion, however, is not definitive, since it must be realized that a clear distinction between wear in compaction and wear in ejection is only possible by the use of split-type dies. Otherwise, it could be maintained that wear observed in the compacting zone might be caused by the "breaking loose" of the compact from the die wall in the initial stage of ejection. The experimental data reveal that the wear resistance of a die material depends on its hardness as well as on its frictional characteristics. For example, it is indicated that the primary reason chromium-plated steel die linings are superior to alloy-steel die linings in molding iron parts, is that the coefficient of friction between chromium and iron is lower than that between steel and iron.

For the linings examined, the following order could be established with regard to wear resistance: (1) cemented carbide, (2) hard-chrome-plated tool steel, (3) high chrome-carbon tool steel, (4) high-speed steel, (5) chrome-nickel steel, and (6) water-hardened, medium-carbon steel.

Expressed in terms of die life, i.e., in terms of the number of operations before the die is worn beyond the largest tolerance permissible, it may be said that cemented carbide is at least ten times superior to hard-chrome-plated tool steel, which, in turn, is approximately five to ten times superior to high-chrome, high-carbon tool steel. Of the steels, the life based on wear resistance follows approximately the ratio 10 : 4 : 4 : 2 with high-chrome, high-carbon tool steel being first, and water-hardened, medium-carbon steel being last.

CLEARANCES AND TAPERS

It must be emphasized, however, that the above-mentioned ratios for wear performance of different materials are only relative, and may be considerably affected by the quality and state of the powder or by details of the die construction—especially with respect to clearances and tapers. This may be made clearer by the following.

A powder of standard commercial quality may all pass through a 100-mesh sieve; one third may pass through a 325-mesh sieve, the particles being smaller than 43 microns, or approximately 0.0017 inch. If the clearance between punch and die is of a similar magnitude, a great many of the fine particles can enter the space. From there, they fall out of the die, possibly causing disturbances by entering sensitive adjustment parts of the press, or may stay in the clearance space, becoming welded to the punch or die by the pressure of the constantly moving punches. In the latter case, even greater harm may result, since wear and friction problems are increased, and mechanical failure during ejection may easily result.

It has become good practice in molding commercial grades of powder to confine the permissible clearance to 0.0005–0.001 inch. The clearance can be enlarged only if relatively coarse powders are used, or if the compacts exceed 2 inches in diameter (or width). A good rule is to hold the clearance to 0.05% of the diameter for parts larger than 2 inches in diameter.

Just as in the case of clearances, the use of draft to aid ejection must be considered by the die designer. Experience with tapers is largely empirical, and usually remains a closely guarded secret of the individual designer. While there is no question that a slight taper materially reduces friction during ejection, its beneficial action is limited to a very small angle, beyond which no benefits result, and the dimensions of the compact may even exceed the permissible tolerance. In determining the proper taper, the designer should remember that its sole purpose is to provide for instantaneous release of stresses imposed on the compact by the elastically overexpanded die. This may be achieved by a minute enlargement of the die cavity in the direction of ejection, and tapers having an angle of between 6 and 10 minutes have been found quite adequate.

On the other hand, there remains the problem of how far the taper can be carried into the die cavity. While it is ideal to compact the powder within the tapered section, this may not always be permissible for reasons of dimensional accuracy. In that event, the compact must be formed in a cylindrical or prismatic section and, upon ejection, pushed through it into the taper. The movement through the nontapered section involves, of course, considerable friction and wear due to overstressing of the compact. This is the reason why considerable pressures are often necessary in the initial stage of ejection. In practice, this method should be used only if small quantities of parts are to be produced in one die. Otherwise molding should be performed within the tapered section, and a sizing operation should be used to obtain the required accuracy.

FINISH OF DIE WALLS

The finish of the die walls is of great importance, both from the standpoint of reducing friction and of producing a good surface on the molded part. A mirror finish on the die wall is usually desired, and can be obtained by lapping or polishing with polishing rouge; this operation must be carried out by lapping in the direction of the stroke. Even if the die walls are nitrided or electroplated with hard chromium, the high polish is essential.

TYPICAL DIE CONSTRUCTIONS

Experimental Molding Dies

In order to be able to determine satisfactorily the pressing factors and characteristics of the powder, and their effects on sintering and subsequent operations, it has become standard practice to use experimental dies. The final design of high-speed production dies is based primarily on the results of thorough studies with experimental dies, though previous experience with allied problems and general judgment may be of considerable help, and may, in certain cases, obviate the necessity for the construction of expensive preliminary tools. Present knowledge of the mechanics involved in the compaction of metal powders is too scant to permit the establishment of definite rules to aid the powder metallurgist in designing his dies. The approach has remained an empirical one; we have learned merely to recognize and compensate for the important factors, such as fill volume of the mix, degree of compaction for a fixed pressure, counteracting the abrasive quality of the powder by lubrication, effective ejection without

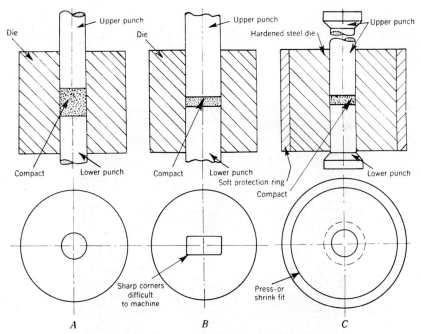

Fig. 123. Schematic designs of solid, single-piece dies: *A*, cylindrical form; *B*, rectangular form; and *C*, reinforced die and punches.

destruction of the compact, dimensional changes during sintering, and allowance for subsequent sizing or coining. Therefore, the need exists for experiments in preliminary, simply constructed dies, and it is always recommended if sufficient time is available for tooling-up. Although the additional expense is sometimes considerable, the approach assures safety and may well make for greater eventual economy when costly mistakes in the production die can be avoided, and when improvements derived from the preliminary experiments can still be incorporated in the final design.

Typical Molding-Die Constructions

Depending on which of the three demands (contours, strength, or satisfactory translation of the molding force throughout the compacting powder) is dominant, the choice of die design may go to one of the following typical constructions: (*1*) solid, single-piece; (*2*) solid, multiple-segment; (*3*) split-segment; (*4*) floating; or (*5*) multiple-cavity.

SOLID SINGLE-PIECE DIES

Dies made from a single block of material constitute the simplest construction. They are particularly adapted to cylindrical shapes, such as thin disks or pellets. The die cavity can easily be drilled, machined, and hand-lapped (or ground) from ordinary engineering materials or hardened

Fig. 124.　Rectangular hard metal die, reinforced by
soft steel shrink-ring.

steels. Complimentary punches can readily be made by machining, grinding, or lapping round bar stock. A sketch of such a construction is shown in Figure 123*A*. The same construction can also be used for simple prismatic shapes, such as cubes or rectangular parts (Fig. 123*B*), although it is rather difficult to machine very sharp corners. The chief merit of this construction lies in its simplicity and economy. The construction of such dies is a question of days—if not hours—against much longer periods necessary for building more complicated units. On the other hand, where hardened steel dies are necessary for high-pressure compression work, the powder metallurgist must be on guard against excessive stresses which may cause bursting of the die. In such cases, the hardened die block may be surrounded by a soft shell that will prevent fragmentation (Figs. 123C and 124). Punches, on the other hand, can usually not be safeguarded in this way, and the only safety lies in keeping them as short and bulky as possible.

SOLID MULTIPLE-SEGMENT DIES

For compacts having rectangular or other noncylindrical cross sections, dies are generally made in several parts. In this manner grinding is facilitated and the individual, hardened steel segments can be held in a soft block by a press- or shrink-fit, or with wedges or set screws. Figure 125 shows a multiple-segment die clamped together by a shrink-ring,[4] and

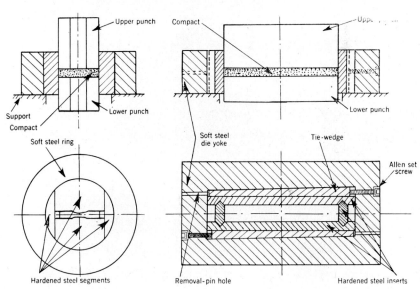

Fig. 125. Schematic design of solid, multiple-segment die with shrink-ring (according to Wulff[4]).

Fig. 126. Schematic design of solid multiple-segment die with wedge yoke.

Figure 126 illustrates how die inserts can be held together with wedges. Although this die arrangement permits ejection in the direction of pressing, the die inserts can be removed for cleaning or regrinding by loosening the wedges with the aid of pins.

[4] J. Wulff, in J. Wulff, *Powder Metallurgy*. Am. Soc. Metals. Cleveland, 1942, p. 248.

SPLIT-SEGMENT DIES

The principle of removing the inserts after compaction may be applied regularly in those cases where, because of undercuts or projections, ejection of the compact is not possible. Moreover, most hard powders form compacts too fragile to withstand the sudden stress releases which occur in successive layers during ejection. Even if nonmetallic binders are added to the powder—laminated compacts may result. In addition, in the case of

Fig. 127. Collapsible die for molding refractory metal and carbide slabs in dismantled condition (courtesy of Hydraulic Press Manufacturing Co.).

especially abrasive powders, the walls of the die cavity become badly scored during ejection. Consequently, dies are taken apart after pressure release, and compacts are removed manually. This practice is common in the molding of tungsten, tantalum, and molybdenum briquettes used for sheets or wire drawing and also for tungsten carbide slabs and bars for tool tips. Figure 127 is a photograph of a dismantled, collapsible die used in this work. The design in Figure 128 is suitable for pressing ingots, and the one in Figure 129 for molding wire bars. Because of the great length of these bars, very large die yokes would have to be built, and it is simpler to insert the die segments into a vise which is closed by the power of the press. Molding in dies of this type has been confined to cases where the finished product is an expensive one and the labor and time required

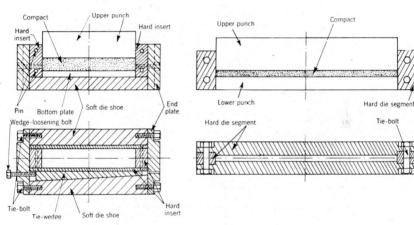

Fig. 128. Schematic design of collapsible-segment die for molding ingots.

Fig. 129. Schematic design of collapsible-segment die held in power vise of press for molding bars.

for assembly and disassembly amount to only a small part of the total costs. It is obvious that mass production by this method cannot be contemplated.

FLOATING DIES

In the foregoing examples, no particular reference has been made to the problem of pressure distribution. It has been shown in the previous chapter that relatively flat compacts can be pressed into a fairly uniform agglomerate of particles by applying the pressure in one direction only. In that case, the density may not vary greatly in different layers, if the die is supported during the operation. The situation is quite different for compacts of greater height, where pressure application from two opposite directions becomes necessary to obtain a fairly uniform density throughout the longitudinal section of the compact. As presses providing for pressure from two opposite sides are not always available, the floating-die principle constitutes an ideal substitute of general usefulness. In fact, there are but a few cases where a floating die cannot give a performance equal to that of a stationary die inside a press operating from two directions.

The floating die principle is amazingly simple. The die rests on a float, made of springs, rubber pads, or oil cushions. As soon as the pressure applied in one direction becomes great enough to agglomerate the powder particles into a coherent mass, part of the pressure is translated to the walls of the die cavity. Frictional forces became so great that they over-

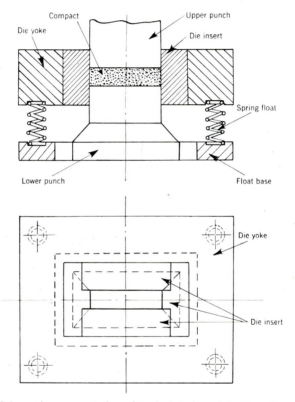

Fig. 130. Schematic representation of typical design of floating die assembly.

come the spring tension. Upon further compression, the die is carried along with the punch that exerts the pressure, and is forced over the complimentary punch which remains solidly supported, in much the same manner as a glove is pulled over a finger. Interparticle friction balances the rate of compression of each complimentary punch, so that densification of the compact progresses equally from the opposite ends. A typical design of a floating die for rectangular bars is shown in Figure 130, while in Figure 131 a photograph is shown of such a die in actual service. Pneumatic attachments permit vibratory and jolting-type dynamic mold-

Fig. 131. Floating die assembly with pneumatic vibrator, jolting table, and static follow-up piston, set in hydraulic laboratory press for experimental dynamic-type molding of bar-shaped test specimens (courtesy of Sintercast Corporation of America).

ing. Experimental dies of similar construction for molding tensile specimens are shown in Figures 132 and 133. In Figure 132 a die used by Eilender and Schwalbe[5] is reproduced; the die sketched in Figure 133, as reproduced from Wulff,[6] is of similar design to the one shown in Figure 130.

[5] W. Eilender and R. Schwalbe, *Arch. Eisenhüttenw., 13,* 267 (1939).

[6] J. Wulff, in J. Wulff, *Powder Metallurgy.* Am. Soc. Metals. Cleveland, 1942, p. 251.

MULTIPLE-CAVITY DIES

In plastics and die casting, the possibilities of simultaneously molding several identical forms in multiple-cavity dies were explored successfully and developed economically. In molding plastics the forming of the part is combined with the curing and setting of the plastics, and in die casting with a slow solidification of the casting—both being slow operations. Since in these fields, the unit pressures are low, the press capacities

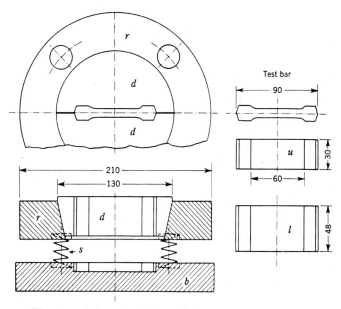

Fig. 132. Schematic design of floating-type laboratory die for molding tensile test specimens (according to Eilender and Schwalbe[5]). All dimensions in mm. *d*, die; *r*, steel ring; *u*, upper punch; *l*, lower punch; *b*, bottom plate; *s*, float springs.

also remain low in spite of the use of multiple-cavity dies. In the field of powder metallurgy, however, the speed of ordinary presses is usually satisfactory and often produces compacts at a faster rate than can be accommodated during sintering. If the speed is not satisfactory, the use of rotary presses is preferred to the use of multiple-cavity dies which would involve an increase in the required press capacity. In addition, the use of multiple-cavity dies would introduce difficulties arising from nonuniform powder feeding and particle size classification, and from unequal ejection pressures caused by erratic friction condition at the die walls. Only in the rare instance where soft powders—possibly mixed with plastics—are pressed into very simple shapes, as in the case of molding magnetic cores, have multiple-cavity dies found use in powder metallurgy.

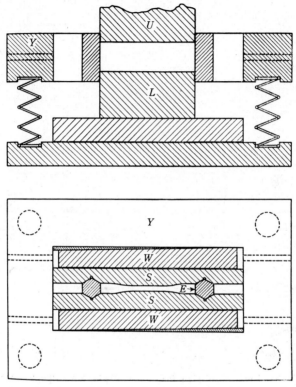

Fig. 133. Schematic design of floating-type experimental die for molding tensile test specimens (according to Wulff[6]): U, upper punch; L, lower punch; E, S, hardened die inserts; W, wedges; Y, die yoke.

Mass-Production Molding and Coining Dies

In discussing production dies and die assemblies, we must distinguish between dies used solely or primarily for molding of powders and dies used for coining or sizing of sintered parts. In compacting powders, greater fill depths and, consequently, higher dies are required than for coining. In pressing porous bearings, sizing pressures are either less than, or about equal to those employed in compacting, and the stresses acting on the dies are similar in both cases. This is not the case for the pressing of dense structural parts where coining pressures frequently have to be higher than the initial pressures applied on the powder. Coining dies for these parts must be built much stronger, and punches must be properly reinforced to permit longer use.

Information on production dies and die assemblies has been scant, and only a few designs have been published in recent years. In Figure 134 a

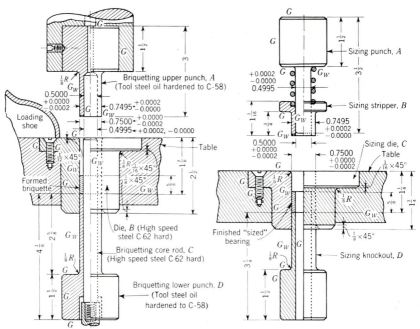

Fig. 134. Tool and die design for production molding of plain cylindrical bearings (according to Langhammer and Smith[7]): G designates hardened and ground surface; G_W designates hardened, ground and superfinished surface.

Fig. 135. Tool and die design for production sizing of plain cylindrical bearings (according to Langhammer and Smith[7]): G designates hardened and ground surface; G_W designates hardened, ground and superfinished surface.

production die assembly for the molding of plain cylindrical bearings is shown.[7] The die part, B, is stationary, forming the outside diameter of the bearing, and the inside diameter, formed by the core rod, C, is also stationary. The movable die elements are the upper punch, A, which transmits the pressure and forms the upper end of the compact, and the

[7] A. J. Langhammer and M. F. Smith, *Metal Progress, 41*, No. 3, 335 (1942).

lower punch, *D*, which forms the lower end of the bearing and ejects the compact from the die after compression. A sizing die assembly is shown in Figure 135. The die, *C*, which controls the outside diameter is stationary, as is the stripper attachment, *B*. Again, the movable parts are the upper punch, *A*, and the lower punch, *D*. The upper punch sizes the inside diameter of the bearing, while the lower punch, also known as "knockout," performs the ejection. The sintered bearing is first placed on the die, and then forced into it by the punch which sizes the bore of the bearing while descending.

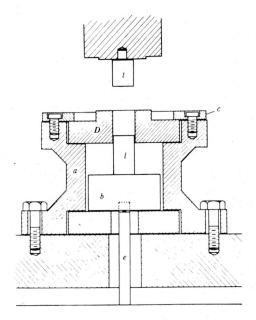

Fig. 136. Schematic design of production die assembly for molding and coining of simple shapes (according to Seelig[8]): *D*, die; *c*, top plate; *a*, adaptor; *b*, pressure block; *e*, ejector pin; *l*, lower punch; *t*, top punch.

Figure 136 is a schematic design of a production die assembly for unidirectional compression.[8] The die can be used both for molding and coining, but has not incorporated all features recognized as necessary for uniform pressure and density distribution in the compact. The die is made of several hardened steel inserts, held together by a tie ring, and set in a cast steel adaptor, *a*. The die is carefully centered over the ejector pin (*e*)

[8] R. P. Seelig, *Metals & Alloys, 12*, No. 6, 744 (1940).

with the aid of a pressure block (*b*) and the lower punch (*l*), and is fastened in position by tightly clamping a top plate (*c*) to the adaptor. The top punch (*t*) is accurately centered above the die and then tightly fastened in the punch adaptor. The punch adaptor is screwed onto the moving slide of the press, while the ejector rod is fastened to the ejector of the press. The lower punch and pressure block are designed so that upward ejection of the longest possible compact is possible. During compression, the pressure is not borne by the ejector, but by a pressure plate below the pressure block which bridges the ejector hole. The entire die assembly is bolted to a heavy base plate.

In operation, the die cavity is filled with powder from a funnel arrangement and leveled with an automatic slide, so that the powder is always flush with the top of the die. Sometimes the die can also be fed with automatically preweighed powder charges. Compaction is obtained only from above by lowering the press slide until a predetermined height or a definite pressure is reached. The pressure is then released, and the slide reverses, while the ejector removes the compact upward—out of the die. The ejected briquettes can be removed from the top of the die either automatically or manually.

A production die assembly embodying the floating die principle is shown in Figure 137, while the die itself with a set of punches is shown in Figure 138. The design detailed in Figure 139 shows certain improved features over that illustrated in Figure 136. The adaptor is replaced by a float, consisting of a base ring into which four dowel pins are screwed and which serve as guide posts for the die. Closest fit of the pins and bearings of the die is essential so that the floating ability will not be impeded. The floating ring holds four additional pins around which strong springs are placed. These springs provide the floating action by yielding during compression when the friction between compact and die walls counteracts the spring tension.

The merits of the floating die principle have been discussed before. A further improvement of the design may be seen in the particular manner in which the hardened, tool-steel inserts are placed in the shrink ring. A tapered fit makes it possible to re-use the same inserts many times. If the die walls have worn beyond their dimensional tolerances, they can be reground at the inner faces. The assembly of inserts can be brought back to the required dimensions simply by regrinding the joints and by placing the inserts in the tapered fit, in a position slightly below that formerly occupied. The top of the die can easily be made flush again by grinding the necessary amount from the top of the shrink ring. In this manner, a die of the type shown in Figure 138 has been overhauled repeatedly without replacement of the inserts or any other part of the die, and at a pressure exceeding 50 tsi, it proved possible to make more than 2 million successful pressings.

Fig. 137.   Floating-type production die assembly in actual operation (courtesy of American Electro Metal Corp.).

Fig. 138.   Production die and punches of Figure 137, disassembled (courtesy of American Electro Metal Corp.).

A comprehensive discussion of die assemblies and tooling for molding and coining of powder metallurgy parts on a mass production scale

has very recently been presented by Dalby,[9] and the reader interested in
this problem is advised to consult the original paper for a close study of
the many interesting details of tool and die construction.

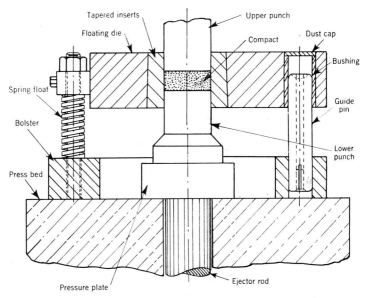

Fig. 139. Schematic design of production die assembly for
molding and coining of dense structural parts (courtesy of American Electro Metal Corp.).

Summary

Any die to be used in powder metallurgy must be tailor-made since
it must be accurate enough to guarantee the required dimensional tolerances. It must also be strong enough to withstand the great forming or
coining pressures used in the operation. As a consequence, die cost is of
major importance and, next to raw material costs, is probably the most
decisive economic factor. High die costs can be diminished by constructing
the die of a wear-resistant material to assure long die life and a minimum
of maintenance and repair. If production is sufficiently large, cemented
carbide die linings are very satisfactory from the standpoint of wear
resistance, although initial costs are higher. Die linings plated with hard
chromium are also satisfactory for many applications, and have the advantage of being less expensive and capable of being replaced.

[9] W. E. Dalby, *Proc. Fourth Annual Spring Meeting of Metal Powder Association,* Chicago, April 15–16, 1948, p. 36.

As for the strength of dies to be used in powder metallurgy, no mathematical approach exists that takes into account all factors inherent in the compaction process. Thus, die design is, in most of its aspects, still in an empirical stage.

There exist many typical die designs, both for compacting powder and coining sintered compacts. Single-piece or multiple-segment dies are used for most experimental and production work. Very fragile compacts are best made in split-segment dies. Dies operating on the floating principle assure more uniform density distribution in compacts while not requiring expensive double-action presses.

CHAPTER XI

Presses

In the preceding chapters, the molding phase in powder metallurgy has been discussed with respect to the behavior of a powder under pressure, the design possibilities of powder metallurgy parts, and lastly, the types of mold constructions. We may now move on to a consideration of the type of machine that is able to exert the necessary forming pressures and perform the required cycles of operation.

There are a number of ways in which pressure can be exerted upon a plunger to compact a metal powder. In extreme cases, a hammer blow might be sufficient to compress a very thin layer of a soft powder, making the mass sufficiently coherent for further handling. However, where an appreciable density is also required, more pressure is necessary than can be exerted by hand. Therefore, all metal powder molding machines of industrial importance are power driven. The necessary pressures are generally supplied by presses, although several processes use rolling mills for the compression of powders, a method used most advantageously where flat layers and long or continuous shapes are to be formed—as in the case of certain linings and bearings.[1] Although presses are integral parts of metal-forming and metal-working industries, we are now concerned only with their significance in connection with powder metallurgy. Those readers interested in studying the considerable literature on presses may refer to the recent texts by Crane[2] and Hinman.[3]

Adopting a functional approach to presses for powder metallurgy, we may state (on the basis of our preceding studies) that the following requirements must be fulfilled: (*1*) Pressure must be applied in the desired order of magnitude. (*2*) Pressure must be applied in the desired direction. (*3*) Length and speed of the press stroke must be controllable. (*4*) Length and speed of the ejection stroke must similarly be controllable. (*5*) The die fill must be adjustable. (*6*) In two-directional pressing, the press strokes must be adjustable for the purpose of synchronization. (*7*) In

---

[1] U. S. Pat. 2,158,461.
[2] E. V. Crane, *Plastic Working in Presses*. Wiley, New York, 1943.
[3] C. W. Hinman, *Press working of Metals*. McGraw-Hill, New York, 1941.

multiple-punch arrangements, the press strokes must facilitate sectional powder transfer.

In general, it may be said that almost any press construction can be adapted for use in powder metallurgy. As in other molding and metalworking operations, however, it is obvious that no one press and accessory combination can cure all the production problems in powder metallurgy. Therefore, economy as well as technical performance must be considered in choosing the right press for the job. The selection of a press depends on a number of factors connected with both the quantity and quality of the desired product; questions of density required in the pressed part, depth of the part to be compressed, cross-sectional area and total surface of the part, as well as the type of die to be used, are just as important for the decision as are questions concerning production rates, press capacity, or flexibility in accommodating different jobs. Fundamentally, the more complex the die, the simpler is it possible for the press construction to be. On the other hand, simple dies require more complicated presses for all except the simplest molding problems. The presses which now are designed primarily for this work are either of the mechanical type (for high production rates) or of the high-tonnage, hydraulic type—with modifications for rapid plunger strokes. Slow-motion, hydraulic presses, as generally employed in the field of plastic molding, are used in the metal powder field only where end-product costs are sufficiently high to warrant the amount of time spent.

Both of these types of presses can be made to operate by exerting pressure in either one or two directions, either opposite or perpendicular to each other. The last principle is generally available only in hydraulic presses, although any of these presses is adaptable for compaction of powders, or for recompression of sintered compacts. Only the details of the particular design may make one machine more suitable for compacting and another for coining.

MECHANICAL PRESSES

Mechanically operated presses are used to a large extent in metalworking practices. For stamping, forming, and coining operations of all kinds, many press constructions are being used, including such different types as simple arbor presses, eccenter presses, crank presses, and cam-operated or toggle (knuckle-joint) presses. The widespread use of the mechanical press may be attributed mainly to the important features of high-speed production rates, flexibility in design, simplicity and economy in operation, and relatively low investment and maintenance costs. In view of these facts, it is only natural that mechanical presses found early applications in powder metallurgy. In fact, we may recollect that Wollaston's classic molding work with platinum powder, which may be considered the beginning of modern powder metallurgy, was carried out with a simple mechanical toggle press. (See Figure 2, p. 24.) There arose, however, a serious difficulty when mechanical presses were tried for the molding of powdered metals into denser compacts—as for tungsten bars in wire production. Here, high unit pressures were to be applied on forms having large cross-sectional areas, and conventional eccentric or crank presses were not powerful enough for the task. On the other hand, production rates were low, so that the high production rates inherent in mechanical presses could not be exploited. Consequently, the more powerful though slower moving hydraulic press had to be used for all work requiring high forming pressures, and only with the rise of the porous metal bearing industry in the 1920's did the mechanical press re-enter the picture.

Bailey[4] gives an interesting account of the development of the high-speed automatic mechanical press in the field of powder metallurgy. The need for automatic molding presses led first to the use of standard automatic compressing machines then being used in many industries. With the addition or change of a few features, the standard presses readily became adaptable to the molding of bearings, bushings, or brushes. During the last 25 years, several basic press designs have been developed for specific use in powder metallurgy. Details of actual operating procedures have been standardized, and important auxiliary functions, such as powder feeding and removal of compacts, have been accommodated.

Tonnage Capacity

In recent years, the press power of these mechanical presses has been increased greatly, and fully automatic molding presses are today available with capacities of up to 150 tons.[5] These tonnages are still small in comparison to high-power mechanical presses of the knuckle-joint type, having press capacities of up to 2000 tons.[6] In making these comparisons, however, questions of press rate, powder fill, and ejection stroke must be taken into consideration. As the tonnage of the press increases, the rate of pressing tends to decrease (if uniform and sound products are to be maintained). Consequently, a definite power limit exists for fully automatic mechanical presses, beyond which they lose their chief advantages of production speed and simplicity in design and operation. Where high-tonnage capacity is required, the hydraulic press is preferred.

[4] L. H. Bailey, in J. Wulff, *Powder Metallurgy*. Am. Soc. Metals, Cleveland, 1942, p. 271.
[5] Products of Kux Machine Co., Chicago, Ill.
[6] Products of E. W. Bliss Co., Brooklyn, N.Y.

386 PRESSES

Production Capacity

The main feature of the fully automatic mechanical press is the high production rate which depends on a number of factors independent of the particular design, such as fill depth, pressure stroke, and ejection stroke, and, to a lesser degree, flow and deformability of the powder and cross section of the mold cavity. It is obvious that a mold for a $1/2$-in. diameter compact to be pressed to a height of $1/2$ in. can be filled more rapidly, and the compact compressed and ejected in shorter time than for a compact 2 in. in diameter and 2 in. high. While, in the first case, a small eccentric press of approximately 10-ton capacity could produce between 50 and 100 compacts per minute, molding of the larger compact would require a large toggle-type or cam-operated press of at least 60-ton capacity that would normally produce not more than 15 compacts per minute.

Where shapes of compacts and fill conditions are sufficiently simple to permit high production speeds, presses with a number of dies arranged on a rotating table have been used successfully for briquetting metal powders. These presses, generally known as rotary-type tablet machines, are widely used for pharmaceutical purposes, where relatively low unit pressures are sufficient. Their adaptation for very small bushings, sleeves, and other mass production articles was simple, and many of these presses are now used by the porous bearing industry, producing from 300 to 500 compacts per minute—depending on the individual model. The total tonnage of most of these presses ranges from 2 to 30 tons, although special models of up to 100-ton capacity have been built.[7]   Their output may still exceed 500 pieces per minute, and duplex models—forming two compacts with each pair of punches at each revolution of the head—have yielded normal production rates of 1500 tablets per minute. Special models are making nonmetallic industrial products at rates up to 4000 units per minute.[8]   On the other hand, the larger rotary models can produce nonmetallic parts up to 6 inches in diameter at the still remarkable production rates of 40 to 60 per minute.

Basic Constructions

Because of the variety of types of mechanical presses suitable for metal powder compaction, a brief review of the individual designs appears appropriate. The following basic constructions can be distinguished: (1) eccentric and crank presses, (2) toggle and knuckle-joint presses, (3) cam-type presses, and (4) combination type presses.

[7] Products of F. J. Stokes Machine Co., Philadelphia, Pa.
[8] C. Hardy, *Symposium on Powder Metallurgy*. ASTM, Philadelphia, 1943, p. 5.

Eccentric and Crank Presses. Eccentric-type, single-punch machines are available in a closely related series of tableting machines—the individual model varying chiefly in the size of pieces produced. A typical

Fig. 140. 20-ton double-action eccenter press with cam-driven knock-out. Press arranged for applying 20 tons pressure simultaneously from top and bottom, with 2⅝-in. fill and for handling pieces up to 3 in. in diameter, complete with core-rod attachment, adjustable gibs, guided lower plunger, spring pulldown, hydraulic pressure equalizer, and variable-speed drive with combination clutch and brake. (Courtesy of F. J. Stokes Co.)

construction is shown in Figure 140. Pressure is usually applied from above by means of an adjustable eccentric. The basic design is simple and practical, permitting precise adjustment of the upper punch for pressure, thickness and density control. The lower plunger is well guided and adjustable to control the weight of the powder fill, which, in turn, governs ranges of thickness and density for the compact. Ejection of the compacts is made possible by a cam and lever system, making the press a combination type in its true sense. The output rate is controlled by adjustment of a variable-speed drive, and the presses are also adaptable for single-stroke operation. In the following table, extremes of capacity are given for this type of tableting press.[7]

Type (Stokes)	Eureka	R
Maximum output per min....................	100	15–45
Maximum tablet diameter, in................	$1/2$	3
Maximum depth of fill, in..................	$7/16$	$2^5/8$
Pressure, tons..............................	$1^1/2$	20
Power for drive, hp.........................	$1/4$	3 (5)

Toggle and Knuckle-Joint Presses. For pressures in excess of 25 tons, the simple eccentric-type press becomes too weak, and stronger constructions must be used. Toggle and knuckle-joint presses offer all the advantages of strength; besides being ideal for coining and sizing work, they are well suited for molding larger pieces of relatively simple shape. In addition to the heavier pressures attainable, the toggle action has two outstanding advantages. There is quick action during idle motion of the punches, while, during final compression, a smooth squeeze of the compact allows better flow of the metal particles into all sections. Moreover, as the punches approach the final compression point, the toggle allows multiplication of the applied pressure without requiring an increase in power input. As in the case of eccenter and crank presses, production rates are adjusted by variable-speed drives, while single-stroke operation is also possible.

Presses of this type can easily be constructed as "double-action" presses, with one punch slide moving from above and another from below, or with one slide moving within the other. A press of the latter type,[6] illustrated in Figure 141, can be used for molding and sizing flange-type porous metal bushings. The outer slide can be timed so as to dwell for sizing of the bushings and to withdraw during stripping. A more powerful single-action, toggle-type press, shown in Figure 142,[7] is really a combination press, since the election or "knock-out" movement of the lower punch is facilitated by cams and levers. The incorporation of a floating die table as a special adaptation of the toggle design permits simultaneous pressure from above and below—up to the rated capacity of the machine. The float may be regulated to obtain regions of lowest density in the center of the compact. This is conveniently accomplished by a regulated downward movement of the table at a rate half of the relative travel of the upper punch.

The floating die table system assures the reduction of adverse wall-friction effects and permits transmission of the full pressure of the upper

Fig. 141. Duplex toggle press with one slide moving within another (courtesy of E. W. Bliss Co. and E. V. Crane).

punch through the compact to the lower punch, which remains stationary throughout the act of compaction. The system, furthermore, makes possible the forming of certain types of protrusions on the compact, e.g., flanges, ridges, etc. Special tool members which are supported in the die utilize the table motion. In this case, however, it becomes necessary to provide an adjustable mechanical stop for the table to assure that positive pressure is exerted by the table upon the protrusions.

Fig. 142. Single-action toggle press with cam-driven knock-out. Press designed for 75 to 100 tons pressure, with floating die table, 4-in. depth of fill, and arranged for pieces up to 4 in. in diameter; also including core-rod equipment, supported die table, clamp-ring holder, guided lower plunger, adjustable gibs, spring pulldown, 15-HP variable-speed motor drive with clutch and brake (courtesy of F. J. Stokes Co.).

One of the latest press constructions of the floating die table type (built by Stokes[8a]) provides "dual pressure" both from top and from bottom, and appears to be particularly suitable for the production of pre-

[8a] F. J. Stokes Machine Co., Philadelphia, Pa., Bulletin No. 488.

forms from iron and other metal powders. The main improvement of the construction over the above-mentioned type lies in a *power-controlled* movement of the table, at a rate exactly half that of the upper punch movement. Compacts produced by this "Dual-Pressure Preformer" excel by their even density and high uniformity through the entire thickness, and their size or weight can be larger than that obtained on the previously described machines due to a more effective distribution of the pressure. Operation of the press is simplified by two combination adjustments, one controlling the die fill (weight of the compact), as well as the ejection stroke, and the other controlling both the pressure applied to the preform and that applied to the die-table movement. Another interesting feature is a shuttle-type feed which permits the handling of large and irregularly shaped preforms up to 6 in. in length and 4 in. in width, or 4 in. in diameter with a corresponding die fill of $2^5/_8$ in. The capacity of the press is up to 60 strokes a minute at the rated tonnage of 75 tons.

Further modification and improvement of the previously described floating die system have resulted in a press of utmost flexibility which is finding increasing use in the porous bearing industry, particularly with those manufacturers who include flange-type bushings in their manufacturing schedules. This press is essentially a single-slide machine with either crank or toggle action and with conventional cam-operated feed shoe and ejector mechanism. Its distinguishing feature lies in the refinement of the table float construction, which allows for *controlled* downward and upward motions as well as adjustable mechanical limiting stops in both directions. The control of the table movement consists of timing, speed of travel, and pressure regulation. The chief advantage of such a press construction lies in more effective control of the density distribution in flange parts and the direction of the low density area toward the center of the compact, since the problems of proper timing and coordination of punch movements, which are so difficult to solve in conventional cam-operated or eccenter presses, is practically eliminated. An added advantage is obtained in this press construction by the delayed upward action of the floating die table at the completion of the pressure stroke. This feature eliminates compact breakage at the intersection between the flange and the hub portion of the compact, since the die table is held in its pressing position until the ejector begins to move the compact out of the die cavity, at which time both the lower punch and the floating die table travel upward together at the same speed.

Although many units embodying some or all of the above-described special features have been built by powder metallurgy fabricators for their own use, none of the press manufacturers is known so far to have designed or built presses of this type, except for the simpler floating die-type toggle press mentioned before.

In the following, typical capacity specifications are given for the press shown in Figure 142.

Type (Stokes)	280	280-A
Strokes per min............................	15–50	15–50
Maximum tablet diameter, in..............	4	4
Maximum depth of fill, in.................	2	3 or 4
Pressure, tons............................	80–100	80–100
Power for drive, hp.......................	$7^1/_2$–10	10–15

Cam-Type Presses. Presses incorporating cam and lever systems constitute a large proportion of the mechanically operated machines. While they are generally not quite so powerful as the toggle-type presses, they excel in versatility of design and performance. Basically, three different designs can be distinguished, each satisfying a particular need in powder metallurgy molding work. They are: (1) rotary-type presses, (2) single-punch presses, and (3) multiple-punch presses.

It has already been shown that when rotary presses are used, very high production rates are emphasized—shapes and fill conditions necessarily remaining very simple. In the rotary press, the various cams operating the punches are rigidly fixed in a circle, and the revolving head carries the punches through the required motions. Even at high speeds, pressure is applied on the pieces more slowly with such a machine than with an eccentric-type press. Pressure is applied simultaneously from both top and bottom, with the resulting compacts being equally dense and hard on the upper and lower edges. Another advantage of the rotary-type press lies in its more satisfactory feed, since more time can be allowed for the fill of each die because of the particular construction which provides filling of each individual die on the revolving head while it is several stations ahead of the briquetting punch. In addition, each die cavity is assured a uniform fill, since it is first overfilled, and then the excess is automatically removed and returned to the feeding device. The knock-out stroke is performed immediately after compaction has been completed and the top punch has been withdrawn. Besides a set of dies which may vary from 6 to 60 in number, the press must contain a complementary set of upper and lower punches. Fast movement of these punches while not under load contributes materially to the high production rates. In certain models "double compression" is possible—two applications of pressure to form a single compact. This has been found particularly advantageous in handling certain

TABLE 39

Capacity Specifications of Rotary Presses[a]

Type	B-2	BB-2 (a)	BB-2 (b)	D-3	DS-3	DD-2 (a)	DD-2 (b)	DDS-2	DDS-2 Spec. 1	#230	#210
Maximum output per min..	300–500	1200	1500	325	275	700	1000	300	150	65 (90)	60
Maximum tablet diam.. in..	5/8	5/8	1/2	31/32	13/16	13/16	15/16	13/16	13/16	2 1/2	6-in. square
Maximum depth of fill, in..	11/16	11/16	11/16	13/16	11/16	11/16	13/16	2	3	4 1/4	1 3/8–3
Number of dies............	16	27	33	16	15	23	31	23	16	10	6
Pressure, tons............	2 1/2	2 1/2	2 1/2	7	10	15	15	15	15	30	100
Power required, hp........	2	2	2	2	3	5–10	5–10	5–10	10	15	30

[a] Manufactured by F. J. Stokes Machine Co., Philadelphia, Pa.

powdered products where the air is squeezed out of the powder in the first compressing cycle, and then, without ejecting the charge, pressure is applied a second time to finish the forming of the piece. In Table 39, capacity specifications are given for a number of Stokes rotary-type presses, and Figure 143 is a photograph of one particular construction.

Fig. 143. Rotary-type cam-driven tableting press. Press used for applying up to 15 tons pressure, with 23 punch stations, 3-inch die fill, core-rod bushings, upper adjustable pressure roll, pressure equalizer, and variable-speed drive for operating at 100–300 RPM. (Courtesy of F. J. Stokes Co.)

Cam-operated, single-punch presses are machines in which two cams control the movement of upper and lower punches independently, and in any relation to one another to produce a wide variety of shapes, including many which cannot be made on other types of presses. The infinite variety of motions possible is an advantage that gives cam-operated machines a broad scope of usefulness in many branches of metal powder molding. The cams in these machines are so arranged that by means of inserts or adjustments the motion of the punches with respect to each other as well as to the cycle can be changed according to the individual requirements. Pressure can be applied both to the top and bottom of the compact, either simultaneously or separately, or from the top alone. The timing of the press cycle can be arranged so that the upper and lower punches press the material alternately, facilitating air escape or compaction of two-level parts. In other words, the motions determined to be most satisfactory in producing a part of uniform density and strength can be readily accomplished on these machines. Thin-walled pieces (involving filling around core rods), pieces requiring deep die-fills, difficult shapes, and poorly flowing powders require extra feeding time. The cam-operated machine is the only mechanical press that makes extra feeding time available without slowing the production rates. In fact, it is possible, by simply holding the upper punch out of the way for about $90°$ of the movement of the controlling cams, to double production on cam-operated presses over that possible on other types. A further contribution to uniform density and strength can be achieved by using a proper relationship between the flow rate of the powder and the speed of compression (regulated by inserting the proper cams). If desired, a longer dwell on the piece at full compression can be obtained in the same way. The capacity ranges of simple cam-operated briquetting presses of one manufacturer follow:

Type (Stokes)	G	S	P-1	P-2
Strokes per min..................	15–45	12–36	8–24	8–24
Maximum tablet diameter, in.....	$1^1/_2$	2	$2^3/_4$	$2^3/_4$
Maximum depth of fill, in........	3 (4)	$6^1/_4$ $(4^1/_4, 3^1/_2)$	$6^1/_4$	8
Pressure, primary upper punch, tons	12	30	60	60
Pressure, primary upper punch, tons	$2^1/_2$	5	10	10
Power required, hp..............	3 (5)	10 $(7^1/_2)$	20	20

Cam-operated presses can also be constructed to permit compression with multiple upper and/or lower punches. The molding of complicated parts with varying sectional thicknesses or protruding lugs, flanged pieces, parts with concentric projections, or pieces having holes only part way through requires individually moving and individually pressing punches. From Chapter IX it is seen that only a "differential" method of applying pressure (while a certain part of the powder is transferred within the die) will give a uniform density throughout the cross section of such multiple-section parts. For this purpose, modern, multiple-punch presses are equipped with a number of cams operating the telescoping segments of upper and lower punches; such multiple-punch presses have recently been constructed by the manufacturers of mechanical tableting presses. One particular press[9] is equipped with two individual upper and two individual lower punches, as well as with a third lower punch which can be used either as a movable low-pressure punch or as a stationary core rod. The press is rated at a 50-ton cumulative pressure, with a $5^1/_2$-in. die fill, and a maximum compact diameter of 5 in. A press operating on a similar principle, but of considerably stronger construction,[10] is shown in Figure 144. It also has two telescoping upper, and three telescoping lower punch arrangements. On the outer, upper punch a pressure up to 85 tons can be applied, whereas the inner, upper punch can operate at pressures up to 40 tons. The lower, outer punch withstands up to 40 tons pressure, while the intermediate, lower punch can apply 30 tons. The innermost punch can operate at 15 tons or can be used as a movable core rod with a pull-down pressure of 15 tons. The part is ejected by locking the two outer, lower punches and by using them jointly for the knock-out operation; in this position the punches deliver a maximum pressure of 70 tons. The press can accommodate compacts up to 6 in. in diameter, and powder-fill depths up to 6 in Fill uniformity can be obtained by using an overfill device, which returns the excess of powder to the feed shoe before compaction. Fully automatic production rates of such a machine can vary from 8 to 24 strokes per minute.

Combination Mechanical Presses. Except for the cam-operated presses, all mechanical tableting presses are actually combination machines, that is, they combine several of the above-mentioned principles. To be exact, all tableting presses have certain cam-operated motions. The simplicity in construction and the relative ease with which cam-controlled motions can be adjusted to particular needs make them especially adaptable to feed-shoe motion, knock-out, and lower punch press features— more so, since lower pressures are generally satisfactory for these operations. Therefore, in order to keep a simple and economical construction, practically all eccentric and toggle presses are equipped with cams for the auxiliary motions rather than with eccentric or toggle drives.

[9] J. Kux, *Machine Design, 14*, No. 10, 59 (1942).
[10] Product of F. J. Stokes Machine Co., Philadelphia, Pa.

Fig. 144. Multiple-action, cam-operated, full-automatic press. Press used for applying up to 85 tons pressure simultaneously from top and bottom, with two separate upper punch movements and two lower punch movements, plus a third movement arranged for use as a movable lower punch in addition to a stationary core rod, or as a movable core rod; also with supported die table, over-fill device, and built-in clutch and brake. (Courtesy of F. J. Stokes Co.)

397

Operating Principles

Mechanical presses employed for briquetting (or for the simpler sizing or coining operations) may be classified according to their mode of operation: (*1*) single-action presses, exemplified by eccentric and toggle presses in general with stationary or floating die table; (*2*) double-action presses, working in a single direction, illustrated by toggle presses having two telescoping upper punches; (*3*) double-action presses, working with single punches in two opposite directions, as eccentric, cam-operated combination presses with stationary die table; and (*4*) multiple-action presses, working with telescoping punches in two opposite directions, as cam-operated, multiple-punch presses.

In the case of single-action presses (both of the stationary die table and floating die table types), the bottom of the die cavity is fixed during compression by the stationary lower punch, which does not participate in the compression action and acts only as knock-out punch. The die cavity is filled with powder in partial vacuum in a uniform and reproducible manner by leading a feed shoe over the cavity before the lower punch recedes from the preceding ejection. Thereafter, pressure is applied from above by a single punch moving downward to a definite stop. This position yields the maximum pressure and may be adjusted very accurately. The lower punch is also movable, but only to eject the part. The pressure supplied by the cam and lever arrangement usually lies below the values practical for molding purposes. Like the pressure-applying upper punch, the position of the lower punch can be adjusted at the fill, pressing, and knock-out points. Its position in the first instance controls the fill powder of the die cavity, and consequently the weight of the compact; the height and density ranges of the compact can be set at the same time. The accuracy of height and density, of course, are controlled by the pressure stroke of the upper punch, while the adjustment of the knock-out position of the lower punch serves to clear the compact from the die cavity. During the next filling motion, the feed show pushes the previously formed compact to an incline, from where it is dropped into a collector or picked up by a conveyor.

About the same procedure is followed in double-action presses, whose pressure-applying punches operate in one direction. The sole difference lies in a duplex compression motion from above. In the case of double-acting presses applying pressure in two opposite directions, the procedure is different. Here the bottom of the cavity is formed by the upper surface of a second punch, which moves upward during the downward movement of the upper punch. In this way, the powder is compressed quite equally between the two opposing punches. At their respective ultimate positions, the two punches exert the maximum pressure. The lower punch generally ejects the compact by continuing its upward motion after compression, and after the pressure has been released by the withdrawal of the upper punch. The lower punch may be hollowed out to provide for either a concentric or eccentric, stationary core rod. This feature permits the forming of ring-shaped compacts, bushings, or other parts with a center hole— without changing the procedure.

With multiple-action presses involving telescopic upper and lower punches, the operations remain basically the same, although of course, individual motions will be dependent on the nature of the part to be pressed. In general, the metal powder must be fed into the die cavity by means of a hollow feed shoe, the fill volume being predetermined on the basis of the apparent density of the powder. The depth of the die cavity, set by the fill positions of the lower punch segments, is calculated carefully; when the feed shoe fills the die completely, smoothing the fill surface flush with the top of the die, the required amount necessary for forming the compact to desired density and weight must be obtained. When starting the compression cycle, the individual segments of the upper and lower punches move with different speeds to different positions while initial compression of some sections is initiated. By this action, part of the powder is transferred from its original position (flush with the top of the die) into regions where material is needed to build up heavier cross sections (while the density remains uniform). After the material has been completely transferred, all punch segments continue their movement at the same rate—compressing the compact uniformly—while retaining a satisfactory density distribution. In this connection the reader is referred back to Figure 102 in Chapter IX, which shows a practical example of this procedure. In order to control the pressure of the different punches, their stroke must also be calculated, and its relationship to the powder-fill depth considered in the light of the characteristics of the powder—especially apparent density and plasticity. Punches will always move between the same positions, once they have been determined. If the powder characteristics remain uniform, especially compressibility and apparent density, reproducible results can be obtained both for the weight of the compacts (as determined by the "volume control" of the powder fill) and for the thicknesses and density (as determined by the "stroke control" of the pressures). The various operational steps of a multiple-action press are illustrated in the series of photographs in Figure 145, picturing the molding of armatures as described in Chapter IX, with the press shown in Figure 144.

Fig. 145. Four principal stages in the molding of armature bodies from iron powder in the press that is shown in Figure 144 (courtesy of American Electro Metal Corp.).

In closing the discussion of mechanical presses, two important points must be emphasized: (*1*) the necessity for uniform fill, and (*2*) the need for special care in adjusting the presses. The success of the molding operation and, to a great extent, the success of all subsequent operations depend on how uniformly the cavity can be filled by the constantly repeated motions of the press; in many cases the powders can be produced or treated to give no serious difficulties in this respect. But peculiar features of the press design, or lack of space or time, may often restrict the flow of the powder. Feed hoppers are conventionally constructed in the form of a funnel, creating a bottleneck near the point of transition to the feed shoe. Feed shoes, on the other hand, often have narrow passages in order to make room for upper punches and punch holders. When fluctuations in humidity and atmospheric pressure are added to this difficulty, there is every reason to expect flow and fill disturbances. In these instances, vibrators are usually attached to the feed shoe, hopper, or feed chute to remedy this effect. But this is not always sufficient, or it may even introduce other problems, since constant vibration may tend to increase the tendency of powder to pack and arch in the narrow passages, aside from the possibility of classification or segregation of the powder. In such cases, enlarging of the passages and slight heating of the powder in the hopper with electric bulbs have often solved this problem. A widely used expedient consists of a so-called overflow mechanism. During the filling operation, the volume of the die cavity is increased by dropping the lower punch below its necessary fill position, allowing an excess of powder to enter the cavity. Thereafter, while the feed shoe is still above the cavity, the lower punch moves into its correct position, and the excess powder is pushed back into the feed shoe.[11]

Movable core rods, acting on a similar principle, can also be employed to overcome feeding difficulties.[12]   Other proposals to remedy the situation suggest feeders having stirrers built in wire cages,[13] or having a vacuum applied to the feed hopper to eliminate air pockets and stratifications in the powder,[14,15] and at the same time to reduce the moisture content of the powder—thus improving flow characteristics.

In regard to the adjustment practices on mechanical presses, one must always keep in mind that these machines are rigid structures which cannot endure overloads. The pressures applied by each punch cannot usually be

[11] Product of Kux Machine Co., Chicago, Ill.
[12] U. S. Pat. 1,607,389.
[13] German Pat. 716,118.
[14] U. S. Pats. 2,198,612 and 2,259,465.
[15] Products of the F. J. Stokes Machine Co., Philadelphia, Pa. See also *Mech. Eng., 69,* **928** (1947).

read on gages; and in order to avoid breaking of the frame, all adjustments must be made with greatest caution. It is always advisable to start
with underfills and underloads, and to approach gradually the desired end
positions and strokes of the punches. It is only natural that these adjustments become more difficult when the press construction is more complicated. While in simple, single-punch presses the adjustment possibilities
are confined to the fill and knock-out setting of the lower punch and the
pressure-stroke setting for the upper punch, a complicated press of the
multiple-action cam type involves no less than eleven standard adjustments (two for the pressure stroke of the upper punches, and one for each
of the three lower punches, the fill, pressure stroke, and knock-out positions). To this can be added an overflow adjustment, in addition to any
possible changes of cam inserts for the molding of differently shaped parts.
For reasons of economy, the extremely complicated adjustment process of
these presses demands large production runs of the same part. It also
raises a definite barrier against further expansion in the use of this type
of press.

HYDRAULIC PRESSES

Modern hydraulic presses display superiority over their mechanical
counterparts in respect to high power, safety against overload, and flexibility in performance. It is because of these features that the hydraulic
press has found its place in powder metallurgy.

Single-action hydraulic presses with rams moving upward slowly
(between 1 and 10 in. per minute) are used extensively in the plastics and
ceramics industries. A commercial utilization of these conventional designs
is achieved only by the application of multiple-cavity dies. According to
the size of the piece and the number of cavities, the tonnage of the presses
may vary from 50 to over 1000 tons. Although the construction of these
presses and their operation are simple—inasmuch as the presses apply
pressure only in one direction—auxiliary operations such as feeding and
ejection must be accomplished by complicated die attachments, such as
loading sheets, etc.

Since multiple-cavity molding has met with little success in powder
metallurgy (see p. 374), the conventional, slow-acting hydraulic press has
found few uses in the field. Small models are used in many laboratories,
but results obtained by the slow-motion compaction are not necessarily
comparable with those which can be obtained in high-speed industrial
machines. Since the beginning of the century, high-tonnage hydraulic
presses of the slow-moving type have been used for the compression of
large slabs, briquettes, or rods, which after sintering must be reduced to

small parts by machining, or to wires and sheets by customary metal-working practices. But in general, the antiquated, slow-acting hydraulic press has given way to the modern, powerful, quick-acting hydraulic machine; various interesting designs of these machines are at the disposal of the powder metallurgist. It is this type of press—with its many auxiliary motions and features, its inherent simplicity of adjustments, and its versatility—that will now be discussed in more detail.

Capacity of Hydraulic Presses

It has already been mentioned that hydraulic presses are generally more powerful, but yield considerably lower production rates than comparable mechanical presses. While the capacity of the fully automatic mechanical tableting machine stretches from 5 to 100 tons, the field of application for hydraulic production presses starts at about 100 tons and ranges up to 5000 tons. We must realize that such heavy presses cannot yield a high production output even if all modern means of stroke acceleration are used in the construction. In addition, the rate of compression for larger compacts must be quite low because the speed with which pressure can be applied to the powder is limited by a number of factors, particularly the resistance of trapped air, particle regrouping, and plastic flow of the metal. For example, a press of 1000-ton capacity would be able to compress at 50 tsi a briquette with a cross section of 20 sq. in. and a possible thickness of about 2 in. The mass required to yield a briquette of approximately 85% of full density would weigh about $9\frac{1}{2}$ pounds in the case of copper powder. A comparatively long period of time must be allotted for the uniform filling of this mass in the die cavity and compaction to a high density, for the particles must be allowed to group themselves so as to permit the escape of trapped air. However, if full advantage is taken of all the modern improvements which high-power, quick-acting hydraulic presses have to offer, remarkable production rates can be obtained. Thus, for example, by keeping at a minimum the idle stroke of the press ram and the dwell, output rates of 7 to 10 pressings per minute have been accomplished with hydraulic presses of 200- to 500-ton capacity, on parts of about 5 sq. in. cross section weighing about $\frac{1}{2}$ pound.

In view of this basic relationship between tonnage and output rate, a tendency has developed lately to employ the heavy hydraulic presses primarily for: (1) the forming of heavy briquettes of uniform density as an intermediate stage in the shaping of small parts or tool bits, or for the production of wires and sheets; (2) for molding friction materials; and (3) for sizing, repressing, and coining sintered compacts into final shape.

In the first instance, relatively large powder masses are involved in each compression cycle. Therefore, slow output rates of about one to two minutes for each briquette are still economical, especially when costly raw materials (*e.g.*, tungsten or tungsten carbide) are involved. In the case of friction materials, the large lateral dimensions of the parts to be pressed necessitate the use of heavy, high-capacity presses. With regard to coining, output rates can be increased substantially because the compression stroke can be kept smaller than for powder compaction. Deformation of the metal in the direction of compression generally represents only a fraction of the distance that the punch must travel to produce compaction of a powder. In coining operations, the idle stroke can also be reduced, since less space is required to introduce and remove a sintered piece than is needed to operate a powder feed shoe over the die cavity. The speed with which pressure is applied can also be increased in coining, since the problems of trapped air and particle reorientation are absent; the possible speed, however, is limited by its effect on dimensional accuracy.

Basic Construction and Operating Principles

The fundamental difference between the use of hydraulic and mechanical presses for compression of powders lies in the fact that, in most cases, mechanical presses work to a definite stop before the movement of the piston is reversed, whereas hydraulic presses usually work to a definite pressure, although there exists the possibility of reversing hydraulic presses by stroke. Accordingly, mechanical presses produce compacts of equal *volume* (thickness), whereas hydraulic presses produce compacts of equal *density*, but not necessarily of equal thickness. It is apparent that in the latter case the powder will always be subjected to the same maximum pressure. If, however, the powder varies in apparent density or plasticity, or if the die cavity is not filled uniformly, considerable variations in the thickness of the compact may result in the case of the hydraulic press.

Single-Action Presses. As for mechanical machines, hydraulic presses can be classified as single-, double-, and multiple-action units. Single-acting presses consist of a downward-moving main ram propelled by a main pump which forces oil into the main cylinder. Usually, a secondary pump forces oil into a smaller cylinder which moves a smaller piston upward and acts as ejector. Generally, the force behind the ejector is kept to at least one-tenth of the maximum pressure obtainable on the main ram. During compression, the press slide bears on punch adaptors which, in turn, force the punch into the die cavity. The die either rests solidly on the bed of the press or on a floating adaptor which rests on the bed of the press,

The bottom of the die cavity is formed by a lower punch which transmits the pressure to a wide pressure block that bridges the ejector pin. In most constructions the ejector cylinder is not strong enough to withstand the

Fig. 146. 200-ton, fast-cycle hydraulic press for molding and coining powder metallurgy parts (courtesy of E. W. Bliss Co.).

pressure of the main ram. The ejector cannot be used for upward compression, and merely delivers the knock-out stroke after compression has been completed and the top slide retracted. The ejector must be permitted to withdraw to its original position before the new compression cycle is begun.

Many presses of this type have proved valuable in powder metallurgy work, as, for example, a 200-ton, fast-cycle "Hydro-Dynamic" molding press with bottom (ejector) cylinder built by Bliss,[16,17] as shown in Figure 146. This press is typical of the fast-acting, self-contained hydraulic type. The oil tank is placed on top of the press with variable delivery pump and motor at the rear. Limiting positions of the quick advance, the pressing stroke, the quick return, and the knock-out are conveniently adjustable, and pressure and position stops are equipped to control the work stroke; the speed of the complete cycle is adjustable over a limited range. The press can operate on a full-automatic, semi-automatic, or hand-controlled cycle. In semiautomatic operation, the movements of the press are stopped after the cycle has been completed, and the operator must start the cycle again. In full-automatic operation, one cycle follows the other until the operator stops the press. For powder molding work, the full-automatic operation is not practical, since the press does not provide for automatic powder feed or for work removal. The press operates most advantageously on a semiautomatic cycle (*i.e.*, completing one complete press and knock-out operation at a time) and a press operator must be at the press constantly to handle the feeding and removal operations. Filling of the die cavity can be accomplished either by the "volume-control" method typical of mechanical presses (except that the powder has to be fed and leveled off by hand, instead of by feed shoe and hopper) or by the so-called "weight-control" method when great precision fill is required. Here, the powder allotment for each compact is weighed on an automatic scale, controlled by a photoelectric cell. If a precision weight scale is used, an accuracy of $\pm1/4\%$ can easily be obtained—comparing favorably with the customary volume-control method. An advantage of the weight-control method lies in its independence of fluctuations in the apparent density of the powder. The method allows approximately six 100-gram weighings per minute. A photograph of this equipment in conjunction with a 200-ton Hydro-Dynamic Bliss press is shown in Figure 147.

Other presses of similar construction but of different capacities are built by the leading press manufacturers of this country. One press built by Bliss can exert a pressure up to 350 tons and allows up to 45 strokes per minute. In this type of press the speed of the ram approaching the powder is 500 in. per minute, while the speed during actual compression is only 23 in. per minute. The speed for the return stroke is again as high as the preliminary speed.

[16] Product of E. W. Bliss Co., Brooklyn, N. Y.
[17] E. V. Crane and A. G. Bureau, *Trans. Electrochem. Soc., 85,* 63 (1944).

Another press[18] of particular interest has been developed recently by Hydropress. This press (Fig. 148) has a capacity of 500 tons, an

Fig. 147. Molding of iron powder parts in the 200-ton hydraulic press shown in Figure 146. The powder allotment for each cycle is automatically weighed by a photoelectrically controlled, exact-weight scale. (Courtesy of American Electro Metal Corp.)

ejector force of 50 tons, and an output rate exceeding 10 pieces per minute when in full-automatic production. It has two novel features, a full-automatic, powder-feed device, and a semiautomatic, coining-feed device. The powder-feed device, consisting of feed shoe and hopper—as in

[18] Product of Hydropress, Inc., New York, N. Y.

Fig. 148. 500-ton, fast-cycle, self-contained, oil-hydraulic press for molding and coining powder metallurgy products. Press is equipped with removable powder and coining feed attachments for semiautomatic or full-automatic operation; selector switch to swing electro-hydraulic pressing cycle from molding to coining operation and vice versa; two-way, variable delivery pumping unit; electronic timer for dwell action control; automatic 50-ton capacity ejector; and stationary core rod (courtesy of Hydropress, Inc. and American Electro Metal Corp.).

mechanically operated presses—is organically integrated into the press; the press can undergo speed changes during compaction in order to allow for an escape of trapped air. The coining-feed device consists chiefly of a carriage shuttling back and forth over the die and dropping a piece into the cavity; after compression, the coined piece is pushed ahead by the

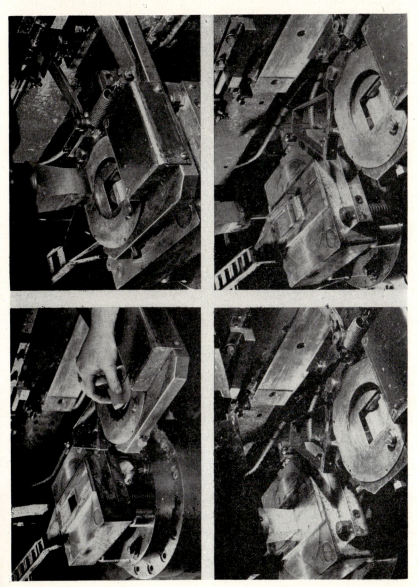

Fig. 149. Four principal stages in the joining of flat sintered iron products in the press shown in Figure 148 (Courtesy of American Electro Metal Corp.).

carriage before another is dropped into the die. During the act of compression an operator is required to deposit a new piece into a receptacle inside the carriage. Figure 149 shows the four principal stages in the coining cycle for a flat iron pole piece. For simple shapes the coining-feed device could be made full-automatic by supplying pieces to the sliding carriage from a magazine. Other improvements of this particular design include an independently actuated ejector ram which, if required, can apply a pressure up to 50 tons simultaneously with the action of the main ram, and so achieve a limited double action. The ejector is hollow to accommodate a core rod which can also withstand a 50-ton pressure. The main press movement is actuated by a two-way, variable-delivery pump, credited with giving a particularly smooth press operation without shocks during the reversal of the action, as well as a very accurate pressure control, because the pump "short-strokes" itself as soon as the preset pressure is reached. A dwelling action can be controlled by a standard electronic timer.

Multiple-Action Presses. Hydraulic press types described above require complicated floating die sets to get uniform compression from above and below. Such die constructions are not always feasible; and especially in cases where very high bushings or multiple-section parts, *e.g.*, flanges, are to be formed, presses acting in two opposite directions are preferred. Of course, presses of this type lose some of the compactness of construction which has marked most recent hydraulic-press creations. Presses applying high pressures from two opposite directions are extremely tall and may cause difficulties when being erected in an average factory building. A 200-ton, double-action hydrodynamic press (Bliss) measures over 20 ft. in height against 12 ft. for the single-action press of equal capacity. Of course, the lower part of the press can be placed in a pit, but this necessitates a foundation on the ground floor. Placing such a press horizontally is not practical for powder molding operations, but would be acceptable for coining purposes.

In the hydraulic press field, double-action usually signifies the independent movements of two rams in one direction. Presses of this type are generally used in deep-drawing work and involve one operating slide inside another. This method makes possible the molding of flange-type parts. Usually, the two telescoping pistons act in a downward direction, but presses are also being built that have this arrangement for upward-moving punches; here, the inner piston acts as a movable core rod, making the arrangement particularly adaptable for closed-end bushings, cup-shaped bodies, etc. Presses embodying this double-action feature both on top and bottom are also being constructed. They are usually referred to as "duplex, double-acting" presses, and are built up to a total pressure of 5000 tons (Bliss). Figure 150 shows a 200-ton hydraulic press[18a] designed by the Hydraulic Press Manufacturing Co. for the forming of parts from

Fig. 150. 200-ton, fast-cycle hydraulic press with two hydraulic ram assemblies for molding powder briquets (courtesy of Hydraulic Press Manufacturing Co.).

metal powders. This press ("Fastraverse") is equipped with two hydraulic ram assemblies, permitting the application of high pressure from two different points. The press cycle is entirely automatic, including loading

[18a] Product of The Hydraulic Press Mfg. Co., Mt. Gilead, O.

Fig. 151. 500-ton self-contained oil-hydraulic double-action powder briquetting press with top and bottom rams, and of open four-column construction. Press has capacity of 500 tons on upper ram, 475 tons on lower ram, 70 tons on ejector, 16-in. stroke of upper ram, 8-in. max. fill height, 12½-in. daylight, 100-HP motor for upper ram pump, 60-HP motor for lower ram pump, and a powder feed capacity of 30 cu. ft.; allows 7 cycles p.m. based upon 8 × 8-in. max. die cavity, 4-in. fill height, 1½-in. height of compact, and ⅛-in. compression with each ram under high pressure (courtesy of Hydropress, Inc.).

of the die with powder, briquetting, and ejecting the finished part from the die. (The entire operating cycle is hydraulically controlled.)

A more powerful double-acting hydraulic press has recently been built by Hydropress,[18] for the briquetting of metal and carbon powder ingots (Fig. 151). Full-automatic production of briquettes at a rate of seven per minute is accomplished by modern push-button and electronic controls, and variations in thickness and density of the briquettes are kept to closest tolerances. A stationary die bed receives the powder charge from an hydraulically operated feed carriage. The powder is then subjected to pressure from both below and above by the two main rams acting in unison and in opposition. Pressure (500 tons maximum from top and 475 tons maximum from bottom) is maintained for a period which can be adjusted by timing relays and can range from 0 to 30 seconds. When the pressure is released, the lower ram ejects the finished briquette (70 tons maximum ejection pressure) which is then moved forward by the feed carriage as it advances to start the new cycle.

In our discussion of multiple-action hydraulic presses, constructions which involve action in two directions perpendicular to each other must be included. While the main pressure is applied in a vertical direction, usually by means of a downward-moving ram, secondary pressure of considerable force is applied in a horizontal direction. The prime purpose of the latter action is to close by force split dies employed in molding fragile briquettes of large dimensions which cannot be ejected from the die by force, but must be removed manually after the pressure is released and the die segments have been opened. Naturally, operating cycles of this kind are very slow, but the high cost of the end product may justify this procedure. Several different models of this particular construction have been publicized in recent years; Watson-Stillman[19] developed a press with a 400-ton vertical ram and a 300-ton horizontal ram for the briquetting of metal powder preforms. The machine has a mold space measuring 30 in. in length by 6 inches in width and allows a powder fill of 7 in. The pressures are adjustable within close limits, and the main pump is driven by a 30-hp. motor; the advance of the rams is actuated by solenoid-operated pilot valves. Operating speeds are 238 in. per minute in downward and return movement of the vertical ram, and 22 in. per minute for the horizontal ram; the vertical pressing speed is 17.5 in. per minute.

In a 750-ton press built by the Hydraulic Press Mfg. Co.,[20] the downward-acting force is similar to the horizontally acting one. This press is used for pressing of tungsten carbide briquettes, 12 in. long, $1^1/_8$ in. wide, and $^1/_4$ in. thick. A press constructed on a similar principle, but of considerably greater power has been built by the same manufacturer[21] for the forming of larger briquettes from tungsten, titanium, and tantalum carbide powders, from which cutting tools, dies, and inspection gages are

[19] Product of The Watson-Stillman Co., Roselle, N. J.
[20] Product of The Hydraulic Press Mfg. Co., Mt. Gilead, O. See also *Iron Age, 153*, No. 12, 76 (1944).
[21] M. T. Muirhead, *Metal & Alloys, 19*, No. 2, 368 (1944).

fabricated. The press is also advertised for the molding of other metal powders requiring the application of high pressure from two different points. It exerts a downward-acting force of 1500 tons and a horizontally acting force of 1000 tons; the unit—like all modern hydraulic presses used in powder metallurgy—is completely "self-contained." This term refers to an arrangement where the hydraulic system, including the hydraulic pumps which generate the operating pressure, is mounted together with the oil storage tank either on top of the press or attached next to its base.

Fig. 152. 3000-ton hydraulic press providing compression from top, bottom, and side for briquetting refractory metal powders (courtesy of Baldwin Southwark Div., The Baldwin Locomotive Works).

The creation of Baldwin-Southwark[22] shown in Figure 152 may perhaps be considered the ultimate in power and versatility in hydraulic presses of this kind. Known as a "U-D" type (uniform density), this press has a capacity of 3000 tons and is used to press tantalum and tung-

[22] Product of Baldwin Southwark Division of the Baldwin Locomotive Works, Eddystone, Pa. See also *Iron Age, 152*, No. 26, 24 (1943).

sten briquettes at pressures up to 80 tsi, the briquettes being 30 in. long, $2^1/_2$ in. wide, and $1^1/_4$ in. thick. It provides compression from above and below, as well as from the side. The upper and lower rams oppose one another, each with a capacity of 3000 tons. The metal powder is thus compressed into a briquette of more uniform density than would otherwise be possible. A 3000-ton side cylinder supplies a tremendous force for clamping the split die. The machine provides a bed width of 24 in. and a daylight of 30 in. It is operated by a 50-hp. motor; the weight of the total assembly is nearly 150 tons.

An impressive acceleration of hydraulic-press operation can be achieved by the use of compressed air. Certain presses employed in the manufacture of friction parts are equipped with a "quick traverse," actuated from an air-pressure-loaded hydraulic accumulator, which can bring the mold into position up to about 60 times as fast as is possible with an ordinary hydraulic pump.[23]

Pulsating Presses. A rather recent development in the field of hydraulic presses adapted to powder metallurgy deserves special mention, since its trend is opposite to the "high-tonnage, low-production-rate" principle, which has been mentioned as characteristic for this type of pressing machinery.

The Denison Engineering Company[23a] has brought on the market a line of standard, fast-acting, flexible presses, ranging from 4 to 12 tons capacity, which are credited with compacting rates that match those of the mechanical single-punch presses. The same company has most recently engineered a 75-ton capacity press which operates on the same principles. A particular feature of these units consists of a *pulsating* pressure application which is designed to obtain over-all higher and more uniform densities in the compact. This pulsating action is obtained by the combination of *interrupted* hydraulic pressure on the upper ram, coupled with a *continuous* hydraulic pressure of smaller magnitude on the lower ram. During the interval, while the upper, pulsating pressure is approaching zero, the lower, constant pressure ram will tend to lift the compact slightly (thereby reducing the die-wall friction) until it is forced down once more against its mechanical stop by the greater pressure of the oncoming upper ram.

Another type of operational cycle has gained favor in compacting ceramic materials and may conceivably have beneficial influence in compacting metal powders as well. This cycle commences with the

[23] Anonymous, *Materials & Methods, 23*, No. 3, 729 (1946).
[23a] The Denison Engineering Co., Columbus, O.

hydraulic or pneumatic operation of the feed shoe, which is followed by a fast traverse downward movement of the upper ram, slowing down just in time before entering the die cavity to prevent undue powder loss by dusting. After the upper ram has closed the die cavity, the lower ram starts moving upward, thereby precompressing the compact until it reaches a position where it exerts its maxium preset pressure. Thereafter, the upper ram starts moving again, forcing the compact and the lower ram downward against its hydraulic resistance until maximum pressure is finally exerted by the top ram against the mechanical stop provided by the lower ram. The ejection cycle operates in a similar way to what has been described previously in other presses.

COMBINATION PRESSES

In addition to mechanical and hydraulic presses, specific models have been developed which combine both operational principles. However, since constructions of this type have only limited applications in powder metallurgy, they will be discussed briefly in this text.

The prime purpose behind these combination designs is the desire to combine the operational safety of the hydraulic press with the quick-acting motion of the mechanical machine. Accordingly, these combination constructions are basically mechanical presses to which hydraulic mechanisms have been attached for the purpose of absorbing overpressures, or of equalizing the applied pressure from stroke to stroke to yield compacts of more uniform density. In general, these combination units have received only limited approval from powder metallurgists. There has been ample ground for criticism, since these machines combine not only certain basic features of the two systems, but also some of their faults. The main objections have been in connection with oil losses (incapacitating the hydraulic system) which are accentuated by the high rate of strokes achieved by mechanical presses. Additional critism has concerned complications in construction and operation, as well as appreciable increase in maintenance.

Several manufacturers have incorporated hydraulic devices in mechanical tableting presses. Stokes Machine Co. constructs a 100-ton toggle press with a hydraulic excess-pressure-release device,[24] built into a compact unit as shown in Figure 142. The same manufacturer equips his standard 20-ton, eccentric-type presses with a detached hydraulic device combining pressure equalization and excess pressure release.[24]

The Kux Machine Company[25,26] constructs presses of 50- and 150-ton

[24] Product of F. J. Stokes Machine Co., Philadelphia, Pa.
[25] Product of Kux Machine Co., Chicago, Ill.
[26] J. Kux, *Machine Design, 16,* No. 11, 135 (1944).

capacity with a combined mechanical-hydraulic operating mechanism. The 50-ton press is shown in Figure 153, and Figure 154 shows a sectional view of the 150-ton press operating at 10 cycles per minute. These machines combine the speed of mechanical presses with the uniform

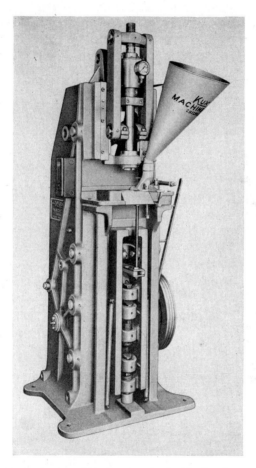

Fig. 153. 50-ton, combination mechanical-hydraulic press for molding metal powders (courtesy of Kux Machine Co.).

pressures of hydraulic presses. The upper punch is operated by a mechanical toggle linkage, whereas the lower punch is raised by a hydraulic ram; the pressure applied to the tablet by this ram is registered on a dial gage so that it can be controlled exactly. Once set, equal pressure is exerted upon every tablet regardless of slight variations in fill, and the

densities of all tablets are thus alike and uniform. The combined mechanical-hydraulic machine is furnished with a self-contained hydraulic unit,

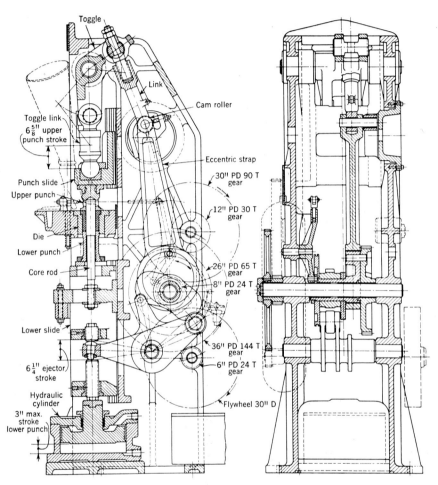

Fig. 154. Sectional schematic drawing of 150-ton, combined mechanical-
hydraulic press with cam-controlled eccentric linkage for 60-degree dwell to permit
action of hydraulic ram on lower punch (according to Kux[26]).

mounted and guarded inside of its oil storage tank. During the compression cycle, the upper punch enters the die first with a definite stroke,
applying a partial compression to the powder from above. As soon as the

upper punch comes to a stop, the lower ram is raised by hydraulic pressure, which applies a predetermined pressure on punch and punch holder, the latter resting solidly on the ram. After the desired pressure is obtained, the hydraulic ram recedes to its starting position, while the upper punch is withdrawn in an upward direction. The knock-out of the compact is achieved with the aid of a cam-lever system which raises the lower punch holder above the hydraulic ram; and the knock-out stroke is timed in order to prevent interference with the withdrawing upper punch.

A 345-ton briquetting press, operating with a triple hydraulic lower motion and a mechanically actuated slide, has recently been constructed for the molding of large parts with irregular cross sections.[27] The press provides a floating die table of 120-ton capacity, a core rod that can exert a pressure of 60 tons, and an ejector that can press with 75 tons. These features, and the fact that the slide can be adjusted for a 5- to 10-in. stroke at 6, 9, or 18 strokes per minute, make the machine one of the most powerful and, at the same time, most flexible briquetting units ever built for powder metallurgy operations.

In closing this discussion on combination designs, a few words should be said about those constructions which are basically hydraulic presses but embody some mechanical features. Devices installed to facilitate the feeding of powders or parts or for the proper removal of the formed objects lie in this group. Of particular interest are installations of "mechanical stops," consisting of heavy pressure blocks placed next to the die, or of hardened steel collars placed around the uprights. Adjustments in the height of these stops can be made readily by adding or subtracting thin steel plates or shims of different thicknesses. The primary function of these stops is to arrest the press ram at a definite, predetermined level, so that the compact is pressed to a definite and reproducible size. When the applied pressure is sufficiently rated to compact the powder to a level below that of the mechanical stops, the ram will touch these stops with every stroke, some excess pressure being absorbed by the stops. In consequence, the compacts may vary in density if the powder characteristics or the fill are not uniform. On the other hand, one of the chief advantages of the mechanical press, namely the formation of uniformly sized pieces, is transferred to the much stronger and safer hydraulic machine. This feature is of particular interest where such presses are to be employed for final sizing and coining operations.

[27] Product of E. W. Bliss Co., Detroit, Mich. See also *Iron Age, 160*, No. 4, 71 (1947).

420

Fig. 155. Battery of heavy mechanical and hydraulic presses for production of powder metallurgy products (courtesy of American Electro Metal Corp.).

TABLE 40
Powder Metallurgy Presses and Press Manufacturers[a]
(Schumacher and Souden[28])

Operating principle	Pressures, speeds, and other features			Type of work for which well-suited
	Single-acting	Double-acting	Other types	
Hydraulic (including oil - hydraulic and oil-electric)	Production pressures run from 500 tons to 5000 tons or more, with speeds available according to pressure and part-size. A Hydropress machine, for example, rated at 500 tons operates at 10 cycles/min. A 35 - ton lab press works at 30 in. per min.	Pressures from below 75 tons (a Kux toggle type, for example) up to 5000 tons (Bliss) are available. Stroke speed of a Kux 150-ton press is 20/min.; of a 600 - ton Hydropress, 2/min. Ram speeds of an HPM 750-ton press are 350 in./min. closing and 20 in. /min. pressing.	Duplex and duplex double-acting presses reach 5000 tons pressure (Bliss). HPM has several double-action machines with horizontal and vertical rams whose pressures are 60/-45 up to 1500-/1000 tons and speeds 6/6 to 114/4 in./min. Hydropress's quadruple-action press is rated at 350/-250 tons and 8 cycles/min.	*Single-acting:* Good for simpler shapes in various sizes; suitable for coining and sizing. *Double-acting:* General purpose presses, and for higher - pressure, more complicated jobs. Can be used for carbides, sizing, dense parts, etc. *Others:* Duplex widely used for thick shapes. Right angle actions popular for carbides and other work.
Mechanical (toggle, rotary, straight side, knuckle-joint, etc.)	Pressure ratings from 10 to 100 tons for one type of inclinable press to 25 to 1500 tons for a straight side and knuckle-joint press (both made by Bliss). The speeds are 6 to 150 strokes /min.	Pressures from 15 tons to 140 tons, and stroke speeds of 6 to 1000/-min. (the last a Kux 25-ton rotary press) are available. Stokes, Kux and others have wide selection of presses of various ratings classified according to depth of die fill, from $1^3/_8$ up to 8 in.	Typical duplex presses are rated at 100–600 tons and 4 to 30 strokes per min.	*Single-acting:* Good for general pressing of porous parts (not too complex in shape), preforming, sizing, etc. *Double-acting:* Most diversified type, used on general work, porous parts, intricate shapes, very high production jobs, dense parts, repressing, sizing, etc.

[a] Press manufacturers: Baldwin-Southwark Div., Philadelphia. Birdsboro Steel Fdry. & Machine Co., Birdsboro, Pa. (Bliss) E. W. Bliss Co., Brooklyn, N. Y. Fred S. Carver Co., New York. Clearing Machine Corp., Chicago. Arthur Colton Co., Detroit. Continental Machines, Inc., Minneapolis, Minn. Denison Engineering Co., Columbus, O. Chas. F. Elmes Engg. Works, Chicago. A. B. Farquhar Co., Ltd., York, Pa. Hydraulic Machy., Inc., Dearborn, Mich. (HPM) Hydraulic Press Mfg. Co., Mt. Gilead, O. (Hydropress) Hydropress, Inc., New York. (Kux) Kux Machine Co., Chicago. Lake Erie Engg. Corp., Buffalo, N.Y. Schloemann Engg. Corp., Pittsburgh (Stokes) F. J. Stokes Machine Co., Philadelphia. Watson-Stillman Co., Roselle, N. J. R. D. Wood Co., Philadelphia. Zeh & Hahneman Co., Newark, N. J.

422

Summary

Presses which now are designed primarily for forming parts from metal powders may be of the mechanical, hydraulic, or combined mechanical-hydraulic types. For lower pressure ranges, mechanical presses are usually more advantageous, whereas hydraulic presses are used extensively for pressures above 150 tons. Because of the fact that high-tonnage presses cannot be made to act as quickly as mechanical ones, the advantage to be gained from the use of automatic feeding is not as great. For this reason, mechanical powder feeds have not been provided on most hydraulic machines. Manual feeding is often satisfactory and, although expensive, provides for much greater safety in operating costly machines and die equipment.

The choice of the most suitable press depends, therefore, on the particular features which are most important for the work on hand. Of course, flexibility in application and performance of the machine is generally required for competitive reasons. A survey of powder metallurgy presses and press manufacturers is reproduced from Schumacher and Souden's manual[28] (Table 40), which describes briefly the various designs and characteristics as an aid in the selection of the most suitable press.

Figure 155 shows a battery of heavy hydraulic and mechanical presses in actual operation in the powder metallurgy plant of American Electro Metal Corporation. The press in front is a 500-ton, automatic, single-action universal press for powder compaction and coining (Hydropress); the second is a 200-ton, single-action "Hydro-Dynamic" press (Bliss); the third is a full-automatic, cam-operated, multiple-action, multiple-punch press of 100-ton total capacity (Stokes); the last is a full-automatic, single-action toggle press with floating die table and hydraulic excess-pressure release (Stokes).

[28] E. E. Schumacher and A. G. Souden, *Metals & Alloys, 20,* No. 5, 1342 (1944).

CHAPTER XII

Compression of Powders under Heat

PRINCIPLES OF HOT-PRESSING

Purpose and Functions

In the preceding chapters our discussion of the compression of metal powders has been confined to operations which occur at room temperature. The beneficial effect of heat on the plasticity of metal particles (*i.e.*, their property of sustaining permanent deformation without rupture) has already been mentioned briefly in connection with friction phenomena. Although it has not been established that frictional heat alone is sufficient to increase particle plasticity to any appreciable extent, many metal powders definitely display improved deformability if subjected to controlled elevated temperatures. Moreover, considerably denser and more coherent agglomerates of particles can be obtained by compaction at elevated temperatures than would be possible in operations at room temperature. The effects of temperature on plastic deformability of particles are not surprising in view of the fact that, in general, the resistance of metals to deformation decreases with rise in temperature; this, incidentally, is the reason that forging, rolling, drawing, and extruding of metals are usually done while the metal is hot. At the same time, it becomes apparent that, with increased plasticity of the metal, many of the characteristics of initial powder so important in ordinary molding lose their significance. This is particularly true of particle size and shape, and also of the tendencies of some powders to classify, segregate, or form striations during compaction.

If we now recall that bonding between particles is increased by improved plasticity of the metal (Chapter VIII), an appreciable consolidation of the structure, coupled with a substantially increased cohesive strength of the entire body, can be expected in hot-pressed compacts. For the purpose of creating a network of contact points where links between particles can be established, the heat acts in the same direction as the pressure, either adding to the effects of the latter or supplanting them in part.

Apart from the fact that hot-pressing improves the compactibility of a powder, another point of view can be taken. Basically, powder metallurgy works on the principle of applying both pressure and heat to a powder. Pressure, while molding the particles into a definite shape, has the important function of making this shape coherent by creating strong particle-contact bonds. Heat, on the other hand, while being allowed to act only in a moderate way to maintain the formed shape (*i.e.*, below the mushy state or fusion temperature), has the equally important function of consolidating and really homogenizing the structure by allowing for a sufficient atomic mobility and "place interchange" to transform the original pattern of agglomerated particles into a recrystallized, uniform grain structure. In order to obtain these desired effects, it is immaterial whether pressure and heat are applied in subsequent steps (as done for practical reasons in the vast majority of cases), or in a simultaneous operation, as in hot-pressing. In other words, if pressure and temperature conditions are properly chosen, hot-pressing can combine both molding and sintering, the two principal processing operations. Since heat and pressure act in conjunction, modifications in either of these two principal operations (if combined) are quite possible. In fact, hot-pressing may be considered intermediate between ordinary cold-pressing and sintering.

The temperatures at which the advantages of hot-pressing become evident are those at which the processed metals begin to exhibit marked plasticity; therefore, they depend on the recrystallization temperature and melting point of the metal. In the instance of aluminum and similar low-melting metals, compacting at room temperature might in a sense be considered to be a hot-pressing operation, whereas, for refractory metals such as tungsten, compaction at 1500°C. (2730°F.) should still be classified as "cold" work. The designation "hot" in connection with pressing has therefore only a relative meaning.

Application of Pressure and Heat

It is obvious that a number of ways can be devised to exert pressure on a powder compact under heat. Depending on whether the pressure or the heat is the more important factor, we may distinguish between hot-pressing and pressure sintering. In the latter operation, a furnace treatment is implied, whereby the powder or compact—passing through the heating cycle—is subjected to static pressure, usually of moderate degree. Several processes of this kind have been patented in the hard metal field[1] and for iron powder applications,[2] with pressures up to 66 tsi. In most

[1] French Pat. 793,020.
[2] U. S. Pat. 2,341,860.

cases pressure is exerted by dead weights, requiring large furnace chambers; compacts must be small to permit an effective unit pressure.

In contrast to this procedure, hot-pressing involves molding of compacts under heat, with one compact at a time being produced. Techniques of this nature always involve the use of a mold to give the compact its shape, and a machine to exert the necessary pressure. The facilities for applying the heat, although basic, can be adapted to the design features of press and die. Hot-pressing of metals in powder form usually involves considerable difficulties; since the powder is heated in a storage container, the greater atomic mobility in the particle surface regions tends to weld the particles together and to impede the flow of powder into the die cavity. Moreover, a large surface of hot metal is exposed, and the danger of oxidation becomes acute. Nevertheless, several procedures involving heated powders have been suggested and a hot-pressing apparatus operating on this principle has been patented recently by Koehring.[3]

To facilitate feeding of the die cavity, the practice of preforming the powder has been employed with great success. Although the additional cold-pressing step complicates the procedure, the time saved during feeding, the increased protection against oxidation, and the general easing of the entire hot-press operation are decidedly advantageous. Preforming pressures may be very low, just sufficient to compact the powder into a body that can be handled safely. For such powders as iron, copper, or copper alloys, preforming pressures of 5 to 10 tsi have been found quite satisfactory. It is advantageous in other instances to apply higher pressures in order to obtain denser preforms; this is especially necessary when the apparatus and method do not provide adequate protection from oxidation.

The three methods of heating metal powders or preforms and dies for pressing are: (1) convection heating, (2) induction heating, and (3) resistance heating.

Convection heating involves heating of the powder or preform as well as of the die, either independently or by heating the die alone and transferring heat into the powder or preform. The heat is supplied by a suitable electrical resistance furnace surrounding the die and placed inside the working space of a hydraulic or mechanical press. For low temperature work, the die may be simply heated with blow torches or gas burners. Sometimes referred to as the "hot-die" method, it has been used by many investigators particularly in their earlier work.

Induction heating, which also applies the principle of heating the die externally, has been used in many of the more recent procedures. The

[3] U. S. Pat. 2,362,701.

chief advantage of this method lies in quicker heating and cooling of the die, eliminating the necessity for ejecting the hot-pressed compact. This is especially advantageous in experimental work, where the danger of oxidation during cooling (after ejection from the hot die) must be eliminated. The same method (of cooling the compact within the die) has been widely practiced in hot-pressing of cemented carbide materials in graphite molds, where a series of hot-press dies can be operated by one high-frequency generator, the high cost of the material permitting slow production rates.

Resistance heating is, perhaps, the most interesting method. Heat is generated either by passing a current of high amperage through the powder or preform, or by passing the current through a graphite or metal mold. Many methods and apparatus based on resistance heating have been suggested in the patent literature, among which a process by Jones is particularly interesting.[4] Some procedures even use a heavy-duty, electrical resistance-welding machine to supply pressure and heat,[5] with special tungsten-base electrodes as dies. A disadvantage of resistance welding, however, may be seen in its implied restriction of the compact's shape. Only small cylindrical bodies can be heated uniformly between the electrodes. Although bar-shaped compacts could be compacted lengthwise, it would be necessary to have a special die construction and a nonconducting material for the mold. Pieces of irregular contours could probably not be hot-pressed by this method at all.

Pressure–Temperature–Time Relationship

The selection of the best method of applying heat and the most suitable apparatus and procedure depend largely on a thorough understanding of the relationship between pressure and heat, as well as of the time element involved. Obviously, the pressure to be used in this kind of work must lie within a certain practical range, the limits being determined not only by the temperature of the compact, but also by the hot strength of the die material and by the tendency of the hot-pressed powder to extrude between the moving and stationary parts of the mold assembly. Other factors, such as friction during pressing and ejection, and gaseous reactions with the compacted material, may also play an important role in determining the best pressure conditions.

PRESSURE–TEMPERATURE

Let us first consider the pressure range and the pressure–temperature relationship in connection with the forming of completely dense bodies.

[4] Brit. Pats. 530,995 and 530,996.
[5] U. S. Pat. 2,355,954.

In hot-pressing, obviously, the upper pressure limit does not exceed the ultimate pressure needed for cold-pressing of powders. This, however, may be so high that die materials may fail before the necessary pressure for complete densification can be reached. The lowest conceivable pressure

TABLE 41

Pressure–Temperature Relationship on Density of Hot-Pressed Iron

Material	Pressure, tsi	Temperature °C.	°F.	Density, g./cc.	Density (vs. theoretical), %	Observer
Copper.............	100	250	482	8.87	99.3	Trzebiatowski[a]
	100	300	572	8.87	99.3	
	100	400	752	8.90	99.6	
	50	300	572	8.77	98.3	Goetzel[b]
	50	400	752	8.90	99.6	
	50	500	932	8.91	99.7	
	25	400	752	8.63	96.6	
	25	500	932	8.74	97.9	
Brass...............	50	300	572	8.19	98.7	Goetzel[c]
55 copper, 45 zinc..	50	500	932	8.30	100.0	
	25	500	932	8.24	99.3	
65 copper, 35 zinc..	59	300	572	8.12	96.0	
	50	500	932	8.44	99.6	
	25	500	932	8.35	98.6	
60 copper, 40 zinc..	5	700	1292	(8.38)	100.0	Pratt[d]
	5	800	1472	(8.38)	100.0	
Iron................	30	500	932	7.47	95.0	Schwarzkopf and Goetzel[e]
	30	600	1112	7.87	100.0	
	18	600	1112	7.50	95.4	
	20	700	1292	7.87	100.0	
	11	700	1292	7.44	94.5	
	10	780	1436	7.58	96.5	Henry and Cordiano[f]
	10	800	1472	7.85	99.7	Schwarzkopf and Goetzel[e]
	5	800	1472	7.51	95.5	

[a] W. Trzebiatowski, Z. physik. Chem., A169, 91 (1934).
[b] C. G. Goetzel, Trans. Am. Soc. Metals, 28, 909 (1940).
[c] C. G. Goetzel, Trans. Am. Soc. Metals, 30, 86 (1942).
[d] W. N. Pratt, Symposium on Powder Metallurgy. ASTM, Philadelphia, Pa., 1943, p. 49.
[e] P. Schwarzkopf and C. G. Goetzel, Iron Age, 148, No. 10, 37 (1941).
[f] O. H. Henry and J. J. Cordiano, Trans. Am. Inst. Mining Met. Engrs., 166, 520 (1946).

for hot-pressing work is given by the pressure requirement for complete densification of a material heated to a temperature infinitely close to its melting point. Since liquid metals are completely dense under normal conditions, the lowest possible pressure can be considered to approach atmospheric pressure. It is therefore clear that the pressure range needed for complete densification is very wide indeed. For practical purposes, it

TABLE 42
Time–Temperature Relationship on Density of Hot-Pressed Electrolytic Iron

Iron powder	Density (per cent vs. iron)	Pressure, tsi[a]	Pressure, tsi, at hot-pressing temperature of						
			500°C (932°F.)	600°C (1112°F.)	700°C (1292°F.)	800°C (1472°F.)	900°C (1652°F.)	1000°C (1832°F.)	1100°C (2012°F.)
Swedish sponge, −100 mesh	90[b]	(75)[c]	21	12	8	4	2½	(4)	(3)
	95	(125)	27	18	11	6	(4)	(6)	(5)
	100	(200)	(75)	(40)	(20)	(12)	(8)	(10)	(9)
Electrolytic, −100 mesh	90	(100)	19	10	(6)	(3)	2½	(4)	(2)
	95	(150)	(35)	19	(10)	(5)	(4)	(6)	(4)
	100	(225)	(60)	30	(20)	(10)	(8)	(9)	(7)
Hydrogen reduced, −325 mesh	90	(125)	24	10	6	3	1½	—	—
	95	(200)	34	17	10	4	—	—	—
	100	(300)	(70)	30	20	10	—	—	—

[a] Pressing performed at room temperature.
[b] Percentage density of Swedish sponge iron compacts related to 7.85 g./cc., but may actually be higher due to irreducible impurities.
[c] Figures in parentheses represent estimated values. Full density based on assumption that no elastic springback in compact is encountered.

is important to establish working values for pressure. These values will depend, to a considerable degree, on the type of material used. In some metals and compositions, such as iron, copper, and many copper alloys, the range of plasticity before the melting point is reached is wide, so that comparatively low temperatures are satisfactory for hot-pressing. The same applies to those compositions in which one or more molten constituents form during the hot-pressing operation. In other metals the plasticity range is very narrow, requiring precise temperature control for hot-pressing work. However, irrespective of the degree of ductility of the particular material affecting the numerical relationship between temperature, pressure, and time, there exists for each temperature a minimum pressure at which complete densification is obtained almost instantaneously (e.g., by dynamic hammer blow or by very brief, static pressure application).

Several investigations have dealt with the pressure–temperature relationship for different materials, and some data are reproduced in Tables 41 and 42. Most interesting, perhaps, is the recurrent experience that, when comparing pressures needed to obtain a certain predetermined density, the application of high temperatures permits pressures of only 10% of those necessary in cold-pressing. Pratt,[6] for example, reports that 60–40 brass powders can be compacted to 100% density at 700° or 800°C. (1290° or 1470°F.) with a pressure of only 5 tsi. Jones[7,8] reports similar results with brasses varying from 90–10 to 50–50 copper–zinc compositions. When heated to about 200°C. (360°F.) below the respective melting point of the alloy, only 5 tsi were necessary for complete densification, even in the absence of a reducing atmosphere. In the case of an 80–10–10 copper–tin–lead bronze, Pratt obtained a density as high as 7.8 g./cc. (92% of theoretical) at a temperature of 750°C. (1380°F.) for as low a pressure as 300 psi. The author found that for straight tin bronzes containing from 5 to 20% tin, completely dense compacts could be obtained by pressing at 500°C. (930°F.) at 25 tsi.[9] Similar conditions, applied to brass compacts containing 15, 25, and 35% zinc, also resulted in complete densification.[10] For a 55–45 copper–zinc composition heated to 500°C. (930°F.), only 15 tsi were necessary to achieve compaction to full density. For pure copper compacted at 500°C. (930°F.), the ideal

[6] W. N. Pratt, Symposium on Powder Metallurgy. ASTM, Philadelphia, Pa., 1943, p. 49.
[7] W. D. Jones, Metal Ind. London, 56, 69 (Jan., 1940).
[8] W. D. Jones, Metal Ind. London, 56, 225 (Mar. 1940); also Brit. Pats. 530,-995 and 530,996.
[9] C. G. Goetzel, Trans. Am. Inst. Mining Met. Engrs., 161, 580 (1945).
[10] C. G. Goetzel, Trans. Am. Soc. Metals, 30, 86 (1942).

density was approached at 25 tsi; the same result was obtained when the compact was pressed at 50 tsi at 400°C. (750°F.).[11] Sauerwald[12] obtained perfectly dense compacts by heating to 810°C. (1490°F.) and applying 24 tsi. Trzebiatowski[13] reached a density of 8.9 g./cc. when hot-pressing a very fine copper powder at 400°C. (750°F.) at 100 tsi. For very fine gold powder a density of 19.11 g./cc. was achieved with the same high pressure of 100 tsi at 300°C. (570°F.).

Turning now to the less plastic iron, the pressure–temperature relationship shifts upward, of course. For example, it is impossible for iron powders to be compacted to bulk density at room temperature with pressures up to 50 tsi, while at 600°C. (1110°F.) it is possible to reach ideal density with a pressure of only 30 tsi; at 700°C. (1290°F.) the necessary pressure decreases to 20 tsi, and at 800°C. (1470°F.)—to 10 tsi.[14]

TIME ELEMENT

The foregoing examples probably suffice to give an indication of the range in which pressures and temperatures can be expected to lie for most of the metal powders important in practice. No conclusive picture can be obtained, however, without due consideration of the time element, since plastic flow of metals is a function of time, as well as of temperature and force. Whereas from a practical standpoint the time during which the hot compact is to be kept under pressure should be as short as possible, there is evidence that densification and physical properties of such compacts can be greatly improved by prolonged application of pressure and heat. Henry and Cordiano[15] have stated that increasing temperature and time (at a particular temperature) under sustained pressure tend to increase the number and area of interparticle contacts as well as the amount of plastic deformation, having a cumulative effect. Experimenting with electrolytic iron powder, they found, e.g., that a density of 7.58 g./cc. (96.4% of theoretical) could be reached after applying a pressure of 10 tsi for 50 seconds at 780°C. (1435°F.). The same density could be obtained with the same pressure at only 700°C. (1290°F.) if applied for 450 seconds. Table 43 shows the significant trend of the time–temperature relationship. Apparently, an extended pressure dwell at a given temperature gives results equal to those obtainable for short dwells at considerably higher temperatures, or presumably at higher pressures. Of course, a practical evaluation

[11] C. G. Goetzel, *Trans. Am. Soc. Metals, 28,* 909 (1940).
[12] F. Sauerwald, *Z. Metallkunde, 21,* 22 (1929).
[13] W. Trzebiatowski, *Z. physik. Chem., A169,* 91 (1934).
[14] P. Schwarzkopf and C. G. Goetzel, *Iron Age, 148,* No. 10, 37 (1941).
[15] O. H. Henry and J. J. Cordiano, *Trans. Am. Inst. Mining Met. Engrs., 166,* 520 (1946).

of these findings depends largely on commercial aspects, since production rates for hot-pressing are (even without extended dwells) much lower than for cold-pressing and sintering. While the compression and ejection phases, in hot-pressing, need not be more time consuming, a certain period must be allowed in most procedures for the establishment of temperature equilibrium before pressure can be exerted. This phase alone may consume several minutes, so that an added period of perhaps several minutes for pressure dwell may condemn the entire process as impractical. Instead, it appears desirable that the process allow for a minimum of time to be

TABLE 43

Time—Temperature Relationship on Density of Hot-Pressed Electrolytic Iron
(Henry and Cordiano[15])

Temperature °C.	°F.	Time, sec.	Density, g./per cc.	Density (vs. theoretical), %
500	932	50	6.31	80.2
		150	6.38	81.1
		450	6.71	85.3
600	1112	50	6.70	85.2
		150	6.89	87.5
		450	7.05	89.6
700	1292	50	7.32	93.0
		150	7.52	95.6
		450	7.58	96.4
780	1436	50	7.59	96.5
		150	7.71	98.0
		450	7.76	98.6

spent in reaching temperature and pressure equilibrium, e.g., by employing high static pressures in the pressing of presintered hot compacts in pre-heated dies.

Controlled Atmospheres

The mechanical procedure of hot-pressing is complicated appreciably by the introduction of a reducing atmosphere. Most processes reported have not dispensed with this precaution, especially where loose powders have been heated before compression.[15a] Elimination of harmful oxide films must be the primary objective in all manipulations involving heat, because the contacts between metallic surfaces must be safeguarded under all circumstances. This has been recognized by most investigators, and elaborate precautions for controlling the atmosphere in the die have been devised accordingly. While hydrogen has been used conveniently by Sauerwald, Trzebiatowski, and others, carbon monoxide produced from graphite dies has also been found feasible. In other cases certain additions to

[15a] U. S. Pat. 2,362,701.

the powder were sufficient to reduce oxide surface films or to prevent excessive oxidation of the particles. Strauss[16] suggests the addition of 5% titanium hydride to nonferrous alloy powders that develop a liquid phase of about 10% by volume at a soaking temperature of 400°C. (750°F.) prior to hot-pressing at 650°C. (1200°F.). Jones[17,18] even went so far as to hot-press copper-base and iron-base compacts with no protective atmosphere at all, and achieving unexpectedly good physical properties with some compositions. In most cases any oxide film formed on the particle surfaces apparently did not appreciably prevent sintering or diffusion of the particles. The films probably were punctured sufficiently during compaction to permit the establishment of metallic surface bonds.

In discussing the question of oxidation, Jones holds that the great majority of his test results appear to show that internal oxidation does not always detract from the qualities of the product as seriously as in the case of cast alloys. He even assumes that the presence of oxide is actually beneficial in certain cases. Unquestionably, the good physical properties obtained with mixtures of 0.3% carbon steel powders and 5% of iron–carbon–phosphorus eutectic (1.3% carbon and 10.2% phosphorus) hot-pressed in air at 1000°C. (1830°F.), and of a variety of copper-base alloys manufactured by hot-pressing powders in air, indicate that this point is well taken. On the other hand, these few examples are insufficient to disregard the predominantly bad experiences encountered during normal sintering practices, where oxides have usually proved very harmful. The dangers inherent in the oxidation of the particle surfaces are well known, and without exaggeration it may be said that the entire powder metallurgy technique is drastically influenced by the struggle against the bad effects caused by oxygen. Therefore, a method of retaining a neutral or reducing atmospere in the compacting zone of the die assembly must be considered to be basic for all hot-press work of a general nature. Of course, considerable modifications would have to be made in the apparatus to allow for operation under such conditions, and several techniques, to be discussed later, have been developed which show that such modifications are quite practicable.

The Material Problem

Closely connected with the problem of atmosphere is that of the material for the die and punch construction. If, for instance, a material such as graphite can be used, the problem of atmosphere loses its importance. In fact, a desire to surface carburize the compact during hot-pressing

[16] H. L. Strauss, Jr., *Steel, 118*, 18 (1946).
[17] W. D. Jones, *Metal Ind. London, 56*, 69 (Jan., 1940).
[18] W. D. Jones, *Metal Ind. London, 56*, 225 (Mar., 1940).

may make the use of graphite molds mandatory. Graphite dies have found applications in the hot-pressing of tungsten carbide mixtures at very high temperatures, but the mold materials used for this purpose are so weak and easily abraded that only very low compacting pressures are permissible. Even the best grades of graphite do not withstand more than 2 to 3 tsi static pressure, and fail at even lower pressures if subjected to repeated stress. This points to one of the greatest problems in hot-pressing: the strength of die materials.

Ceramic dies made from materials such as steatite, apatite, quartz, or magnesium silicate have also been suggested, but no experience with these materials as hot-press dies has yet been reported. It appears doubtful that the hot-strength, heat-shock resistance, or lubricating properties of these ceramics can be developed to a degree making them superior to those of graphite.

In view of the relatively high specific pressures necessary for forming dense compacts, metallic dies are preferable to graphite or ceramic materials, but the limited hot strength of most metallic materials adaptable for die making restricts the hot-pressing temperature considerably. Little has been made public about experiences with hot-press die materials. In discussing the use of steel dies, Jones[17] comes to the conclusion that the steels used would not be suitable for temperatures above 500°C. (930°F.). However, Henry and Cordiano[15] have proved that a high nickel–chrome alloy steel can be used for repeated hot-pressing at 10 tsi up to a temperature of 780°C. (1435°F.), and in some of the author's work, high-speed steel dies withstood 10 tsi at 800°C. (1470°F.) for several hundred consecutive operations without appreciable wear.[19] In discussing the work of Henry and Cordiano, Wulff reported the successful use of Stellite up to 800°C. (1470°F.).[15] For temperatures exceeding 800°C. (1470°F.), there appears to be no ordinary metallic material available. It has been suggested that tungsten-base alloys, or even cemented carbide dies should be tried, but the performance of such materials has not yet been made public. Cemented carbide tools, it appears, would have to be properly reenforced to prevent bursting.

However, there remains a single approach where reasonably high compacting pressures can be obtained at even higher temperatures than those mentioned above—retaining the idea of using steel dies. For the sake of a uniform temperature distribution in the compact, the methods described so far generate heat in the die; the heat is carried into the compact by convection. If, however, means are found of confining the heat

[19] C. G. Goetzel, in J. Wulff, *Powder Metallurgy*. Am. Soc. Metals, Cleveland, 1942, p. 395.

developed in the compact to the die cavity without heating the entire die, reasonably high compacting pressures can be obtained at even higher temperatures than previously mentioned. This would involve preheating the compact directly in the die, or in an independent furnace, before placing it in the mold, and then pressing the hot compact in a cold die. The operation must then be carried out so quickly that no appreciable cooling of the compact occurs before compression. The operation would thus resemble a drop-forging technique, where a preformed hot compact is pressed in a relatively cold die. Since loose powder cannot be transferred easily from a furnace to the die, a cold preforming of compacts would probably have to precede the operation. In practice, therefore, such a technique would have to be divided into three distinct phases: (*1*) cold-pressing of the powder into a compact rigid enough for handling, (*2*) heating of the compact to sintering temperatures without pressure, and (*3*) pressing of the compact while still hot in a forging operation. In such a process, temperature gradients in the compact and in the die (which cannot be prevented entirely) make the method less accurate and less reproducible. If the production rate is low, nonuniformity of the structure of the compact will result from the cooling effects near the surface. This may be particularly objectionable in the case of steels. On the other hand, if the production rate is high, temperature differences will be less pronounced in the compact, but more so in the die. The interior regions of the die will gradually heat to temperatures which may be beyond safety with regard to rigidity, strength, hardness, or wear resistance. Moreover, the entire assembly will be subjected to a state of nonequilibrium of temperature which may cause stress concentrations and may impair dimensional accuracy because of uncontrollable expansion.

Mechanical Interferences

In addition to the problem of die strength, several other factors are troublesome in hot-pressing: (*1*) the wear of the die material, (*2*) the tendency of the compact's surface particles to weld on die walls and punch facings, and (*3*) the tendency of very plastic metals to extrude into the clearance space between the die walls and the moving punches.

DIE WEAR

During compression, metal powders may exhibit extraordinarily high frictional forces. Even if pressing occurs at room temperature, more stringent conditions are experienced than in the pressing of plastics. This is why additions of lubricants become necessary for most cold-molding processes involving metal powders. Unfortunately, these lubricants (with the exception of graphite and mica), volatilize at such low temperatures that their use for hot-pressing work is impossible. Graphite, however, retains its excellent antifriction properties even at very high temperatures. In powder form, it can be mixed with the metal powder, but becomes effective in hot-pressing only if added in appreciable quantities (e.g., 1–5%). This amount may be prohibitive in many cases where homogeneous, high-quality alloys are desired. Large graphite additions may be objectionable particularly in iron-base materials. Aside from the possibility of forming low-melting eutectic compositions, substantial diffusion of the carbon into the iron would cause replacement of many of the lubricating graphite particles by abrasive cementite—thus counteracting the lubrication effects. Flake or spherical graphite powder can, of course, be placed in immediate contact with the die and punch facings by brushing or spraying. A very effective method of lubrication involves the application of colloidal graphite suspended in water or oil (Aquadag, Oildag[19a]) either to the die walls or to the surface of the preformed compact.

Finely divided mica has proved satisfactory as a lubricant for hot-pressing at moderate temperatures and particularly for hot-repressing.

SURFACE WELDING

Just as serious as die-wall abrasion is the problem of welding of the surface of the hot compact to the facings of punches and die cavity. If high pressures and temperatures are applied, diffusion between the compact's metal and the die metal can hardly be prevented, especially if a long dwell is allowed. This problem is particularly serious at the facings of punches serving as electrodes for direct heating. Powdered graphite is usually not sufficient to prevent the tendency of the electrodes to weld to the compact. In that case one has to resort to inserts between the punch and the compact. Solid disks from graphite or refractory metals, especially tantalum, have been tried successfully. Of course, the problem varies with the nature of the metal involved, and also with the required temperature, time, and pressure. For processes working on the direct-heating principle, Jones[17] and others[19b] recommend tungsten–copper electrode tips as a wear- and weld-resistant material for punches. Apparently, this material does not have to be supplemented by nonwelding powder surface additions unless circumstances become extremely difficult.

It has also been found that surface-welding and galling effects on the walls of hot-press dies can be reduced materially if compaction is

[19a] Products of the Acheson Colloids Corp., Port Huron, Mich.
[19b] U. S. Pat. 2,355,954.

obtained by impact blows. Instantaneous compaction prevents welding by gradual abrasion, and tends to break up any solid contacts formed on the facings of the die.

PLASTIC EXTRUSION

The tendency of certain metals—if made very plastic by high temperatures—to extrude between the punches and dies also interferes seriously with smooth operation of the hot-press technique. Especially in the case of certain brasses, aluminum alloys, and other low-melting metals, this situation is so serious that one pressing may be sufficient to block any further movement of the punches, making it necessary to dismantle the entire hot-die assembly. Obviously, lubricants or other foreign additions cannot help to eliminate this situation and the only remedy lies in pressing very ductile alloys at lower temperatures and higher pressures than would otherwise be feasible. For all hot-pressing work, it is advisable to fit the punches into the die as closely as possible. Since the compaction cycle is slower than in cold-pressing, no particular consideration need be given to space for escape of trapped gases. In cases where clearance between punches and dies cannot easily be held to the desired minimum, split dies held together in a water-cooled vise may solve the problem.

Ejection

It has been noted that, in hot-pressing, the problem of wear is particularly serious during ejection if the compacts are allowed to cool appreciably during removal from the die. Therefore, it is desirable to eject the part while still hot and relatively plastic. Since metal oxides are strong abrasives, the surface of the compacts should remain protected by a controlled atmosphere at least until the compact leaves the die. In fact, for the sake of a smooth surface finish, the compact should remain surrounded by a protective medium until completely cold. Besides gaseous protection in a cooling chamber, burying in an inert powder mass or quenching in water will prevent oxidation.

ACHIEVEMENTS OF HOT-PRESS TECHNIQUE

Having surveyed the fundamentals of the hot-press technique with its attending operational difficulties, we shall now discuss its advantages. Products of the cold-pressing technique are generally porous, requiring an unproportionally high energy input for complete densification. This method is therefore ideally suitable where porosity is an essential characteristic. Porous bronze bearings, metallic filters, and low-strength functional parts are typical examples. Of course, it is quite possible to manufacture nonporous articles in this manner, although this involves expensive and time-consuming manipulations. Tungsten, tungsten carbide, heavy metal, and refractory metal base contact materials are examples of fully dense materials obtained by cold-pressing. But in all of these cases densification is connected with great volumetric contraction caused by high-temperature sintering or by subsequent working. This change in dimension, a *sine qua non,* involves the loss of one of the greatest advantages of the process, namely, the molding of the powder into a form of precise dimensions.

The only general method of producing absolutely nonporous compacts of precise dimensions in any alloy is the hot-press method. All alloys or powder mixtures which are capable of being sintered can be produced by this method, eliminating many limitations peculiar to cold-pressing. The versatility and adaptability of hot-pressing to many molding problems have proved a great asset and have induced leading men in the field to develop the technique into a practical procedure in spite of the many and serious obstacles. Of equal importance, of course, is the fact that, with the hot-press technique, materials can be developed whose physical properties and structural quality far exceed anything that can be achieved by the conventional cold-pressing and sintering practices.

Practical Advantages

The practical advantages of the hot-press method have been aptly summarized by Pratt.[20]   Hot-pressing would generally be a substitute for sintering, and, accordingly, handling and losses due to breakage of the green compacts would be minimized. Closer size control is possible, since volumetric changes, caused by the shrinkage forces or by gas evolution, occur in the hot-press die. Subsequent sizing or coining operations thus become superfluous. Moreover, the hot-press method can easily yield required heavy weights or high densities without repressing or subsequent working. As far as is known, physical characteristics obtained by using this method are better, and comparable in many cases to those of wrought materials. Thanks to the high density, hot-pressed parts will respond readily to surface treatments, plating, or hardening. Even complicated parts with re-entrant angles may possibly be hot-pressed if the alloy is sufficiently plastic and contains a liquid constituent, as in the case of brass and bronze powder mixtures. In general, the hot-press method is less sensitive to the powder's origin and to particle characteristics. As pointed out by Comstock[20a] hot-pressing offers the possibility of utilizing the advan-

[20] W. N. Pratt, *Symposium on Powder Metallurgy.* ASTM, Philadelphia, Pa., 1943, p. 49.
[20a] G. J. Comstock, *Communication at Powder Metallurgy Colloquium,* New York University, April 26, 1946.

438

Fig. 156. Microstructures of electrolytic copper powder compacts hot-pressed in hydrogen atmosphere, before and after annealing: A, pressed at 400°C. (750°F.), at 50 tsi (×120), and B, same, after annealing at 750°C. (1380°F.) for 1 hour (×120).

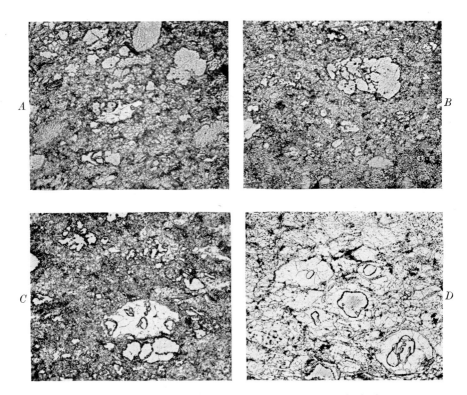

Fig. 157. Microstructures of brass compacts hot-pressed in hydrogen atmosphere from alloy powders at 500°C. (930°F.) at 50 tsi (×120). Powders produced by sintering at 450°C. (840°F.) for 2 hours. Particle structure shown in Figure 16. A, 85–15 Cu–Zn alloy; B, 75–25 Cu–Zn alloy; C, 65–35 Cu–Zn alloy; and D, 55–45 Cu–Zn alloy.

tages of those prealloyed powders which, on account of their extreme hardness, exhibit poor compactibility at room temperature.

Lacking pores, the internal structure of hot-pressed compacts is more uniform and consolidated than in their sintered counterparts. As evidence,

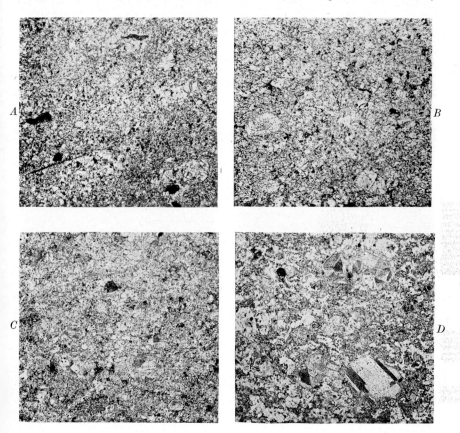

Fig. 158. Microstructures of bronze compacts hot-pressed in hydrogen atmosphere from alloy powders at 500°C. (930°F.) at 50 tsi (×120). Powders produced by sintering at 550°C. (1020°F.) for 1 hour. Particle structure shown in Figure 17. *A,* 95–5 Cu–Sn alloy; *B,* 90–10 Cu–Sn alloy; *C,* 85–15 Cu–Sn alloy; and *D,* 80–20 Cu–Sn alloy.

a series of photomicrographs are shown in Figure 156 for hot-pressed copper, in Figure 157 for hot-pressed brasses, in Figure 158 for hot-pressed bronzes, and in Figure 159 for hot-pressed iron. All photographs have a substantially dense structure of agglomerated particles in common, with signs of recrystallization becoming clear at or above 400°C. (750°F.) for the nonferrous materials. No marked alloying effects are noticeable up to 500°C. (930°F.). For powder mixtures, the period of time allowed for alloying during hot-pressing is evidently too short. If the powders are prealloyed at approximately the same temperature for a prolonged period

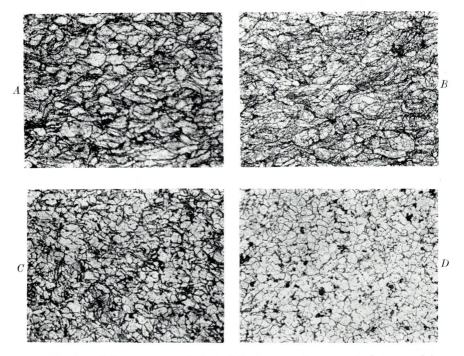

Fig. 159. Microstructures of electrolytic iron powder compacts hot-pressed in hydrogen atmosphere at different temperatures and pressures in alloy steel dies (A and B) and in graphite dies (C and D) (× 140). A, pressed at 500°C. (930°F.) at 30 tsi; B, pressed at 600°C. (1110°F.) at 30 tsi; C, pressed at 900°C. (1650°F.) at 2½ tsi; and D, pressed at 1100°C. (2010°F.) at 1½ tsi.

of time, the additional, brief heat treatment causes little change. Recrystallization of the iron begins at 600°C. (1110°F.); at higher temperatures, grain growth gives the structure an entirely new pattern of polygonal, ferritic grains.

If sufficient time is allowed for plastic flow, there is little variation in density or hardness of hot-pressed pieces from one section to another. Losses by evaporation—so serious in the case of zinc during sintering of brass—are considerably reduced, and because of the lower temperatures used, it may even be possible to hot-press certain compositions with a molten constituent which cannot be produced otherwise—the liquid constituent being lost under normal sintering conditions.

Physical Properties

Since the physical properties of metal powder compacts will be discussed in detail elsewhere (Chapter XXXI, Volume II), it is sufficient (on the basis of a few examples) to indicate here how the density, hardness, and tensile properties of hot-pressed compacts are affected by the collective action of pressure, temperature, and time.

COPPER AND GOLD

Trzebiatowski's results[21] with hot-pressed copper and gold are reproduced in Figure 160. With a uniformly applied pressure of 100 tsi, density, hardness, and specific resistance are shown as a function of the hot-press temperature. The density of the compacts rises rapidly with increasing pressing temperature, and attains a maximum of 8.9 for copper and 19.11 for gold. These values differ by less than one per cent from those determined by means of precise lattice parameter measurements (copper, 8.937; gold, 19.29). At higher pressing temperatures, the densities tend to fall (according to Trzebiatowski this is attributable to increased plasticity of the metal extruding slightly between the press punches and the die cavity). Some gas, moreover, may have been absorbed by copper, or water vapor by gold. Hardness values show an initial increase to the remarkable figures of 190 for copper and 165 for gold. Above

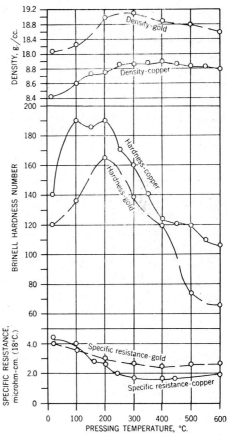

Fig. 160. Density, hardness, and resistivity as a function of temperature of pressing for copper and gold powder compacts hot-pressed at 100 tsi (according to Trzebiatowski[21]).

[21] W. Trzebiatowski, *Z. physik. Chem.*, A169, 91 (1934).

200°C. (390°F.) the hardness begins to diminish because of annealing effects. These high hardness figures could not be interpreted satisfactorily by Trzebiatowski. X-ray investigation disclosed a gradual disappearance of the work-hardening structure, and at 350°C. (660°F.) the Ka doublet is once again resolved. At this temperature, however, the hardness of copper is still above 150 Brinell, and at 600°C. (1110°F.) still above 100, leading to the conclusion that lattice deformation is only partly responsible for the exceptional hardness. Trzebiatowski could not prove that the fine particle size used in this work could account for these effects. A microscopic examination in addition to x-ray work gave no evidence of grain growth, and no unusual phenomena were observed with regard to specific resistance. With both copper and gold, the specific resistance—measured on the cooled, finished compact—decreased continuously with increasing temperatures up to about 300°C. (570°F.) and then remained approximately constant. The temperature coefficient of resistance increased up to 300°C. (570°F.) and then also remained constant.

Inspired by Trzebiatowski's fundamental research, the author (using commercial grades of copper powder) extended the investigation of hot-pressed copper compacts to include coarser powders and lower pressures.[22] The characteristics of the principal powder were as shown in Table 46 (page 448). The powders were preformed at 2 tsi at room temperature into pellets $5/_8$ inch in diameter and $5/_8$ inch high. They were then dropped into a hot die, and, while protected by hydrogen, were hot-pressed at various temperatures and pressures (for details of apparatus and procedure, see page 475, Chapter XIII).

In addition to density and hardness measurements, the compact was subjected to a static compression strength test to determine the ductility of the metal by cold rolling and compression. For an electrolytic powder, all passing through a 100-mesh sieve and nearly one-half of it passing through a 325-mesh sieve, densities of 8.9 were obtained for pressing temperatures of 400° and 500°C. (750° and 930°F.). The peak hardness value was reached only for compacts pressed at 50 tsi; at 300°C. (570°F.) a Brinell figure of 101 was measured. This value could be raised to 114 Brinell if a finer powder, all passing through 325 mesh, was compacted under the same conditions. These figures are considerably below the abnormally high hardness values recorded by Trzebiatowski, but may be considered more representative for powders of commercial (coarse) grades.

In Figure 161, density, hardness, cold-rolling reduction, and compressive properties are shown as a function of the compacting temperature.

[22] C. G. Goetzel, *Trans. Am. Soc. Metals, 28,* 909 (Dec., 1940).

(The greater the compacting pressure the higher the curves in the diagram.) But only the highest pressure (50 tsi) and temperatures (400°C. (750°F.) minimum) used yielded sufficiently solid compacts to permit a

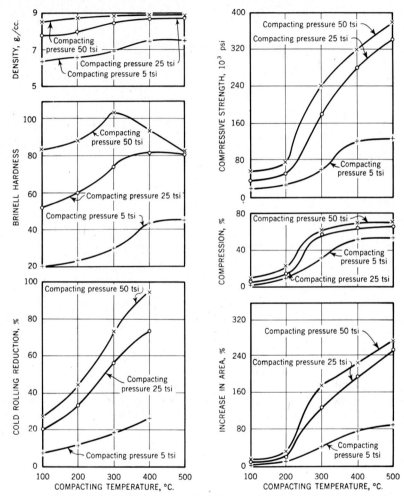

Fig. 161. Density, hardness, rollability, and compressive properties as functions of temperature of pressing for copper powder compacts hot-pressed at 5, 25, and 50 tsi.

direct reduction into wire size (95%) without intermediate annealing (*i.e.*, sintering). For the compressive properties too, each curve refers to a definite compacting pressure, with the highest pressure giving the highest position in the diagram. The maximum load which the specimens could take before showing signs of fracturing at the surface was used for computing the compressive strength. The tendency of the compressive properties to increase with rising temperatures—first slowly, then more rapidly, and then slowly again—is seen from the S-shaped appearance of the curves. At a pressure of 50 tsi and temperatures above 400°C. (750°F.), or at a pressure of 25 tsi at 500°C. (930°F.), the values equal those of cast copper. Figure 162 shows a series of compacts hot-pressed under different conditions, after compression to the point where edge failure was observed.

TABLE 44

Tensile Properties of Hot-Pressed Copper Compacts

Hot-press temperature		Pressure, tsi	Yield point, psi	Tensile strength, psi	Elongation, %	Observer
°C.	°F.					
610	1130	24	—	33,600	—	Sauerwald[23]
715	1319	24	—	30,600	—	
810	1490	24	—	29,600	—	
800	1472	5	27,000	27,000	4	Jones[24]
950	1742	5	11,000	30,000	60	

If density, hardness, rollability, and compressive properties are plotted for different temperatures against the compacting pressure (Fig. 163), a series of curves results which show tendency of the values to increase with rising pressure—the curves having a positive slope. Usually, the increase is more pronounced at lower pressures and the curves have concave-upward shapes, with some curves approaching the "S" shape found in the graphs of Figure 161. With the exception of the hardness curve for 500°C. (930°F.), which tapers off above 25 tsi and is then crossed by the 300°C. (570°F.) curve, the curves in the diagrams lie higher as the press temperature is elevated.

These data on the physical properties of hot-pressed copper powder are supplemented by some early experiments by Sauerwald[23] and by more recent tests by Jones,[24] reproduced in Table 44. Sauerwald tested the tensile strength of compacts hot-pressed at a constant pressure of 24 tsi at various temperatures—all above those employed in the investigations just mentioned. He found that the tensile strength reached 33,600 psi for a hot-pressing temperature of 610°C. (1130°F.), but decreased slightly with higher temperatures; for 810°C. (1490°F.) the tensile strength was

[23] F. Sauerwald, Z. Metallkunde, 21, 22 (1929).
[24] W. D. Jones, Metal Ind. London, 56, 225 (Mar., 1940); also Brit. Pats. 530,995 and 530,996.

445

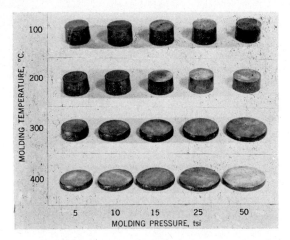

Fig. 162. Series of electrolytic copper powder compacts hot-pressed at different pressures and temperatures after compression test to point of failure.

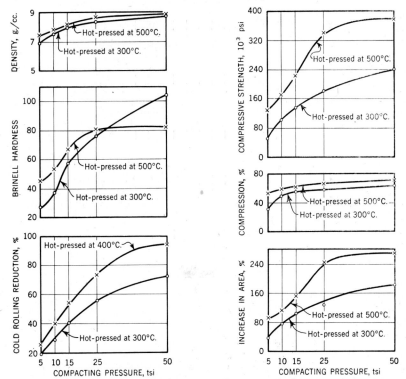

Fig. 163. Density, hardness, rollability, and compressive properties as a function of pressure for copper powder compacts hot-pressed at different temperatures.

TABLE 45
Tensile Properties of Hot-Pressed Copper Compacts[a] (Jones[24])

Type of raw material	Composition, %						Hot-press temperature		Yield point, psi	Tensile strength, psi	Elongation, %	Brinell hardness
	Copper	Zinc	Tin	Nickel	Silicon	Iron	°C.	°F.				
Prealloyed powder	70	30	3.2–4.5	—	—	—	900	1652	17,000	30,000	21	—
	82–85	9–15		—	—	0.2–1.2	870	1598	25,000	43,000	46	—
Mixture of individual metal powders	90	10	—	—	—	—	900	1652	17,000	30,500	22	—
	80	20	—	—	—	—	900	1652	18,000	37,000	34	—
	70	30	—	—	—	—	800	1472	22,000	38,000	16	—
	50	50	—	—	—	—	775	1427	21,000	21,000	0	110
	95	—	5	—	—	—	700	1292	26,000	35,000	9	114
	95	—	5	—	—	—	800	1472	24,000	45,000	47	114
	93	—	7	—	—	—	800	1472	24,000	47,000	75	—
	91	—	9	—	—	—	800	1472	30,000	42,000	17	—
	85.7	10.56	3.66	—	—	—	900	1652	18,000	38,000	53	—
	89.25	—	5.35	4.24	0.79	—	850	1562	32,000	45,000	13	—
	88.55	—	5.4	5.3	0.63	—	850	1562	44,000	52,000	5	—
	83.1	10.3	2.51	3.97	—	—	900	1652	18,000	39,000	32	—
	28	25	—	45	2	—	900	1652	18,000	42,000	0	—

[a] All compacts were first preformed at 8 tsi at room temperature, then hot-pressed at 4.5 tsi.

only 29,600 psi. Although details as to the powder characteristics or time of pressing were not given, the decrease in tensile strength may be attributed to annealing and grain growth, the latter becoming apparent at 700°C. (1290°F.). The lack of figures for elongation or reduction of area makes it difficult to evaluate these data for any technical purposes. Jones, pressing with only 5 tsi, but employing still higher pressing temperatures, found tensile strength values of the same magnitude as those reported by Sauerwald for a pressure of 24 tsi. The elongation of compacts prepared in this way amounted to no less than 60%—a very remarkable finding, since hot-pressing was carried out without provisions for a protective atmosphere!

BRASSES AND BRONZES

Encouraged by the results obtained with the pure metals, gold and copper, the hot-press experiments were extended to the industrially important field of copper alloys. Jones[24] explored the possibilities of hot-pressed brasses, bronzes, and more complex copper alloys. Alloy powders as well as mixed powders were used as starting materials. Spherical alloy particles, all passing through 300 mesh, were obtained first by casting ingots and then by mechanically pulverizing them to fine size. Each alloy powder was then hot-pressed at 5 tsi at a temperature which was 200°C. (360°F.) less than the melting point of the alloy concerned. In the tests with the mixed powders, the method employed was slightly different, involving: (1) appropriate mixing, (2) preforming at room temperature at 8 tsi, and (3) hot-pressing at only 4.5 tsi at various temperatures. Perfectly nonporous alloys were reportedly obtained with this procedure.

The most important results of Jones' investigation are reproduced in Table 45. With regard to the experiments with prealloyed powders, tensile test results were unexpectedly low—except for two compositions. These exceptions were a simple 70–30 brass and a copper–zinc–tin alloy, analyzing 82–85% copper, 3–4.5% tin, about 1% iron, and balance zinc. All other alloys, including copper–nickel and particularly the high tensile complex brasses, were reported to show very low tensile strengths and little or no elongation. A possible explanation for these results may be found in the fact that, even at high temperatures, the pressures employed by Jones were probably too low to overcome the internal rigidity of the alloy particles. It is also possible that insufficient time was allowed during hot-pressing for complete annealing of the strained structure of the mechanically disintegrated particles. Finally, formation of oxide films during the high-temperature treatment may have further aggravated the resilience of the particles.

Table 45 shows that results obtained with compacts hot-pressed from these powder mixtures can compare favorably with engineering requirements. For example, Jones was able to obtain 37,000 psi tensile strength and 34% elongation with a simple 80–20 brass composition; a tensile strength of 45,000 psi and 47% elongation with a 95–5 bronze; a tensile strength of 47,000 psi and an elongation of 53% with a composition containing 85.7% copper, 10.7% zinc, and 3.7% tin. Attempts made by Jones to ascertain whether these results could still be improved substantially by alloying additions were unsuccessful. As shown in Table 45, no appreciable

*TABLE 46

Physical Properties of Copper and Zinc Powders Used for Hot Pressing

Property	Electrolytic copper	Atomized zinc
Purity, %	99.5	98.5
Impurity, %		
Oxygen(hydrogen loss)	0.0915	1.14
Iron	Trace	—
Lead	0.023	—
Antimony	Trace	—
Tin	Trace	—
Zinc	Trace	—
Silica	0.011	—
Carbon } Sulfur }	0.057	—
Grease	0.029	—
Nitric acid, insoluble	0.068	—
Apparent density, g. per cc	2.50	2.55
Flow, sec./50 g.	30	33
Screen analysis, fraction %		
+100 mesh	0	0
−100 +150	5.05	0
−150 +200	18.18	9.88
−200 +250	2.30	4.54
−250 +325	23.14	24.58
−325	50.50	60.95

improvements, either in tensile strength or in elongation, could be recorded, causing Jones to conclude that the physical metallography of alloys hot-pressed from powdered ingredients is radically different from that of cast materials. On the other hand, according to Jones, the success with the more simple brasses and bronzes hot-pressed from individual metal powders may possibly be caused (to a certain extent) by the fact that the zinc and the tin are (for a period at least) in the liquid state during the heat treatment. This circumstance may facilitate the dislodging of oxide films formed in the absence of a protective atmosphere. This explanation, though apparently quite plausible, does not take into account the fact that good physical properties have also been observed in some compacts hot-pressed from prealloyed powders where no molten constituent existed during the hot compression.

The investigation of hot-pressed binary copper alloys was carried further (by the author) by including a pressure and temperature variable. Alloy powders were again compared with mixed powders, and the physical properties of the original copper and zinc powders are given in Table 46; those of copper and tin powders are given in Table 47. In order to prevent strain-hardening effects, the alloys were obtained by diffusion during a preliminary sintering treatment. In the case of brass compositions, the mixed powders were heated in hydrogen for 2 hours at 450°C. (840°F.)[25]; in the case of bronzes, the treatment was carried out at 550°C. (1020°F.)

TABLE 47

Physical Properties of Copper and Tin Powders Used for Hot-Pressing

Property	Electrolytic copper	Electrolytic tin
Purity,[a] %	99.5	99.5
Apparent density, g./cc	2.52	2.33
Screen analysis[b]		
+100 mesh	0	—
−100 +140	6.44	—
−140 +200	19.19	—
−200 +230	3.03	—
−230 +325	23.77	0
−325	47.50	100
Particle-size distribution[c]		
0 − 5 microns	0	29
5 −10	2	33
10–20	15	27
20–30	14	7
30–40	12	3
40–50	10	1
50–75	23	0
75–100	19	—
Over 100	5	—

[a] Impurities for copper same as given in Table 46.
[b] U. S. Standards.   [c] As determined by microscopic count.

for 1 hour.[26]   Photomicrographs of some of these alloy powders are shown in Figures 16 and 17 in Chapter III. These treatments did not result in noticeable sintering of the particles, and subsequent disintegration of the cake did not impose appreciable strain on the particles. Although a preforming pressure of 2 tsi was found sufficient for all mixed powders, it had to be raised to 5 tsi for the prealloyed brass powders, and 8 tsi for prealloyed bronze powders. This increase in preforming pressure can be explained by the increased resistance of these powders to plastic deformation. The hot-press temperatures were kept lower than in Jones' work in

[25] C. G. Goetzel, Trans. Am. Soc. Metals, 30, 86 (1942).
[26] C. G. Goetzel, Trans. Am. Inst. Met. Mining Engrs., 161, 580 (1945).

450

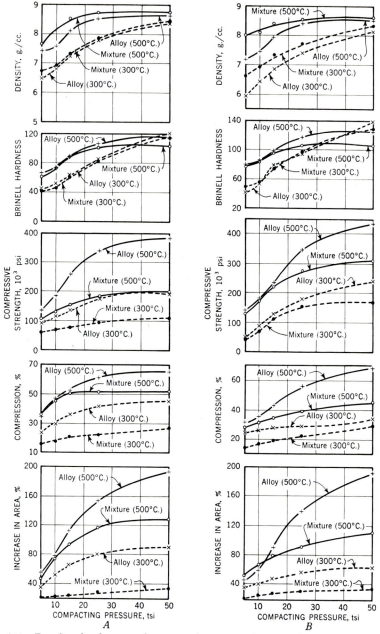

Fig. 164. Density, hardness, and compressive properties as a function of pressure for brass compacts hot-pressed at 300°C. (570°F.) and 500°C. (930°F.): A, 85–15 Cu–Zn alloy composition; B, 75–25 Cu–Zn alloy composition; C, 65–35 Cu–Zn alloy composition; and D, 55–45 Cu–Zn alloy composition.

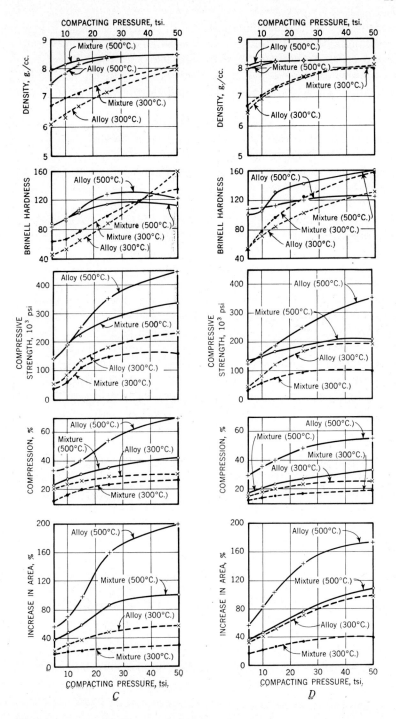

order to facilitate the manipulations and to allow for sufficient die strength in working with specific pressures up to 50 tsi. Most pressings were carried out at 300°C. and 500°C. (570° and 930°F.), respectively, on cylinders approximately $^5/_8$ inch in diameter and $^1/_2$ to $^5/_8$ inch high. The effects of oxidation were eliminated by hot-pressing all compacts in a hydrogen atmosphere.

In the diagrams of Figure 164, the density, hardness, and compressive properties of four brasses are shown as a function of the compacting pressure—the individual graphs representing alloys containing 15, 25, 35, and 45% zinc. In comparing the curves with those drawn for pure copper (Fig. 163), it will be seen that the general trend has not been changed by alloying with zinc. All properties tested rise with increasing pressure, the increase being more drastic for low pressures than for high. The curves for a press temperature of 300°C. (570°F.), have the familiar concave upward shape; those for 500°C. (930°F.) are frequently "S" shaped and, in general, lie above those for 300°C. (570°F.), with the exception of the hardness values, where an overlapping of the curves above 25 tsi is apparent for all compositions. The inferiority of compacts hot-pressed from alloy powders when compared with those made of mixed copper and zinc powders—found by Jones—could not be affirmed generally. The density values of all compositions and the hardness values of the 55–45 copper–zinc composition were found to be higher when mixed powders were used as starting materials. In all other instances, compacts made from previously alloyed powders showed superior physical properties. This may be attributed, at least in part, to a more diffused microstructure (Fig. 157); the structure of these hot-pressed materials, however, still lacked uniformity and a continuous matrix of polygonal-shaped crystals. For high molding pressures, the microstructure as shown in the micrographs of Figure 157, appeared to be substantially dense without the widespread pattern of porosity so familiar in all cold-pressed and sintered alloys. Accordingly, the density reached peak values closely approaching normal densities of commercial brasses of corresponding composition; 8.75 g./cc. for 85–15 brass, 8.6 for 75–25 brass, 8.5 for 65–35 brass, and 8.25 for 55–45 brass.

In the case of the lower pressing temperature of 300°C. (570°F.), very high hardness values were obtained for these compositions (pressed at high pressures): 120 Brinell for the 85–15 composition, 135 for the 75–25 alloy, 150 for the 65–35 alloy, and 160 for the 55–45 composition. These values are comparable to those of severely worked wrought brass, exceeding, in certain instances, the peak data reported for hot-pressed copper. If finer powders were to be used, it may be assumed that the hardness could be increased still further, reaching perhaps a Brinell figure of 200. Compressive strength figures were also slightly higher than corresponding values for pure copper, but the values for compression and increase in area remained fairly low, indicating that the temperatures used during hot-pressing were insufficient to anneal the structure and to render the compact sufficiently ductile to endure severe cold deformation.

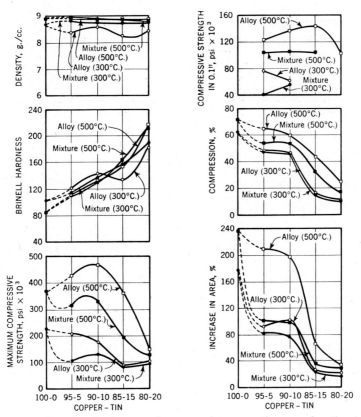

Fig. 165. Density, hardness, and compressive properties as a function of the tin content for bronze compacts hot-pressed at 300°C. (570°F.) and 500°C. (930°F.) at 50 tsi.

In considering the hot-pressing experiments with bronze compositions, it may be seen from the diagram of Figure 165 that, as a result of the alloying with tin, density, hardness, and compressive properties are slightly higher than those obtained with the copper–zinc compositions. The curves in Figure 165 represent data for compacts hot-pressed at 300° and 500°C.

(570° and 930°F.) at the highest compacting pressure used (50 tsi) for four alloy compositions containing 5, 10, 15, and 20% tin. Increased amounts of tin have their expected effects on microstructure (Fig. 158) and on physical properties; the obscurity of the grain structure is understandable in view of the processing method used. Hardness values increase steadily to values exceeding 200 Brinell and compressive properties—

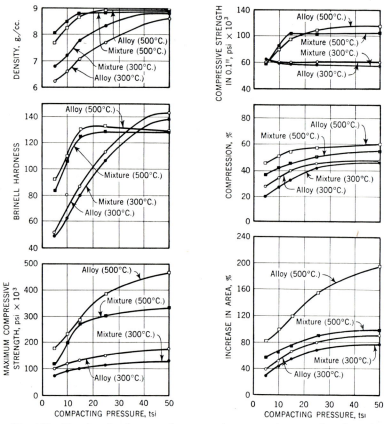

Fig. 166. Density, hardness, and compressive properties as a function of pressure for 90–10 Cu–Sn bronze compacts hot-pressed at 300°C. (570°F.) and 500°C. (930°F.).

attaining peak values in low-tin alloys—drop markedly for the tin-rich compositions.

In the case of the 90–10 composition, the data are plotted against the compacting pressure in Figure 166, the curves assuming the same trend and relative locations as already noticed for the brasses. The density increases with molding pressure regardless of composition. Compacts which have been hot-pressed at 300°C. (570°F.) approach theoretical values only at high pressures (50 tsi) while compacts which have been pressed at 500°C. (930°F.) reach the theoretical limit at lower pressures (25 tsi). Again, as in the case of brass, hot-pressed bronzes are denser when made from the powder mixtures (especially for the compacts pressed at 300°C.; 570°F.), indicating increased plasticity and effective cementing caused by the liquid tin or tin-rich phases. The high hardness values, in many cases comparing favorably with severely work-hardened bronzes, may be attributed to the dense structure, and also, in a measure, to severe straining of the particles when compressed at the high pressure below the recovery temperature range. All compacts become harder with increasing pressures. When compressed at 500°C. (930°F.), hardness maxima are reached at a pressure of 25 tsi, while hardness values remain approximately constant up to 50 tsi. For compacts hot-pressed at 300°C. (570°F.) the values are generally lower because of lower density; only the more ductile low-tin compositions yield higher hardness values when strained severely at 50 tsi. High compressive properties can be obtained only in materials hot-pressed at or above 500°C. (930°F.), since all compacts hot-pressed at 300°C. (570°F.) are naturally poor in strength and ductility. Figures for compressive strength, compression, and increase in area grow to maxima above 25 tsi with rising molding pressure. As indicated by the improved data for compacts prepared from prealloyed powders, values become better as the structure of the compacts become more uniform.

In comparing mixed powders with prealloyed powders (for hot-pressed brasses and bronzes), the author found (except for minor exceptions) certain definite advantages in using prealloyed material for work at relatively low temperatures. Considerable improvements in hardness, strength, and ductility are obtainable through increased homogenization of the different alloys. Prealloying, on the other hand, increases the internal rigidity of the individual particles, causing increased resistance to compression and slightly less dense compacts. At the higher temperatures where annealing becomes more effective, resistance to compression and plastic deformation is overcome, at least in a measure, and complete consolidation of the structure becomes possible. These results are not in agreement with the experience reported by Jones, who employed considerably higher temperatures; substantial annealing of the work-hardened pulverized alloys should presumably have been accomplished at these temperatures. It is doubtful, however, whether a comparison between the two investigations would have much meaning, since the particle structures of the alloy powders used were quite different, and Jones' experimental procedure differed considerably from that used by the author. Jones also used considerably higher hot-pressing temperatures and very low molding pressures. In addition, he failed to establish comparative data on compacts pressed under exclusion of oxygen. It is quite possible that heavy oxide films were formed around most alloy particles at the high temperatures used; this, of course, tended to weaken the grain boundaries materially, resulting in poor tensile strength values despite high densities.

IRON AND STEELS

The high physical properties obtained with nonferrous materials led to a general belief that the hot-press method would also produce iron and steel compacts which would be stronger and tougher than otherwise possible. This was supported in part by Sauerwald's classic experiments.[27] At various temperatures he found the following tensile values for hot-pressed iron powder compacted at 24 tsi:

Press temperature	Tensile strength
610°C. (1130°F.)	25,000 psi
715°C. (1320°F.)	37,000 psi
810°C. (1490°F.)	50,400 psi

During recent years, the systematic studies of the properties of hot-pressed metal powders have been extended into the field of ferrous materials. Information on the density, hardness, and tensile properties was sought for a variety of iron powders, as well as for a number of ferrous compositions. The main difficulty experienced when adapting one of the formerly used hot-press methods to the higher melting materials was that an appreciable increase in plasticity of iron occurs only at temperatures which are already dangerously high for metallic die and punch materials. Whereas a temperature of 500° or 600°C. (930° or 1110°F.) is sufficient to make copper completely dense even at relatively low pressures, higher temperatures and pressures are necessary for iron and steels. Because of these difficulties, the shape of the compacts has been limited either to simple cylinders—suitable only for metallographic, density, or hardness testing—or to simple small bars from which tensile specimens—often below standard specifications—had to be machined.

The consolidation of hot-pressed iron powders of various origin was studied by Schwarzkopf and Goetzel.[28] Structure, density, and hardness were examined on compacts hot-pressed from Swedish sponge iron (−100

---

[27] F. Sauerwald, Z. Metallkunde, 21, 22 (1929).
[28] P. Schwarzkopf and C. G. Goetzel, Iron Age, 148, No. 10, 37 (1941).

mesh), electrolytic iron powder (–100 mesh), and hydrogen-reduced, fine iron powder (–325 mesh). Various press temperatures ranging from 500° to 1200°C. (930° to 2190°F.) were employed. In the low-temperature region, including 800°C. (1470°F.), high-speed steel dies were used, while graphite molds were used for higher temperatures. Corresponding to the limited hot strength of the die materials, the maximum applied pressures were reduced from 50 tsi for 500°C. (930°F.) to $2^1/_2$ tsi for 900°C. (1650°F.) and above. Metallographic examination showed complete densification of the particle agglomerates for low temperatures and high pressures. At temperatures above 900°C. recrystallization and grain growth had caused a change to a normal structure consisting of fine-grained ferrite after cooling; insufficient pressure, however, left a certain amount of porosity (photomicrographs of electrolytic iron compacts are shown in Figure 159). In Figure 167, curves for the different raw materials are given, in which density and Brinell hardness are plotted against compacting pressure (on a semilogarithmic scale). Each curve represents a different press temperature. It is apparent from these graphs that, in the case of iron, too, high temperatures in conjunction with low pressures, as well as lower temperatures with correspondingly higher pressures, can lead to dense compacts (see also Table 41). The latter method, however, appears to be preferable in view of the generally short life of nonmetallic molds. The finer, hydrogen-reduced (Fig. 167C) and purer, electrolytic (Fig. 167B) powders apparently yield better densities and hardness values than the coarser and less pure sponge iron (Fig. 167A).

Of particular interest is the fact that distinct minima in density and hardness values could be observed for all three types of raw material at 1000°C. (1830°F.). It is believed that the marked drop in density between 900° and 1000°C. (1650° and 1830°F.) (at identical pressures) has some connection with the phase change in iron, indicating that the compacts are more resistant to deformation when entering the austenitic condition than when still in the ferritic condition. It is known that if a metal exists in a new polymorphic form at the higher temperature, the hardness may increase and the formability decrease suddenly at a certain temperature—as occurs in gamma iron above 910°C. (1670°F.).[29] The apparent decrease in hardness in the powder compacts can be explained by the loosening of the structure.

As already stated, the effects of temperature or pressure cannot be evaluated independently with any degree of success. It is quite apparent that for each given press temperature, including room temperature, there

[29] G. E. Doan, *The Principles of Physical Metallurgy*. McGraw-Hill, New York, 1935. p. 111.

458

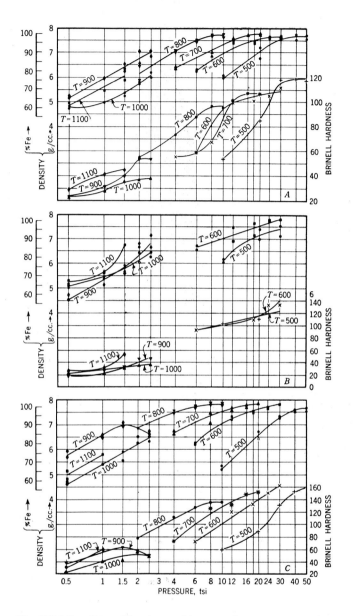

Fig. 167. Density and hardness of iron powder compacts as they vary with compacting temperatures, T (in °C.), and pressures: A, compacts from Swedish sponge iron powder; B, compacts from electrolytic iron powder; and C, compacts from hydrogen-reduced fine iron powder.

exists a pressure at which the compact becomes completely dense. While for iron at room temperature, the necessary compacting pressure may well be 200–300 tsi, this pressure limit decreases rapidly with rising compacting temperatures, especially when plasticity of the metal improves at temperature ranges above the recovery and recrystallization zone. Although, of course, absolute values depend to a great extent upon the nature, history, and properties of the powder used, great plasticity of the powder at room temperature (Swedish sponge iron) does not necessarily mean that the same powder would yield optimum properties at lowest possible pressures and temperatures.

TABLE 48

Physical Properties and Chemical Analysis of Electrolytic Iron Used in Hot-Pressing Experiments (Henry and Cordiano[30])

Screen analysis	Fraction, %	Element	Amount, %
+100 mesh	1.0	Oxygen (hydrogen loss)	0.34
−100 +150	10.5	Carbon	0.005
−150 +200	21.5	Manganese	0.002
−200 +250	11.0	Nickel	0.008
−250 +325	13.0	Silicon	0.003
−325	43.0	Sulfur	0.004
		Phosphorus	0.001

Apparent density, 2.4 g./cc.; flow, 46 sec./50 g.

TABLE 49

Tensile Properties of Hot-Pressed Iron Powders (Henry and Cordiano[30])

Temperature °C.	°F.	Time, sec.	Density, g./cc.	Tensile strength, psi	Elongation in 1 inch, %	Brinell hardness, 500-kg. load
500	932	50	6.31	26,200	0.0	50
		150	6.38	25,500	0.0	51
		450	6.71	39,800	1.0	63
600	1112	50	6.70	36,900	0.5	62
		150	6.89	40,800	1.0	77
		450	7.05	48,800	2.0	80
700	1292	50	7.32	47,800	1.0	90
		150	7.52	57,300	12.0	95
		450	7.58	57,500	27.0	100
780	1436	50	7.59	54,100	22.0	101
		150	7.71	52,400	32.0	93
		450	7.76	52,900	37.0	96

The tensile properties of hot-pressed iron have been investigated by Henry and Cordiano[30] who employed relatively low temperatures and a constant pressure of 10 tsi in molding an electrolytic iron powder. The powder, whose properties are given in Table 48, was preformed cold into

[30] O. H. Henry and J. J. Cordiano, Trans. Am. Inst. Mining Met. Engrs., 166, 520 (1946).

$3\times^3/_8\times^5/_8$-in. bars at 20 tsi prior to introduction into the hot die. After a period of 10 minutes had elapsed—believed sufficient to achieve temperature equilibrium—the pressure was applied for the predetermined time intervals of 50, 150, and 450 sec. (see Table 49). In Figure 168, the density, hardness, tensile strength, and elongation values are plotted against compacting temperature for the three time intervals. All the curves rise with increasing temperature. In all cases the longest period gives the highest curve, indicating that sufficient time for plastic deformation must

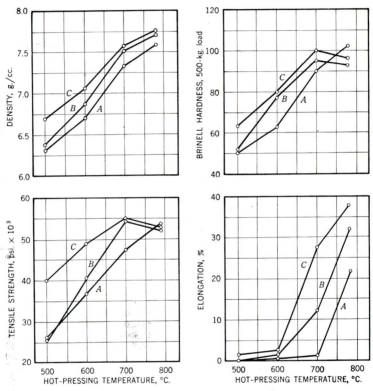

Fig. 168. Effect of pressing temperature on density, hardness, tensile strength, and elongation of iron compacts hot-pressed at 10 tsi, at different dwelling times (according to Henry and Cordiano[30]). Pressing time: A, 50 sec.; B, 150 sec.; and C, 450 sec.

be allowed if optimum properties are desired. In the diagram of Figure 169, the same results are shown as a function of the time of pressure dwell, all curves rising with increasing time. Except for an overlapping of the hardness and tensile values for higher temperatures (indicating effective annealing), the curves take a logical position in the diagram, with those for the higher temperatures lying above those for the lower temperatures. The high values for tensile strength (up to 57,500 psi) and elongation (up to 37%) of Henry and Cordiano exceed even Sauerwald's earlier results, and compare very favorably with fused and annealed electrolytic iron (tensile strength 40,000 to 50,000 psi, elongation 40 to 60%).

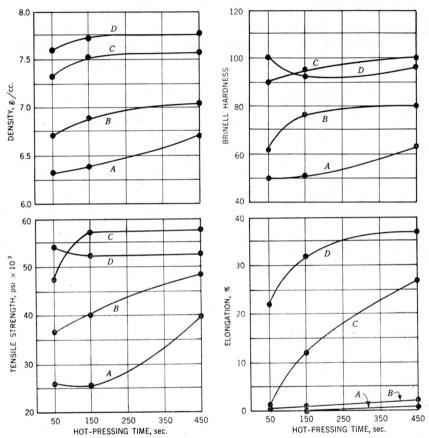

Fig. 169. Effect of pressing time on density, hardness, tensile strength, and elongation of iron compacts hot-pressed at 10 tsi, at different temperatures (according to Henry and Cordiano[30]). Pressing temperature: *A*, 500°C. (930°F.); *B*, 600°C. (1110°F.); *C*, 700°C. (1290°F.); and *D*, 780°C. (1435°F.).

Henry and Cordiano's values have lately been surpassed in Germany by the results of Wassermann's work[30a] with eddy-milled and carbonyl

[30a] G. Wassermann, *Metallforschung, 2,* 129 (1947).

iron powders. Deviating considerably from the previously described procedures, the powders were presintered in the uncompressed state at 850°–1000°C. (1560°–1830°F.) and the resulting sinter-cake was transferred while hot into the hot-press die. After a final normalizing treatment, a metal of homogeneous structure and fine grain size was obtained, which had the following excellent physical properties: 7.5–7.84 g./cc. density, 130–140 Brinell hardness; 51,000–64,000 psi tensile strength; 27–35% elongation; and 62–73 ft.-lb. (10.5–12.6 cm.kg./mm.^{2}) impact strength.

Several investigations were performed with hot-pressed steel and high-carbon iron alloys. Sauerwald and Kubik[31] preformed powdered white cast iron at 24 tsi into compacts which were then hot-pressed at the same pressure for 30 seconds at 800°C. (1470°F.). After this treatment, the compacts were only about half as hard as the original material. This disappointing result may be attributed to an insufficiently high temperature for rendering the brittle, cast-iron particles plastic enough to yield a completely solid structure. It is quite possible that at higher press temperatures, 1000°C. (1830°F.) perhaps, the compacts would have attained the same hardness as the original cast iron.

The author developed a hot-pressing technique whereby small rectangular specimens could be produced in graphite molds in a resistance welding machine (see Chapter XIII). A protective atmosphere of hydrogen was introduced into the die cavity in experiments with compacting loose powders; the carbonaceous atmosphere produced by the mold material was found satisfactory for hot-repressing iron and steel compacts which had been previously formed and sintered. Test specimens obtained by direct hot-pressing electrolytic iron powder of 100-mesh grade (35%, −325 mesh) at 1100°C. (2010°F.) and 2 tsi displayed an average density of 7.07 g./cc., a tensile strength of 37,000 psi, and an elongation of 7%. When 0.5% graphite was mixed with the same iron of uniformly distributed pearlite and ferrite grains resulted, whose density powder, and the same procedure employed, a 0.33% carbon steel material sity averaged 6.96 g./cc., tensile strength 49,000 psi., and elongation 3%. Far superior physical properties, however, could be obtained when the powders were first cold-preformed (at 20 tsi), and then sintered (at 1100°C. (2010°F.) for 15 min.) and cooled before being hot-repressed at 1100°C. (2010°F.) at 2 tsi. Pure iron thus treated possessed an average density of 7.47 g./cc., tensile strength of 41,000 psi, and elongation of 15%. Analogous values for 0.33% carbon steel specimens, produced in an identical manner, were an average density of 7.44 g./cc., tensile strength

[31] F. Sauerwald and S. Kubik, Z. Elektrochem., 38, 33 (1932),

of 50,000 psi, and elongation of 14%. (For additional physical properties of these materials, see Table 63, Chapter XVIII.)

More recent studies with hot-pressed steels and iron–carbon alloys have been made with compacts which were first preformed and presintered to a high temperature, and then compressed in a cold die while still hot. Koehring,[32] reporting very interesting results with several ferrous compositions, prefers to call this procedure a forging operation. But inasmuch as his procedure closely resembles that reported by Jones,[33] it appears appropriate to mention Koehring's results in connection with hotpressing. Koehring investigated five different materials, namely, (1) ground steel chips of $1/_8$-inch size; (2) decarburized comminuted iron–carbon alloy (powder "A"), (100%, –100 mesh; 50%, –325 mesh); (3) same powder "A" to which 0.60% carbon was added in form of graphite; (4) reduced iron powder "B" (from millscale), 100%, –100 mesh; 50%

TABLE 50

Physical Properties of Hot-Forged Bars Made from Various
Kinds of Iron and Steel (Koehring[32])

Composition of bar[a]	Density, g./cc.	Tensile strength, psi	Yield point, psi	Elongation in 2 inches, %	Reduction in area, %	Hardness, Rockwell B
Ground steel chips.........	7.79	54,400	38,100	11	13	59–74
Iron powder "A"..........	7.78	55,200	37,800	14	13	61–68
Iron powder "A" with 0.25 to 0.35% carbon.......	7.81	73,300	51,200	23	32	74–85
Iron powder "B"..........	7.30	39,900	33,000	3	3	67–82
Iron powder "B" with 0.25 to 0.35% carbon.......	7.40	44,700	46,900	4	5	66–74

[a] Iron powder "A" made by decarburizing finely ground iron–carbon alloy; iron powder "B" made by reducing iron oxide.

–325 mesh, and (5) reduced iron powder "B" to which 0.60% carbon was added. In all cases bar-shaped specimens were pressed and were heated uniformly to 1100°C. (2010°F.) for one hour; the pure iron compacts were placed in hydrogen atmosphere, and the other compacts were placed in a box filled with carbon. Specimens were forged into specimens of high density; then, after hot compression, the carbon-containing specimens were cooled slowly in lime. The results of the physical tests are shown in Table 50. The test bars made from the ground steel chips have the tensile strength and yield point to be expected from a wrought material of this composition. The elongation and reduction in area values are very low, however, probably because of high oxide content of the material. The specimens made from the decarburized powder displayed a cleaner struc-

[32] R. P. Koehring, in J. Wulff, Powder Metallurgy. Am. Soc. Metals, Cleveland, 1942, p. 304.
[33] W. D. Jones, Metal Ind. London, 56, 69 (Jan., 1940).

ture, less often interrupted by discontinuities due to oxide inclusions. The mechanical properties of this powder, with a final carbon content of about 0.3%, approximate those of a steel of similar composition. The bars made from the reduced iron powder have inferior mechanical properties and lower density than those made from the other materials. Whether the low values for tensile strength and elongation obtained with this kind of powder are caused by the sponginess of the particles or by insufficient reduction (resulting in an excessive oxygen content) is not apparent from the data available, but Koehring attributes the failure of this material to the latter cause.

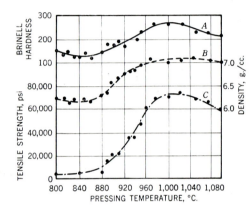

Fig. 170. Influence of pressing temperature on physical properties of powdered gray cast iron hot-pressed at 8 tsi (according to Jones[34]).

A, Brinell hardness
B, density
C, tensile strength.

Jones[34] extended his research into the ferrous field by experimenting with several iron–carbon compositions. Powdered gray cast iron of –80 mesh size (analyzing 3.16% C, 1.13% Si, 0.58% Mn, 0.126% S, and 1.154% P) was preformed into small compacts at 8 tsi, and then heated in the die to a temperature of 800–1100°C. (1470–2010°F.) in an electric furnace (without a protective atmosphere). Immediately after reaching the desired sintering temperature, the die was withdrawn from the furnace and the bars were hot-pressed in air at the same pressure of 8 tsi. Figure 170 shows the density, hardness, and tensile strength of compacts produced in this manner as a function of press temperature. Below 800°C. (1470°F.) the tensile strength is still very low, amounting to less than 6000 psi for a density of 6.25 g./cc. From 880° to 975°C. (1615° to 1785°F.), however, there is a marked improvement in mechanical properties, and a continuous change in microstructure—involving the appearance of temper carbon, the disappearance of cementite, and the general formation of pearlite. The improvement is particularly impressive for the

[34] W. D. Jones, Foundry Trade J., 59, 401 (1938).

tensile strength. For a temperature of 975°C. (1785°F.) hot-pressed compacts from pulverized cast iron have quite remarkable properties: a density of 7.1 g./cc., a hardness of 270 Brinell, and a tensile strength of 72,000 psi. Above 975°C. (1785°F.) no further improvement is found in any of the properties examined, and a tendency of the values to decrease—especially tensile strength—becomes apparent. At 1100°C. (2010°F.) the microstructure consists only of ferrite, carbon nodules, and some phosphide eutectic. According to Jones, the surprisingly strong coherence of the annealed, unalloyed, cast-iron compacts may be attributed chiefly to the formation of a liquid phase during sintering. At approximately 960°C. (1760°F.) a liquid iron–carbon–phosphorus eutectic is formed, containing 1.3% carbon and 10.2% phosphorus, which under the influence of pressure may dissolve the oxide films at least partially, thus facilitating an intimate bond between the individual powder particles. As in the case of the brasses and bronzes, there is reason to believe that the liquid phase causes a favorable reaction during the hot-pressing of iron powders under oxidizing conditions. As a further example of the unusually high physical properties that can be achieved in this manner, Jones[34] cites an experiment in which a hot-pressing was made of a mixture of mild steel powder (containing 0.3% carbon) with 5% of the above-mentioned eutectic (1.3% carbon, 10.2% phosphorus, balance iron). The pressing was effected in air at 1000°C. (1830°F.), presumably at the same pressure (8 tsi) as used in Jones' other experiments. Thereafter, there followed slow cooling to room temperature and machining into test specimens. The tensile strength obtained with this material reached 106,000 psi without elongation, and the hardness was 230 to 270 Brinell. Values of this order must be considered extremely favorable; if one excepts the case of refractory alloys containing tungsten, this is probably the highest tensile strength obtained in any alloy made from powders in the annealed condition.

CEMENTED CARBIDES

The fact that hard metal compositions produced by the conventional cold-pressing and sintering method always display a certain degree of porosity caused Hoyt,[35] in 1930, to study the properties of hot-pressed tungsten carbide – cobalt mixtures. When produced by sintering, hard metal compositions do not surpass a Rockwell A hardness of 92, but with hot-pressed material a hardness of 95.6 Rockwell A can be reached. Pressures employed by Hoyt were extremely low, ranging from $1/2$ to possibly 2 tsi; periods necessary for the heat treatment were appreciably shorter

[35] S. L. Hoyt, *Trans. Am. Inst. Mining Met. Engrs., 89,* 9 (1930).

than customary for sintering of the cold-pressed material. With regard to
the pressure–temperature relationship, Hoyt found satisfactory results if,
for a measured temperature of 1350–1400°C. (2460–2550°F.), a pressure
of 0.65 tsi was used; for a temperature above 1400°C. (2550°F.), a pressure of 0.45 tsi gave equally good results. From Hoyt's observations, it
appears unlikely that the pressure accelerates the rate at which cobalt
dissolves tungsten carbide, although such an effect is possible on account
of the intimacy of contact involved. On the other hand, if tungsten carbide
is not previously formed, and a mixture of the three elements, tungsten,
carbon, and cobalt (in the proper proportion), is hot-pressed, the condi-

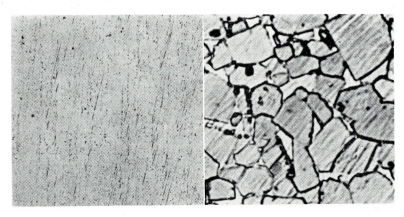

Fig. 171. Microsection and microstructure if hot-pressed cemented
tungsten carbide (according to Hoyt[25]): left, as polished; (×65) and right,
etched (×2500).

tions are such that tungsten takes up carbon rapidly, and the end product
is of standard quality. Hoyt concluded that hardness, strength, toughness,
and cutting properties of hot-pressed, cemented carbides are generally
superior to the same properties of cold-pressed and sintered material.
The photomicrographs of Figure 171, displaying the structure of a hot-
pressed, cemented tungsten carbide clearly indicate the dense structure of
the hot-pressed material.

Many years later, Meyer and Eilender[36] confirmed Hoyt's findings in
a systematic study of different hard metal compositions. It was found that
the greatest possible hardness can be obtained in compacts subjected to
pressure and sintering temperatures simultaneously. This may not be explained entirely by greater density. As in copper-base alloy mixtures, the
liquid constituent (cobalt in the hard metal compositions) is forced into

[36] O. Meyer and W. Eilender, Arch. Eisenhüttenw., 11, 545 (1938).

practically all pores and between the interparticle interstices. Moreover, some of the cobalt may even ooze out of the compact, evaporating at the surface, while an alloy somewhat poorer in cobalt remains. This increment in hardness was observed in all well-known hard metal compositions, including some developed from tantalum carbide.[37]

Summary

If metal powder molding is performed together with application of heat, compacts result whose properties are very different from those obtainable by the simple compaction of the particles at room temperature. If temperatures and pressures selected are sufficiently high, and a sufficient period of time is permitted for their action, increased plastic deformation of the powder particles results in completely consolidated compacts whose structures are recrystallized and may be stress-released by a simultaneous or subsequent annealing treatment. The properties of such compacts are dependent upon the amount and rate of diffusion which, in turn, depends on the number and area of the interparticle contacts as well as on the amount of plastic deformation to which the compact is subjected. Increasing temperature and time at temperature under sustained pressure tend to increase each of the above effects individually, as well as all of them collectively.

The application of heat during the compressing operation introduces a processing method that may be readily adapted for specific requirements in the compact. If, for example, the structure must be entirely free of voids, this may be achieved by the proper selection of any one of the variables (pressure, heat, or time) in relation to the other two, such as high pressures at low temperatures for a certain period of time, or lower pressures for a shorter time at very high temperatures. The solution of the problem of selecting the best combination may depend on the plasticity of the metal to be molded, on its capacity for alloying with the die material, on the desired grain structure, or on the die materials or lubricants available. On the other hand, a controlled porosity may be achieved by the use, at fairly high temperatures, of pressures about one-tenth those necessary at room temperature.

The achievements of hot-pressing combined with a suitable atmosphere may be summarized on the basis of results from a variety of experiments: (1) reduction in gas content, (2) avoidance of oxide, (3) accuracy of size, (4) high density, (5) high tensile strength and hardness, and (6) great elongation.

[37] L. Molkov and A. V. Chochlova, *Redkie Metal.*, *4*, No. 1, 10 (1935).

Although attractive properties are obtainable with this procedure, the process has many attendant problems. (*1*) Heat developed must be confined effectively to the die cavity and the compact. (*2*) Die materials must be developed which are capable of withstanding without deformation or failure sufficiently high pressures and temperatures to render the compact plastically deformable. (*3*) A high-temperature medium must be used which is capable of acting as a lubricant or separator to prevent excessive die wear or welding of the compact to the punch facings or die walls. (*4*) Provisions must be made for surrounding the compact (when above room temperature) with a reducing or neutral atmosphere. (*5*) Means of ejection and controlled cooling of the hot-pressed compact are needed.

CHAPTER XIII

Hot-Pressing Apparatus and Techniques

Our discussion of the compaction of metal powders by the combined effects of pressure and heat has so far been confined largely to the process of consolidation and to the properties in bodies thus formed. Since problems connected with the techniques of hot-pressing have been mentioned only in so far as they had immediate bearing on the mechanism of compaction or on the achievements of this method, a more general study of the different apparatus and techniques employed, both for experimental and industrial purposes, is now in order. Moreover, such a study is indispensable in any successful attempt to design a soundly engineered machine for mass production hot-pressing.

EARLY DEVELOPMENTS

The development of hot-pressing of metal powders is best described in the patent literature. As early as 1870, in a search for a material for journal boxes, Gwynn[1] experimented with hot-pressing of tin powder mixed with a small amount of petroleum still residue as binder. In 1904 Axelrod[2] used high pressures in conjunction with a moderate temperature of about 500°C. (930°F.). The pressure was sufficient to strain the metal particles beyond the elastic or flow limit, thereby compressing the particles into a solid body.

In 1912, for the first time, it was suggested in a German patent[3] that refractory metal powders, e.g., boron, tungsten, or tungsten carbide, should be heated by direct current so that pressure could be applied to the powder after high temperatures were reached. In a slightly modified form, the same principle was disclosed in the United States in 1917[4] for the production of alloys of tungsten and molybdenum with carbon, boron, titanium, iron, etc. Several years later, a German patent[5] described an apparatus

[1] U. S. Pat. 101,863.
[2] U. S. Pat. 863,134.
[3] German Pat. 289,864.
[4] U. S. Pats. 1,343,976 and 1,343,977.
[5] German Pat. 356,716.

that made possible simultaneous heating and pressing or hammering of alloys of any desired metal and metalloid combination for bearing materials; other patents issued in Germany in 1926 and 1927[6] suggested the production of hard metals by hot-pressing. In the United States, the hot-pressing of tungsten at 2400°C. (4350°F.) at a pressure of approximately 6 tsi was disclosed in 1930 by Pirani,[7] who suggested wear-resistant pressure plates containing more than 50% tungsten carbide. At the same time, Smith[8] described a method whereby iron powder associated with other, lower melting metals was hot-pressed at a temperature high enough to fuse the lower melting phase, but below the melting point of the iron. Taylor,[9] in the same year, disclosed a method whereby two dissimilar metal structures were welded together by superimposing the metals in powder form in a nonconducting mold, and by simultaneously applying pressure to the powder while a current was passed through them. In 1934, Kempf[10] disclosed a method for producing high-silicon alloys of aluminum by hot-pressing coarse powders at pressures varying from $12^1/_2$ to 75 tsi. The temperature was kept in a range from 200°C. (390°F.) to that just below the fusing point of the lowest melting constituent. Since 1930, many patents have been issued here and abroad, in most cases describing modifications of hot-pressing processes already disclosed in the earlier patent literature. They refer either to methods employing direct current heating, or heating of the powder through the die by convection or induction. To the former belong, for instance, several processes patented in the middle 1930's by Kratky,[11] who produced cutting tools by placing a shaped block of hard carbides between two carbon electrodes above a matrix and below a hammer die. When the necessary temperature was reached, the hot, hard metal block was driven into the matrix and densified by a dynamic blow with the hammer die. To the same class belongs also a process developed by Gillet and Dayton,[12] who pressed powders at predetermined pressures between washers at the end of electrodes through which a high-density current was passed. A similar process developed by Jones[13] will be described below in greater detail. Methods involving means of heating other than direct electric current have been described by Stout[14] for low melting

[6] German Pats. 497,558 and 504,484.
[7] U. S. Pat. 1,747,133.
[8] U. S. Pat. 1,775,358.
[9] U. S. Pat. 1,896,853.
[10] U. S. Pat. 1,944,183.
[11] U. S. Pat. 2,089,030; also Aust. Pats. 137,815 and 146,381, and Brit. Pat. 434,830.
[12] U. S. Pat. 2,149,596.
[13] Brit. Pats. 530,995 and 530,996.
[14] U. S. Pat. 1,913,133.

materials, and by Goldenzweig[15] for cemented carbides. In the latter case a powdered mixture of carbides and binders was introduced into a carbon mold and heated in a furnace under vacuum to at least 1600°C. (2900°F.). After fusion of the binder, a slight pressure of 0.1 tsi to 0.5 tsi was applied. To this group, also, belongs a process by Comstock,[16] who obtained dense articles for ornamental purposes by rubber stamping a presintered base (*e.g.*, from copper or iron) with an adhesive, such as glycerine, sprinkling with a precious metal powder, and hot-pressing at 1 to 2 tsi at a temperature above that used for presintering. The method is also applicable to the production of mosaic patterns of many metallic bodies of different compositions,[17] such as combinations of silver and copper. The silver and copper were first individually hot-pressed at approximately 0.25 to 0.6 tsi at a temperature of 200–500°C. (390–930°F.) and, after assembling into the mosaic, were consolidated further by a second hot-pressing—this time at a temperature of about 700°C. (1290°F.) and at a pressure of $1^1/_2$ tsi.

EXPERIMENTAL APPARATUS AND PROCEDURES

Hoyt's Technique

The first detailed account of a practical hot-pressing procedure was given in 1930 by Hoyt[18] in connection with his fundamental work on hard metal carbides. (See also Chapter XII.) The mold, constructed to permit direct electrical heating, was provided with a top and bottom plunger for applying pressure. The heating was then accomplished either by passing a current through the plungers, using top and bottom electrodes, or by passing current through the mold directly by the use of side electrodes. The procedure that followed was impressively simple. After tungsten carbide and cobalt powders were mixed by ball milling, the loose powder mix was poured into the mold on top of the bottom punch level, with the die surface. Top punch was inserted, the mold placed in a press, and pressure applied to the top plunger. The heating cycle was then carried out while the charge was kept under pressure. Hoyt reported pressures ranging from $1/_2$ to "many" tsi, with resulting shrinkage and substantial consolidation of the compact while under heavy pressure. Heating time was not specified, but was declared to be very short and much less than

---

[15] French Pat. 793,020.
[16] U. S. Pat. 2,126,737.
[17] U. S. Pat. 2,162,701.
[18] S. L. Hoyt, *Trans. Am. Inst. Mining Met. Engrs., 89,* 9 (1930).

ordinary sintering periods of the cold-press process. Temperatures were reported to be about 1400°C. (2550°F.)—presumably at the charge. Because of cooling effects, temperature measurements made on the graphite mold with an optical pyrometer showed a temperature differential between surface of the mold and the charge proper of as much as 200°C. (360°F.). Direct measurements of the charge temperature were conducted through a small hole drilled at the center of the mold.

Sauerwald's Procedure

Apparatus. The technique of Sauerwald and Kubik,[19] dating back to 1929—when the first results were published on experiments with hot-pressed iron and copper (see Chapter XII)—was described in detail in 1932 in connection with compaction experiments on pulverized, white cast iron. Sauerwald's experiments are remarkable inasmuch as they constitute the first systematic attempt to hot-press metal powders in a protective atmosphere, and at pressures and temperatures that imply stresses very close to the limit possible for iron-base die materials.

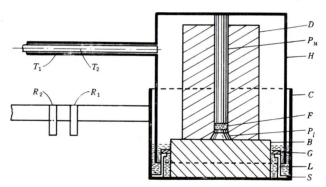

Fig. 172. Schematic cross section through apparatus used by Sauerwald and Kubik[19] for hot-pressing powder compacts as a reducing atmosphere: D, die; P_u, upper punch; H, hood; C, iron cup; F, preform; P_l, lower punch; B, steel base; G, guide ring; L, lead seal; S, support; T_1 and T_2, telescoping hydrogen pipes; and R_1 and R_2, steel bars.

Figure 172 shows a cross section of the experimental apparatus used. The die, D, and the upper and lower punches P_u and P_l, respectively, surrounding a preformed pellet, F, were machined from "heat-resistant" material. They were placed on a perfectly plane steel cylinder, B, which

[19] F. Sauerwald and S. Kubik, Z. Elektrochem., 38, No. 1, 33 (1932).

was centered by a loosely fitting guide ring, G, plus support, S. The entire assembly was then placed in a thick-walled iron cup, C, and a hood, H, of similar gage material was placed on top. A liquid lead seal, L, prevented access of the outside atmosphere. Two telescoping pipes, T_1 and T_2, facilitated the introduction of a continuous stream of hydrogen during the experiment. The entire apparatus could be transported, even when at peak temperature, with the aid of steel bars, R_1 and R_2, fastened to the cup.

Technique.   The powder allotments were chosen to yield final compacts approximately $5/8$ in. in diameter and $1/4$ in. high. After weighing, a pellet of slightly greater height was preformed in a hand-operated, hydraulic press at room temperature with a pressure of 24 tsi. Thereafter, the die, punches, and compact were placed on the round steel base, resting on the cup which was kept at a temperature of approximately 500°C. (930°F.). After the hood was placed into position, the hydrogen hose line was connected, and the assembly was placed into a gas-fired, muffle furnace kept at the desired temperature of 800°C. (1470°F.). The temperature of the hot-press apparatus was measured to an accuracy of ±10°C. with a separate iron-constantan thermocouple which was submerged in the liquid lead bath between the hood and the cup. After the apparatus was heated in the furnace to the temperature at which hot-pressing was to be carried out, 15 minutes were allowed to pass in order to obtain temperature equilibrium. Thereafter, the hydrogen and thermocouple connections were severed and the apparatus was quickly transferred back into the press. A pressure of 24 tsi was applied immediately and exerted for 30 seconds while the temperature of the compact was presumably maintained close to 800°C. (1470°F.) on account of the relatively large mass of the die assembly. After pressure release, the entire apparatus was moved back into the furnace, and hydrogen and thermocouple connection were installed anew. The apparatus was then furnace-cooled until the lead seal had solidified, the hydrogen cut off, and the hood removed. Ejection of the compact followed air cooling to room temperature.

Trzebiatowski's Procedure

The most important contribution to the experimental exploration of hot-pressing has been made by Trzebiatowski.[20] His procedure vastly improved previous techniques, since it combined a method of heat supply to the die during the act of compression (used by Hoyt) with a reducing atmosphere introduced into the die cavity (as provided for in Sauerwald's work) with the atmosphere being permitted to surround the compact

[20] W. Trzebiatiwski, *Z. physik. Chem.*, A169, 91 (1934).

uninterruptedly through the periods of heating, drawing, compression, and cooling to room temperature.

Apparatus.   The experimental apparatus built by Trzebiatowski is sketched in Figure 173. It was placed in a 13-ton hydraulic press and

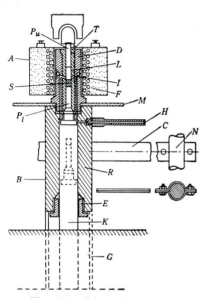

secured to its vertical posts, N, with the aid of clamps, C. Basically, the apparatus consisted of two parts, the actual hot-press die assembly, A, and a receiving compartment, B. The two parts of the equipment were screwed together, but insulated against heat transfer by means of mica and asbestos, M. The die proper, D, was machined from nickel-chromium, corrosion-resistant steel and contained a replaceable die liner, L, made of a s p e c i a l heat-resisting hardened steel, out of which the punches, P_u and P_l, were also made. Heat was supplied to the die by a low-voltage, resistance tube furnace, F, whose temperature was determined by a thermocouple, T, situated a few millimeters from the compact. Through the inlet, H, hydrogen was passed into the receiver and led into the die cavity through small inlet channels, I. The lower punch, P_l, was supported by a plunger, K, which fitted closely into the receiv-

Fig. 173. Schematic cross section through apparatus used by Trzebiatowski[20] for hot-pressing powder compacts in a reducing atmosphere: A, hot-press die assembly; B, receiving compartment; E, stuffing box; H, hydrogen inlet; I, hydrogen inlet channels; M, insulation; and N, press column.

ing chamber, R, and was provided with a stuffing box, E. A steel tube, G, was used for raising the apparatus during ejection.

In this apparatus, temperatures up to 600°C. (1110°F.) on the specimen, S, could be maintained. The maximum permissible pressure was governed by the strength of the die materials at this temperature, and could reach 100 tsi. Under these optimum conditions, however, the die and particularly the punches required frequent replacement.

Technique.   Trzebiatowski employed the following technique. The die was heated to the required temperature with the punches in their proper position while a slow stream of hydrogen was allowed to pass through the die opening. Copper and gold powders were cold-prepressed at a low pressure into cylindrical compacts having a slightly smaller diameter and a greater thickness than the final size (10 mm. in diameter and 4–6 mm. high.) The preformed compact was placed on top of the lower punch, P_l, and the preheated upper punch, P_u, was immediately placed on top. After the preformed compact was left in position for 5 minutes to attain the necessary temperature equilibrium, pressure was applied, the maximum load being reached in 30 seconds—and maintained for 1 minute to allow for completion of the plastic deformation of the compact. Ejection of the compact was then initiated by raising the apparatus and placing the spacer tube, G, beneath it. The piston, K, was pushed out of the receiver, R, by increased hydrogen pressure. By inserting spacer blocks under the press head, the compact and the lower punch, P_l, could be pushed into the receiving chamber by applying only slight pressure. After cooling to room temperature under hydrogen, the compact was withdrawn, and the lower punch repositioned in the die on top of the piston, K. The entire operation did not require more than 10 minutes, including the period necessary for attainment of temperature equilibrium in the specimen and punches.

Author's Procedures

Depending on the type of material to be experimented with, the author initiated the development of several different apparatus and techniques. For the hot-pressing of copper and its alloys, a procedure similar to Trzebiatowski's was used, whereas for the hot-pressing of iron and steels, the process was modified considerably.

METHOD FOR HOT-PRESSING COPPER AND ITS ALLOYS

The apparatus and technique employed in this work[21] constituted only a modification of Trzebiatowski's method of applying highest possible pressures on preformed compacts at elevated temperatures in the presence of a reducing atmosphere. The same method of applying the required temperature with a low-voltage, electrical resistance furnace surrounding the die was adopted. The most important modification was the introduction of a quench bath for rapid cooling and acceleration of the entire operation. In order to conserve die materials, pressures and temperatures were kept below those used by Trzebiatowski. The same die and tools

---

[21] C. G. Goetzel, *The Influence of Processing Methods on the Structure and Properties of Compressed and Heat-Treated Copper Powders, Dissertation.* Columbia University, New York, 1939, p. 10.

were used for many experiments, and no appreciable wear of the die could be noticed even after several thousand compressions under various conditions within the maximum temperature (500° C.; 930F.) and pressure (50 tsi) limits.

The experiments conducted in this manner produced compacts of copper, brass, bronze, and alloys of copper with nickel, manganese, and

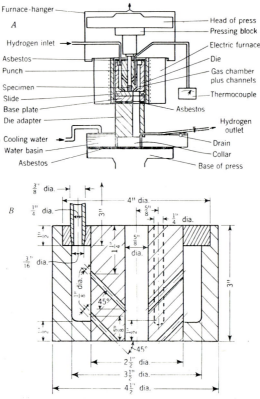

Fig. 174. Schematic cross section through apparatus used by author for hot-pressing copper and copper alloy compacts in a reducing atmosphere: *A*, assembly, and *B*, die.

silicon, formed either from powders in the elemental state or from prealloyed powders. The final size of the compacts was $5/8$ in. in diameter and $1/2$ to $5/8$ in. high.

Apparatus. The details of the apparatus used are shown in the cross-sectional view in Figure 174. It could be operated at temperatures up to 500°C. (930°F.), and was fitted for compression in a reducing atmosphere. Pressure, varying from 5–50 tsi, was applied by means of a 20-ton, Carver hand-operated hydraulic press. Heat was applied by an electric furnace surrounding the die, which was built from 18–4–1 tungsten–chromium–vanadium high-speed steel to retain its hardness up to 500°C. (930°F.). Base plate, slide, and punch were built from the same material as the die. All these parts, with the exception of the base plate, were hardened. The base plate rested on a soft steel adapter and, in turn, on a steel collar in a water basin. The base plate, slide, and adapter had eccentric holes of such diameter as to permit the specimens to be dropped into the quenching bath. The temperature could be maintained constant within ±10°C. (±18°F.) by a rheostat connected in series with the furnace. The temperature was measured by an iron–constantan thermocouple, the junction of which was brought as close to the specimen as possible.

Technique. The process of hot-pressing was carried out in the following manner. The furnace was heated to the desired temperature and the die purged with hydrogen. The specimens were compressed (cold) at 2 tsi into pellets of slightly reduced diameter, so that they could easily be dropped into the cavity of the hot die. The cavity was then closed with the preheated punch, on top of which loading blocks were set to fill the gap between the punch and the head of the press. To insure a uniform temperature, the pellets were kept in the die for 5 minutes before applying pressure, a time assumed to be sufficient because of the large mass of the die and the small mass and better conductivity of the copper compacts. After this period, the load was applied, left for 1 minute, and released. Then the furnace was lifted, and the slide pushed back by means of a specially constructed hook, bringing the opening of the slide directly underneath the compressed compact. Slight pressure ejected the compact into the hydrogen-filled opening, and by pulling the slide into its original position the hot compact was dropped through the bore hole of the adapter into the water bath. The entire operation took 8 to 10 minutes per specimen.

METHOD FOR HOT-PRESSING IRON

Since thermoplastic conditions of iron are quite different from those of copper or copper-base alloys, certain changes in hot-press apparatus and procedure[22] were implied. Above all, higher temperatures and pressures were found necessary for consolidation, imposing severe stresses on the available die materials. The process appeared further complicated by

[22] C. G. Goetzel, in J. Wulff, *Powder Metallurgy.* Am. Soc. Metals, Cleveland, 1942, p. 398.

a chemical reaction between the compact's surface and the die material if the latter was graphite. Therefore, a maximum temperature of 1130°C. (2065°F.) could not be exceeded, lest the compact liquefy at its surface, causing loss of shape and operating difficulties due to extrusion.

Apparatus. Other modifications in apparatus consisted primarily in the use of graphite dies for all high-temperature work, the use of induction heating, and the replacement of the quench tank by a double-compartment receiver fitted with a sand-catch bucket. The details of the apparatus are given in the sketch in Figure 175; a photograph of the entire hot-press equipment is shown in Figure 176.

Graphite dies, in conjunction with plungers made either of graphite or of cemented carbide, were used only at temperatures ranging from 1130°C. (2065°F.) down to 900°C. (1650°F.), whereas for work at lower temperatures down to 500°C. (930°F.), 18–4–1 tungsten–chromium–vanadium high-speed steel was used for both die and plungers. High-frequency current heated the die to the required temperature, and the heat was transferred to the compact and punches by convection. A thermocouple was placed close to the specimen. The compacts were surrounded by hydrogen during the operations of pressing, ejection, and cooling. The size of the specimens after pressing was $3/4$ in. in diameter and $3/4$ in. high. Hydrogen entered the die cavity through an inlet tubing. To eliminate the possibility of oxidation of the compact during cooling the gas is fed into the two-chamber receiver by a second supply line. The pressure, applied by means of a 25-ton, double-speed hydraulic jack built into a press framework, was varied between 10 and 50 tsi where the high-speed steel die was used and was kept between $1/2$ and $2^1/_2$ tsi where pressings were made in graphite molds.

Technique. The die was heated to the desired temperature while the entire system was purged with hydrogen. The powders were premolded at room temperature at pressures up to 10 tsi into cylinders of slightly undersized diameter. They were then dropped into the hot die on top of the bottom punch, P_l, and the cavity was immediately closed by inserting the top punch, P_u. In order to attain temperature equilibrium in the compact a 5-minute period was allowed to elapse. Pressure was exerted, and the maximum load was applied for 1 to 10 minutes, depending on how much time was necessary for the completion of the plastic flow. After the pressure was released, the base plate was slid into position so that the opening coincided with the die cavity. The compact and lower punch were then ejected downward through the opening in the slide into the first compartment of the receiver. After the slide was moved into its original position, the top punches were removed and a duplicate bottom punch and

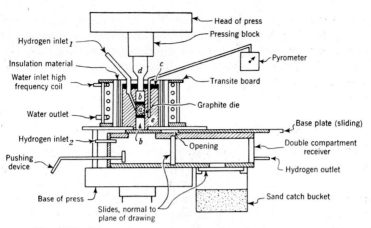

Fig. 175. Schematic cross section through apparatus used by author for hot-pressing iron compacts in a reducing atmosphere: *a,* specimen; *b,* carbide punches; *c,* graphite cover; *d,* spacer punch; and *e,* die.

Fig. 176. Experimental hot-press equipment consisting of die assembly of Figure 175, hydraulic jack, press frame, and high-frequency converter (courtesy American Electro Metal Corp.).

a new preform introduced into the die. The hot-pressed compact and the original bottom punch were first moved into a second compartment of the receiver with the aid of a pusher and additional slides, and then dropped into the catch bucket filled with sand. The sectioning of the receiver was necessary to avoid the formation of explosive mixtures of hydrogen and air in the cooling compartments.

METHOD FOR HOT-PRESSING STEEL

Further modifications in apparatus and technique became necessary when bar-shaped steel compacts were hot-pressed. The direct current method of heating was found best for this purpose, and split molds, held together by a vise, were used to facilitate removal of the compacts. The compact's tendency (at very high temperatures) to weld to the die and extrude between the die and the punches would have caused great difficulties during ejection from solid dies.

Apparatus. A sketch of the apparatus is shown in Figure 177. It resembles a resistance-welding machine in its principal functions. Heat was supplied by passing a low-voltage current (approx. 6–8 volts) through water-cooled electrodes, E_1 and E_2, forming a vise. The current (15,000 amp.) was passed sideways into graphite blocks, G, which served as spacer blocks for a split mold, D, made of graphite or of special cemented carbide. A preformed compact was pressed between a lower punch, P_l, and an upper punch, P_u, on which pressure was exerted by means of the hammer, H. The punches were made of the same material as the mold, *i.e.*, graphite or cemented carbide; the hammer was machined of a block of sintered molybdenum. Pressure was applied from a downward-acting hydraulic press, J, having a two-way piston connected with a four-way valve and a high-speed Vickers pump. The press permitted quick motion of the ram and a maximum pressure of about 1 ton. Provisions were made to introduce hydrogen through an opening, H, between two sections of the die. The resulting specimens were $2 \frac{1}{4}$ in. long, $\frac{3}{8}$ in. wide and $\frac{3}{8}$ in. high.

Technique. The procedure was as follows: The compact was preformed cold at about 10 tsi. It was then placed in the die between the two punches, and the die was assembled and placed between the graphite spacers in the vise. After closing the vise and purging with hydrogen, current was passed through the assembly. The rise in temperature was followed by optical pyrometer measurements. Within one-half to one minute the die was brought to the desired working temperature (1125°C. (2055°F.) in the case of graphite dies, and up to 1250°C. (2280°F.) when cemented carbide dies were used), whereupon the maximum pressure ($1\frac{1}{2}$

tsi) was applied with a quick blow of the hammer. The current was then shut off, and the mold was cooled within a few minutes to below red heat

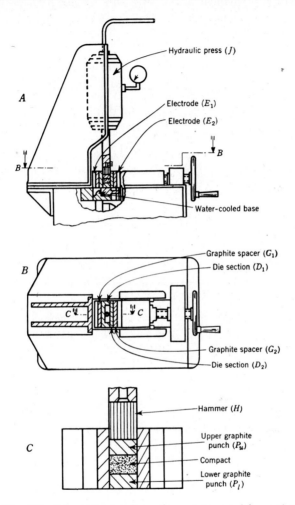

Fig. 177. Schematic drawings of apparatus used by author for hot-pressing steel compacts in a reducing atmosphere (according to U. S. Pat. 2,133,495; courtesy of P. Schwarzkopf and American Electro Metal Corp.). *A*, elevation of machine and section through vise and die; *B*, downward view; *C*, section through die assembly, enlarged.

with the water-cooling system. The hydrogen supply was then disconnected, the vise opened, and the mold taken apart to remove the specimen. The entire operation did not last longer than 5 minutes.

Henry and Cordiano's Procedure

The main improvement in the procedure for hot-pressing iron powder developed by Henry and Cordiano[23] is in the use of a closed die permitting the hot-pressing of bar-shaped compacts 3 $\times$ $^5/_8$ $\times$ $^3/_8$ in., out of which tensile test specimens could be machined.

Apparatus. The details of the apparatus are reproduced in Figure 178. The die, upper and lower punches, and sliding block were made of heat-resistant, high nickel–chromium steel, whose analysis and properties at elevated temperatures are given in Table 51. The die block was

TABLE 51

Analysis and Properties of Punch and Die Steel Used in Hot-Press Apparatus by Henry and Cordiano[23]

Chemical Analysis			
Element	Per cent	Element	Per cent
Carbon	0.33	Nickel	19.70
Manganese	0.49	Chromium	8.31
Silicon	1.16	Iron	Balance

Properties at Elevated Temperatures				
Temperature		Tensile strength, psi	Proportional limit, psi	Elongation, per cent in 2 inches
°C.	°F.			
20	70	112,500	45,200	—
205	400	102,000	45,600	21.0
480	900	90,800	40,100	20.0
595	1100	80,000	35,500	19.7
650	1200	74,000	35,800	19.3
705	1300	65,400	26,200	18.5

machined from cast iron, and the chute was fabricated from sheet steel. The entire die assembly was enclosed in an electric furnace heated with a nichrome wire coil.

Technique. To facilitate handling during hot-pressing, the iron powder was precompacted cold at 20 tsi into 3 $\times$ $^3/_8$ $\times$ $^5/_8$-in. specimens. The die lubricant used was made by dissolving 1 g. of stearic acid in 20 cc. of carbon tetrachloride. To provide high-temperature lubrication, the hot-pressing punches and die parts were rubbed with flake graphite. Before the furnace was brought to temperature, the system was purged with dry nitrogen in order to prevent oxidation of the die parts and to pro-

[23] O. H. Henry and J. J. Cordiano, *Trans. Am. Inst. Mining Met. Engrs.*, *166*, 520 (1946).

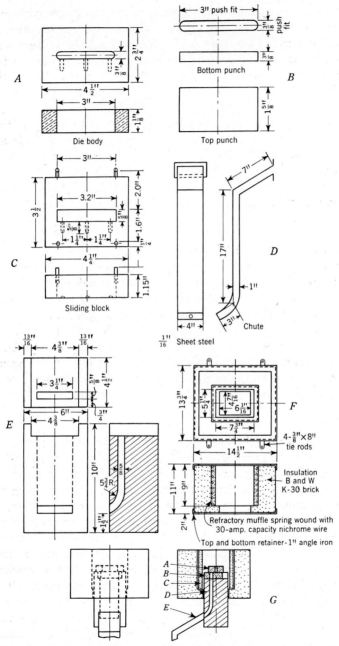

Fig. 178. Schematic drawings of apparatus used by Henry and Cordiano[23] for hot-pressing iron compacts in a reducing atmosphere: *A*, die body; *B*, punches; *C*, sliding block; *D*, chute; *E*, die block; *F*, electric resistance furnace; and *G*, die assembly.

vide a means for purging the hot die cavity with hydrogen without form-ing an explosive mixture. When the furnace had reached the desired temperature, purified hydrogen was passed through the system to replace the nitrogen, and the escaping hydrogen was ignited and allowed to burn. The temperature was measured by means of a Chromel–Alumel thermo-couple inserted through the top of the die assembly during the heat-up (through a hole in the cover). Initially, the tip of the thermocouple was in the die cavity and rested approximately $3/8$ in. from the top of the com-pact. The loosely pressed compact was placed in the heated die and left to soak for 10 minutes. A pressure of 10 tsi was applied for the predeter-mined time interval—50, 150, or 450 seconds. The specimen was ejected and, after it had cooled, the hydrogen was replaced by nitrogen and the specimen removed.

INDUSTRIAL PRACTICES

In view of the fact that properties obtainable from hot-pressed metal powders are, in general, far superior to the properties of cold-pressed and sintered powders, and that there exist many procedures which could be followed to obtain the desired results, the question of the industrial possi-bilities of the method must, of course, receive primary consideration. There are many problems to be solved toward this end. The proper choice of the die material is probably of the greatest importance, but other problems remain, such as maintaining a reducing atmosphere around the compact, and achieving a smooth functioning of the ejection mechanism. Working at high temperatures with high pressures is, in itself, not espe-cially convenient, although it does not differ much from hot-press forging, drop forging, and other hot-working practices.

On the other hand, these conventional metal-working practices are well established and economically sound. They all involve the plastic de-formation of solid slugs or ingots from considerably less expensive raw materials than even the cheapest metal powders. Thus, the high price of powders, coupled with the necessity of using a protective atmosphere, makes industrial hot-pressing of the technically important metals (iron, steel, brass, etc.) hardly competitive, except under the following condi-tion. (1) Extremely inexpensive raw materials are used, such as fine brass or cast iron shot. (2) An inexpensive atmosphere, such as cracked pro-pane or natural gas, is used for protection against oxidation. (3) A dur-able metal of high hot strength, such as Stellite or special cemented car-bide, is used for punches and dies, so that very close dimensional toler-ances and excellent surface finishes can be obtained. (4) A soundly engi-neered machine is designed to permit continuous operation for the production of parts of various sizes and shapes by simply changing the timing and the dies.

If these conditions are fulfilled, if physical properties are stabilized at their high level, and if such operations as sintering or coining are eliminated, it would seem possible for hot-press products to compete not only with cold-press products, but also with some articles that are generally machined.

To date, hot-pressing has found only limited applications in industry. It is used for the manufacture of some friction materials, such as brake linings and clutch facings[24] (where it may be designated as pressure sintering); for some bushings made from waste materials, such as cast iron and steel chips[25,26]; for the production of some diamond tools[27]; and for the production of cemented carbide products.[28] In Germany during the war, hot pressing of cemented carbides was developed to a high degree of perfection in the mass production of armor-piercing bullet cores and tips[29] (for details see Volume II, Chapter XXII), and a hot-press used for this purpose is shown in Figure 179. Since these applications are specialized, the machinery used is suitable only for the production of the particular article in question. A few of these techniques will be discussed later, together with some methods suggested for more general use.

Hot-Press Methods for Cemented Carbides

A commercial method of producing hot-pressed carbide parts has have described in general terms by Engle.[28] The equipment used in this process is of modern design and quite complicated; a photograph is shown in Figure 180. The procedure generally followed involves: (1) placing the powder in an enclosed graphite mold, and (2) heating the mold and its contents to sintering temperature while sufficient pressure is imposed in one direction to overcome the forces which cause shrinkage in the other two directions when a cold-pressed compact is heated freely. Pressures between 400 and 2500 psi are commonly used. The molds are heated electrically from within the mold itself (by resistance), by a resistance furnace, or by high-frequency current. Oxidation is prevented by the introduction of a hydrogen atmosphere which also serves to prevent undue deterioration of the graphite mold.

[24] C. S. Batchelor, *Metals & Alloys, 21*, No. 4, 991 (1945).

[25] Anonymous, *Steel, 108*, No. 21, 76 (1941).

[26] A. F. Macconochie, *Steel, 110*, No. 11, 98 (1942).

[27] U. S. Pat. 2,228,871.

[28] E. W. Engle, in J. Wulff, *Powder Metallurgy*. Am. Soc. Metals, Cleveland, 1942, p. 442.

[29] G. J. Comstock, *Iron Age, 156*, No. 9, 36A (1945).

More recently, it has been reported[30] that the process can produce very large pieces, such as carbide die nibs which are too large for sintering in ordinarily available furnace equipment. Parts of this kind weigh

Fig. 179. Hot-press used in Germany for production of cemented carbide armor-piercing bullet cores (courtesy of Industrial Powder Metallurgy Laboratory, Stevens Institute of Technology).

up to 50 pounds, having cross sections up to 100 sq. in., outside diameters up to 20 in., and thicknesses up to 8 in. The replacement of ordinary graphite by "graphite-base" compositions (probably containing clay or

[30] Anonymous, *Metal Progress*, *45*, No. 4, 681 (1944).

other refractory oxides) has increased the strength and life of the molds and has also improved the surface finish of the product. The mold consists of an outer case, a hollow inner core or bottom plunger having a corresponding flange at its bottom to hold the powder in place, and a ring-shaped top plunger to squeeze down the carbide powder between the case

Fig. 180.   Hot-press for production of cemented carbides shown in operation (courtesy of Carboloy Co., Inc.).

and the core. When very large-sized parts are being formed, the powder-filled mold should be preheated in a portable electric resistance furnace in a hydrogen atmosphere. After the preheating temperature is reached, the furnace, together with the mold, is pulled by a convenient cable and pulley arrangement from its stand-by position to a location directly underneath the press platen. Usually, the preheating furnace draws current during the entire process of hot-pressing; the heating condition is thus actually continuous, including presintering and sintering during pressure application. The equipment for this operation consists of a 100-ton hydraulic press with water-cooled platens; the construction (by the National Machinery Co., Tiffin, Ohio) resembles the so-called press-welding type of machine. The necessary heat is provided by electrical resistors placed to attain sintering temperature; it is supplied through a 750 kva. transformer and controlled by General Electric electronic contactors.

Another process by which cemented carbides are produced by simultaneous application of pressure and heat is described by Baëza.[31] The powder is precompacted cold at about 10 tsi into a slug of the same mass as the finished product. The size of the preform roughly corresponds to the final size, but has a simpler shape to help the molding process. The prepressed compact is placed over the die in the bed plate of a press (in which it is to be molded into the final form) and is insulated from the base by a thin sheet of mica. The compact is heated (preferably in a vacuum in order to prevent oxidation) by having two electrodes pass a current through it sidewise. For small pieces, the time required to heat the compact to near sintering temperature may be less than one minute. Heat may also be supplied by means of a high-frequency induction coil. On reaching the required temperature, the press ram moves a plunger until it contacts the compact, which is then forced into the die carrying the thin sheet of mica with it. The compact is compressed into its final form at about 30 tsi and is ejected. Except for stripping the mica, no other finishing operations are necessary. Even though the hot compact is in contact with the walls of the die for only a short time, considerable die wear takes place because of the high pressures used. However, water cooling of the dies and electrodes has partly reduced the wear effects. The method is said to give particularly good results in producing cemented carbide tool bits, and also is supposed to be equally adaptable to the production of iron and steel products of great density, high tensile strength, and hardness equivalent to that of hard steels.

During the war, the hot-press method was extensively used in the German hard metal industry.[29] Electrical resistance heating of spring-clamped graphite molds was the preferred method of heating, and mechanical as well as hydraulic systems applied the necessary pressure. The excessive wear of the graphite dies during mass production work was made economically tolerable by employing exchangeable cartridges— usually cylinders or other tube forms broached from graphite—which were pressed into heavier graphite mold containers. In order to shorten the hot-press punch movements, the powder mixtures were first cold-

---

[31] W. J. Baëza, *A Course in Powder Metallurgy*. Reinhold, New York, 1943, p. 74.

pressed into temporarily reinforced graphite molds under heavy hydraulic pressure. The production cycle could thus be accelerated to such an extent that pieces weighing from 50 g. to 1 kg. could be produced within 3 to 7 minutes.

Hot-Press Methods for Iron- and Copper-Base Metals

In the manufacture of cemented carbides, the product is of sufficiently high value to make it economically feasible to destroy the graphite mold after the compression even if new graphite molds are required at the rate of a dozen or more per hour. However, when materials of lower market value are to be made by this method, a continual change of molds is impossible. For this reason graphite molds are unsuitable, even if they can withstand 20 to 50 compressions (as has been found possible with special clay-containing compositions). Graphitic dies are also undesirable when iron powders are to be hot-pressed, since surface carburization of the compact is not easily controlled. Consequently, all methods suggested for industrial hot-pressing of metals other than cemented carbides use metallic dies.

Stellite and cemented multicarbides provide the ideal materials for construction of the dies due to their high hot hardness and creep resistance, although their extreme brittleness demands very careful operation. Where conditions do not permit the use of this material, high-speed steels and high-alloy tool steels with good hot hardness, toughness, and creep resistance may be chosen. These metallic die materials, of course, impose a definite limit on the pressing temperature but, on the other hand, allow the application of much higher pressures than are possible with graphite molds. A very interesting procedure with industrial possibilities has been developed by Jones,[32] who with the aid of metallic dies hot-pressed a variety of copper-base metal powders[33] and suggested the use of the same apparatus for the hot-pressing of iron-base materials. The method involves a technique similar to that for drop-forging providing for the preliminary compaction of the powder at slight pressures, a brief heat treatment of the compacts at temperatures well above the desired hot-press temperature, the pressing in a die preheated to a temperature several hundred degrees below the working temperature, and then the actual drop-forging into a cold die. The apparatus constitutes the first design operating on a continuous basis. It would be of even more commercial significance, however, if modifications in design permitted introduction of a protective atmosphere around the hot compact.

[32] W. D. Jones, *Metal Ind. London*, *56*, 69 (1940).
[33] W. D. Jones, *Metal Ind. London*, *56*, 225 (1940).

Jones' apparatus is reproduced in the sketch of Figure 181. He describes the operation as follows. The apparatus is disposed under three presses; the cold-pressing die is on the left-hand side and is provided with the usual feed hopper loading into and completely filling the die cavity. Compression by one or by both rams squeezes the powder to a compact, and the rams then withdraw. It must be imagined that the heating chamber shown at the right of the diagram occupies the central position

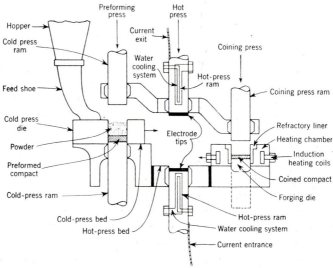

Fig. 181. Schematic cross section of multiple press that combines preforming, hot-pressing, and cold-coining (according to Jones[32]).

which it can do by sliding along the surface of the coining-press bed. In this position the heating chamber is then covered by the cold-pressing die, which slides along the surface of the cold-press bed, so that at this stage of the operations both the heating chamber and the cold-pressing die are centered under the hot-press rams. The cold-compressed compact is pushed into the heating chamber by the upper ram, and rests there, supported by the lower ram. The upper ram then retracts to allow the cold die to withdraw to the left, and the ram finally returns to the heating chamber. The heating chamber is provided with a refractory liner, and the diagram indicates that induction heating coils can be substituted. Pressings of simple shapes, however, are more conveniently heated by direct passage of electric current through the compact; in this case the apparatus is provided with the electrical portion of a resistance welder, with the rams becoming electrodes. (They are provided with cavities for water cooling.) The current enters at the bottom lead, passes through the compact, and leaves at the top lead, the compact being maintained under slight pressure during the heating period. In electrical sintering of this type it has been difficult to overcome the tendency of the electrodes to weld to the compact; a successful solution is to use tungsten–copper electrode tips, and only under difficult circumstances must this material be supplemented by nonwelding powder surface additions. On attainment of the sintering temperatures, the compact, now appreciably consolidated, must be finally treated by forging into a cold die. This die is situated at the right, and it must be supposed that the heating chamber, now containing the hot compact, has moved to a position over it (as indicated in the diagram). Before appreciable cooling occurs, the walls of the segmented liner open slightly and release the compact, which then falls and is simultaneously knocked into the die below by the coining-press ram.

Crane and Bureau[34] have suggested, for an improved industrial hot-pressing of metal powders, a procedure similar to common progressive forging in double-crank presses. The method is based on a combination of closely progressive heating and dense-pressing of powder preforms in a so-called "Transfer Feed" press, which provides a series of stations, close together and enclosable, with the necessary movements. Protection in transit by flame curtains, or enclosure with Pyrex windows, plus circulation of a reducing atmosphere, prevent oxidation of the, charge. Powder briquetted at the first station, or preforms delivered to the first station by a conventional friction dial feed, are transferred from one position to the next by suitably shaped fingers on the feed bars. The initial stations consist of an automatic furnace with refractory, ceramic die blocks and resistance heating contacts, or induction heating coils where great speed is required. In the pressing (and possible coining) station, high-temperature-resistant forging steels with hot-oil circulation for temperature control are used. Thus, the difficulty in finding die materials that will stand up both to the high heating temperature and also to the high forming pressures without loss of dimensional accuracy has been overcome by a clever construction and layout of operating functions, which permits the use of available materials.

In all the procedures mentioned so far, it was considered essential to preform compacts from powders at room temperature. As long as the metal in its highly dispersed form remains cold, the danger of oxidation is small and, in addition, its flow characteristics remain favorable for high production rates. Nevertheless, it was suggested that the metal powder be

[34] E. V. Crane and A. G. Bureau, *Trans. Electrochem. Soc.*, *85*, 63 (1944).

heated before the pressing operation; Koehring[35] developed an interesting process on this basis. The process involves an apparatus for the continuous production of agglomerated charges of powdered metals by the progressive heating of a portion of the charge to a temperature at which agglomeration occurs. A certain portion of the agglomerate is then cut off and dropped into a die where it is hot-pressed into the final form. Simultaneously, another portion of the powder is advanced toward the furnace for heating to the agglomeration temperature. The charge, which is ejected from the furnace, consists of a measured quantity of powder that is agglomerated and has been brought to the required pressing temperature. This charge need not necessarily be very close in shape or size to the finished article; in fact, to facilitate the operation, it is better to make the charge somewhat undersize in diameter. Where high production rates are desirable, induction heating is suggested as the best means of heating the powder. The entire apparatus, consisting of the powder chamber, the pressing chamber, and a cooling chamber, is maintained in a protective atmosphere to prevent oxidation in the loose powder, in the agglomerate, or during compaction or cooling.

A hot-press suitable for direct molding of small cylinders from alloy powders has recently been developed by Strauss.[36] The machine operates on the principle of resistance heating of the graphite molds, and is adaptable for both laboratory and production work; it appears particularly suitable for the hot-pressing of small contact points from copper- or silver-base compositions.

Wassermann[36a] has recently described a hot-pressing practice of industrial significance which resembles conventional die-forging methods. The technique is considerably simplified by eliminating the initial cold-pressing operation, and can be employed on the production of blind-end bushings and similar parts from iron powder. The required amount of powder (e.g., eddy-milled or carbonyl iron) is first poured into a graphitized iron crucible, and heated to 850°–1000°C. (1560°–1830°F.) in a hydrogen atmosphere for about 1 hour. The resulting iron cake is removed from the crucible and immediately placed into the die of a friction-driven screw press, where it is hot-pressed into the final shape with one or two strokes. According to Wassermann, this method of hot-pressing sintered cakes is especially applicable to those metal powders which, on account of their particle shape, are cold-pressed only with difficulty, but which display good sintering properties (as is the case with carbonyl iron powders).

[35] U. S. Pat. 2,362,701.
[36] Anonymous, *Product Engineering, 17,* No. 11, 93 (1946).
[36a] G. Wassermann, *Metallforschung. 2,* 129 (1947).

It would undoubtedly lead too far afield to cite further examples of hot-press machines and methods suitable for production work. The patent situation in the hot-press field is quite complex, and probably all of the basic ideas have been patented. It may be sufficient to conclude that it appears quite possible to develop a machine that is capable of producing parts of very good physical properties and excellent dimensional accuracy on a commercial basis, if the following conditions are met: (*1*) proper choice of raw materials, (*2*) selection of a durable material for dies and punches, (*3*) automatic transfer of powder and compact, and (*4*) proper precautions taken against harmful oxidation effects.

Hot-Press Extrusion

Hot extrusion is closely related to hot pressing. In a few instances, standard extrusion processes have been applied to powdered metals. This has been done in spite of the fact that in such a process one of the main advantages of powder metallurgy, namely, the direct molding of completed shapes to close dimensional accuracy, is lost at least in part. Extrusion of profiled bar stock, even if kept to closest cross-sectional tolerances, invariably necessitates a cut-off operation to produce small prismatic shapes. The accuracy in the longitudinal direction depends entirely on the efficiency of the cut-off mechanism.

Where the raw material and end products are costly, extrusion of powdered metals has found some industrial applications. For instance, refractory metal-base contact materials can be worked into profiled stock.[37,38] The mixture of tungsten and silver or copper powders is first pressed into an ingot which is sintered at a high temperature, usually close to the melting point of the lower melting metal. The hot ingot is then transferred into the extrusion press, where it is subjected to a hot-press and extrusion operation. A reducing atmosphere is maintained in the furnace during the transfer and in the extrusion die chamber. If the extrusion temperature is kept sufficiently high by preheating the extrusion dies, and if the plastic low-melting metal is present in the composite in a sufficiently large proportion, *i.e.*, at least 25% by weight, the material can be extruded at pressures not exceeding 50 tsi, in spite of its highly refractory nature. In fact, Schwarzkopf was able to apply this method to compositions which contained an appreciable amount of tungsten carbide.[39] He was also successful in obtaining solid and dense steel alloys by hot extrusion of ingots that had previously been densified by pressing

[37] R. Kieffer and W. Hotop, *Pulvermetallurgie und Sinterwerkstoffe*. Springer, Berlin, 1943, p. 321.
[38] Brit. Pat. 484,996; French Pat. 823,238.
[39] U. S. Pat. 2,148,040.

and sintering.[40]   The pressures necessary for extrusion ranged from 60 to 100 tsi. Alloys fabricated in this manner were plain carbon steels and nickel, chromium, molybdenum, and tungsten alloy steels. Extruded hard metal products of various shapes (rods, spirals, tubes) have recently been made commercially available by the Carboloy Co., Inc.[41]   Details of the production process have not been disclosed.

A process with interesting possibilities was developed by Stout[42] and Tyssowski[43] in connection with the production of pure, oxygen-free copper, designated as "coalesced" copper. A relatively coarse copper powder is obtained by the electrodeposition of brittle cathodes, followed by stripping and washing in usual manner. The powder is then cold-briquetted at about 10 tsi to approximately 80–86% density. The briquettes are heated in a controlled atmosphere to 870–910°C. (1600–1670°F.), where they are purified and deoxidized. They then pass directly to the extrusion press through a chamber retaining the controlled atmosphere. Coalescence and extrusion are accomplished in a hydraulic extrusion press of conventional design, exerting a pressure up to 38 tsi on the plunger. Actual pressures during extrusion vary from 15 to $26^1/_2$ tsi for normal sizes—up to the press maximum for small cross sections. The sizes and cross sections of the extruded products as obtainable by this method are limited only by the dies used and by certain economic considerations. Tyssowski found it uneconomical to extrude an article of much below 0.5-in.^{2} cross-sectional area, or much above 4.5 inches in diameter. But within this wide size range, round, rectangular, and especially profiled shapes could be readily produced—the extruded metal being of highest purity, of a uniform recrystallized structure, and of excellent physical properties.

Summary

Experimental hot-pressing has led to a variety of individual techniques which have differed primarily in the types of die and in the methods of heating employed; most procedures, however, have provided a protective atmosphere around the hot compact and have required preformed specimens for the sake of speedier and safer operation.

Solid dies were preferred for work at lower temperatures and higher pressures, as in hot-pressing of copper, copper alloys, and iron, whereas split dies were often more suitable for work at very high temperatures, e.g., above 1200°C. (2200°F.) in the hot-pressing of cemented carbides

[40] U. S. Pats. 2,205,865 and 2,225,424.
[41] E. W. Engle, in J. Wulff, *Powder Metallurgy*. Am. Soc. Metals, Cleveland, 1942, p. 442.
[42] H. H. Stout, *Trans. Am. Inst. Mining Met. Engrs.*, *143*, 326 (1941).
[43] J. Tyssowski, *Trans. Am. Inst. Mining Met. Engrs.*, *143*, 335 (1941).

and steels. At these temperatures, the tendency of the compact to weld to the die and to extrude between the punches and the die, make a regular ejection procedure very difficult, if not impossible. Although direct resistance heating was preferred for very high temperature work, resistance or high frequency furnaces were found satisfactory in the other instances.

High-speed steels and heat-resistant tool steels have been employed effectively as die materials (up to approximately 800°C., 1470°F.). Above this temperature, graphite was found satisfactory only for very low pressures (not exceeding $2^1/_2$ tsi); higher pressures necessitated the use of special cemented carbide dies and tools.

Hot-pressing of metal powders in industry today is used only on a limited scale. The most important application is in the production of very hard or very large cemented carbide parts. For the production of less refractory materials, such as bushings from bronze powders or from steel chips, or friction materials, the method is used to a lesser extent. Its widespread application for ordinary iron-base and copper-base parts, however, awaits the construction of a carefully engineered, fully automatic, hot-pressing machine capable of molding inexpensive raw materials in a protective atmosphere into high-density, high-strength parts of close dimensional accuracy. The need for such a machine is well recognized, and numerous attempts have been made in recent years to overcome the inherent difficulties—especially with regard to the die materials, but none of the methods so far disclosed appears to be applicable to more than one particular use.

The adaption of conventional extrusion practices to metal powders is possible, and hard metals and refractory base metals, as well as lower melting metals, such as steels or copper, can be produced in this manner. The products obtained by extrusion have excellent physical properties, comparing favorably with ordinary wrought material. The hot-extrusion process defeats one of the primary gains in powder metallurgy, namely, the direct molding of complete shapes to accurate size, but permits the production of shapes that are too long for the application of standard powder metallurgy procedures.

CHAPTER XIV

Principles of Sintering

Sintering, the mechanism by which solid bodies are bonded by atomic forces[1] through the application of pressure and/or heat, represents the fundamental process in powder metallurgy. In its larger sense this process includes such well-known phenomena as welding, brazing, soldering, seizing, or sticking of metals. No suppositions concerning temperature, pressure, or phase changes are necessary, and it may be possible for sintering to occur at room temperature without the application of pressure. While a liquid phase may be present during sintering (as in soldering), it is not a definite necessity; the conception that at least two substances or two phases of one substance are necessary to facilitate sintering[2] has been proved erroneous by the recent history of the sintering technique. The concept that allotropic changes are responsible for the sintering process[3] is equally untenable since metals sinter regardless of whether or not they undergo a phase change at elevated temperatures. It is conceivable, however, that the volume changes connected with the transformation may affect the sintering process to a certain extent, e.g., through pressure exerted by neighboring particles.

In accordance with this definition of the process there are five different methods of sintering: (1) sintering with pressure alone, equivalent to the molding of powders—and treated in that section; (2) sintering with heat alone, called thermal sintering or molding without pressure—to be discussed briefly in connection with the determination of sintering temperatures, and with shrinkage phenomena; (3) sintering by simultaneous application of pressure and heat, called hot-pressing—and covered in that section; (4) and (5) sintering of compacted powders with or without the production of a liquid phase, constituting the technically important sintering operations—and to be treated as such in the following analysis.

In the technical sintering of metal powders, several different processes can be distinguished: (1) sintering of homogeneous powders, i.e., mono-

[1] P. E. Wretblad and J. Wulff, in J. Wulff, *Powder Metallurgy*. Am. Soc. Metals, Cleveland, 1942, p. 36.
[2] K. Endell, *Metall u. Erz, 18*, 169 (1921).
[3] R. C. Smith, *J. Am. Chem. Soc., 123*, 2088 (1923).

metallic powders, solid solutions, or intermetallic compounds; (2) sintering of heterogeneous powders, *i.e.*, polyphasic (polymetallic or metal–nonmetal) powders without the formation of a liquid phase; (3) sintering of heterogeneous powders with the formation of a liquid phase; and (4) sintering of powders, followed by impregnation with a liquid metal in which the powders may be either in the loosely heaped condition or compressed into compacts. Before discussing these individual processes and their various conditions and effects, it appears advisable to study some of the basic aspects of the sintering mechanism in the homogeneous systems that are common to all of these processes. In the course of this discussion, those processes which involve heterogeneous systems and, especially, intervening gaseous or liquid phases will be treated as variations of the fundamental process.

MONOMETALLIC SYSTEMS

Sintering Forces

A discussion of the sintering process in its widest sense involves fundamental aspects of powder metallurgy, for instance, the basic concepts of adhesion of particles by the action of surface cohesive forces, the atomic bonding of particle surfaces with or without the aid of pressure, or the bonding phenomena related to increased plasticity at elevated temperatures. Inasmuch as this part of the text is primarily concerned with the technological aspects of the powder metallurgy process, it may be sufficient to stress the theoretical implications only to that extent necessary for a full understanding of the subject. Additional treatment of the theoretical aspects of the sintering process can be found in Chapter XXXV, Volume II.

FORCES ACTING BETWEEN TWO SURFACES

For the formation of a sinter bond between two powder particles, there must be utilized the same forces as cause the coherence of the smallest building units (atoms) of any solid body in general. The coherence of these units in a solid body leads to the conclusion that a certain attractive force is effective. On the other hand, a definite limit is set for this attraction, and a further reduction of the normal distances of the atoms requires external compression force. In other words, the unlimited approach of the atoms is prevented by repulsive forces. The difference between attractive and repulsive forces constitutes the resultant force in which we are interested. Following the procedure given by Kieffer and

Hotop,[4] we express the two forces as a function of the atomic distance, r, by the formula:

$$K(r) = a/r^l - b/r^m$$

where $K(r)$ is the resultant force, and a and b are constants. The power, l, should be smaller than the power, m, meaning that with increased distance, r, the repulsive force should decrease more rapidly than the attractive force. With rising temperatures, the movement of the atoms increases the repulsive forces, since the distance at which the atoms are at equilibium increases with rising temperatures. The thermal repulsive force can again be expressed as inversely proportional to a certain power, n, of the distance, where n is smaller than m or l. The complete expression for one particular temperature, T, can then be written as:

$$K(r) = a/r^l - b/r^m - c(T)/r^n$$

In Figure 182 the three members of the formula and their sum are sketched graphically. For small distances, the curve for the resultant force

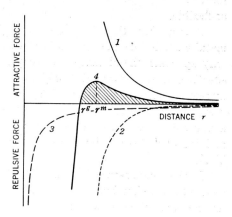

Fig. 182. Schematic diagram of the attractive and repulsive forces and the resulting curve for the interatomic force (according to Kieffer and Hotop[4]).
Curve 1, a/r^l
Curve 2, b/r^m
Curve 3, $c(T)/r_n$
Curve 4, $K(r)$.

shows a repulsion which decreases rapidly with increasing distance. At the point r_g it becomes zero and equilibrium is reached. At greater distances, an attractive force comes into play, which passes a maximum at distance r_m, and decreases thereafter. At great distances and especially at high temperatures the resultant force again becomes repulsive, the value, however, remaining insignificantly small. Of importance to our problem is only the region where the resultant force is of attractive nature, as

[4] R. Kieffer and W. Hotop, *Pulvermetallurgie und Sinterwerkstoffe*. Springer, Berlin, 1943, p. 62.

indicated by the cross-hatched area in Figure 182. An elevation of the temperature causes a shift toward a larger distance, r (to the right of the diagram), of the equilibrium distance, r_g, as well as the distance, r_m, of maximum attraction. The sphere of influence of an atom can extend to ten to twenty times the atom's radius.

These observations on the action and on the regions of influence of the forces relate only to free metallic surfaces. They must now be translated to the process of sintering metal powder particles. For the sake of simplicity it may be assumed at the moment that the surfaces of the particles are smooth and free of sorbed substances, particularly gases. Then the attractive forces would cause a coherence of the particles even if the powder were only loosely heaped. This baking or caking effect, of course, becomes greater as the total particle surface (where intimate contacts affect the action of the attractive forces) increases. Jones[5] has indicated by a number of examples that where large, smooth surfaces of solid metals or materials, such as mica, are available, the part of the surfaces enabling intimate contacts is relatively large and the adhesion forces become remarkably noticeable, especially if the material is pliable or flexible.

It is obvious that, in practice, the irregular surface contours of the metal particles are the reason that in loosely heaped powders the surfaces only approach the distance necessary for the attractive forces to become effective. If the powder is compressed, the surfaces contributing to the attraction become enlarged when the powder is more deformable. Thus it is possible to obtain, even at room temperature, compacts which possess a marked cohesive strength. This phenomenon is the basis for all cold molding of metal powders, and has been referred to in detail in Chapter VIII.

If the temperature is now raised, the maximum attractive force is reduced somewhat by the above-mentioned increase in repulsive forces. At the same time, however, the surfaces contributing to the attraction become considerably enlarged, primarily because of improved plasticity of the metal. This second effect outweighs by far the first; this is true for loose powders, as well as for compacts. The increase in surface areas contributing to attraction becomes particularly significant when the powder is compressed at elevated temperatures, as discussed in Chapter XII. Since plasticity of many metal powders increases rapidly with rising temperatures, moderate temperatures are frequently sufficient to achieve complete and intimate contact between all particles. In the absence of interfering phases, such as surface films or gases, complete sintering or welding along the total surface of the particles is possible in hot-pressed compacts,

[5] W. D. Jones, *Principles of Powder Metallurgy*. Arnold, London, 1937, p. 4.

resulting in polycrystalline materials whose properties resemble metals obtained by common practices involving fusion.

SURFACE TENSION FORCES

The attractive forces manifest themselves in the tendency of the particles to reduce their surface. Since the decrease in surface diminishes the surface energy, a more stable condition is reached. To enable this process to take place, however, a certain atomic mobility is required. This, in turn, means that sintering is favored by increasing temperatures, and its rate increases rapidly above the temperature range at which recrystallization becomes evident.

That sintering forces can be aided materially by surface tension forces, has been emphasized by Balke.[6] Using gold as an example because it is not subject to oxidation, Desch[7] observed that if it is melted to form a liquid bead, the surface of the gold will appear perfectly smooth. When it freezes, however, the surface becomes wrinkled because of contraction during solidification. After the bead is subjected to a temperature several hundred degrees below the melting point of gold, the wrinkles disappear within a few hours, indicating that there is a powerful force operating along the surface. Furthermore, Desch showed that small triangular crystals of the same metal (formed by reduction of gold chloride by aldehydes) will become completely rounded when heated to a temperature considerably below the melting point. At 900°C. (1650°F.) the corners gradually round off and the triangles eventually assume the shape of perfect little spheres. The metal apparently becomes sufficiently plastic at the high temperature, so that forces operating along the surface can smooth the surface and stretch it out.

From these examples Balke has drawn a theoretical picture explaining the sintering forces in a pressed powder compact heated to temperatures below the melting point. When two particles are pressed together so that at certain points the surface of one is in contact with that of the other (intervening surface films having been ruptured or otherwise eliminated), there will be a deep crevice which is gradually filled up by the action of the surface tension forces. Balke compares this action with that of a "zipper." This action fans out (in a circle) from the point of contact, and as adjacent circles eventually meet, they may trap solid or gaseous impurities. Provided these impurities remain of minor nature, the zipper action continues throughout the entire mass and shrinkage occurs on a large scale, being more pronounced with greater surface energy (meaning

[6] C. C. Balke, *Iron Age, 147*, No. 16, 23 (1941).
[7] C. H. Desch, *J. Am. Chem. Soc., 123*, No. 1, 280 (1923); also *The Chemistry of Solids*, Oxford Univ. Press, London, 1934.

smaller particles). General experience tends to corroborate Balke's explanation that finer powders cause greater shrinkage. It is true that grain growth may also accompany shrinkage, but it is hard to believe that grain growth alone could account for this phenomenon, or that the vapor emanating from a metal (below its melting point) could fill in the crevices or pull two particles together. Surface tension forces, apparently constitute an important factor in causing the shrinkage and in aiding the sintering forces to produce a solid metal, while elevated temperatures simply allow the forces to operate within a reasonable time interval.

DIFFUSION AND PLASTICITY

If two particles form contact at sufficiently high temperatures, the atoms in the surface layer become so mobile that they enlarge the contact area, thereby being grouped into one of the two differently oriented lattices. Usually this lattice then grows at the expense of the other. The interchange of atoms, called diffusion (or self-diffusion in monometallic systems), is more pronounced as the sintering temperature becomes higher. In fact, the interchange of the atoms reaches a considerable magnitude at temperatures very near the melting point, so that a solid union of neighboring crystals and extensive grain growth can frequently be observed.

In the diffusion of bulk metals the activation energy should be lowered and the speed of diffusion increased by internal stress dislocations, atomic voids, and small grain size, while the activation energy should be raised and the speed of diffusion decreased by interfering phases, such as impurity films at the surfaces. Diffusion at temperatures below the recrystallization temperature of the metal would most likely be slow and would not involve the movement of atoms over long distances. According to Wretblad and Wulff[7a] the sintering process, which involves the transport of an appreciable amount of material, cannot be regarded solely as a diffusion process. The microscopic process of diffusion should rather be considered to be superimposed upon the macroscopic process of elastic deformation, with both facilitating the formation of new sinter bonds.

The surface energy acting in the interstices between contacting particle surface areas involves considerable stresses. If the temperature is high enough that these stresses exceed the elastic strength of the metal, plastic deformation should remain until the stresses are relieved, resulting in a flow of metal from the adjacent contact surface into a crevice or cavity which produces new sinter bonds, as indicated in the diagram (Fig. 183). At the same time, however, metal surrounding the cavity, but away

    [7a] P. E. Wretblad and J. Wulff, in J. Wulff, *Powder Metallurgy*. Am. Soc. Metals, Cleveland, 1942, p. 44.

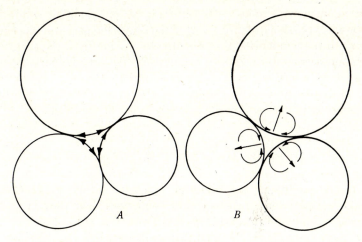

Fig. 183. Schematic representation of a two-dimensional pore (according to Wretblad and Wulff[7a]): *A*, arrows represent direction in which surface atoms would tend to diffuse at elevated temperatures, based on differences of chemical potential; *B*, arrows represent movement of material due to plastic deformation caused by surface stresses greater than the cohesive strength of the material at the specific temperature.

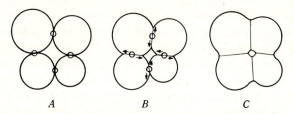

Fig. 184. Schematic representation of progressive spheroidizing and contraction of a pore through action of surface-tension forces (according to Rhines[7b]): *A*, compacted powder particles at room temperature—only contact points "0" exist; *B*, compacted particles at higher temperatures—contact points have widened laterally and corners of shrinking cavity have rounded; *C*, compacted particles at very high temperatures—contact areas have further widened and diminished cavity has completely spheroidized.

from the contact point, would move in the opposite direction (possibly away from the cavity) to relieve the stresses. Surface tension forces will produce rounded cavities of diminishing size, as indicated in Figure 184.

Sintering, accentuated by plastic deformation, should increase with time at any one temperature up to a certain saturation value. Raising the temperature, other factors remaining the same, should initiate further

[7b] F. N. Rhines, Communication on "Principles of Sintering," *Powder Metallurgy Colloquium*, New York University, May 24, 1946.

sintering until, with time, a constant value should again be attained. The increase of the bonded area with increasing temperature depends, of course, on the assumption that the stresses arising from the surface configuration decrease at a rate less rapid than the cohesive strength of the material. However, it appears improbable that complete bonding of the total available surface in a compact can be accomplished by this process at a temperature below that of incipient fusion of the metal.

EFFECTIVENESS OF THE SINTERING FORCES

Jones[8] discussed the theoretical aspects of the sintering process in great detail, paying particular attention to the forces that cause sintering and to the conditions that facilitate or obstruct their range of action. His interesting interpretations can be reproduced here only in part, and the reader is advised to consult his text for further study.

According to Jones, the conditions which give rise to the sintering of a metal powder compact are identical with those which condition the cold welding of massive materials. Considerable emphasis must be placed on a very rigid distinction between the sintering *forces* and factors that condition their *effectiveness*. The actual forces that effect sintering are the normal, effective cohesive forces which determine the theoretical strength of a metal. It has already been shown that these forces *decrease* with temperature. Hence, the tendency to sinter, if it were conditioned only by the magnitude of the sintering forces, would decrease with temperature. But in normal sintering practices, obstructions to sintering decrease much more rapidly with rising temperatures than the magnitude of the sintering forces, and thus adhesion and the cohesive strength of the aggregate grow. The principal factors obstructing the effectiveness of the sintering forces are given by Jones as (*1*) incomplete contact between surfaces, (*2*) reduced contact due to surface films, (*3*) reduction in sintering forces by chemical action of gases, and (*4*) insufficient plasticity of particles.

With regard to the incompleteness of contact, it has been shown in Chapter VIII that external pressure can correct this situation and that high compacting pressures materially increase the strength of the compacts. Similarly, it has been shown that compression of powders can also overcome the obstacles formed by surface films. During compaction, the particles are moved and distorted, favoring abrasion of the film. Furthermore, the chemical reactivity of the films increases with temperature, and the chance of dissipation by volatilization, segregation, or solution is therefore enhanced. The reduction of the sintering forces by the chemical action of gases will be considered in connection with the general effects of gases in powder compacts.

Lack of plasticity may generally be considered as obstructing the in-

[8] W. D. Jones, *Principles of Powder Metallurgy*. Arnold, London, 1937, p. 55.

crease of contact areas. Macaulay,[8a] for example, demonstrated that two freshly cleaved surfaces of mica seize only when one surface is flexible. The more plastic the metal, the larger are the areas in contact, and the more able are the sintering forces to pull the surrounding area into contact. The most important effect of temperature in aiding sintering, therefore, is only a matter of causing an increase in the plasticity of the material. The fact that oxides can be sintered densely only at temperatures very close to the melting point, while a much larger temperature range is available for metals, can be laid to the greater plasticity of metals in general, as compared with ceramic oxides, which become plastic only near the melting point.

Skaupy,[9] in discussing Jones' view, remarks that it is possible, in principle, to obtain strong and dense bodies by sintering powders which do not possess plastic deformability. This case exists if particle size distribution is such that after sufficient compaction all free surfaces come into effective range of the surface attraction forces. Since oscillation of the surface molecules increases with rising temperature, the sintering is aided by the approach of the vibrating molecules to the adjacent free surface. This consideration would be in agreement with Sauerwald's[10] statement that "within this critical temperature range, the forces of adhesion between metal surfaces increase considerably."

Sintering Phenomena

EFFECTS OF TEMPERATURE AND TIME

In the sintering process, time and temperature act in the same direction. The longer the time of heating the better the mechanical coherence of a compact for a given temperature; the higher the temperature the better coherence for a given period of heating. However, the effect of time generally becomes less marked at higher temperatures. As a matter of fact, at temperatures approaching the melting point of the metal, a maximum of the compact's cohesive strength may sometimes be reached in as short a time as one second.[11] In monometallic compacts, which are free of the disturbing influence of gases, the effectiveness of sintering is simply a matter of plastic flow. Jones[12] cites the example of two metals of different plasticity, namely, tin and bismuth. At given pressure and temperature, the softer metal (tin) will give a greater increase in density in a given time than the harder metal (bismuth).

It has previously been shown that the sintering forces as such are not markedly affected by temperature and that, strictly speaking, the expres-

[8a] J. M. Macaulay, J. Roy. Tech. Coll. Glasgow, 3, No. 3, 353 (1935).
[9] F. Skaupy, Metallkeramik. Verlag Chemie, Berlin, 1943, pp. 80, 97.
[10] F. Sauerwald, Metall u. Erz, 21, 117 (1924).
[11] W. Engelhardt, Z. Metallkunde, 34, 12 (1942).
[12] W. D. Jones, Principles of Powder Metallurgy. Arnold, London, 1937, pp. 4, 59.

sion "sintering temperature" has no fundamental significance. In practice it amounts to a temperature at which certain changes associated with sintering occur—as recorded by the particular method of testing or observation. It must be realized that the "sintering temperature" will vary with all the disturbing influences already mentioned, as well as with the sensitivity of the measuring device. It may be mentioned, however, that the greater part of the experimental work done in connection with the sintering of metal powders has involved determinations of a "sintering temperature." The sintering process is not bound to a definite temperature; the lower limit is usually governed by the fact that, for practical reasons, the time cannot be extended indefinitely. The upper temperature limit for the process is determined either by the complete (or substantial) fusion of the compact or by certain other changes such as evaporation, allotropic transformations, excessive gas solubility, etc.

The review of the various methods of investigating the sintering temperature, as given by Jones, refers, in most cases, not to a temperature useful for technical purposes, but to the temperatures at which sintering in its true sense becomes apparent. Of course, a distinction must then be made between the behafior at elefated temperatures of loose and compacted powders. In this connection the investigations of Schlecht, Schubardt, and Duftschmid,[13] Hedvall,[14] and Garre[15] are significant and should be studied by those particularly interested in this matter.

Of special interest is the determination of the beginning of sintering of a loose powder mass by the arrested stirrer method, as developed by Tammann and Mansuri.[16] The powder is stirred with a paddle and heated at the same time. The paddle is rotated by a friction drive; the stirrer stops at a certain temperature. The particular temperature measured depends on a large number of factors, e.g., stirring speed and force, rate of heating, grain size and shape of the powder, and volume of the powder. The results are quite remarkable, some of them being reproduced in Table 52. Comparison with the results obtained by Smith [17] by heating 150-mesh powders in hydrogen (carbon dioxide in the case of palladium) and determining the onset of sintering by prodding with a wire and observing change in color and appearance of luster (Table 53), shows that the latter come much closer to values that are useful in practical work. It will be noticed that the stirrer experiment indicated a maximum variation of only 29°C. (52°F.) for coarser particles between so widely differing metals as iron and tin! Even though there is an appreciably greater variation in the figures with finer powders, the significance of these data in the study of the

[13] L. Schlecht, W. Schubardt, and F. Duftschmid, Z. Elektrochem., 37, 485 (1931).
[14] J. A. Hedvall, Z. physik. Chem., 123, 33 (1926).
[15] B. Garre, Z. anorg. allgem. Chem., 161, 152 (1927).
[16] G Tammann and Q. A. Mansuri, Z. anorg. allgem. Chem., 126, 119 (1923).
[17] R. C. Smith, J. Am. Chem. Soc., 123, 2088 (1923).

sintering mechanism in powder compacts remains highly problematical. It cannot be denied, however, that these results permit certain conclusions as to the behavior of loose powders in such treatments as annealing or reduction.

TABLE 52

Temperature at Which Stirrer Stops in Heated Powder Mass[a]

(Tammann and Mansuri[16])

Metal	Melting point		Grain diameter			
			0.3–1.0 mm.		Less than 0.3 mm.	
	°C.	°F.	°C.	°F.	°C.	°F.
Silver............	960	1760	146–148	294.8–298.4	136–141	276.8–285.8
Antimony........	632	1170	126–130	258.8–266.0	248–252	478.4–485.6
Tin..............	232	450	134–138	273.2–280.4	—	—
Copper..........	1084	1983	144–148	291.2–298.4	296–302	564.8–575.6
Iron.............	1530	2786	144–155	291.2–311.0	146–148	294.8–298.4
Zinc.............	419	786	141–150	285.8–302.0	203–215	397.4–419.0
Lead.............	327	620	141–138	285.8–280.4	138–133	280.4–271.4
Cadmium........	321	610	149–152	300.2–305.6	—	—
Cobalt..........	1480	2696	126–134	158.8–273.2	—	—
Aluminum.......	657	1214	—	—	196–202	384.8–395.6

[a] Metal powder heated in air at 10°C. per minute.

In a modified stirring test (two strong magnets rotate outside a closed tube containing the powder under different atmospheric conditions and act on an armature mounted on a vertical stirring shaft), Durst[17a] found that sticking requires considerably higher temperatures in vacuum than in air, with hydrogen taking an intermediate position. For silver (44–74μ), the temperatures were 185°C. (365°F.) in vacuum, 146°C. (295°F.) in air; for carbonyl iron, 210°C. (410°F.) in vacuum, 140°C. (284°F.) in air; for copper (44–74μ), 360–372°C. (680–702°F.) in vacuum, 310–316°C. (590–601°F.) in hydrogen, and 210–212°C. (410–414°F.) in air. If any of these powders were heated to below sticking in vacuum but above sticking in air, and air were let into the system, the stirrer stopped immediately. Also, powder sticking in air loosened up in vacuum, and the same sample would undergo several of these reversible changes. Durst concludes from these observations that the adherence of the powder in the stirring test might be caused by films rather than by contact between metallic surfaces.

Although it has been frequently stated in connection with the optimum sintering temperature range for metal powders that $2/3$ to $4/5$ of the absolute melting temperature ($a = 0.67$ to 0.8) is a good approximation for the maximum temperature, while the minimum temperature corresponds to that of apparent recrystallization, it must of course be realized

[17a] G. Durst, *Metal Progress, 51,* No. 1, 97 (1947).

that such a statement is a generalization, and that a closer definition for each individual material is necessary.

The density of a metal powder compact, and with it properties such as tensile strength and elongation, tend to increase with increasing sintering temperature and time. However, the effect of these sintering condi-

TABLE 53

Sintering Temperatures of Various Metals[a] (Smith[17])

Metal	Temperature °C.	Temperature °F.	Metal	Temperature °C.	Temperature °F.
Platinum, black precipitate..500		930	Iron, reduced.250, 750, 850		480, 1380, 1560
in vacuo 500		930	in vacuo .. 750, 850		1380, 1560
Palladium, black precipitate.600		1110	Iron precipitate 750		1380
Silver precipitate............180		355	(not heated above 800°C., 1470°F.)		
Gold precipitate.............250		480	Lead precipitate· 200		390
Aluminum filings............200		390	Nickel filings... 650		1200
Cobalt filings 700		1290	precipitate .. 700		1290
Cobalt precipitate...........200		390	Molybdenum		
Copper reduced.............500		930	filings 800		1470
Copper precipitate 250		480			

[a] These were obtained by heating loosely piled 150-mesh powders in hydrogen (palladium in carbon dioxide). The onset of sintering was detected by change in color, appearance of luster, and by prodding with a fine wire.

tions on the physical properties is influenced considerably by the compacting pressure applied previously. Thus, powders only loosely heaped or slightly compressed usually show a rapid increase in density and me-

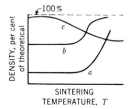

Fig. 185. Schematic representation of the density of loose and compacted metal powders as a function of the sintering temperature (according to Kieffer and Hotop[18]): a, noncompacted powder; b, powder compacted at intermediate pressures (about 30 tsi); and c, powder compacted at very high pressures (about 200 tsi).

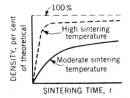

Fig. 186. Schematic representation of the density of compacts as a function of the sintering time for different temperatures (according to Kieffer and Hotop[18]).

chanical properties, while powders pressed at very high pressures frequently show a *decrease* in density with increasing sintering temperature, expansion of the compact being caused by stress releases and gas evolution. In the diagrams of Figure 185 the density is shown schematically as a function of temperature and in Figure 186 as a function of time for different degrees of powder compaction.[18] The effects of a rising sintering

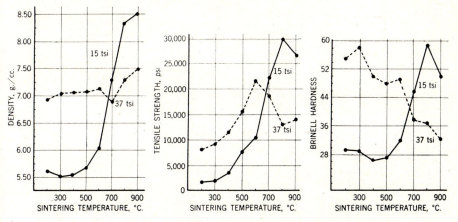

Fig. 187. Effect of sintering temperature on density, tensile strength and hardness of copper powder compacts (according to Sauerwald and Kubik[20]).

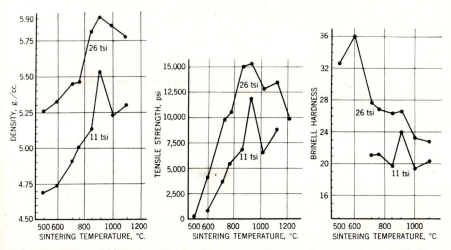

Fig. 188. Effect of sintering temperature on density, tensile strength, and hardness of iron powder compacts (according to Sauerwald and Kubik[20]).

[18] R. Kieffer and W. Hotop, *Pulvermetallurgie und Sinterwerkstoffe.* Springer, Berlin, 1943, pp. 95, 97.

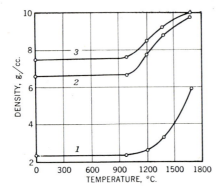

Fig. 189. Effect of sintering temperature on the density of noncompacted and compacted molybdenum powder (according to Grube and Schlecht[21]):
1, loosely filled
2, pressed at 24 tsi
3, pressed at 43 tsi

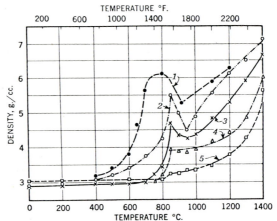

Fig. 190. Density changes of loose powders of various origins and particle-size distributions when annealed 24 hours in hydrogen at temperatures up to 1400°C. (2550°F.) (according to Libsch, Volterra, and Wulff[23]): 1, carbonyl iron powder (1 to 10 μ); 2, electrolytic iron powder (1 to 30 μ strained); 3, electrolytic iron powder (1 to 30 μ preannealed at 730°C. (1350°F.) for 2 hr.); 4, electrolytic iron powder (preannealed at 730°C. (1350°F.) for 2 hr.); and 5, reduced magnetite (preannealed at 730°C. (1350°F.) for 2 hr.). Size distribution of the electrolytic iron powder (4) and reduced magnetite (5) is 66%, 100 to 200 mesh; 17%, 200 to 325 mesh; and 17%, —325 mesh.

temperature on copper and iron compacts have been studied by Sauerwald and his coworkers,[19,20] and some of the results are reproduced in the diagrams of Figures 187 and 188, showing density, tensile strength, and hardness as a function of the sintering temperature. Similar data for nickel and molybdenum have been reported by Grube and Schlecht,[21] the curves for

[19] F. Sauerwald and E. Jaenichen, Z. Elektrochem., 30, 175 (1924).
[20] F. Sauerwald and S. Kubik, Z. Elektrochem., 38, 33 (1932).
[21] G. Grube and H. Schlecht, Z. Elektrochem., 44, 367 (1938).

molybdenum being reproduced in Figure 189. In the case of iron, density data obtained by Eilender and Schwalbe,[22] by Libsch, Volterra, and Wulff,[23] and by Heath and Stetkewicz[24] are in agreement in showing a marked increase of the density from about 400° to 850°C. (750° to 1560°F.) followed by minima in the range between 850° and 1000°C.

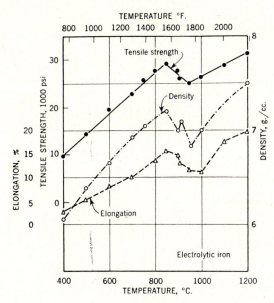

Fig. 191. Physical properties of electrolytic iron compacts as influenced by sintering temperature (according to Libsch, Volterra, and Wulff[23]).

(1560° and 1830°F.), and by a further increase at higher temperatures. The minima are related to the $a \rightarrow \gamma$ transformation of iron. Figure 190 shows these effects for loosely heaped powders of various makes, and Figure 191 for a compacted electrolytic iron.[23] In Figure 192 a group of curves is shown for different compacting pressure.[24] The decrease in density of highly compressed compacts with rising temperatures has also been found by Trzebiatowski,[25] who experimented with very soft copper and gold powders. As Figure 193 illustrates, the density of both metals decreases markedly above 300°C. (570°F.) if the compacts were previously pressed at 200 tsi, while the same powders pressed at 40 tsi exhibit the increase in density usually found.

[22] W. Eilender and R. Schwalbe, *Arch. Eisenhüttenw.*, *13*, 267 (1939).
[23] J. Libsch, R. Volterra, and J. Wulff, in J. Wulff, *Powder Metallurgy*. Am. Soc. Metals, Cleveland, 1942, p. 379.
[24] C. O. Heath. Jr., and J. D. Stetkewicz, *Metal Progress*, *48*, No. 1, 73 (1945).
[25] W. Trzebiatowski, *Z. physik. Chem.*, *B24*, 75, 87 (1934).

512

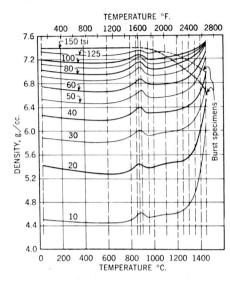

Fig. 192. Relation of sintering temperature to density of iron powder cylinders compacted at various pressures (according to Heath and Stetkewicz[24]). All samples sintered in hydrogen for 1 hour. Broken vertical lines designate temperatures of tests.

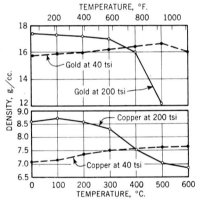

Fig. 193. Density of gold and copper compacts as influenced by compacting pressure and subsequent heat treatment (according to Trzebiatowski[25]).

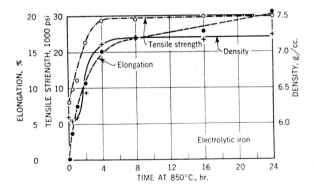

Fig. 194. Physical properties of electrolytic iron compacts as influenced by sintering time (according to Libsch, Volterra, and Wulff[23])

For a temperature of 850°C. (1560°F.), the effect of sintering time on density and tensile values of an iron powder pressed at 33 tsi is shown in Figure 194. After half an hour, the density rises and approaches a saturation value after four hours; the initial loss in density may be due primarily to expansion of gas. At higher temperatures, Libsch et al. report that the saturation of the density curve occurs in less than four hours.

SHRINKAGE

The increase in density of a loose or compacted powder with rising temperatures or prolonged heating times is brought about by a reduction of the pore volume of the mass and with it, the over-all volume. If this phenomenon, generally referred to as shrinkage, occurs at the start of sintering in its larger sense, this is truly accidental. However, while sintering continues at higher temperatures, there is a steady decrease in porosity

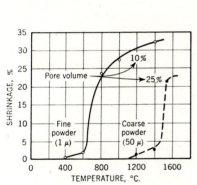

Fig. 195. Effect of sintering temperature on shrinkage of two types of iron powders compacted at 1.6 tsi (according to Dawihl[26]).

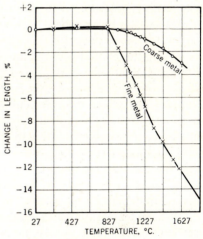

Fig. 196. Change in length of pressed tungsten bars sintered at various temperatures (according to Smithells, Pitkin, and Avery[29]).

and since this is impossible without simultaneous material movement into the voids, it follows that the continuation of sintering must involve a corresponding decrease in over-all volume (usually commencing at the surface of the compact and growing toward its center). The effect of sintering temperature on shrinkage is shown for two types of iron powders in Figure 195, according to Dawihl.[26]   The powders had an average

[26] W. Dawihl, Stahl u. Eisen, 61, 909 (1941).

particle size of 1 μ and of 50 μ, respectively, and were compressed slightly at about 1.6 tsi. The minimum porosity values reached were 10% and 25%, respectively.

The possible reason for the shrinkage has already been mentioned. When two particles touch and start to sinter at a particular point, the two surfaces are drawn closer together, and the sintered area gradually increases, spreading outward from the first point of contact. Once two particles have sintered at a particular point, one may regard the shrinkage forces simply as surface tension forces that operate in the direction of minimizing the surface areas and which have been shown to be effective in conditioning the shape of metal particles.[27]   The spheroidizing and contraction effects that the surface tension forces produce on an interstice formed by four adjacent spherical powder particles have been schematically presented at the beginning of this chapter (Fig. 184). Surface tension forces, however, are more likely to cause a rapid variation in external shape with angular than with spherical particles.[28]   Similarly, very small particles would be influenced to a greater degree by surface tension forces than would larger particles.

A definite relation between fine and coarse powders is revealed during shrinkage. Smithells, Pitkin, and Avery[29] experimented with tungsten powders of varying fineness, the coarser powder having a maximum size of 3.5 μ, the finer about 0.6 μ. A number of bars were heated in pure dry hydrogen at different temperatures; the results of subsequent volumetric measurements are shown in Figure 196. Both the rate of shrinkage and the total shrinkage were found greater for the fine powder than for the coarse. This increasingly marked effect on shrinkage with increasing dispersion of the powder can generally be observed and is true for most metals, including iron, nickel, copper, etc.

Next to temperature and powder dispersion, shrinkage is most affected by the density of the powder mass, *i.e.*, whether it is loosely or heavily compacted. The higher the initial density, the smaller is the rate of densification during the sintering treatment, and the smaller also are both the rate of shrinkage and the actual shrinkage. On the other hand, powders subjected to so-called "thermal sintering" (without the aid of external pressure) naturally exhibit the effect of shrinkage forces to the greatest extent. A good example of the ability of the shrinkage forces to cause compaction has been given in the case of iron. In studies with finely divided carbonyl iron which was loosely filled in porous ceramic tubes and heated

[27] C. H. Desch, *J. Am. Chem. Soc.*, *123*, No. 1, 280 (1923); also *The Chemistry of Solids*. Oxford Univ. Press, London, 1934.

[28] C. C. Balke, *Iron Age, 147*, No. 16, 23 (1941).

[29] C. J. Smithells, W. R., Pitkin, and J. W. Avery, *J. Inst. Met., 38*, 85 (1927).

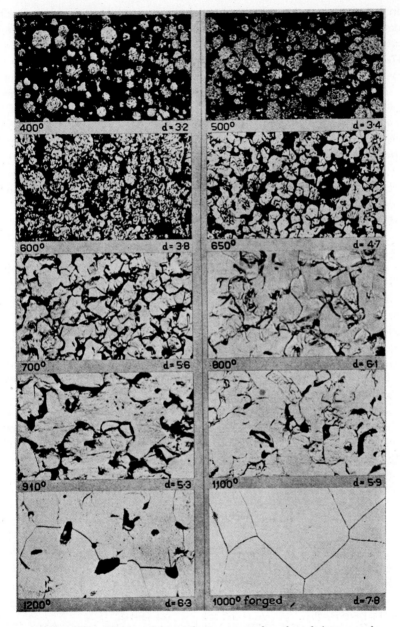

Fig. 197. Microsections and structure of carbonyl iron powder sintered at various temperatures without application of pressure ($\times 500$) (according to Schlecht, Schubardt, and Duftschmid[20]). Temperatures are given in degrees centigrade; d denotes density.

in hydrogen for 24 hours, Schlecht, Schubardt, and Duftschmid[30] found that shrinkage began at temperatures around the recrystallization temperature of carbonyl iron powder, and increased rapidly to about 800°C. (1470°F.), where the porosity was relatively small. Over the $a \rightarrow \gamma$ transformation point, the porosity increased, possibly because of the contraction which accompanies phase change. With increasing temperature, porosity once again decreased, and after sintering at 1000°C. (1830°F.), the mass could be hot-forged into a solid ingot. The microsections of the sintered rods are reproduced in Figure 197. The density of the sintered material varied in accordance with the figures shown under the photomicrographs.

In a study of a more recent date by Deslisle,[31] in which shrinkage data were collected on a variety of loosely packed iron powders that were thermal-sintered, it was confirmed that shrinkage is more pronounced with finer particle size and higher temperature, with carbonyl iron powder giving the highest figures. Results similar to those obtained with iron have also been obtained using nickel carbonyl powder. Hamprecht and Schlecht[32] heated powders varying in size from 0.5 to 5 μ in a refractory vessel in a reducing atmosphere to 1200°C. (2190°F.). Under these conditions shrinkage occurred to the extent of 60 to 70% of the initial volume. A powder of apparent density of from 2 to 3 was sintered into a porous mass having a density of from 6.5 to 7.5 g./cc., which was suitable for hot-forging into an ingot of a density of 8.85.

Other factors influencing shrinkage are chemical reactions and crystallization. The effect of chemical reactions is demonstrated by the increased shrinkage of compacts made from metals which form oxides reducible by hydrogen.[33] It is known that the presence of larger amounts of water vapor in the reducing atmosphere increases shrinkage according to the reaction:

$$\text{Metal} + H_2O \rightleftharpoons \text{Metal oxide} + H_2$$

This reaction can take place in both directions simultaneously on account of minor external disturbances of the equilibrium, where the more active superfine particles grow (via the oxide) into larger crystallites requiring less volume. Similar effects can also be found with carbon dioxide according to the reaction:

[30] L. Schlecht, W. Schubardt, and F. Duftschmid, Z. Elektrochem., 37, 485 (1931).
[31] L. Deslisle, Trans. Electrochem. Soc., 85, 123 (1944).
[32] G. Hamprecht and L. Schlecht, Metallwirtschaft, 12, No. 1, 281 (1933).
[33] R. Kieffer and W. Hotop, Pulvermetallurgie und Sinterwerkstoffe. Springer, Berlin, 1943, p. 99 ff.

$$\text{Metal} + CO_2 \rightleftharpoons \text{Metal oxide} + CO$$

Finally, crystallization phenomena within and between the particles may affect shrinkage, especially at higher temperatures. Polycrystalline particles may thereby be transformed into single crystal particles which, in turn, may grow with neighboring particles into still larger grains requiring less space. In this process small interstices between the original particles may be completely isolated and may finally be closed entirely by the surrounding metal. In Figure 198 a photomicrograph of a copper compact

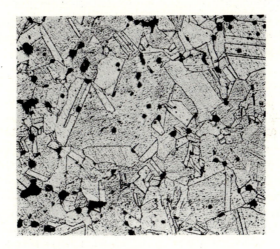

Fig. 198. Microstructure of electrolytic copper compact sintered in hydrogen for 16 hours at 900°C. (1650°F.) after compaction at 45 tsi (×200).

after sintering for 16 hours at 900°C. (1650°F.) shows a number of small pores completely surrounded by grown crystals. The total reduction in volume of the compact after this treatment amounted to 8%.

In concluding the discussion on shrinkage it may be emphasized that excessive shrinkage of the type just mentioned is not desirable. Since they depend on so many factors and are therefore difficult to control, shrinkage effects generally constitute a serious problem in the manufacture of articles of predetermined shape and size. Although the above-mentioned general rules apply in most cases, each metal, and sometimes each powder shipment, will require specialized shrinkage control, and the fact that problems of this nature have been successfully overcome in certain fields does not permit the conclusion that this can be achieved in every case.

SWELLING CAUSED BY GAS EVOLUTION

Our analysis of the various sintering phenomena has so far concerned itself only with the ideal case where only one phase is present in the compact, and where the particles have a surface free of foreign films. This, however, is never the case in practice; the attracting forces act not only on neighboring metal particles, but also on the gas molecules that are sorbed at the metal surfaces. As a matter of fact, even in a vacuum, adsorbed and absorbed gases and films of oxides must be taken into account. Therefore in its true sense, a homogeneous monometallic system does not exist, except in high-vacuum in the metallurgy of sintering. Influences of gases are discussed in this part of the chapter in order to facilitate the understanding of the various sintering phenomena and their grouping according to practical importance. We thus leave for study under the heading of heterogeneous systems only combinations of several solid phases, solid and liquid phases, and solid and gaseous phases that do not escape during the heat treatment.

Jones[34] devotes a large part of his treatise to the influence of gases and vapors on the properties of compacts and on their behavior during pressing and sintering. He distinguishes between the following effects: (*1*) A volume effect—by which are understood the influences on porosity and density, and the accompanying properties. (*2*) A physical effect—by which are understood influences on the adhesion forces between the metal particles. (*3*) A chemical effect—by which are meant mostly chemical reactions between the metal surfaces and the gaseous or oxide films. The complexity of this topic permits only a brief review here. The reader is advised to consult Jones,[34] the papers by A. Sieverts, a bibliography of which is given by McBain,[35] and also by Skaupy,[36] and the important contributions by Hüttig[37] and collaborators (see also Chapter IV). Hüttig's theory of partial processes of sintering connects the sorptive capacity of powders for dyes with six distinct temperature stages, as follows:

(*1*) Below $a = 0.25$, the adsorptive capacity drops in accordance with a "covering" effect as a consequence of adhesive forces which intensify with rising temperature.

[34] W. D. Jones, *Principles of Powder Metallurgy*. Arnold, London, 1937, p. 38 ff.
[35] J. W. McBain, *The Sorption of Gases and Vapours by Solids*. Routledge, London, 1932.
[36] F. Skaupy, *Metallkeramik*. Verlag Chemie, Berlin, 1943, p. 94.
[37] G. F. Hüttig, *Kolloid-Z.*, *97*, 281 (1941); *98*, 6, 263 (1942); *99*, 262 (1942),

(*2*) With values of a from 0.23 to 0.35, the adsorptive capacity rises again in accordance with an activation due to molecular rearrangements in the surfaces (caused by surface diffusion), and in accordance with an evolution of gases adsorbed in the surfaces.

(*3*) With values of a from 0.33 to 0.45, a pronounced decrease of the adsorptive capacity takes place because of a stabilization of the surface after completion of the molecular rearrangements in the surface regions.

(*4*) With values of a from 0.37 to 0.53, no appreciable change in adsorptive capacity occurs. Activations are caused by molecular rearrangements in the crystal interior (lattice diffusion effects) and by expulsion of gaseous constituents from the lattice, but the effects of these processes on the surface properties as determinable by the adsorption of dyes are insignificant.

(*5*) With values of a from 0.48 to 0.8, the adsorptive capacity falls off in accordance with deactivation caused by lattice diffusion and a decrease of surface due to "collective" recrystallization.

(*6*) Above $a = 0.8$, no record was made of the effect on the adsorptive capacity by presumptive new activations preparatory to fusion.

In good agreement with these tests were additional experiments made by Hüttig and Bludau on the degassing of iron powders of various sources and treatments.[38] The gases contained in eight different samples were found to vary in quantity with the method of preparation of the powder and also with the previous history of the material. In all cases, however, maxima in the rate of gas evolution were found at approximately 200°C. (390°F.) $(a = 0.26)$, 400°C. (750°F.) $(a = 0.37)$, and 710°C. (1310°F.) $(a = 0.55)$. These maxima are explained (*1*) by a desorption at 200°C. (390°F.) of the gases held on the surface, (*2*) by degassing due to a loosening of the lattice which permits diffusion to take place at 400°C. (750°F.), and (*3*) by a more thorough opening of the lattice due to the closeness of the a transformation at 710°C. (1310°F.).

Experiments made by Hüttig and Arnestad[39] on the rusting of fritted iron powders in relation to previous heat treatments are also of interest. Three samples of powdered iron, electrolytic iron, Armco iron, and wrought iron were subjected to identical heat treatments and were then compared as to oxidizability and speed of reaction in ferric chloride solution. After preheating in hydrogen, electrolytic iron powder showed a minimum oxidizability at 200°C. (390°F.) $(a = 0.26)$, pulverized Armco iron at 400°C. (750°F.) $(a = 0.37)$, and pulverized wrought iron at 780°C. (1435°F.) $(a = 0.58)$. These same samples showed maximum oxidizability at 355°C. (670°F.) $(a = 0.35)$, 530°C. (985°F.) $(a = 0.44)$, and 850°C. (1560°F.) $(a = 0.62)$, respectively. Similar minima and maxima were noted in the reactions of the same powders with ferric chloride.

[38] G. F. Hüttig and H. H. Bludau, *Z. anorg. allgem. Chem.*, *250*, 36 (1942).
[39] G. F. Hüttig and K. Arnestad, *Z. anorg. allgem. Chem.*, *250*, 1 (1942).

Probably the only powders entirely free of gases are those prepared and comminuted in vacuum. All others may contain gases or develop them durinɡ heating, either by simple evolution or by liberation through chemical reactions. The surface of the particles alone can adsorb considerable quantities of gases, the initial covering by gases usually occurring extremely rapidly. Among the gases absorbed during the sintering process, hydrogen is particularly important, since it is soluble in most metals but generally does not disturb sintering. On the other hand, the persistence of a gas film, and particularly the resulting reactions between the gas and the metal or its oxide—causing compounds to form—constitute the major problem in the sintering process. Here, however, it must be remembered that such reactions may completely remove interfering films so that the attractive forces between expansive free surfaces can operate.

Trzebiatowski's[39a] work on highly compressed copper and gold powders (see Fig. 193) showed that an increase in the sintering temperature may lead to a *decrease* in density instead of an increase. This decrease may be attributed to the removal of adsorbed and absorbed gases (especially water vapor) which, during excessive evolution, may inhibit the formation of sinter bonds, or even disrupt already formed linkages—especially in more densified surface regions. If lower pressure is used and the density of the compact remains low, the gases can escape from the compact through interconnected pores before the commencement of shrinkage, and no destruction of the sinter bonds occurs.

During sintering, gases develop from: (1) gases dissolved in the metal particles, (2) absorbed gas or vapor films, (3) air or other gases trapped between the particles during pressing, (4) chemical reactions during the heating of the compact, and (5) purposely added gas or vapor-evolving substances (e.g., hydrides or volatilizers). Individually or jointly, these different sources can be responsible for the large quantities of gas that can be observed (under certain conditions) during the annealing of powders as well as during the sintering of compacts. In general, gases originally dissolved in the solid particles or gases produced by a chemical reaction (e.g., reduction of oxides) may either be liberated during heat treatment, thus acting like trapped gases, or may remain in the powder, acting as a foreign phase. Adsorbed gases as well as solid surface films retard sintering in both a direct and an indirect manner. They affect surface diffusion and inhibit plastic deformation, which, as previously shown, arises from surface energy forces.

As shown by Ruer and Kuschmann,[40] the amount of gas absorbed

[39a] W. Trzebiatowski, *Z. physik. Chem.*, B24, 75, 87 (1934).
[40] R. Ruer and J. Kuschmann, *Z. anorg. allgem. Chem.*, 154, 69 (1926); 166, 257 (1927); 173, 233 (1928).

by powders even at room temperature is quite large. A copper powder, obtained by the reduction of cupric oxide at 750°C. (1380°F.) with hydrogen, and subsequently subjected to a vacuum treatment at 440°C. (825°F.), gained as much as 0.01% by weight after being exposed to the open air for 70 hours (see Table 9, Chapter IV). Similar results were obtained with reduced iron, which showed a weight gain of 0.023% after 20 hours.

Durau and Franssen,[41] when milling copper powder in high vacuum at room temperatures, observed a small immediate adsorption of nitrogen, carbon monoxide, ethane, and ethylene, and a larger adsorption of oxygen and carbon dioxide. Adsorbed water, of course, can contribute directly to an increase in volume upon heating apart from any chemical reaction with the metal. Although the amount of adsorbed water is small in most cases, the values are appreciable in the case of gold. According to Trzebiatowski, it is difficult to remove completely the rather tenacious films from the powder.

Amount of air trapped during compression of a powder depends largely on the properties of the powder, and also on the type of die construction used (as discussed in Chapter X).

Gases are frequently formed as a result of a chemical reaction within the compact during the sinter treatment. For example, if copper and iron compacts containing a certain amount of oxides are sintered in hydrogen, water vapor is formed; similarly, carbon dioxide is formed as a reaction product between carbon and oxygen in carbonyl metal powders, usually at a temperature below 700°C. (1290°F).[42] If these gases can escape before marked sintering and solidification occur, they are harmless, and may, in certain cases, even be helpful in orienting the shape and pattern of pores (e.g., interconnected pores for the purpose of oil impregnation). If, however, the gases begin to escape only at higher temperatures, the gas pressures are usually quite large, frequently causing a marked increase in size and a greater number of pores. The compacts then grow considerably and may display other undesirable marks such as surface blisters, fissures, or even laminary cracks. Obviously, the higher the compacting pressure of a powder, the more strongly are the particles united during pressing, and the stronger do they become welded during sintering. Thus occluded gases have little chance to escape, especially if the pores of the compact's interior are no longer connected with the surface by fine channels. A typical case is the formation of blisters and the loosening of the structure during sintering in hydrogen of oxygen-containing com-

[41] F. Durau and H. Franssen, Z. Physik, 89, 757 (1934).
[42] F. Duftschmid, L. Schlecht, and W. Schubardt, Stahl u. Eisen, 52, No. 2, 845 (1932).

pacts which are highly compressed. Here, the relatively small hydrogen atoms diffuse easily into the interior of the compact, whereas the larger molecules of water vapor formed by the reduction cannot pass to the outside through the lattice interspaces, but only through pores, thus leading to a swelling of the sintered body.

Metallographic Phenomena

POROSITY IN SINTERED COMPACTS

The most apparent difference between the microstructure of a metal made from powders and one obtained by the fusion method is the presence of pores in the sintered structure. This porosity may constitute a marked obstacle in the development of satisfactory physical and chemical properties. Variation in the forming pressure affects the density of the powder packing. The resulting porosity, which may be designated as "primary," is largely due to incomplete compaction. Figure 199A shows this by means of an electrolytic copper powder compacted at 45 tsi. This kind of porosity can be controlled to a certain extent by changing basic conditions such as type and particle size of powder, compacting pressure, or shape of the compact. Pores present within the particles would also fall within this category. During sintering, excessive shrinkage and grain growth may cause reduction in size or change in shape (spheroidizing) of these pores which, due to their continuity and originally angular shape, often have a detrimental effect on most properties of the final compact, (e.g., density, strength, hardness, toughness, ductility, and electrical conductivity). Grain growth and diffusion during heat treatment may also serve to increase separation between individual pores if certain prerequisites are fulfilled.[43] Some of these requirements are (1) a powder whose size distribution, particle shape, and purity are favorable for close packing (2) a compacting pressure sufficiently high to cause close packing of the particles but allowing for the escape of gases during heating; (3) a sufficiently high temperature and sufficient time to permit marked grain growth to take place; and (4) an atmosphere that does not create additional pores by chemical reactions with metal or with impurities.

In the case of chemical reactivity of the atmosphere during sintering, or when sintering in vacuum, a new type of porosity may be created by the spontaneous evolution of the gases. This porosity, which may be

[43] C. G. Goetzel, *The Influence of Processing Methods on the Structure and Properties of Compressed and Heat Treated Copper Powders, Dissertation,* Columbia University, New York, 1939, p. 45,

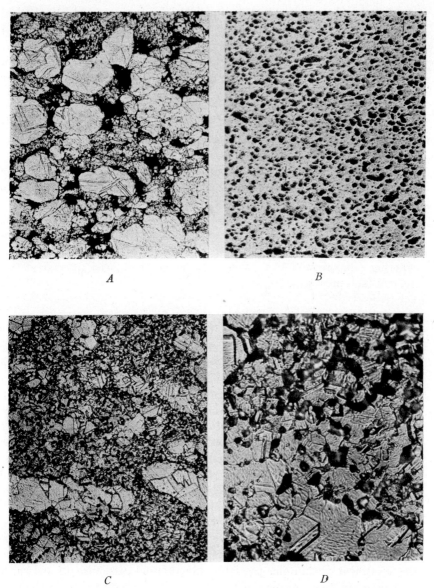

Fig. 199. Porosity in compacted and vacuum-sintered electrolytic copper: *A*, after compaction at 45 tsi (×150); *B*, after vacuum-sintering for 16 hours at 750°C. (1380°F.) (as polished, ×150); *C*, after vacuum-sintering for 16 hours at 750°C. (1380°F.) (etched, ×150); and *D*, after vacuum-sintering for 16 hours at 750°C. (1380°F.) (etched, ×575).

designated as "secondary," is distinguished by the pronounced spherical shape of the pores, as shown in the photomicrographs (Fig. 199B–D) of copper compacts sintered in vacuum.

As already stated, fine powders are able to adsorb larger amounts of gases and trap more air during pressing than coarser grades. If these adsorbed and occluded gases are emitted during heating at atmospheric pressure, they can escape gradually, especially if heating occurs slowly and sufficient time is allowed for complete gas evolution. After this degassing process the porosity is reduced by diffusion and shrinkage. If these precautions are not taken and, particularly, if the sinter-treatment is carried out in vacuum, a large quantity of gases is evolved immediately in an explosivelike fashion at the beginning of the treatment, whereby a myriad of minute, spherical pores within the particles (Fig. 199B), or even within the individual crystallites (Fig. 199D) are formed. Such "secondary" pores are apparently not closed easily or reduced by grain growth or diffusion and, as a result, the density is decreased with rising sintering temperature, just as if the compact had been pressed at extremely high pressures.

RECRYSTALLIZATION AND GRAIN GROWTH

As mentioned before, sintering may increase considerably above the recrystallization temperature. This crystallization usually precedes grain growth; the term is generally accepted as implying a new arrangement of crystals, starting from newly formed nuclei and absorbing the old crystal arrangement. This change in the arrangement of the crystallites takes place within the individual particles, with recrystallization usually commencing at the highly stressed surfaces where interparticle contacts have been formed during pressing. Under certain conditions, recrystallization in powder compacts leads to a larger grain size than the initial grain size within the powder particles—constituting actual grain growth. This is clearly visible in the photomicrographs of Figures 199A and 200A, where an electrolytic copper powder is shown after compression at 45 tsi and after heating to 600°C. (1110°F.) for 16 hours. While the outside contours of the individual particles remained practically unchanged by the treatment within the particles, the original structure of outward-spreading grains—so typical for electrodeposited powders—was changed by recrystallization into an equiaxed structure of entirely new appearance. Grain growth, taking place within the particles before it occurs across the particle boundaries, is also shown in the photomicrograph of Figure 200B for a compact of the same copper powder as in Figure 199A, but sintered for half an hour at 750°C. (1380°F.). Wretblad and Wulff,[44] as well as Jones,[45] hold that such internal grain growth is probably of little practical importance in sintering, but other investigators, particularly Trzebiatow-ski[46] and Bal'shin[47] hold a contrary view.

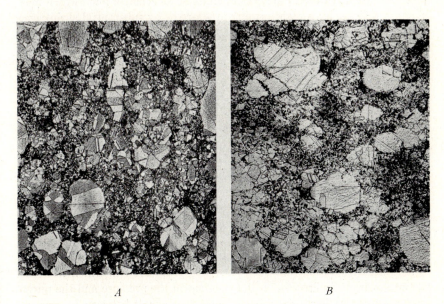

A   *B*

Fig. 200. Microstructure of hydrogen-sintered electrolytic copper after compaction at 45 tsi: *A*, sintered for 16 hours at 600°C. (1110°F.) (×120); and *B*, sintered for ¹/₂ hour at 750°C. (1380°F.) (×120).

So far, we have considered only polycrystalline particles—*i.e.*, each particle consisting of an aggregate of crystallites—which represent the most frequent case in powder metallurgy practice. If a powder particle consists only of a single crystal—which often occurs in the case of fine powders—the crystal cannot grow without a supply of material from the outside. Single-crystal particles can grow only at the expense of each other. According to Jones,[45] it is also possible for 2 particles of identical orientation to sinter together with a disappearance of the boundary. Even if only touching at one point, they could nevertheless be regarded as one

[44] P. E. Wretblad and J. Wulff, in J. Wulff, *Powder Metallurgy*. Am. Soc. Metals, Cleveland, 1942, p. 51 ff.
[45] W. D. Jones, *Principles of Powder Metallurgy*. Arnold, London, 1937, p. 79.
[46] W. Trzebiatowski, *Z. physik. Chem.*, *A169*, 91 (1934).
[47] M. Yu Bal'shin, *Vestnik Metalloprom.*, *16*, No. 17, 87 (1936).

particle. In general, however, grain growth, particularly in the case of single-crystal particles, can occur only after the randomly oriented particles touch each other at a number of contact points. This means that grain growth can commence only after sintering has proceeded to a considerable extent. In fact, the porosity must be sufficiently low (which may be effected either by a high forming pressure or by substantial shrinkage) to permit the ready coalescence of the structure of one particle into another. Jones[45] holds that one may consider the phenomenon of grain growth as the increase in particle size by interassimilation, becoming thereby, like recrystallization, an "extra-sintering" phenomenon. If the particle boundaries are now weaker than the grains themselves, which is usually true for the original boundaries in powder compacts, recrystallization, particularly if associated with moderate grain growth, results in greater strength and ductility of the material than is possible in the unrecrystallized compact.

During the sintering of monometallic powders, however, especially if of high purity, there is often a tendency for extremely large grains to form. According to Wretblad and Wulff[44] such large grain size is no longer of advantage, for the strength of the compact then decreases, since grain boundaries are generally stronger than the grains themselves at any temperature below recrystallization. Sauerwald originally asserted in several of his publications[48-51] that such excessive grain growth, which he terms "spontaneous grain growth," is independent of the pressure and the purity of the powder. It has always been found to occur at between $2/3$ and $3/4$ of the absolute temperature of the melting point. This remarkable statement is explained by Sauerwald on the basis of Tammann's theory of recrystallization[52] founded on the presence of nonmetallic impurity films between the individual crystallites which must first be destroyed by cold-working before grain growth occurs. According to Sauerwald, these foreign films are missing in "synthetic metal bodies," and grain growth can occur even without preceding cold deformation. Although based upon a number of thorough investigations, this concept is not well established and, as a matter of fact, has been severely criticized by other investigators, notably Smithells, Pitkin, and Avery,[53] Trzebiatowski,[54] Dawihl,[55] and Bal'shin,[47] who maintained that recrystallization within each particle has

[48] F. Sauerwald and S. Kubik, Z. Elektrochem., 38, 33 (1932).
[49] F. Sauerwald, Lehrbuch der Metallkunde. Springer, Berlin, 1929, p. 21.
[50] F. Sauerwald, Z. anorg. allgem. Chem., 122, 277 (1922).
[51] F. Sauerwald, Z. Metallkunde, 16, 41 (1924).
[52] G. Tammann, Lehrbuch der Metallographie. 3rd ed., Voss, Leipzig, 1923, p. 98.
[53] C. J. Smithells, W. R. Pitkin, and J. W. Avery, J. Inst. Metals, 38, 85 (1927).
[54] W. Trzebiatowski, Z. physik. Chem., B24, 75, 87 (1934).
[55] W. Dawihl, Stahl u. Eisen, 61, 909 (1941).

not received sufficient consideration. The weight of the evidence presented by these authors forced Sauerwald himself[56] to abandon his original theory. Smithells, Pitkin, and Avery found for tungsten that grain growth, as observed by metallographic methods, begins only at 1227°C. (2240°F.) for a compacting pressure of 8 tsi, while it begins at 927°C. (1700°F.) or lower for a pressure of 32 tsi. Dawihl, studying the effects of pressure on grain growth with x-ray photographs, found that very high pressures lower the temperature of commencing grain growth considerably, while the grain size obtained is smaller for very high pressures than for medium pressures. The temperatures for the commencing grain growth for sintered iron are relatively high in comparison with fused iron. In fused iron a recrystallization temperature of 800°C. (1470°F.) corresponds to a one per cent deformation. Taking this as a basis for comparison, the pressures applied by Dawihl to the compacts (55 tsi) would result in only a very small deformation of the metal structure. Finally, the temperature of grain growth is found by Libsch, Volterra, and Wulff[57] to be largely dependent on the purity of the powder (iron), and the extent of such growth also dependent on the initial particle size distribution of the powder.

One may assume that Sauerwald's temperature for unprovoked grain growth represents the temperature at which the particle boundaries give way completely to the boundaries of the newly formed crystals as shown, for instance, in the photomicrograph of Figure 199C, for a copper powder compact sintered for 16 hours at 750°C. (1380°F.). Apparently, the spontaneous grain growth should be regarded as a pure temperature effect—the temperature being specific to the metal concerned. Other explanations advanced have been that the phenomenon is an indication that the external application of pressure does not cold-work the particles in a compact, or that the particles of the powders used by Sauerwald were already in a condition of maximum cold work, but none of these explanations have been proved. As a matter of fact, Sauerwald and Holub[58] have shown that this phenomenon follows the same general lines with powders that were cooled slowly from the melting point. In another paper, Sauerwald and Jaenichen[59] presented the view that crystallization and grain growth phenomena do not depend on the average distance between particles, but only on the nearest distance of approach, and this distance should not depend very much on pressure.

Jones[59a] holds that this phenomenon of spontaneous grain growth has

[56] F. Sauerwald, Kolloid-Z., 104, 144 (1943).
[57] J. Libsch, R. Volterra, and J. Wulff, in J. Wulff, Powder Metallurgy. Am. Soc. Metals, Cleveland, 1942, p. 379.
[58] F. Sauerwald and L. Holub, Z. Elektrochem., 39, 750 (1933).
[59] F. Sauerwald and E. Jaenichen, Z. Elektrochem., 31, 18 (1925).
[59a] W. D. Jones, Principles of Powder Metallurgy. Arnold, London, 1937, p. 81.

a certain practical value. While the temperature of grain growth in previously sintered compacts (as with cast metals) can be lowered by cold work, such as coining, the temperature corresponding to the initiation of such grain growth cannot be raised above that of spontaneous grain growth in the initial powder compacts. In this way an upper temperature limit is given for heat treatment of the compacts, and above which those qualities harmed by grain growth will start to disappear. Sauerwald's[60] investigation of the grain growth temperature of sintered iron subjected to various kinds of cold work is interesting in this connection. Iron compacts, pressed at 10 and 32 tsi, were sintered in hydrogen for half an hour at 750° and 1070°C. (1380° and 1960°F.). The treatment at the lower temperature showed no grain growth; at the higher temperature a coarser grain structure resulted, but the initial pressure had no influence. These compacts were reduced 60% by coining at 65 tsi and were afterward annealed for half an hour in hydrogen at a series of increasing temperatures. Grain growth in both types of specimens commenced at approximately 750°C. (1380°F.), as against 1070°C. (1960°F.) as the temperature of spontaneous grain growth. Similar results were obtained with copper. Compacts pressed at 32 tsi and sintered at either 620° or 820°C. (1150° or 1510°F.) were deformed 60% by coining, and 97% by cold-rolling. The first indications of a change in structure were detected after annealing to 420°C. (790°F.), while visible grain growth was observed microscopically at 520°C. (970°F.). Cold deformation in this case reduced the temperature of grain growth from 720°C. (1330°F.) to approximately 420°C. (790°F.).

Kieffer and Hotop[61] affirmed that grain growth in sinter metals is a function of recrystallization, analogous to like phenomena in deformed (or nondeformed) metals obtained by fusion. This implies that the term "recrystallization" is used in its widest possible sense, i.e., any structural change in fused and solidified bodies, with or without deformation, except changes in phase or chemical composition; crystallites may thereby form with definite grain boundaries, or the grain boundaries may merely be shifted. The changes that the structure undergoes at higher temperatures are not ended with this recrystallization. Growth of the crystallites follows; this growth is more pronounced with higher temperatures to which the recrystallized metal is heated. The conclusion drawn from this fact is that the newly formed structure (i.e., recrystallized) is at the start, not yet quite stable, and that it tends to develop into greater stability by the growth of the crystallites. Accordingly, the term "recrystallization" in-

[60] F. Sauerwald, Z. Elektrochem., 29, 79 (1923).

[61] R. Kieffer and W. Hotop, Pulvermetallurgie und Sinterwerkstoffe. Springer. Berlin, 1943, p. 99 ff.

cludes two consecutively functioning but partly interconnected phenomena, namely the formation of a new structure and the growth of the crystallites.

With regard to the crystallization phenomena in sintered ("synthetic") metals, it may be concluded that, in the aggregates of crystals in the thermodynamically unstable powder particles, heat treatment will always cause grain boundary shifts and grain growth phenomena, which may be considered to be recrystallization phenomena. In addition to the internal stresses, always present in the particles that form the grains, lattice distortions caused by the compacting pressure add secondary recrystallization effects. In most sintering processes, chemical reactions caused by a reducing atmosphere promote increased grain growth with recrystallization.

NATURE OF SINTERED BOUNDARY

Although the present conceptions of the recrystallization and grain growth phenomena have been clarified above, it appears appropriate to elaborate to a greater extent on the structure of the sintered compacts with particular reference to the nature of the grain boundaries which are formed without solidification from the liquid state. This problem has interested a number of investigators, and has recently been discussed in detail by Jones,[62] Comstock,[63] Skaupy,[64] and Kieffer and Hotop;[65] study of these references is recommended.

In cast metals, when the metal passes from the liquid to the solid state, the atoms arrange themselves to form small crystallites. Upon slight supercooling of the melt, numbers of atoms freeze to form nuclei (centers of crystallization) of the crystalline phase, appearing spontaneously at various points in the liquid. Adjacent atoms join with each other at each of these points, instantaneously releasing their energy of liquid motion as latent heat of fusion at the boundary of the growing crystal. They abandon their random motion and adopt relatively fixed positions about which they oscillate. A crystal lattice is formed in this manner, the nuclei growing by the orderly coherence of other atoms from the liquid. During this growth other nuclei are formed. The rate of nucleus formation and the rate of crystal growth are dependent on the rate of cooling. Finally, the growth of each crystal is obstructed by contact with its neighbors, while

[62] W. D. Jones, *Principles of Powder Metallurgy*. Arnold, London, 1937, p. 72 ff.
[63] G. J. Comstock, *Metal Progress, 35*, No. 4, 343 (1939); *35*, No. 5, 465 (1939); *35*, No. 6, 576 (1939).
[64] F. Skaupy, *Metallkeramik*. Verlag Chemie, Berlin, 1943, p. 5 ff.; also *loc. cit.*, 1st ed., 1930, p. 7.
[65] R. Kieffer and W. Hotop, *Pulvermetallurgie und Sinterwerkstoffe*. Springer, Berlin, 1943, p. 72.

the remaining liquid between the crystals becomes exhausted; between the abutting crystals, now called grains or crystallites, nonsymmetrical boundaries are formed. The orientation of the grains in the solidified metal is random, their form and size depending on the nature of the metal and on the conditions during cooling. Frequently, grain boundaries are also formed in the solid state, particularly if phase changes occur upon further cooling of the cast metal. Mechanical working followed by annealing is another cause for the formation of new grain boundaries (recrystallization), and finally, ordinary heating may have a similar effect (germination, "Sammelkristallisation").

While the properties of the crystallites in cast metals are incurred primarily by the process of transformation during and after solidification, the structure of sintered ("synthetic") metal bodies is largely dependent on the degree of dispersion of the powder and on the possible presence of nonmetallic matter. Furthermore, the sintering process frequently causes definite changes in the crystallites that may even counteract the influence of the original particle size distribution.

According to Skaupy,[64] the size and especially the boundaries of the crystallites can have a deciding influence on the resulting properties of a sintered metal. Whereas at 100m temperatures fractures in metals are intracrystalline for the most part, suggesting that the boundary has a greater effective strength than the crystal, tungsten and probably most other sintered metals suffer intercrystalline fractures—their mechanical properties being decreased with an increase in boundary area. Skaupy's view caused Jones[62] to discuss in great detail the nature of grain boundaries in sintered metals, and particularly to bring up the question of whether the grain boundary in a solidified cast metal can be considered identical with that of a sintered metal, in order to permit certain conclusions to be drawn as to the different physical and mechanical properties of the two types of material. Even under the assumption that it would be possible to bring two metallic surfaces at room temperature into complete contact so that the atoms of the two surfaces could form uninterrupted links, it is improbable *a priori* that such a sintered interface would be identical with a grain boundary between two crystals. Certainly the order of the atomic arrangement in the boundary area is different in the two cases. However, with rising temperature the atomic mobility would eventually increase, causing a new arrangement to take place at the contact areas, which would tend to equalize the initial differences in the atomic orientation of the cast and sintered metals.

Tammann[66] contends that there exists a clear difference between a

[66] G. Tammann, *Z. anorg. allgem. Chem.*, *121*, 275 (1922).

sintered interface and a grain boundary, as illustrated by a difference in grain growth. Sintered metals display distinct grain growth at certain temperatures, whereas in metals obtained by fusion method without cold working, no grain growth is apparent upon annealing. Electrodeposited metal—where irregular recrystallization phenomena can be observed upon heating—is, however, an exception, as is also the one case of very pure cast aluminum, which was found to suffer spontaneous recrystallization upon reaching a temperature near the melting point.[67] The lack of ductility and toughness in sintered metals has been mentioned as another reason for the assumption of a difference in the boundaries of the two types of metal. But it must be realized here that elongation in sintered metals can be improved considerably by a subsequent cold or hot densi-

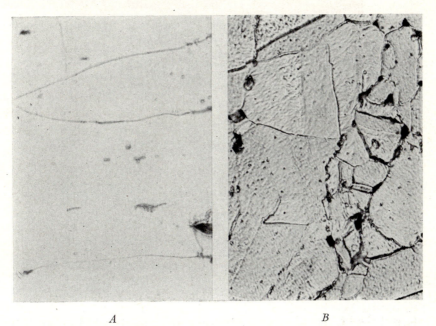

| *A* | *B* |

Fig. 201. Grain boundary structures of fused and vacuum-sintered copper: *A*, cast oxygen-free high-conductivity copper (×100). *B*, electrolytic copper powder vacuum-sintered at 750°C. (1380°F.) for 16 hours after compaction at 45 tsi (×750).

fication and that, with a sufficient degree of working, values of the fused metal can be approached. The low elongation values in simply sintered metals can easily be explained by the presence of pores in the structure and boundaries which impose a tri-axial state of internal stress on the

[67] H. Röhrig and E. Käpernick, *Aluminium*, *17*, 411 (1935).

metal. In Figure 201 the grain boundary structure of vacuum-sintered copper is compared with that of fused copper and shows clearly a difference in the compactness of the boundary.

There is probably some truth in the belief that certain differences exist in the two types of boundary areas because of the different kind, amount, and distribution of impurities. In fact, Tammann maintains that the difference in the behavior of sintered metals in connection with recrystallization phenomena is *entirely* a matter of the differing degree and na-

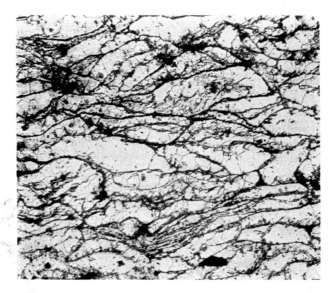

Fig. 202. Microstructure of sintered magnesium (×200) after compaction at 30 tsi and sintering at 600°C. (1110°F.) for two hours in carbonaceous atmosphere, displaying initial magnesium oxide particle boundaries (courtesy of Charles Hardy, Inc.).

ture of the intercrystalline impurity. In fused and solidified metals these impurities can be found chiefly in the grain boundaries, but in sinter-metals these foreign substances are dependent on the history of the metal, being distributed in a more or less disorderly fashion at the grain boundaries as well as inside the crystallites. The nature of the impurities is generally different for fused and sintered metals, since in the former they are primarily introduced during the melting and casting procedures, while in the latter they originate in the history of the powder. Examples are seen in the microstructures of sintered magnesium in Figure 202, and of hot-pressed copper in Figure 203. In both cases the contamination of the boundaries by oxide films is apparent. In spite of these observations, no differences of practical importance may result—as seen in the work by Kieffer and Hotop[68] on iron and nickel, and their alloys with cobalt and

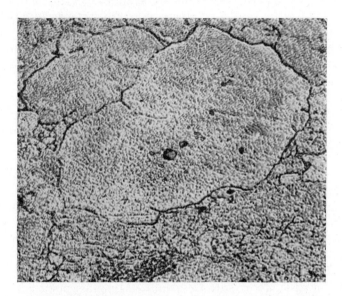

Fig. 203. Microstructure of electrolytic copper powder after hot-pressing at 400°C. (750°F.) at 50 tsi in nitrogen atmosphere, displaying films of Cu_2O at the grain boundaries (×750).

molybdenum—in the physical and mechanical properties of fused and sintered bodies, if the same amount and type of impurities are present in both cases, and if the sinter-metal is subsequently worked to complete density.

MICROSTRUCTURES

Before an analysis can be made of the various physical and mechanical properties of compacts as affected by different sintering conditions, the microstructure of these compacts should be considered further. A number of investigations have dealt with this topic, and the numerous contributions by Sauerwald and coworkers[48-51,58,59,60,68,69] on iron and

[68] R. Kieffer and W. Hotop, *Pulvermetallurgie und Sinterwerkstoffe*. Springer, Berlin, 1943, p. 74.

[69] F. Sauerwald and E. Jaenichen, *Z. Elektrochem.*, *30*, 175 (1924).

copper, already cited repeatedly, as well as the publications by Eilender and Schwalbe[70] on iron, by Goetzel,[71] and by Bier and O'Keefe[72] on copper, and by Smithells and his coworkers[53,73] on tungsten, should be consulted for closer study. In this part of the text it appears sufficient to summarize the different metallographic findings, especially since most of these studies have already been discussed to some extent under the headings of shrinkage, porosity, and boundary conditions.

With the advancement of the sintering process, the powder particles are subjected to a progressively more apparent crystallization, which leads to a more or less intensive grain growth at higher temperatures. In particles that are substantially in a state of structural equilibrium the spontaneous grain growth for the metals investigated has been observed to occur at about the same relative temperature range in reference to the absolute melting point of the metal ($a = 0.67$–0.87). The commencement of grain growth lies at relatively high temperatures in comparison with the recrystallization of fused metals. On the basis of earlier metallographic examinations[48–50,69,74,75] the temperature of the beginning grain growth was believed unaffected by the preceding forming pressure. But more recently, x-ray investigation[75a] of sintered iron and microscopic examinations of sintered tungsten of finest particle size[73] have indicated that the temperature of commencing grain growth may be depressed by the application of very high forming pressures. Figure 204 shows the equiaxed dense grain structure of fully recrystallized tungsten powder after compaction at 100 tsi and sintering at 2000°C. (3630°F.). In addition, other factors influencing the structure must be taken into account. These include the powder's history, purity, particle size, and particle surface conditions, as well as the type of treatment the powder underwent and the atmospheric conditions used during sintering.

With regard to the powder's purity, it appears that metal oxides reducible in hydrogen (cupric oxide, ferric oxide, nickel oxide (Ni_3O_4), tungsten trioxide, molybdenum trioxide) facilitate grain growth because of reversible chemical reactions between the oxides, the metal, the reducing gas, and the water vapor produced. Metal oxides that cannot be reduced

[70] W. Eilender and R. Schwalbe, *Arch. Eisenhüttenw.*, *13*, 267 (1939).

[71] C. G. Goetzel, *The Influence of Processing Methods on the Structure and Properties of Compressed and Heat Treated Copper Powders. Dissertation.* Columbia University, New York, 1939.

[72] C. J. Bier and J. F. O'Keefe, *Trans. Am. Inst. Mining Met. Engrs.*, *161*, 596 (1945).

[73] C. J. Smithells, *Tungsten.* 2nd ed., Chapman & Hall, London, 1936, p. 64 ff.

[74] L. Schlecht, W. Schubardt, and F. Duftschmid, *Z. Elektrochem.*, *37*, 485 (1931).

[75] M. Yu Bal'shin, *Vestnik Metalloprom.*, *61*, No. 17, 87 (1936).

[75a] W. Dawihl, *Stahl u. Eisen*, *61*, 909 (1941).

in hydrogen (aluminum oxide, silicon dioxide, thorium dioxide, etc.) have the opposite effect, hampering crystallization and grain growth.

Fig. 204. Grain structure of tungsten compact after compression at 100 tsi and sintering in hydrogen at 2000°C. (3630°F.) for 30 minutes (×300).

With finer particle size the removal of the last traces of oxygen becomes more difficult. If their oxides are reducible in hydrogen, finer powders display a stronger tendency toward grain growth. It may also be assumed that in finer powders (because of the increased total surface) more discontinuities in the atomic lattice of the surface areas appear, a condition which promotes nucleus formation and grain growth.[61] Particle contours and surface conditions appear to affect the structure of the sintered body in such a way that particles with a rugged, nodular, irregular (unstable) surface tend more toward crystallization than powders whose particles have smoother and more uniform (stable) surfaces, being themselves already in a structural equilibrium. The method of production and conditioning of a powder have a determining influence on the purity, particle size, and particle surface conditions, and thereby indirectly affect the microstructure of the sintered metal.

A reducing atmosphere during sintering, essential for the chemical reactions with the always present traces of oxides, and possibly with other impurities (*e.g.*, carbon), generally promotes grain growth. Inert atmospheres tend to retard grain growth, and the same applies to vacuum, if the oxides have a low vapor pressure. Moreover, vacuum tends to facilitate the formation of a great quantity of microscopic spherical pores within the particles during the early stages of the sintering treatment. An alternate treatment in hydrogen and vacuum makes possible a nearly complete removal of all gases, oxides, and other volatile impurities, thereby promoting grain growth.

Effects of Sintering Conditions on Properties

The preceding study of the sintering phenomena and of the metallographic changes caused by these occurrences enables us now to evaluate briefly the basic trend of the physical properties that can be obtained in monometallic powder compacts by varying the sintering conditions. Since this chapter is concerned solely with the occurrences during sintering and their effects on the properties of the compacts in general, the discussion of the properties obtainable by changing the conditions will, of necessity, be sketchy. A more complete survey of the properties of powder metallurgy products will be given elsewhere. The extent to which density determines other properties, which recently has been studied systematically for iron compacts by Squire,[76] will also be discussed in detail in a following chapter (Chapter XXV, Volume II).

DENSITY

The effects of various sintering conditions on density have already been touched upon during our study of shrinkage and growth phenomena.

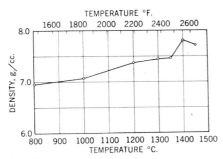

Fig. 205. Density as a function of the sintering temperature for iron compacts pressed at 30 tsi and sintered for 32 hours (according to Kelley[77]).

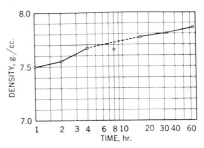

Fig. 206. Density as a function of the sintering time for iron compacts pressed at 30 tsi and sintered at 1425–1450°C. (2600–2640°F.) (according to Kelley[77]).

It has been shown that for loosely packed or lightly pressed powders density increases with rising temperature (as can be seen in Fig. 205 for an iron powder compacted at 30 tsi and sintered for 32 hours, according to

[76] A. Squire, *Trans. Am. Inst. Mining Met. Engrs.*, 171, 485 (1947); see also *Watertown Arsenal Lab. Rept.*, WAL 671/8, 671/12, 671/16, 671/23, 671/31 (1946).

Kelley[77]) while it tends to decrease for powders compacted at very high pressures (Figs. 185 and 193). In the case of a high sintering temperature, the density increases rapidly within a short period of time, followed by a slight increase. Figure 206 shows the increase in density with time for an iron powder compacted at 30 tsi and sintered at 1425–1450°C. (2600–2640°F.) (according to Kelley). For lower sintering temperatures the initial increase in density is usually less rapid and is followed by a long period during which the density increases at a very slow rate (Fig. 186).

While temperature and time are the primary sintering variables that influence the compact's density, there remain other factors worthy of consideration, among them the sintering atmosphere and the rates of heating and cooling. It has already been mentioned that, if the rate of heating is high, treatment in vacuum, for example, may materially decrease the density because of the formation of new "secondary" pores within the crystallites (Fig. 199). A strong reducing atmosphere such as dry hydrogen, on the other hand, may cause a larger increase in density with either rising temperature or with time than possible with a weaker reducing atmosphere, since, as has been shown, reactions between hydrogen and the reducible metal oxide films or inclusions promote grain growth and thereby shrinkage. If the rate of heating is too rapid, a spontaneous evolution of absorbed or trapped gases will disrupt the structure not only in the case of vacuum, but also if the sintering takes place at overatmospheric pressure. A loss in density and the appearance of laminated cracks may result, and only prolonged sintering at a high temperature may again densify the swollen compact through effective shrinkage.

There remain several additional variables which should be noted briefly since they have an indirect influence on the conditions mentioned. In several of the earlier chapters it has been shown that the density of a compact is, of course, a direct function of the closeness of the packing of the particles within a given volume. The packing, in turn, is affected by the particular particle size distribution, as well as by the external pressure applied during compaction. Consequently, the density of a sintered compact is to a certain degree dependent on the particle size and on the compacting pressure, in conjunction with the powder's plasticity. In Figure 207 a schematic diagram summarizes these observations by showing the density of sintered compacts for a soft and a hard metal powder as a function of the pressure.

Finally, the effect of the initial particle size on density remains to be taken into consideration. As shown during our study of shrinkage phe-

[77] F. C. Kelley, in J. Wulff, *Powder Metallurgy*. Am. Soc. Metals, Cleveland, 1942. p. 60.

nomena, finer powder displays shrinkage at a lower temperature, and also results in greater shrinkage than a coarser powder (Fig. 195). Accordingly, density increases with the powder's dispersion if all other factors (pressure, temperature, time, and atmosphere) are kept constant, as shown in Figure 208.

ELECTRICAL CONDUCTIVITY

Electrical conductivity, or specific electrical resistance, respectively, appears to be particularly sensitive to the sintering conditions. The problem has been studied carefully under a variety of conditions and for a

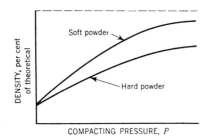

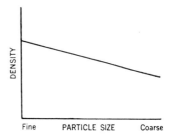

Fig. 207. Schematic representation of the density of sintered compacts as a function of the compacting pressure for hard and soft powders (according to Kieffer and Hotop[77a]).

Fig. 208. Schematic representation of the density of sintered compacts as a function of the initial particle size (according to Kieffer and Hotop[77a]): pressure P; temperature T; and time t are constant.

number of metals by several investigators whose results will now be summarized. The investigations of the electrical conductivity of metal powder compacts cover both pressed and sintered compacts, since smaller compact resistance with closer contact between the particles or crystallites can be established by means of either pressure or heat. Since a study of the conductivity of powders under pressure is of aid in interpreting sintering phenomena at elevated temperatures, it will also be considered here.

The conductivity of pressed powder compacts was first investigated by Streintz,[78] and more recently by Skaupy and Kantorowicz,[79–81] who

[77a] R. Kieffer and W. Hotop, *Pulvermetallurgie und Sinterwerkstoffe*. Springer, Berlin, 1943, pp. 85, 89, 97.
[78] F. Streintz, *Ann. Physik*, *308*, 1 (1900); *314*, 854 (1902).
[79] F. Skaupy and O. Kantorowicz, *Z. Elektrochem.*, *37*, 482 (1931).
[80] F. Skaupy and O. Kantorowicz, *Metallwirtschaft*, *10*, No. 1, 45 (1931).
[81] O. Kantorowicz, *Ann. Physik*, *12*, 1 (1932).

TABLE 54
Specific Resistance Ratio of Powders Compacted at Approximately 15 Tsi Versus Solid Metal (Skaupy and Kantorowicz[79,80])

Metal	Electrical resistance, compact vs. solid metal	Material constant C
Tin..........................	1.1	27
Lead........................	1.7–3.0	16.5–6.3
Graphite.....................	2.3	0.3
Zinc.........................	2.6	34
Silver.......................	3.1–6.9	71.0–71.6
Gold (type "S")..............	7.5	19
Bismuth.....................	7.9	0.6
Gold (type "P")..............	12.0	4.9
Antimony....................	14–29	1.2–0.6
Copper......................	90	6.8
Platinum....................	100	0.6
Nickel (hard)................	120–150	1.9–2.0
Nickel (annealed)............	18–26	9.2–5.3
Iron........................	185	2.1–1.1
Tungsten....................	110–420	—

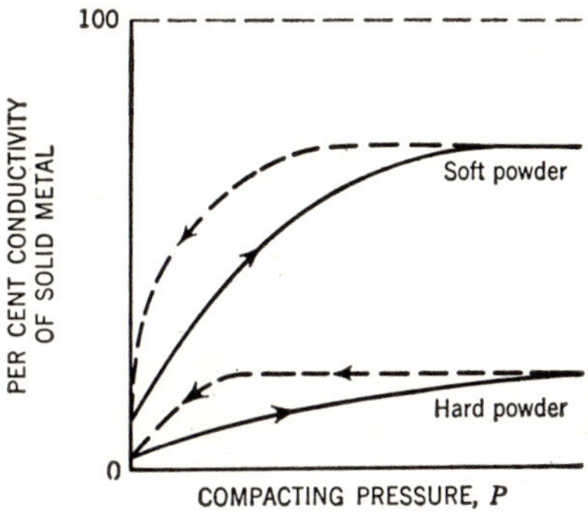

Fig. 209. Schematic representation of the electrical conductivity of powder compacts as a function of the compacting pressure (according to Kieffer and Hotop[77a]).

Fig. 210. Specific electrical conductivity of tungsten powder compacts as a function of the compacting pressure after various repeated pressings (according to Kantorowicz[81]).

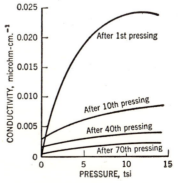

examined the resistivity of a number of materials as affected by particle size distribution and applied pressure. For details of the experimental procedure the reader is advised to consult the original papers. Essentially the following results were observed in these tests:

The resistance of a metal powder under pressure is always greater than that of fused metal. For soft metals (tin, lead, zinc, silver) the difference is smaller than for harder metals (nickel, iron, tungsten, etc.). In Table 54 a comparison is made of a number of metals. The resistance of the compact varies with the time of application of pressure, decreasing rapidly at first and more slowly later on. With rising pressures, P, the resistance, R, may be expressed for a large range of values by the formula:

$$1/R = c\sqrt{P} + C$$

where c and C are constants depending on the material and its history. On the basis of the results of Table 54, the conductivity $(1/R)$ for hard and soft materials can be expressed schematically as a function of pressure by two curves, as shown in Figure 209. The constant c, larger for soft than for hard powders, is a measure of the slope of these curves; while constant C represents the conductivity of the powder in the unpressed state, i.e., for zero pressure. Resistance–pressure curves are not reversible, With diminishing pressure the resistance remains close to the minimum value obtained for the highest pressures and rises gradually at first, often increasing more markedly only at quite low pressures.

TABLE 55

Electrical Resistance[a] and Specific Resistance[b] of Tungsten Powder
(Kantorowicz[81])

Pressure, tsi		Number of pressings							
		1	10	20	30	40	50	60	70
0	R	0.753	0.149	0.193	0.198	0.303	0.408	0.519	0.663
	ρ	10.4	3.35	4.5	5.02	7.8	10.3	13.3	17.3
32.8	R	0.0348	0.0782	0.104	0.136	0.173	0.224	0.298	0.338
	ρ	0.64	1.88	2.6	3.3	4.43	5.74	7.65	8.92
65.6	R	0.0234	0.0578	0.0784	0.0977	0.115	0.145	0.18	0.212
	ρ	0.48	1.41	2.03	2.54	3.2	4.03	5.14	6.06
98.4	R	0.0179	0.0538	0.0688	0.0855	0.0975	0.121	0.14	0.160
	ρ	0.4	1.32	1.77	2.0	2.71	3.36	4.0	4.57
131.2	R	0.0168	0.0447	0.0615	0.077	0.0843	0.104	0.118	0.132
	ρ	0.42	1.18	1.68	2.11	2.37	3.0	3.58	4.20

[a] R, measured in ohms.
[b] ρ, Ω mm.^{2}/meter.

The inverse trend for conductivity is expressed in Figure 209. After such treatment the initial value for resistance before the application of renewed pressure is not obtained again. With most materials (e.g., tungsten, platinum, iron, nickel, copper, silver, antimony, zinc, bismuth, and graphite, but not gold, lead, and tin), the resistances obtained for any given pressure are higher on repressing than on the preceding pressing; this is more pronounced with the harder metals. With repeated pressing, the resistance-pressure curves generally lie higher with each pressing—up to about the seventieth, by which time (in the case of tungsten) the resistance under pressure will be about ten times that at the initial pressing. For tungsten powder, the values for the variations in resistance, R, and specific resistance, ρ, with pressure and repeated pressings are shown in Table 55; in Figure 210 the conductivity is shown as a function of pressure, with different numbers of pressings presented by different curves. The increase in electrical resistance with repeated pressings is dependent upon the duration of pressure, generally being more pronounced the longer the powder is subjected to the pressure. On the other hand, with short high pressure and a soft metal, the resistance-pressure curves obtained after successive applications of pressure may lie below each other. The resistance of a powder compact depends on particle size, being higher with finer particles (within the observed range of 2 to 60 μ). The behavior of tungsten powder freshly reduced in hydrogen is no different from that of a powder previously exposed to the atmosphere. It may therefore be assumed that oxide effects play no role in these results.

The temperature coefficient of resistance of a powder before pressing is strongly negative, but at high pressures it depends on the direction of change of pressure, being negative for increasing pressures and positive for decreasing pressures. Increase of temperature (because of increased sinter bonds) reduces the resistance measured at room temperature.

Skaupy and Kantorowicz's results have been carefully analyzed by Jones.[82] It is only natural to expect a reduction in the resistance of compressed metal powders, since compaction causes an increase in contact points. It is less readily understandable, however, why repeated pressing further increases the resistance. Cold deformation of the particles may be a contributing cause, although in cold-deformed fused metal the increase in resistance is considerably less (not more than 5%). The conclusion must be drawn (with reservations) that powder particles are more severely strained. On the other hand, Jones has (rightly) observed that the increase in resistance by the repeated pressings may be due to oxide films which have formed on account of the frictional heat caused by the repeated compressions of the metal. Oxide films thus obtained may be of greater thickness and tenacity than those that form on the particle surfaces during a simple exposure to air. This reasoning would also explain why this particular phenomenon could not be observed on gold powders, which are always free of oxide films. In the case of lead and tin the deformability of the particles was probably so great that the oxide films could be punctured readily and ample contact areas helped reduce the resistivity.

Trzebiatowski[83] determined the electrical conductivity of compacted copper and gold powders after heating in hydrogen to temperatures up to 700°C. (1290°F.). The original paper should be consulted for particulars of apparatus and procedure. The measurements were made after cooling

[82] W. D. Jones, *Principles of Powder Metallurgy*. Arnold, London, 1937, p. 97.
[83] W. Trzebiatowski, *Z. physik. Chem.*, *B24*, 87 (1934).

to room temperature; the results are shown in Figures 211 and 212. Three regions can be observed: up to 100°C. (210°F.) the resistance increases followed by a decrease up to 300°C. (570°F.), and then by a continuous increase. This phenomenon was much more impressive for copper than for gold, and also more pronounced for compacts pressed at a lower pressure (40 tsi) than at a very high pressure (200 tsi). The decrease that follows above 100°C. (210°F.) is, according to Trzebiatowski, explained by the onset of structural changes (recovery?) ; according to Jones, by the onset of sintering. Evolution of sorbed gaseous films may play a part in this phenomenon. The difference between the curves of the two metals may be explained by the possible presence of oxide films in the copper powder. The increase of resistance up to about 100°C. (210°F.) is not in agree-

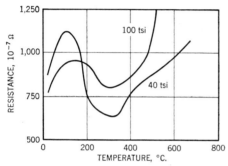

Fig. 211. Variation of electrical resistance of copper compacts with temperature (according to Trzebiatowski[83]).

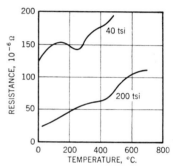

Fig. 212. Variation of electrical resistance of gold compacts with temperature (according to Trzebiatowski[83]).

ment with the general conclusions concerning the negative temperature coefficient of metal powder compacts. One would rather expect a continuous decrease of resistance from the commencement of heating. The final increase in resistance which occurs above 300°C. (570°F.) is explained by Jones by the possibility that sintering is now largely "completed," and the compact now behaves like a normal cast metal. This explanation appears highly doubtful, however, and it must be assumed that at temperatures above those used by Trzebiatowski (where pronounced grain growth takes place) a flattening of the resistivity curves toward a slight normal increase can be expected.

Among other investigations dealing with the conductivity of metal powder compacts, the results obtained by Grube and Schlecht[84] on a highly pressed carbonyl nickel powder are of particular interest. In Figure

[84] G. Grube and H. Schlecht, Z Elektrochem., 44, 367 (1938).

213 the specific electrical resistance is shown as a function of the sintering temperature for compacts pressed at different pressures. The compacts were tested only after they had been sintered at the different temperatures for two hours and had been cooled to room temperature. The specific

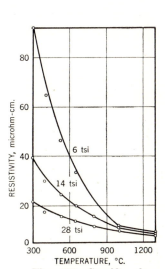

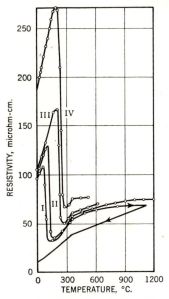

Fig. 213. Specific electrical resistance of carbonyl nickel powder compacted at different pressures as a function of sintering temperature (according to Grube and Schlecht[84]); sintering time, 2 hours.

Fig. 214. Specific electrical resistance as a function of sintering temperature for carbonyl nickel powder compacted and sintered under different conditions (according to Grube and Schlecht[84]): curve I, carbonyl nickel powder pressed 5 minutes at 43 tsi, measured during sintering in hydrogen; curve II, same, measured during sintering in argon; curve III, same, measured during sintering in high vacuum; curve IV, same as III, but pressed at 35 tsi.

electric resistance decreases with rising sintering temperature, rapidly at first, and more slowly at higher temperatures. For lower temperatures the resistance is higher for lower pressures during the forming of the compact. The character of the curves is as expected and is in agreement with tests on copper powders conducted by the author.[85] According to Grube and Schlecht,[84] the total resistance of a sintered metal body consists of the resistance of crystallites themselves and the contact resistance between the

[85] C. G. Goetzel, *The Influence of Processing Methods on the Structure and Properties of Compressed and Heat Treated Copper Powders. Dissertation.* Columbia University, New York, 1939, p. 27.

individual crystallites. Free powder particles are generally surrounded by a layer of sorbed gases—the contact resistances being quite large. The total resistance of a loose or slightly compacted powder is consequently also large. An increased compacting pressure effects a closer approach between the particles and larger areas of contact (partly because of abrasion of the isolating surface gas films), and so reduces the total resistance. The absorbed gases begin to evolve with rising temperature, causing a marked reduction of the electrical resistance. With further temperature increases the resistivity is further decreased because of recrystallization and grain growth which form conducting links.

In order to examine the effect on the electrical resistance of powder compacts of oxide and gas films absorbed by the surface of the particles, Grube and Schlecht also determined temperature–resistance curves on carbonyl nickel powder. The compacts were heated under different atmospheric conditions. The results, reproduced in Figure 214, are in good agreement with the values for copper found by Trzebiatowski.[83] First, the resistance increases with a remarkably high temperature coefficient, especially in the compact treated in high vacuum. According to Grube and Schlecht this increase in resistance is caused by increased resistances between the particles effected by the gas films, which are, of course, greatest in high vacuum. The steep drop of resistance is explained by the evolution of the absorbed gases; the resistance shows a normal increase after all gases have escaped. The gradual flattening of the curves above 600°C. (1110°F.) can be attributed to the start of more pronounced sintering caused by the occurrences during recrystallization already discussed. The compact treated in hydrogen was also measured with temperatures decreasing from 1140°C. (2085°F.) to room temperature; the curve corresponds to that of bulk nickel. In all curves the discontinuity observed in the neighborhood of 360°C. (680°F.) is caused by the Curie temperature of nickel.

In a recent publication, Steinitz[86] shows that the resistivity of iron and steel compacts decreases rapidly after short sintering periods, indicating that the formation of actual metallic bonds between particles starts in the earliest stages of sintering.

HARDNESS

The effect of the sintering temperature on the over-all hardness of a compact (i.e., the hardness of the particles or crystallites plus the interstices) has been studied by several investigators. Trzebiatowski's results[86a] for copper are shown in the diagram of Figure 215, together with data ob-

[86] R. Steinitz, Powder Met. Bull., 1, 6 (1946).
[86a] W. Trzebiatowski, Z. physik. Chem., B24, 75 (1934).

tained by Kieffer and Hotop.[87] Results obtained by Grube and Schlecht[84] on carbonyl nickel powder are shown in the diagram of Figure 216, and the data of Heath and Stetkewicz[88] on hydrogen-reduced iron are given in Figure 217. The remarkably high hardness values obtainable in compacts of very plastic metal powders (*e.g.*, copper or gold) has already been discussed in connection with the molding process. It may be explained by severe work hardening, although some of the values obtained exceed those commonly experienced in work-hardened bulk metal. Trzebiatowski explains the phenomenon by the high dispersion and purity of the powders used. Hardness increases for low compacting pressures with rising sintering temperature (Fig. 215, curve 2); for intermediate and high pressures

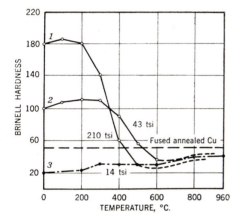

Fig. 215. Effect of sintering temperature on the Brinell hardness of copper compacts compressed at different pressures (curves *1* and *2* according to Trzebiatowski[83], and curve *3*, according to Kieffer and Hotop[87]).

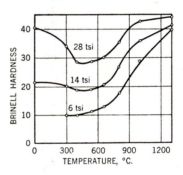

Fig. 216. Brinell hardness of carbonyl nickel compacts compressed at different pressures as a function of sintering temperature for sintering time of 2 hours in hydrogen atmosphere (according to Grube and Schlecht[84]).

increasing sintering temperature first causes a marked decrease in hardness, which is more pronounced for the more deformable copper than for the nickel. For higher temperatures hardness again increases, especially for the more rigid nickel and iron. The steady rise in hardness with temperature for slightly compressed compacts can be explained by increased density during progressive sintering, caused by a greater effectiveness of the adhesion forces and by recrystallization phenomena. For pure mono-

[87] R. Kieffer and W. Hotop, *Pulvermetallurgie und Sinterwerkstoffe.* Springer, Berlin, 1943, p. 113 ff.
[88] C. O. Heath, Jr., and J. D. Stetkewicz, *Metal Progress, 48,* No. 1, 73 (1945).

metallic systems, the maximum hardness values obtained for the highest temperatures cannot, of course, exceed that of the fused metal. In most cases, values of the bulk metal will not be reached because of remaining porosity and possibly extensive grain growth at the higher temperatures. The drop in hardness in highly compressed compacts is undoubtedly caused by a recovery of the metal particles that were strain-hardened during compression. In addition, in the case of highly compressed fine powders, the pronounced decrease in hardness may depend at least in part on the decrease in density due to degassing phenomena. The discontinuities in the curves for iron correspond to those found in the density curves (Fig. 192), and are caused by the $\alpha \rightarrow \gamma$ transformation.

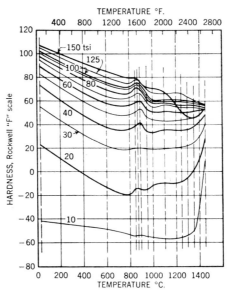

Fig. 217. Relation of sintering temperature to hardness of iron powder cylinders compacted at various pressures. All samples sintered in hydrogen for 1 hour (according to Heath and Stetkewicz[88]). Broken vertical lines designate temperatures of tests.

In Figure 218 the available results are combined in a schematic diagram showing hardness as a function of the sintering temperature—the individual curves representing different compacting pressures. These curves refer only to easily deformable metals, and cannot be applied to very rigid metals such as tungsten or to brittle compounds such as tungsten carbide. In these cases the compacting pressure causes only a small deformation of the particles, if any at all, so that no decrease in hardness at low sintering (annealing) temperatures would be noticeable.

Little information is available about the effect of sintering time on hardness. It may be concluded that the effects of time on hardness conform to those on the density; in other words, that a slight increase in hardness takes place with time. In the case of electrolytic copper powder

Fig. 218. Schematic representation of the Brinell hardness as a function of sintering temperature for compacts compressed from plastic metal powders at different pressures (according to Kieffer and Hotop[87]): curve *1*, low compacting pressure (approx. 15 tsi); curve *2*, medium-high compacting pressure (approx. 40–60 tsi); and curve *3*, very high compacting pressure (approx. 140–200 tsi).

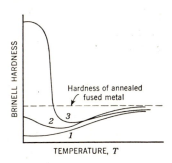

compacted at 25 tsi and sintered at 750°C. (1380°F.) the author[85] could notice an increase in Brinell hardness from 38 for the unsintered compact to 44 after 16 hours of sintering. Studies of the effect of particle size on hardness[85,89,90] have shown that with other factors such as pressure, sintering temperature, time, and atmosphere remaining constant, hardness like density decreases with increasing initial particle size.

MECHANICAL PROPERTIES

The effects of sintering temperature and time on tensile strength and elongation have already been discussed to some extent. Libsch, Volterra, and Wulff[91] investigated these properties for electrolytic iron and found an increase for both rising temperature (Fig. 191) and time (Fig. 194) with a marked discontinuation between 800° and 1000°C. (1470° and 1830°F.). Eilender and Schwalbe,[89] experimenting with "technical" (mechanically comminuted) iron powder, found substantially the same trend in these properties, as can be seen from Figures 219 and 220, where data for tensile strength and elongation are plotted as a function of sintering temperature and time, respectively. The rapid change in tensile properties at lower temperatures is caused by the increasingly effective action of the

[89] W. Eilender and R. Schwalbe, *Arch. Eisenhüttenw.*, *13*, 267 (1939).

[90] P. Schwarzkopf and C. G. Goetzel, *Iron Age, 146*, No. 12, 39 (1940).

[91] J. Libsch, R. Volterra, and J. Wulff, in J. Wulff, *Powder Metallurgy*. Am. Soc. Metals, Cleveland, 1942, p. 379.

adhesive forces; at higher temperatures (above 600°C.; 1110°F.) the increase in strength and ductility is largely due to recrystallization effects. It is particularly interesting to note that, in the case of iron, there is a temporary decrease in tensile strength and elongation in the region between 800° and 1000°C. (1470° and 1830°F.). This phenomenon has been observed in both investigations, and has also been observed on grades of iron powder (finer and coarser) other than used in these experiments.[92] Apparently, the initial particle size has little influence on the character of the strength-temperature curves, although the absolute values appear to increase with particle fineness. Sauerwald and Jaenichen attributed the drop in tensile values above 800°C. (1470°F.) to the volu-

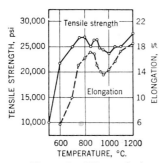

Fig. 219. Effect of sintering temperature on tensile strength and elongation of iron compacts compressed at 43 tsi (according to Eilender and Schwalbe[80]).

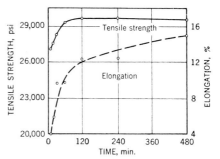

Fig. 220. Effect of sintering time on tensile strength and elongation of iron compacts compressed at 43 tsi and sintered at 800°C. (1470°F.) (according to Eilender and Schwalbe[80]).

metric changes connected with the $\alpha \rightarrow \gamma$ transformation. The contraction and the atomic realignment occurring during the transformation cause a relapse of sintering forces that is counteracted only at considerably higher temperatures. Eilender and Schwalbe[89] maintained that the occurrences during recrystallization affect the tensile properties extensively, with strength increasing during the formation of a uniform, equiaxed, ferritic grain structure, but then decreasing when marked grain growth occurs. The contention that grain growth is the chief contributing factor in the decrease in tensile properties is supported by the fact that all investigations established that this phenomenon commences below the actual transformation temperature (910°C.; 1670°F.). The renewed increase of tensile values at higher temperatures may be attributed to the closer

[92] F. Sauerwald and E. Jaenichen, Z. Elektrochem., 30, 175 (1924).

agglomeration of the crystallites caused by shrinkage. It may be assumed, moreover, that the forces interfering with sintering at higher temperatures become progressively smaller.

The effects of sintering temperature on the tensile strength of carbonyl nickel powder compacted at different pressures have been investigated by Grube and Schlecht[84] and by Kieffer and Hotop.[93] The results are combined in the diagram of Figure 221, which shows, especially for the

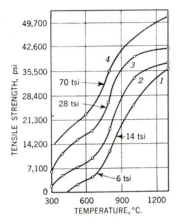

Fig. 221. Tensile strength of carbonyl nickel powder compacted at different pressures as a function of the sintering temperature for a sintering time of 2 hours (curves *1–3* according to Grube and Schlecht,[84] and curve *4,* according to Kieffer and Hotop[93]).

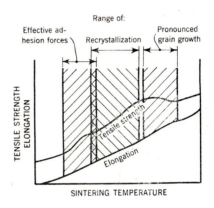

Fig. 222. Schematic representation of the effect of the sintering temperature on tensile strength and elongation of sintered metal powder compacts (according to Kieffer and Hotop[93]).

curves representing the higher compacting pressures, two temperature ranges (300–500°C.; 570–930°F.; and above 650°C.; 1200°F.) where tensile strength increases rapidly with rising sintering temperature. In the lower range the increase in strength may be attributed to the initial effectiveness of the sintering forces through the elimination or reduction of the counter forces, such as oxide films and dissolved or occluded gases. In the upper range increased sintering due to recrystallization phenomena is responsible for the rapid increase in strength of the compact. It is only

[93] R. Kieffer and W. Hotop, *Pulvermetallurgie und Sinterwerkstoffe.* Springer, Berlin, 1943, p. 116 ff.

natural that the application of higher compacting pressures leads to higher strength values at any temperature. The investigations on iron, nickel, and especially copper powders (see Figs. 187 and 188) by Sauerwald and co-workers,[92,94] and studies carried out on copper by the author[85] are substantially in agreement with these findings. In the case of copper, however, a maximum value for the tensile strength is frequently found at intermediate temperatures (900°C.; 1650°F.), above which the strength again decreases slightly—this phenomenon probably being caused by excessive grain growth at very high temperatures.

In Figure 222 the results of the individual investigations are summarized by plotting schematically the tensile properties as a function of the sintering temperature. The curves refer to compacts formed with an intermediate pressure (about 25 tsi); higher pressures would shift the curves upward in the diagram; lower pressures, downward.

The effect of sintering time on the tensile properties of metal powder compacts has been investigated, in the case of iron, by Libsch, Volterra, and Wulff[91] (Figure 194), by Eilender and Schwalbe[89] (Figure 220), and by Squire[94a] (for iron plus graphite); and, in the case of copper, by the author.[85] For a given sintering temperature the tensile strength of iron increases rapidly within the first few hours; thereafter the values remain nearly constant. The curves for elongation have a similar course, except that the initial strong increase is followed by a slower rise that continues at least for a period of 24 hours (Fig. 194). Prolonged periods of sintering apparently do not effect an additional increase in tensile strength, but improve elongation appreciably. No important variation from these trends was found in the case of copper, leading to the conclusion that the relations found between tensile properties and sintering time for iron and copper are essentially applicable to other metals of similar deformability.

The initial particle size has a similar effect on tensile strength and elongation as it has on density and hardness, as can be seen from the work by Eilender and Schwalbe[89] and by Libsch, Volterra, and Wulff.[91] In this connection, the recent investigations by Squire should again be cited. Squire reports[94b] that compacts prepared from electrolytic iron powders, graded to pass a 325-mesh screen, may reach a tensile strength of about 45,000 psi, impact resistance of 60 ft.-lb., the corresponding density of the material being 7.4 g./cc. or 94% of the theoretical value.

Among the other properties examined as a function of the sintering

[94] F. Sauerwald, *Z. Metallkunde, 21*, 22 (1929).
[94a] A. Squire, *Trans. Am. Inst. Mining Met. Engrs., 171*, 473 (1947); see also *Watertown Arsenal Lab. Rept., WAL 671/14* (1946).
[94b] A. Squire, *Trans. Am. Inst. Mining Met. Engrs., 171*, 485 (1947); see also *Watertown Arsenal Lab. Rept., WAL 671/16* (1946).

conditions, yield point and reduction in area,[89] as well as impact strength, compressive properties, and fatigue strength,[85] may be mentioned. Since these properties do not contribute any basically new factors in the analysis of the sintering process, further discussion of them will be dispensed with in this section.

POLYMETALLIC SYSTEMS

Our discussion of the sintering mechanism has so far been restricted to homogeneous systems, and essentially to monometallic powders. As we now study the application of sintering conditions to heterogeneous systems, we must bear in mind that most powder metallurgical products consist of more than one component, and that it is inherent in the process that practically any desired combination can be obtained between different metallic substances, or between metallic and nonmetallic constituents, regardless of the mutual solubility conditions in the liquid or solid state. It is a different problem, however, to determine whether the rules established in monometallic systems for pressing and sintering conditions and properties obtainable can be translated unqualifiedly into polymetallic systems. To facilitate this study it is advisable to distinguish between those systems involving only solid phases, where sintering is accomplished without the aid of liquid cement, and systems where sintering occurs in the presence of a temporary or permanent liquid phase that may constitute a minor or a major part of the entire composition.

Sintering in the Absence of a Liquid Phase

MULTIPLE METALLIC COMPONENTS

The sintering of mixtures of different powders is not very different in principle from the previously analyzed sintering process for single metal powders, since the actual bonding is due, in both cases, to atomic forces. The rate of alloying is determined by the rate of diffusion. In general, the sinter bond that is formed by atoms of different metals will be stronger with greater mutual solubility of the components. Bonding between particles of different metals is also likely if the component metals form a chemical compound. The solubility factor also leads to a more rapid alloying, of course, because of a more rapid rate of diffusion. However, on the basis of the foregoing considerations of the sintering process in single metals, it may be expected that bonding between particles having identical atoms usually occurs before diffusion alloying is noticed, especially for low temperatures. At higher temperatures the two effects may be superimposed—with sintering being greatly improved during the process of homogenization. In this connection it must be noticed that

552

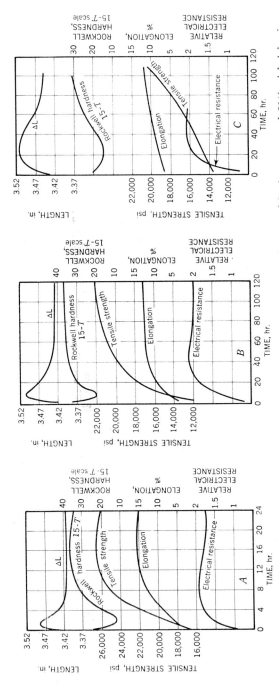

Fig. 223. Effect of sintering time on the physical properties of powder mixtures of 70% copper and 30% nickel having different particle sizes, for a sintering temperature of 980°C. (1795°F.) (according to Rhines and Meussner[96]): A, powder size, −325 mesh; B, powder size, 250 to 325 mesh; and C, powder size, 150 to 200 mesh.

density and tensile properties usually do not approach a saturation value with time (at constant temperatures) as rapidly as in single-component compacts.

Solid Solutions and Compounds. Two groups of alloying compositions can be distinguished in the process of sintering mixtures of several metal powders; these groups are those forming solid solutions or compounds, and those forming heterogeneous structures in which the components are not soluble at all or only soluble to a limited extent with each other in the solid state. The case of the formation of solid solutions will be considered first. Only if the powder particles already consist of the homogeneous alloy composition are the phenomena during sintering analogous to those established for a single metal. Otherwise, the degree of alloy formation by diffusion during sintering becomes of great importance, and phenomena connected with the formation of solid solutions or chemical compounds are superimposed upon the regular sintering phenomena for single-metal compacts. In compacts consisting of a mixture of powders of the pure components or of phases not yet in equilibrium, a change in the physical properties is accomplished by the diffusion between the different metals after a preliminary strengthening of the sinter bonds. Since the degree of diffusion is influenced by such factors as temperature and time of sintering, purity, particle size, and degree of work hardening of the powder, and sintering atmosphere, the prospectives of the resulting alloy bodies also depend on the same factors.

The process of homogenization, *i.e.*, diffusion toward structural equilibrium of compacts of two soluble components, and the effect obtained by interrupting this process before complete structural equilibrium is reached have been studied by Rhines and coworkers.[95,96] By using an isomorphous copper–nickel system, which forms a continuous series of solid solutions, they confined the investigations to one of the simplest possible cases of alloy formation. Rhines and Colton[95] examined the progress of homogenization during sintering of compacted mixtures of copper and nickel powders of different proportions by measuring the electrical resistance of the compacts. Rhines and Meussner[96] concentrated their effort on a 70–30 copper–nickel composition, and measured tensile strength, elongation, and hardness and dimensional changes during sintering in addition to electrical resistance. The interesting results of Rhines and Meussner are reproduced in the diagrams of Figure 223, representing the different properties for

[95] F. N. Rhines and R. A. Colton, in J. Wulff, *Powder Metallurgy.* Am Soc. Metals, Cleveland, 1942, p. 67.
[96] F. N. Rhines and R. A. Meussner, *Symposium on Powder Metallurgy.* ASTM, Philadelphia, 1943, p. 25.

different powder grades as a function of sintering time. As shown in Figure 223*A*, all physical properties, except elongation and electrical resistance, reach maximum values before homogenization is completed. It is assumed that similar maxima would also have appeared in the other two diagrams —with sintering times extended in accordance with the increased coarseness of the particle size. The fact that the maxima constitute diffusion and not sintering effects is indicated by the curves for change in length Δ L which show that the density and volume of the compacts have become stabilized before the maxima are reached.

The interpretation of these results by Rhines and Meussner is based on the concept that, at the time of maximum strength, diffusion has produced alloy layers around the particles in a pattern of "strategic" distribution through the entire structure. During sintering the diffusion process will produce all alloy compositions of the system, and the proportion that forms a physically continuous phase will change with time. The properties of the continuous phase, however, predominate in the over-all properties. Since the copper–nickel system exhibits hardness and strength maxima for a composition approaching the 50–50 proportion, it is to be expected that the compact reaches its maximum hardness and strength when the 50–50 alloy layers form a physically continuous phase. The stage at which most or all of the continuous phase consists of the hardest 50–50 composition need not necessarily be that of completed homogenization, however. As a matter of fact, only in mixtures containing equal parts of copper and nickel powders of high purity will the continuous 50–50 alloy phase coincide with the stage of complete equilibrium. Where the 50–50 phase does not form a continuous network, its effect is considerably reduced; but according to Rhines and Meussner, even in a mixture of 99% copper and 1% nickel the very minor proportion of 50–50 phase should improve physical properties at the early stages of the homogenization process, although a continuous network of the hardest phase can never form. The incompletely homogenized systems can be considered to represent frozen nonequilibria, and the possibility of producing by powder metallurgy such nonequilibrium alloy systems with superior properties may become of considerable practical importance.

As an example of a system of mutual solubility resulting in a compound, iron–graphite may be mentioned. Bal'shin[97] found that the addition of graphite under suitable conditions causes increased strength and density. This, however, requires the formation of a solid solution of carbon in iron and of the compound, iron carbide, at temperatures well below the melting point. The diffusion process is accelerated greatly by an increased

[97] M. Yu Bal'shin, *Vestnik Metalloprom.*, *18*, No. 4, 89, 124 (1938).

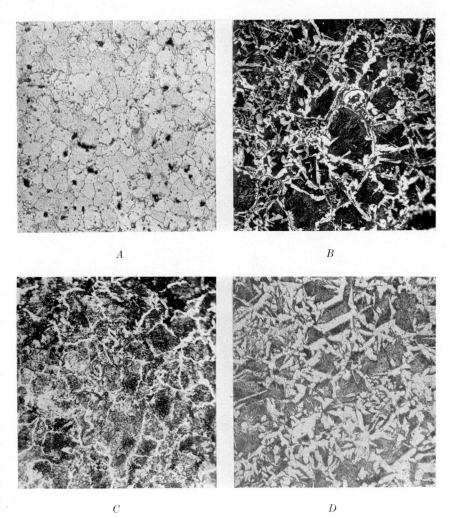

Fig. 224. Microstructures of steel compacts compressed at 50 tsi and sintered at 1250°C. (2280°F.) for 1 hour (×125): *A*, compact from crushed carbon–steel powder decarburized to 0.15% carbon; *B*, compact from mixture of 60% crushed steel, 40% reduced iron powder, containing 0.50% carbon; *C*, compact from crushed cast iron powder decarburized to 0.60% carbon; and *D*, compact from carbonyl iron powder containing 0.55% combined carbon.

dispersion of the powdered graphite and by increased contact areas through the use of high compacting pressures.[94a] Foreign elements (sulfur and phosphorus) may also promote sintering and homogenization.

Carbon-steel structures of normal appearance are obtainable by sintering compacts from incompletely decarburized steel and cast iron or carbon-containing carbonyl iron powders. In Fig. 224 typical microstructures are shown for these materials—after sintering compacts compressed at 50 tsi at 1250°C. (2280°F.) for one hour. By extending the sintering cycle to 44 hours at 1300°C. (2370°F.) it is also possible to produce a normal-type microstructure of a 2% chromium-carbon steel, as shown in Figure 225.

Fig. 225. Microstructure of chrome–carbon steel compact containing 2% chromium and 0.5% carbon. Compact pressed from mixture of elemental powders at 70 tsi and sintered in dry hydrogen for 44 hours at 1300°C. (2370°F.); ×150. (Courtesy of Charles Hardy, Inc.)

Another interesting example of a system of mutual solubility can be found in the austenitic stainless steels. The comparatively slow diffusion rates of chromium and nickel coupled with the former's tendency to oxidize or to produce carbide precipitation in the grain boundaries require extensive sintering cycles and carefully controlled atmospheric conditions. In Figure 226 fully homogenized grain structures are shown for two sintered austenitic stainless steels, but the treatment involved compaction at 70 tsi and sintering in dry hydrogen atmosphere for 44 hours at 1300°C. (2370°F.)

Generally speaking, it may be said that the diffusion in metal powder compacts does not differ basically from that in solid metals. According to Jones,[98] the factors important in connection with the diffusion of metal powders can be summarized as follows. (1) Intimately mixed metal powders interdiffuse much more readily than solid metals, with homoge-

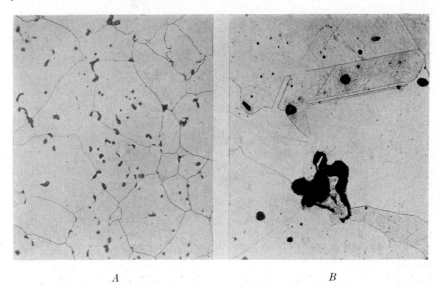

A   B

Fig. 226. Microstructures of austenitic stainless-steel compacts pressed from mixtures of the elemental powders at 70 tsi and sintered in dry hydrogen for 44 hours at 1300°C. (2370°F.); ×120; A, 18–8 Cr–Ni stainless steel; and B, 13–13 Cr–Ni steel containing 1.8% tungsten, 0.3% silicon and 0.4% carbon. (Courtesy of Charles Hardy, Inc.)

nization progressing more rapidly with the use of finer powder. (2) The rate of diffusion increases exponentially with temperature. (3) If the metal powders undergo phase changes during heating, they may be affected by diffusion. (4) The interchange of atoms is affected negatively by all those factors that obstruct adhesion as, for example, poor contact of individual particles due to too large distance or interfering gas or oxide films. (5) Two metals diffuse into each other with the formation of layers of reaction products equal in number to the phases stable in the system at the temperature of diffusion. (6) The rate of diffusion is materially in-

[98] W. D. Jones, *Principles of Powder Metallurgy*. Arnold, London, 1937, p. 112.

creased if small quantities of a liquid phase are present, especially if the liquid phase is able to dissolve oxide films or other interfering impurities. (7) When one constituent is molten the phase to appear first is generally that at the end of the equilibrium diagram corresponding to the molten metal. (8) Otherwise, the same laws apply to the diffusion of metal powders that govern the diffusion of solids.

Nonsolubility of Components. In systems that consist of two or more mutually insoluble components, all characteristics of the sintering of monometallic compacts apply to the same extent, provided that sintering takes place at a temperature at which no liquid phase exists. At higher temperatures, however, recrystallization will occur in each component at its own typical range, and it is possible that the different components may obstruct each other's capacity for recrystallization and grain growth. As examples of this type of multiple-component system a number of technically important combinations used as contact materials may be cited, such as tungsten–copper, tungsten–silver, molybdenum–silver, nickel–silver, copper–graphite, and silver–graphite. It is understood that in all these cases compacts are formed from the individual components, and sintering temperatures do not reach the melting point of the lower melting component.

Comstock has reported[99] high strength values for iron–silver compositions obtained by compression and sintering of a mixture of the two components. The strength and hardness of such composite materials follow the rule of the mixed proportions, based on the properties of the bulk metals, provided that the compact is made substantially free of pores (*e.g.*, by hot forging). Similar results are obtainable with mixtures of iron and copper powders that are sintered below the melting point of copper.[100] The magnitude of the values obtainable for density and tensile strength does not differ much from that of the pure metals, although precipitation hardening effects may have an influence on the strength values. As an example of a system with complete mutual insolubility, copper–graphite compositions may be mentioned. Bal'shin[97] found that the addition of graphite, for such a material, diminishes the shrinkage of copper during sintering, resulting in a lower density and usually in lower mechanical properties. On the other hand, graphite additions effect a more uniform distribution of density, for they act as lubricants during compaction; this may sometimes result in improved strength of the sintered compact. The latter effects are more pronounced with finer graphite powders and coarser copper powders.

[99] G. J. Comstock, *Metal Progress, 35,* No. 6, 576 (1939).
[100] F. Sauerwald and E. Jaenichen, Z. *Elektrochem., 31,* 18 (1925).

INTERFERING SOLID SURFACE COMPONENTS

The last cited combination of copper–graphite constitutes a typical example of a solid phase (graphite) obstructing the previously discussed metallographic changes of the other phase (copper) which would occur undisturbed if the compact consisted only of the pure metal. This mutual obstruction is effected purposely in certain instances, as in the case of tungsten for example, where with the aid of certain foreign additions (thoria or complex silicates), excessive grain growth in the filament is inhibited.

Wretblad and Wulff [101] discuss the interesting problem of solid foreign surface layers and their effect on the sintering process. In practically all powders the particles vary in composition from their surface to the interior, with the surface being covered by one or several of the following nonliquid types of layers: (1) a thoroughly foreign solid phase; (2) an oxide or other chemical compound of the basic metal; (3) a layer characterized by absorbed gases which do not escape during the heat treatment; and (4) a layer not separated from the main phase by any grain boundary, but showing a higher or sometimes lower content of the dissolved substance. Except in the last case, such layers tend to diminish the attractive forces between the particles, as previously shown in the photomicrograph of the sintered magnesium compact (Fig. 202). Moreover, the surface films are an obstacle to the progress of sintering by inhibiting diffusion along and across particle boundaries. Although the layers can be ruptured locally by abrasion during compaction, the remaining part of the films may still seriously diminish the contact between the particles. On the other hand, the existence of a particular surface layer may occasionally promote sintering by the formation of a solid solution or compound (e.g., graphite coatings around iron particles).

Wretblad and Wulff use the existence of surface layers to explain the increase in sintering which generally occurs at the recrystallization temperature, i.e., that particle boundaries in a compact are usually covered by surface films which diminish the attraction between the particles. During recrystallization, new grain boundaries are formed which, as a rule, contain less foreign matter (the latter having been driven off as vapor or dissolved in the main phase, or finally congealed into grains—constituting a new phase). Whereas before recrystallization, two types of boundaries exist, namely the contaminated interparticle boundary and the (clean) boundary between the crystallites within the particles, after recrystallization all boundaries tend to become similar.

[101] P.E. Wretblad and J. Wulff, in J. Wulff, *Powder Metallurgy*. Am. Soc. Metals, Cleveland, 1942, p. 36.

Since recrystallization is connected with the ability of the atoms to overcome diffusion obstacles formed by the boundary areas, the nature and thickness of the layer of foreign substance may have a profound influence on the entire sintering process. Libsch, Volterra, and Wulff,[102] for instance, found that compacts of electrolytic iron, as well as reduced iron oxide—

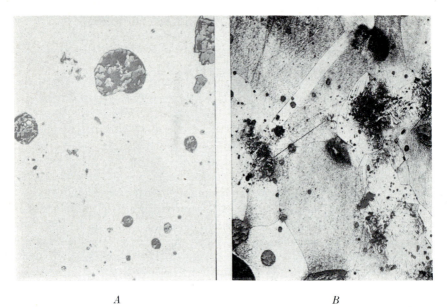

A   B

Fig. 227.   Microsections through polished and etched 18–8 Cr–Ni stainless-steel compact displaying spheroidized oxide inclusions after sintering for 44 hours at 1300°C. (2370°F); ×125; A, polished; and B, etched (Courtesy Charles Hardy, Inc.)

both of high purity—show a very rapid increase in density and strength at 850°C. (1560°F.) (see Fig. 194), whereas compacts from Swedish sponge iron, containing 1.16% insoluble silicates, have a comparable increase in these properties only at 1130°C. (2065°F.).

If the sintering cycle is lengthy, and involves very high temperatures, surface tension forces cause spheroidizing of filmlike or streaky impurities. This is shown in Figure 227, where polished and etched sections of austenitic stainless steel of the 18–8 chromium–nickel type—after sintering at 1300°C. (2370°F.) for 44 hours—are shown. The spheroidized inclusions have their origin in iron and chromium oxide films that adhered to the initial particle surfaces. In Figure 228 a photomicrograph is shown

[102] J. Libsch, R. Volterra, and J. Wulff, in J. Wulff, *Powder Metallurgy*. Am. Soc. Metals, Cleveland, 1942, p. 379.

of an identically treated stainless steel compact that displays a carbide
network constituent which has been precipitated in the grain boundaries
and inside some of the larger grains.

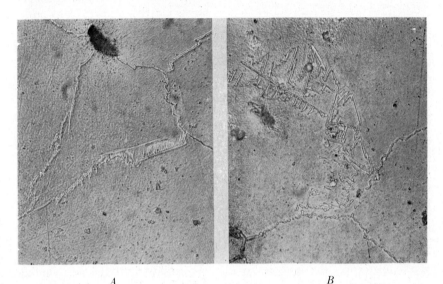

A   *B*

Fig. 228. Microsections through 18–8 Cr–Ni stainless-steel compact displaying
carbide network constituent precipitated out during the sintering treatment at
1300°C. (2370°F.) for 44 hours; ×475; *A*, carbide network constituent in grain
boundaries; and *B*, carbide network constituent inside large grains. (Courtesy of
Charles Hardy, Inc.)

Sintering in the Presence of a Liquid Phase

If the melting points of the components of a system are not very different, as in the iron–nickel, iron–cobalt, or iron–nickel–cobalt systems, the
sintering temperature will always be below the melting point of the lower
melting component. But there are many systems in which a considerable
difference exists between the melting points; in those cases, sintering may
be carried out advantageously above the melting point of the lowest
melting constituent, *i.e.*, while a liquid phase of more or less extensive
proportion is present in the compact. The presence of a liquid phase will
considerably increase the rate of diffusion and thus the rate of alloying.
As long as the major part of the compact is not fused, customary powder
metallurgy methods can be fully employed. To this class of materials belong many of the most important technical sintering materials, such as

the sintered refractory metal carbides, tungsten–copper and tungsten–silver contact materials, and porous copper–tin–graphite bearings. The existence of the liquid phase during the heat treatment of these materials affects the sintering process in all of its aspects, especially if the low-melting part is sufficient to form a continuous matrix.

Homogeneous Alloys. It is again advisable to distinguish those alloys that finally become a solid solution or chemical compound from those that remain in a heterogeneous state. In the case of a homogeneous sinter alloy the liquid phase can exist only temporarily. It is formed at a certain temperature and then absorbed by the base metal, diffusing into a solid solution or compound. Technically important examples are certain binary copper alloys, such as bronzes, brasses or cupronickels. The possible addition of graphite as insoluble constituent may be disregarded in this connection since it does not basically affect relations between the metallic components (acting only as interfering phases in the boundary regions—similar to the pores). The sintering procedure of a 10% tin–bronze usually involves a brief treatment of the compressed powder mixture of copper and tin in a reducing atmosphere at temperatures between 750° and 850°C. (1380° and 1560°F.). After exceeding the melting point of tin (232°C.; 450°F.) this component first exists in the liquid state. With time and rising temperature, the molten tin diffuses rapidly into the surrounding copper particles, so that after the compact reaches the sintering temperature and remains there for a certain length of time a homogeneous solid solution of α-bronze is formed. In Figure 229 this is schematically reproduced according to Sauerwald.[103] Based on colored photomicrographs originally taken by Hall,[104] it shows the actual phase changes that occur during sintering at 810°C. (1490°F.) of a bronze containing 9% tin and 6% graphite. Diffusion between tin and copper was found to commence after one minute at the temperature where tin-rich transitory phases were formed. After 25 minutes, a homogeneous alloy of α-bronze was obtained.

A similar process takes place in the case of sintered permanent magnets of the "Alnico" type, except that here the low-melting aluminum is advantageously pre-alloyed with iron into a powdered master alloy (of 50–50 aluminum–iron ratio) in order to prevent undue oxidation of the liquid metal—which would inhibit diffusion. Sintering of the compacted mixture is usually carried out in dry hydrogen atmosphere at a temperature between 1200° and 1300°C. (2190° and 2370°F.). Upon reaching a temperature of about 1150°C. (2100°F.) the pre-alloyed phase melts and, in the liquid state, diffuses rapidly into the adjacent areas of iron and

[103] F. Sauerwald, *Metallwirtschaft, 20*, 649, 750 (1941).
[104] H. E. Hall, *Metals & Alloys, 10*, No. 10, 297 (1939).

nickel. Here, too, a solid homogeneous phase is obtained within a relatively short time. The high sintering temperature causes appreciable shrinkage, and without subsequent densification, densities of between 97 and 99% of the theoretical values are possible, thanks to the existence of a transitory liquid phase of substantial proportion. (If, for example,

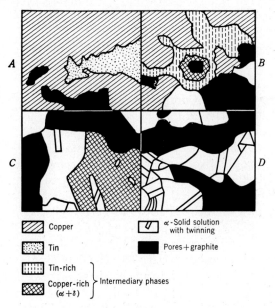

Fig. 229. Schematic representation of the metallographic changes that take place during sintering of a copper–tin–graphite bearing alloy (according to Hall[104] and Sauerwald[103]); A, unsintered mixture; B, after sintering for 3 minutes at 810°C. (1490°F.); C, same, for 15 minutes; and D, same, for 30 minutes.

a composition containing 13% aluminum, 27% nickel, and balance of iron is considered, and pre-alloyed powder of 50–50 aluminum–iron is used, this phase constitutes 26% of the total at the temperature of liquefaction.)

The liquid phase must be kept small enough so that the shape of the compact is not deformed by collapsing or rounding of corners. Generally, however, in those cases where the liquid phase is only temporary, a larger proportion of it is permissible than in compositions where the liquid phase remains throughout the entire treatment at the sintering temperature. The relatively large shrinkage may be explained by the fact that the liquid phase largely removes surface films and irregularities which would otherwise obstruct sintering and crystallization. The rounded grains of the solid metal form closer bonds with the liquid phase because of the increased mobility of a great number of atoms. Surface tension forces of the liquid probably contribute materially to the formation of a more closely knit structure. Once a homogeneous solid solution is formed by this process, further metallographic changes are not different from those experienced with single-metal compacts. The relatively high strength of these materials after sintering may be attributed to the high degree of densification of the structure by close approach of the individual crystallites, as well as to the metallographic changes, such as pronounced grain growth, that take place as sintering progresses.

Heterogeneous and Composite Materials. The conditions, however, are basically different if the sintering of a heterogeneous polymetallic compact containing liquid phase during sintering is considered. The liquid phase may thereby be caused either by the fact that sintering takes place above the melting point of the lowest melting component, or because a low-melting eutectic is formed during the heat treatment. Typical examples of this class of materials are the sintered hard metals and the refractory metal-base contact and heavy alloys, as well as a number of other combinations, notably iron–copper and nickel–silver. In these cases the liquid phase acts as a binder or cement for the solid particles which, without this aid, would sinter into a coherent mass only at very high temperatures—and then probably only into an unsatisfactorily weak compact. The liquid phase, however, can be present only in relatively limited amounts for reasons of preserving the shape of the compact. The sintering process involving a permanent liquid phase will be illustrated briefly by a standard, tungsten carbide–cobalt alloy containing 6% cobalt, and a tungsten-base heavy alloy containing 5 to 6% nickel and 2 to 4% copper.

The sintering of tungsten carbide–cobalt alloys is usually conducted in a reducing atmosphere at a temperature of about 1450°C. (2640°F.). After a temperature of 1400°C. (2550°F.) is reached the pure cobalt disappears and becomes enriched with tungsten carbide—because of limited solubility. During the process the cobalt-rich phase liquefies and a condition arises where the sinter-body consists of about 90% tungsten carbide and 10% of the liquid phase. This is shown schematically in Figure 230A. As long as the temperature is not increased this proportion does not change, even for an unlimited period of time. With increased temperature the amount of the liquid "cementing" phase increases gradually in accordance with the ternary state of equilibrium, while an increasing proportion of the tungsten carbide is dissolved by the melt. In practice, sintering temperatures do not exceed 1500° to 1550°C. (2730° to 2820°F.) because excessive gas evolution causes the formation of surface blisters. Upon cooling of the compact tungsten carbide is again precipitated because of the decreasing solubility of the cobalt in tungsten carbide with falling temperature. This is indicated schematically in Figure 230B. It can generally be observed that the newly precipitated tungsten carbide crystallizes onto the still present, undissolved tungsten carbide particles. This is illustrated

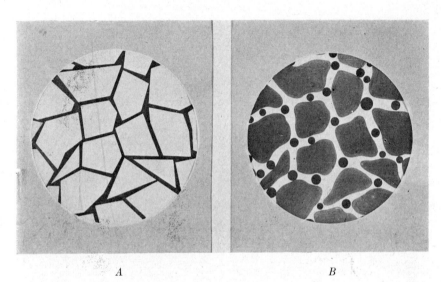

A   B

Fig. 230.   Schematic representation of the grain structure of cemented carbide at sintering temperatures: A, at temperature where cobalt-rich phase liquefies, about 10% liquid phase exists; and B, during cooling from peak sintering temperature, carbide precipitates in form of small crystals adjacent to the larger undissolved carbide grains. (Courtesy Cutanit, Ltd.)

by the microstructure of sintered carbide materials which, as shown in Figure 231, consists of many very small tungsten carbide particles, closely spaced, and separated only by the cementing cobalt network almost completely surrounding each particle. Chemical analysis of the binder after sintering, as well as x-ray studies, have shown that no absolutely free, pure cobalt is present in the finished material, but on resolidification of the liquid phase after sintering, most of the tungsten carbide is precipitated out and less than 1% of it remains in the cobalt at low temperatures. This is considered advantageous from the standpoint of toughness, since iron or nickel, if used as binder, reacts to a much greater extent, forming an undue amount of a brittle constituent. Since the precipitation of the tungsten carbide always occurs on the remaining undissolved particles, and since the smallest particles dissolve most easily, a repeated sintering procedure would result in a diminishing of the finest particles and would thereby cause an increase in the size of the larger tungsten carbide particles. Since this is undesirable from a technical point of view, prolonged periods of sintering are avoided, and the finest possible tungsten carbide powder is used as starting material. In the sintered structure the original

A   B

Fig. 231. Microstructure of cemented tungsten carbide (according to Engle[104a]): A, coarser grain size material containing 13% cobalt; and B, fine grain size material containing 6% cobalt.

fragments of the tungsten carbide crystallites which were produced by ball milling are retained, except for a rounding of the sharp corners and the formation of single-crystal particles. Their form and magnitude are somewhat changed according to the conditions existing during cooling. The advantages of this sintered structure in strength and toughness are apparent, especially if compared with a structure obtainable by fusion, where grains of ten to twenty times the size of the powder particles are crystallized from the melt, and an extremely brittle material results. The introduction of the liquid phase during sintering may be considered an expedient which accelerates the cementing process and improves the toughness of the fun-

[104a] E. W. Engle, in J. Wulff, *Powder Metallurgy.* Am. Soc. Metals, Cleveland, 1942, p. 436.

damentally brittle material. The toughness and strength of the binder, *e.g.*, cobalt, is mainly responsible for these properties, attractive forces between the carbide particles or the formation of a skeleton of carbide particles probably playing only a minor role.[104b] If, for example, the cobalt is chemically dissolved out of the body, the remaining carbide skeleton has but relatively low strength.

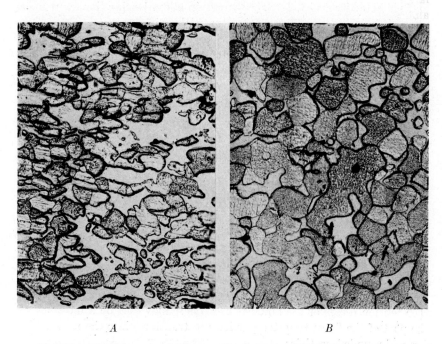

| A | B |

Fig. 232. Microstructure of "Heavy Alloy" (90–6–4 W–Ni–Cu); ×125 : *A*, sintered below the melting point of the binder; and *B*, sintered above the melting point of the binder (courtesy of American Electro Metal Corp.).

Price, Smithells, and Williams[105] established a very similar sintering mechanism for heavy alloys, containing from 87 to 93% tungsten, 4 to 6% nickel, and 2 to 4% copper. The addition of only 5% nickel and 2% copper to very fine tungsten powder resulted in bodies of nearly theoretical density and a tensile strength of over 100,000 psi after sintering for only one hour at 1400° to 1500°C. (2550° to 2730°F.) This treatment led to a linear shrinkage of 17 to 20% and, under best conditions, resulted in a structure practically free of pores. These results are most significant in

[104b] R. Kieffer and W. Hotop, *Pulvermetallurgie und Sinterwerkstoffe*. Springer, Berlin, 1943, p. 134.
[105] G. H. S. Price, C. J. Smithells, and S. V. Williams, *J. Inst. Metals*, **62**, 239 (1938); also *Metal Ind. London*, **59**, 354, 372, 394 (1941).

view of the fact that pure tungsten, even when treated at temperatures closely approaching the melting point of the metal, is still very porous and extremely brittle. The change in the structure with rising temperatures above the melting point of the binder is illustrated by the photomicrographs of Figure 232. The final structure (Fig. 232B) consists of rather large, round, pure tungsten, single-crystal particles, which are surrounded and cemented by a nickel–copper–tungsten alloy. During the low temperature sintering the copper and nickel particles diffuse only partly at first, but the tungsten particles remain unchanged (Fig. 232A). After the melting point of copper is reached, the small part of the copper which is still unalloyed fuses and, with rising temperature, more and more of the copper–nickel alloy is dissolved. At a temperature above 1350°C. (2460°F.) all of the copper–nickel is molten, and is able to dissolve about 18% of the tungsten.[105]   Here, too, the smallest tungsten particles are dissolved first. During cooling of the alloy at least part of the tungsten dissolved at the maximum sintering temperature is precipitated again and, as in sintered tungsten carbides, the precipitate is deposited onto the remaining undissolved tungsten crystallites, resulting in rather large-sized tungsten particles in the final structure. It may be assumed that the comparatively large size of the particles is not entirely due to the precipitation process during cooling, but that during the sintering at the highest temperature fine tungsten particles are continuously dissolved in the liquid phase and then precipitated onto the remaining larger tungsten crystallites by the oversaturated melt. The excessive shrinkage may be explained by the complex solubility conditions during sintering and also by the surface tension forces of the liquid phase. There may also be a tendency on the part of the finest tungsten particles to transform into the more stable form of rounded particles.

The sintering mechanism of the heavy alloys finds an interesting parallel in the high-temperature sintering of iron–copper and iron–carbon–copper alloys[106,107] which may be produced by infiltration of liquid copper into a porous iron or steel skeleton body.[108-111]   The limited solubility of iron into liquid copper at temperature of infiltration causes similar spheroidizing effects on the iron grains—as shown in Figure 232B for the tungsten grains in the heavy alloy. Thus, the copper phase acts not only as a

[106] L. Northcott and C. J. Leadbeater, *Symposium on Powder Metallurgy*. The Iron and Steel Institute, Special Report No. 38, London, 1947, p. 142.
[107] R. Chadwick, E. R. Broadfield, and S. F. Pugh, *Symposium on Powder Metallurgy*. The Iron and Steel Institute, Special Report No. 38, London, 1947, p. 151.
[108] F. P. Peters, *Materials & Methods, 23*, 987 (1946).
[109] E. S. Kopecki, *Iron Age, 157*, No. 18, 50 (1946).
[110] C. G. Goetzel, *Powder Metallurgy Bull., 1*, No. 3, 37 (1946).
[111] C. G. Goetzel, *Product Engineering, 18*, No. 8, 115 (1947).

filler of the pore volume but also as a cementing material which gives strength to the iron body. The characteristic dense structure of such copper-infiltrated iron compositions is indicated in the photomicrograph of Figure 233, representing a composition containing 60% iron, balance copper. The subject of copper-infiltrated iron and steel compacts is discussed in greater detail in Chapter XXV, Volume II.

Jones[112] studied the sintering conditions for a number of other heterogeneous systems, and came to the conclusion that it is not absolutely essential that the melting points of the components be too far apart, as

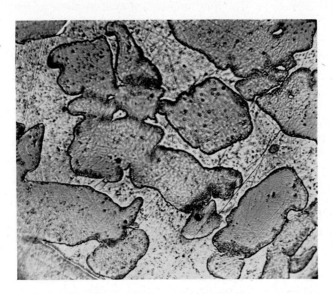

Fig. 233. Microstructure of 60–40 iron–copper composition produced by infiltration of molten copper into a porous iron powder compact at 1100°C. (2010°F.) in hydrogen atmosphere (×750).

long as during sintering very fine particles are in contact with an amount of liquid phase that is capable of dissolving the solid component in part. These conditions are obtainable on a number of systems, and in Table 56 some practical examples are given with the corresponding sintering temperatures.

In conclusion it may be repeated that in this type of sintering the undissolved powder particles are cemented by the liquid phase, and no complete alloying exists. The strength of these composites is dependent on that of the binder, and attractive forces between particles of the matrix

[112] W. D. Jones, *Metal Treatment, 5*, 13 (1939).

material play only a comparatively small role. Important, though, are the forces which form the contact between the particles and the liquid phase, the boundaries between the two phases corresponding to the regular grain boundaries in fused and solidified metals. The contact between the liquid and solid phases must be considered ideal, especially if a certain degree of solubility between the two exists (e.g., in W–C–Co, W–Ni–Cu, Fe–Cu). If such mutual solubility does not exist (e.g., in tungsten–copper), the ideal contact is hardly possible, and the molten phase is not capable of "wetting" the solid particles. Hence, minor alloying additions (e.g., $^1/_2$–1% nickel in the case of copper tungsten) may have to be added to facili-

TABLE 56

Composition and Optimum Sintering Temperatures of
Two-Component Systems[a] (Jones[112])

Alloy		Sintering temperature	
A, %	B, %	°C.	°F.
Lead, 70	Tin, 30	200	390
Tin, 90	Lead, 10	190	375
Lead, 96	Antimony, 4	255	490
Antimony, 70	Lead, 30	300	570
Cadmium, 95	Zinc, 5	275	525
Zinc, 80	Cadmium, 20	275	525
Cadmium, 90	Tin, 10	185	365
Lead, 92	Cadmium, 8	260	500
Gold, 65	Nickel, 35	1000	1830
Copper, 50	Nickel, 50	1275	2325
Copper, 90	Tin, 10	925	1700
Lead, 95	Tellurium, 5	400	750
Aluminum, 78	Magnesium, 22	475	885
Magnesium, 80	Aluminum, 20	450	840
Aluminum, 96	Silicon, 4	600	1110

[a] These systems can be sintered to high density in the presence of a liquid phase

tate the formation of proper contact areas. This latter consideration is of importance if the liquid phase is introduced by impregnating the porous compact by capillary forces. Here, a complete wetting and an evolution of all gases are essential for the production of dense compacts of controlled shape.

Summary

The process of sintering metal powder compacts is based upon atomic bonding between neighboring particle surfaces. A number of characteristic phenomena, such as dimensional changes, take place during sintering.

Shrinkage of the compact is caused by the action of the sintering forces in conjunction with metallographic changes which take place with progressive sintering temperature or time. Swelling of the compact is the result of loosening of the sinter bonds in the early stage of the heat treatment by the eruptive evolution of adsorbed or occluded gases. The rate of heating and the type of atmosphere have a particular effect on this phenomenon.

With rising sintering temperature, the microstructure of the compact undergoes a marked change caused by recrystallization and grain growth, whereby the simply agglomerated powder particles of the compressed compact are transformed into a uniform, equiaxed grain structure which, except for the presence of pores within and between the crystallites, takes on a normal appearance.

The physical and mechanical properties of a compact are affected by the sintering variables in a natural manner. Density, hardness, and tensile properties tend to increase with temperature and time, rapidly at first, and more slowly later. The electrical conductivity changes with sintering temperature in a more complex fashion, a disruption of the compacted agglomerate of particles by spontaneous gas evolution causing an initial decrease which is followed by a rapid increase—indicating the formation of metal-to-metal bonds.

The sintering mechanism of polymetallic compacts is more complex than that of monometallic systems. Two fundamentally different processes must be distinguished. The first refers to sintering of homogeneous or heterogeneous systems in the absence of a liquid phase. Here, the actual sintering processes are not different from those found in monometallic compacts, except that alloying by diffusion occurs at the latter part of the sintering process and that, depending on the degree of diffusion, the properties may be affected materially.

The process of sintering in the presence of a liquid phase may be subdivided into systems in which (1) the liquid phase is only of intermittent character, and (2) the liquid phase remains throughout the high-temperature sintering procedure. In the first instance the liquefaction essentially aids diffusion and homogenization of the alloy, and after solidification the alloy will behave like a homogeneous compact. If, however, the liquid phase is not absorbed by the remaining phase during the treatment, it becomes a cementing material and will govern the strength and other properties of the final product. The particle size of the base metal and the type of contact between the liquid and the solid phase are then of utmost significance. In certain systems partial solubility at the sintering temperature—being facilitated by very fine particle sizes of the base metal—is a distinct aid in the production of dense and strong composite bodies.

CHAPTER XV

Sintering Practices

Practical sintering in powder metallurgy may be linked closely to conventional heat treating of metals and alloys. By heat treatment, one generally understands an operation or combination of operations involving the heating and cooling of a metal or alloy in the solid state for the purpose of obtaining certain desirable conditions or properties; with a few modifications, this definition might well apply to sintering in its practical sense. Strictly speaking, however, the term "solid state" would not be applicable to all cases of sintering. A "substantially solid state" would be a more accurate designation, taking into consideration those compositions which pass through a liquid phase during sintering. The term "compacted state" would have the disadvantage of excluding sintering of noncompacted powders. It is well to keep in mind, besides, that the extremely large surfaces encountered in metal powder compacts (caused by the inherent porosity) require special precautions against chemical reactions, notably oxidation, which are not called for in the ordinary heat treatment of metals and alloys. On the other hand, atmospheric protection, as well as the presence of the multitude of pores, or of nonmetallic inclusions, makes it possible, in certain cases, to apply more drastic heat-treating conditions than would be possible in conventional procedures. For example simple steel powder compacts are sintered advantageously at a higher temperature (between 1000° and 1300°C.; 1830° and 2370°F.) than is necessary (or practical) in the ordinary heat treatment of massive steel of the same composition. The protective atmosphere prevents scale formation at the compact's surface, while the pores or inclusions act as grain growth inhibitors.

With due allowance for the peculiarities inherent in the various individual processes, it may be said that, in general, conventional heat-treating practices have been adopted for the sintering of powdered metals. The different sintering techniques may be classified by certain factors, such as manner of heat application, pressure conditions under which the treatment is carried out, the medium which surrounds the compact during the treatment, conditions that are created by the composition of the compact, and state of the constituents during heat treatment. A brief survey of the various sintering practices based on these variables will follow. In later chapters, the industrially important subjects of sintering furnaces and atmospheres will be treated more fully.

SINTERING BY INDIRECT HEATING

Under sintering by indirect heating may be grouped all those procedures in which the compact to be sintered is heated by convection or radiation from another medium that is heated in a conventional manner (*e.g.*, in a gas-fired furnace or in an electric resistor furnace). The medium may be a liquid, an inert powder pack, or an inert mold; the compact may also be surrounded by a protective atmosphere and heated by radiation from an externally heated muffle or retort, or by direct radiation from the resistor elements.

Sintering by Immersion in Liquids

One of the oldest techniques of powder metallurgy is the submerging of compacts in liquid baths during sintering. The liquid medium permits quick and uniform application of heat to the compact. Upon withdrawal of the part after completion of the sintering operation, the liquid adhering to the compact's surface usually solidifies as a dense film which prevents oxidation on premature exposure to the air.

LIQUID SALT BATH TREATMENT

Most salt baths employed in the heat treatment of metals are also suitable for powder compacts. Depending on the sintering temperature required, the salt baths may be classified as low-temperature, medium- and high-temperature baths. Sodium nitrate, sodium nitrite, and potassium nitrate, either individually, or as eutectic mixtures, may be used for a low-temperature range from 150° to 600°C. (300° to 1100°F.), as in the sintering of compacts from powdered lead, zinc, aluminum, or their alloys. For a temperature range of 600° to 900°C. (1100° to 1650°F.), salts composed of cyanides, chlorides, or chlorides and carbonates, would be suitable. Sodium cyanide (melting at 564°C.; 1047°F.) or a mixture of 50% potassium chloride and 50% sodium carbonate (melting at 585°C.; 1085°F.), for example, would be satisfactory for sintering of copper and copper alloys, while for sintering of ferrous metal compacts a mixture containing 56% potassium chloride and 44% sodium chloride (melting at 660°C.; 1220°F.) would be preferable. If still higher sintering temperatures are required, bath mixtures containing barium chloride, sodium fluoride, borax, silicates, etc., suitable for a temperature range of 900° to

1300°C. (1650° to 2350°F.) with a melting range between 850° and 1050°C. (1560° and 1920°F.) must be selected.

A number of difficulties must be overcome in using salt baths. If a relatively high temperature is required for sintering, a bath of a high-melting temperature must be selected, otherwise losses through evaporation would be too great. It is seldom possible to heat a compact instantaneously to its high-sintering temperature, since the trapped gases would tend to escape in an explosivelike manner, thus disrupting the sinter bonds; heating should be gradual in these instances. Since the compact must be protected against oxidation from the earliest stage of heating, and this cannot be accomplished by the salt as long as it remains solid, the compact has to be heated in another medium to the fusion temperature of the high-temperature bath under protective atmospheric conditions. In practice, this implies heating of the compact first in a low-melting bath, followed by a quick transfer into the high-melting salt. Special precautions against oxidation during this transfer may become necessary.

Probably the most objectionable feature of salt bath sintering is the corrosiveness of the salts. A porous piece, upon immersion into the fused salt, would tend to take up a considerable amount of the fused salt through its interconnected pores, and the difficulties of removing the salt (after cooling) by washing or other methods of extraction are very great. This fact has undoubtedly been the determining factor in discarding the fused-salt sintering method for all practical purposes; only where very dense parts are to be sintered would the method hold hope for practical use.

Sintering by immersion in fused salt has produced interesting results for very dense objects in laboratory work. For particularly fine work, such as medals or engravings, where the slightest distortion would cause the rejection of the part, immersion sintering offers attractive possibilities. It has been reported, for example, that bronze medals have been sintered effectively by being dipped for a few minutes in fused sodium cyanide at a temperature of 675°C. (1245°F.).[1] During the submersion the medals were held in a clamping device while being separated from each other by means of mica sheets. After they had been withdrawn from the bath, the salt was washed off in boiling water; after drying, the medals showed no signs of distortion, and could be electroplated without difficulty. In contrast to these experiments, it has been found that silver compacts sintered in various salt-bath compositions suffered a certain degree of distortion due to volumetric changes.[2] Silver and silver alloy coins were submerged at temperatures between 720° and 770°C. (1330° and 1420°F.) for 10 to

[1] C. Hardy, *private communication.*
[2] C. Hardy, *private communication*, based on Laboratory Report No. 148 of Chas. Hardy, Inc., New York.

30 minutes in such salts as barium chloride, a eutectic mixture of 70% barium chloride and 30% sodium chloride, or a eutectic mixture of 50% barium chloride, 30% potassium chloride, and 20% sodium chloride. Regardless of composition, the coins increased in size during sintering because of spontaneous gas evolution upon submersion. The diameter increased from 3.5 to 8%.

LIQUID METAL BATH TREATMENT

Liquid metals may become additional material for sintering baths. Liquid lead is being used extensively in the heat treatment of steel, and has found some use as a sintering medium, especially in applications where it is essential for the compact to have a high specific gravity. In contrast to immersion in liquid salts, the compacts must be held submerged rigidly in lead on account of the latter's high specific gravity. Liquid lead tends to oxidize strongly, and the surface of the bath must be protected from the atmosphere, since lead oxide is corrosive to most metals. With porous compacts, penetration of lead cannot be prevented, and may present difficulties, since the removal of lead from the pores is practically impossible.

On the other hand, especially for relatively dense parts, there are some advantages in the use of liquid lead baths as sintering media. Not only may lead be melted and heated more rapidly than salt baths—owing to the higher conductivity of the metal—but for the same reason, the liquid lead heats the compact more rapidly and uniformly. Aside from the tendency to oxidize, lead maintains a stable composition, whereas most salts undergo some decomposition during use or when stored, since they are to some extent hygroscopic.

Copper and silver are also very interesting as sintering media. These two metals melt at considerably higher temperatures, so that compacts that have to be presintered to facilitate degassing must first be heated under protective conditions in other facilities. But liquid copper and silver, being less corrosive and having a lower vapor pressure, are easier to work with. Only in the case of copper are precautions necessary against oxidation of the bath surface. The submersion of relatively porous compacts of such metals as tungsten and molybdenum into liquid copper or silver has become an important process in the manufacture of composite contact metals.[3] In this process the compacts are either submerged directly, and sintering allowed to proceed while the liquid metal penetrates the skeleton, or the compacts are presintered under other conditions and infiltration of the liquid bath becomes an additional treatment.

[3] R. Kieffer and W. Hotop, *Pulvermetallurgie und Sinterwerkstoffe.* Springer, Berlin, 1943, p. 206.

One of the disadvantages of the method of sintering in liquid metal baths is the fact that a coating of the solidifying bath metal remains attached to the compact after withdrawal from the bath. It is usually rather difficult to control this coat; mutual solubility between the compact and bath metal may result in surface imperfections which are frequently incompatible with fabrication of finish-molded parts. Although chemical stripping of the coat has eliminated this problem in some cases, the inherent danger of corrosion in many metal combinations has prevented this expedient from becoming a general cure. Therefore, the method of sintering in liquid metal baths, especially of copper and silver, has been largely confined, so far, to the fabrication of blanks—from which individual parts (such as electrical contacts) are obtained by machining.

Sintering in Powder Packs and Solid Molds

Sintering of compacts by burying them in an inert powder or mold provides a very simple technique; it has been in widespread use since the early days of producing parts from powders. If the pack is contained in a tightly sealed box, sintering may be carried out without introduction of a reducing atmosphere. The oxygen from the air which is stored in the packing mass usually reacts with the surface of the compacts at an early stage of the sintering cycle, but is insufficient to cause serious oxidation if they contain a reducing element. Bal'shin and Korolenko,[4] for example, sintered compacted mixtures of iron powder and comminuted cast iron in boxes filled with a pack consisting of 70% sand and 30% graphite by heating the charge to 1100°C. (2010°F.) for 5 hours, and cooling in the furnace. Other materials suitable for packing include cast iron chips and powdered charcoal,[5] which may be mixed with aluminum oxide, beryllium oxide, zirconium oxide, or powdered lime. If the compacts embedded in the pack are made of ferrous powders, the amount of carbon in the pack must, of course, be controlled closely, and of smaller proportion than tolerable for the sintering of nonferrous parts. The diffusion of carbon into an iron compact at higher sintering temperatures is rapid, and a case tends to form readily. One way of inhibiting case formation is to paint the green compacts with a paste of lime or other refractory oxide, which will bake into a solid coat during sintering. A similar method of sintering in a powder pack has been employed successfully in Germany during the war, and is now reported in use in a large powder metallurgy plant in Austria in the production of sintered steel parts, magnets and iron alloys.[5a]

[4] M. Yu Bal'shin and N. G. Korolenko, *Vestnik Metalloprom.*, *19*, No. 3, 34 (1939); also *Metallurgia*, *22*, No. 128, 59 (1940) (in English).

[5] W. J. Baëza, *A Course in Powder Metallurgy*. Reinhold, New York, 1943, p. 107.

[5a] P. Schwarzkopf, *private communication*.

It involves embedding of the parts into a pack consisting of approximately one-half pure iron powder and one-half of a 50–50 prealloyed iron–aluminum powder. To this mixture, up to 5% aluminum oxide powder is added. The pack is contained in heavy, covered, graphite boats, and acts as a getter for the commercial furnace atmosphere (*e.g.*, hydrogen or dissociated ammonia). The graphite casings prevent excessive decarburization of the charge. By proper selection of the packing substance, pack-

Fig. 234. Standard batch-type furnace equipment for heat treating and carburizing that is suitable for pack sintering (courtesy of American Electro Metal Corp., and Leeds and Northrup Co.).

sintering may also be combined with such surface treatments as chromizing, silieonizing, Sherardizing, and the like (See also Chapter XX.) Figure 234 shows a heat-treating and carburizing furnace arrangement that can be used for pack-sintering operations.

Sintering of compacts in solid casings is possible as well as appropriate where complicated shapes are to be treated. Variations in cross section tend to cause deformation and warpage of the part during sintering, and a solid form—acting as a mold—will help more in keeping the shape of the part than would a loose powder pack. Molds suitable for this purpose, consisting of inert ceramics, such as aluminum oxide, are formed to fit the contours of the compact. Allowance must of course be made for the difference in the coefficient of expansion (and shrinkage) of the mold material. The mold is first formed in the pasty state and is then baked at high temperature. If the contours of the compact are such as to prevent its direct insertion into the mold, the latter must be sectioned, and held together during sintering by means of clamps. Sintering of nonferrous compacts may be accomplished in graphite molds, which have the added advantage of providing a reducing atmosphere. Molds of the ceramic type may be sealed sufficiently tight to prevent the access of external oxygen to the compact's surface. Oxidation of the compact by exhausting the oxygen in the air included within the compact may be appreciable and may lead to decarburization effects in steel powder compacts. The use of a protective atmosphere (through which the molds containing the compacts pass) may therefore be necessary.

The use of closed boxes containing powder packs or of sealed molds has been found particularly interesting in the case of brass. Sintering of brass compacts is always connected with a certain degree of dezincification. This is known to be particularly serious in compacts pressed from powder mixtures, with compositions rich in zinc, or if sintering takes place in large chamber furnaces through which charges and atmosphere pass in continuous fashion. Apparently, these zinc losses can be reduced materially when the compact is placed in a mold or a small sealed container where the atmosphere rapidly becomes saturated with zinc vapor and inhibits further dezincification of the alloy.

Sintering in Controlled Atmospheres

BATCH-TYPE SINTERING

In most instances sintering must be carried out under protective atmospheric conditions. The compacts may rest freely on stationary bases which are often stacked in many layers; spacious boxes, retorts, or muffles serve to retain the protective atmosphere. Some furnace construction provide muffles built from the furnace brickwork itself. Where atmospheric requirements are exacting, the entire furnace must be purged and the atmosphere retained in a gas-tight steel housing; where the requirements are more lenient, it is usually sufficient to pass the protective atmosphere over the furnace hearth. Boxes and retorts are generally constructed from high-temperature-resistant metals since they can be gas-proofed more easily than muffles from refractories or brickwork. Stainless steel, nickel–chromium alloys, and pure nickel—because of their high resistance to oxidation and creep—are most suitable for container material. Low-carbon steels are less costly and easier to work with when fabricating the retort, but excessive formation of oxide scales reduces their life considerably if the container is heated in ordinary furnaces under oxidizing conditions.

To eliminate contamination of the charge by oxidation from the outside, the retort must be closed with a cover that remains sealed during the entire sintering and cooling cycle. Only in this way can the protective atmosphere be really effective. This is not an easy task, since most seals loosen with changing temperature. If the entire metal retort is put into the furnace—which may be desirable from the standpoint of temperature equilibrium—it is usually very difficult to find a suitable sealing or gasket material that can withstand the high temperatures required for sintering. Good seals can be obtained by applying refractory cements, but they are usually destroyed by the contraction of the metal while the box is cooling. Surface oxidation of the charge is then a common phenomenon.

An interesting discussion of retort designs suitable for sintering was recently given by Kelley.[5b]

CONTINUOUS SINTERING

If the charges to be sintered are large in quantity, the batch-type operation employing stationary retorts becomes cumbersome and uneconomical. Sintering is then best performed in continuous furnaces of the type used in the metal heat-treating industries. Instead of the retorts containing one batch of compacts at a time, metal muffles retain a steady stream of protective atmosphere while the charge is passed through the muffle. The compacts are usually placed on trays which are propelled either by a roller hearth or a stoker system; they may also be placed directly on a conveyor belt, or may rest on a tray that is carried by the conveyor. In Figure 235 a production layout is shown where molded compacts are conveyed from the press directly to the charging stations of stoker-type continuous sintering furnaces.

There are numerous technical advantages inherent in the continuous sintering procedure. Problems of maintaining good sealings from outside oxidation are less aggravated, since doors or gates can be installed in cool sections of the muffle. A liberal flow of protective gas aids further in the prevention of oxidation caused by outside air infiltrating the furnace

[5b] F. C. Kelley, *Iron Age*, *161*, No. 21, 84 (1948).

chamber. Since the charge travels continuously from room temperature to the maximum sintering temperature, and back to about room temperature in a cooling tunnel, each lateral row of compacts undergoes the same temperature cycle. Thus, a temperature gradient is confined to the cross section of the hearth of the muffle. Continuous sintering also permits flexible charging of the furnace and considerably less manipulation than in batch-type sintering.

Fig. 235. Conveying of compacts to charging stations of stoker-type continuous sintering furnaces (courtesy of American Electro Metal Corp.).

Muffles, trays, and transport equipment subjected to heat must be constructed from high-temperature-resistant materials such as nickel–chromium alloys. The life of these items depends largely on loading and operating conditions, and on the materials' capacity to resist plastic deformation at high temperature.

Sintering under Different Pressure Conditions

Although sintering at atmospheric pressures is the rule, there are certain noteworthy exceptions, and pressures below, as well as above atmospheric pressure may prevail during the treatment.

SINTERING IN VACUUM

A vacuum can be maintained in metallic as well as in ceramic muffles, provided that the system is sufficiently tight. Most metals become permeable to gases at higher temperatures, and it is difficult to obtain a vacuum sufficient for the purpose of sintering in the complete absence of oxygen. But certain nonmetallic materials, such as quartz or sillimanite, are impervious to gases even at very high temperatures (1200°C. (2190°F.) maximum for quartz, 1500°C. (2730°F.) and over for sillimanite), and are therefore excellent muffle materials for vacuum sintering by indirect heating. The best muffle construction provides one closed end and a tightly sealed cover—usually made of rubber—at the opposite, cold end. Sealing wax or plastic cements are used for the vacuum seal. For example, a furnace was used for vacuum sintering of copper compacts that maintained (for 16 hours) a vacuum of less than 1 mm. in a quartz tube which was heated to 900°C. (1650°F.) in a resistance furnace.[6] The arrangement is reproduced in the sketch of Figure 236.

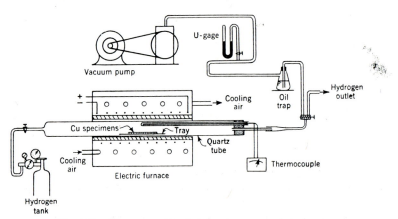

Fig. 236.   Diagrammatic sketch of vacuum-sintering
set-up for laboratory work.

While solidly closed retorts (implying batch treatment) are best for vacuum sintering, it is still possible to sinter continuously in vacuum. The seal must then be made by a liquid, such as lead. A continuous vacuum furnace, for instance, has been developed by the Moraine Products Divi-

[6] C. G. Goetzel, *The Influence of Processing Methods on the Structure and Properties of Compressed and Heat-Treated Copper Powders, Dissertation.* Columbia University, New York, 1939.

sion of General Motors Corporation,[7] in which a copper–nickel sponge-covered steel strip is impregnated with a lead alloy while traveling through the bath; the liquid alloy provides the vacuum seal. The same furnace could undoubtedly be used with only minor changes at temperatures above those needed for the infiltration of the lead alloy.

SINTERING AT PRESSURES ABOVE ATMOSPHERIC

If sintering is to be carried out above normal pressures, two different approaches must be distinguished. The protective gas may be maintained above atmospheric pressure or sintering may take place while the compact as such is subjected to pressure. In the first instance, pressures of from 2 to 5 atmospheres can be obtained in completely closed pressure vessels. Such an arrangement, of course, makes batch sintering mandatory. Sintering above atmospheric pressure has been found helpful in excluding all traces of oxygen that may penetrate the muffle. It is also known that pressures above atmospheric increase the rate of reduction of iron oxide by hydrogen[8] and carbon monoxide.[9] This is especially interesting where the sintering of very sensitive compositions, as stainless steel or chromium, demands the complete elimination of oxygen.

The situation is quite different if a static pressure is only applied locally to the compact in order to achieve greater densification. In this case the atmosphere remains at normal pressure, and no special pressure doors or other seals are necessary. Consequently, pressure sintering can be carried out in batch furnaces and in continuous furnaces. The pressure applied to the compact must be a multiple of that applicable to the atmosphere if substantial gains in the compact's density are to be achieved. This is accomplished in practice either by applying a static load, such as by a stack of trays filled with charges, or by applying spring-loaded pressure—provided the sintering temperature is below the annealing temperature of alloy-steel springs, i.e., below 800°C. (1470°F.). Several processes use the pressure-sintering method; this seems particularly advantageous where bonding to a steel backing is to be obtained in conjunction with the sintering of the powder layer. Best-known examples for this process are the sintering of friction materials[10,11] and of bearings. Figure 237 shows

[7] A. L. Boegehold, in J. Wulff, *Powder Metallurgy.* Am. Soc. Metals, Cleveland, 1942, p. 525.

[8] M. Tenenbaum and T. L. Joseph, *Trans. Am. Inst. Mining Met. Engrs.*, *135*, 59 (1939).

[9] M. Tenenbaum and T. L. Joseph, *Trans. Am. Inst. Mining Met. Engrs.*, *140*, 106 (1940).

[10] E. F. Cone, *Metals & Alloys, 14*, No. 6, 843 (1941).

[11] C. S. Batchelor, *Metals & Alloys, 21*, No. 4, 991 (1945).

a sketch of a furnace arrangement for the sintering of friction materials; pressure is applied to the stack of disks by static load, and sintering and bonding of the powdered material to the steel disks take place simultaneously.

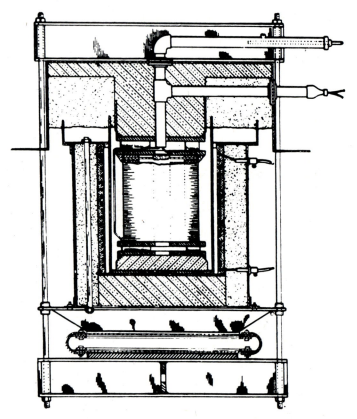

Fig. 237. Diagrammatic cross-sectional view of a sintering furnace showing an arrangement of stacked discs of friction elements with the bell-type heating unit applying downward pressure (according to Cone[10]).

Indirect Heating by High-Frequency Current

Indirect heating of powders and compacts by inducing current into a graphite crucible loaded with the charge is a conventional method in the manufacture of cemented carbides. In this case the charge receives its heat by convection or radiation. Figure 238 shows a graphite crucible heated by high-frequency current, such as is suitable for production of tungsten carbide, or tungsten–titanium carbide powders, at temperatures between 1500° and 2000°C. (2730° and 3630°F.). A similar arrangement would also be adaptable for the sintering of cemented carbide bits, except that in

Fig. 238. High-frequency induction sintering of cemented carbides (courtesy of American Electro Metal Corp.).

the crucible the compacts would have to be separated by graphite spacer disks, fastened to a central graphite trunk (tree-fashion).

SINTERING BY DIRECT HEATING

Direct resistance heating and high-frequency induction heating are the two methods that fall under the heading of sintering by direct heating. They have two advantages over the previously described methods: (1) the efficiency of applying heat to the compact is greater and (2) considerably higher temperatures are obtainable. On the other hand, direct resistance heating can only be effected in one operation on individual bar-shaped compacts; high-frequency heating is possible with a small batch of equal-sized parts.

Direct Resistance Sintering

This method is largely confined to the sintering of refractory metals, where full advantage can be taken of the possibility of obtaining very high temperatures in a reasonable length of time, and where the high market value of the end product permits sintering of individual units. Tungsten, molybdenum, and tantalum rods, and ingots for wires or strips are always sintered by having an electric current pass through them. Electrodes and contact materials are also frequently sintered in this way. The refractory metal powder must be compressed into a shape suitable for conduction of the required current. Bars of square or rectangular cross section whose lengths are about ten to twenty times the width are usually used; the current is then about 1000 amps. per square centimeter.[12] The compacted bar is placed in a clamping device with the top clamp making solid contact with the current lead wire, while the bottom clamp is submerged in a mercury bath to maintain contact while the bar first expands and then shrinks (up to 15% in length) during sintering. Since green bars are very fragile, presintering at 1000°C. (1830°F.) in a muffle furnace usually precedes the main operation. An added advantage lies in the fact that presintering establishes a certain amount of intimate sinter bonds which reduce the very considerable initial resistance of the bar, permitting the use of simpler apparatus and transformers of a narrower voltage range.

The bar and clamping device are enclosed in a tightly sealed sinter bell of such dimensions that the walls do not overheat. Figure 239 shows a picture of two sinter bells in operation. The upper contact clamp, the bell, and the mercury bath are water cooled. Sintering is usually performed in hydrogen, although it may also take place in vacuum, requiring a somewhat more complicated apparatus. Temperature control is maintained by regulation of the current, and visual temperature checks may be made through a sight glass in the bell. The electric current must be adjusted carefully and controlled according to the material to be sintered. Very high current densities may result in an unfavorable grain structure caused by local fusion in the grain boundaries. This difficulty is enhanced by the additives usually employed in lamp filament manufacture. With completed volatilization of these additives and with progressive sintering, the current density can be increased gradually. Heating is continued for about one hour before the final sintering temperature is reached. The maximum

---

[12] P. E. Wretblad, in J. Wulff, *Powder Metallurgy.* Am. Soc. Metals, Cleveland, 1942, p. 425.

current density is selected at about 95% of the current required for fusing the bar. The corresponding temperature is in excess of 3000°C. (5430°F.). Under these conditions, sintering is generally carried on for 15 to 30 minutes. Cooling of the bar in a protective atmosphere follows the break of the electric circuit; its rate is increased by the water-cooling system serving the clamps and the bell.

Fig. 239. Bell-jar equipment for sintering tungsten and molybdenum bars in hydrogen atmosphere (courtesy of P. Schwarzkopf and American Electro Metal Corp.).

The method has also been applied to shapes such as cubes and cylinders of small height.[13] The green compact is put between two graphite slabs serving as electrodes. To prevent distortion or uneven heating, due to imperfect contact, a clamping device holds the graphite blocks tightly to the surface of the compact. The entire arrangement may be enclosed in a protective atmosphere.

[13] C. Hardy, *unpublished notes*.

Induction Sintering

Sintering by high-frequency induction heating is the most modern technique. As in the case of resistance sintering, the heat can be generated exactly where needed, namely in the compact itself. However, it must be taken into consideration that induction heating produces maximum temperatures only at the surface of the compacts. In spite of this limitation the temperature distribution under favorable circumstances can be more uniform than with direct heating methods, thus facilitating sintering and diffusion by more effective heating. Rapid heating and cooling are additional advantages inherent in this method.

In recent years methods and apparatus have been developed which permit direct induction heating of the compact itself.[14,15] This requires that no metal or graphite be placed into the high-frequency field; the choice of a crucible and a container to retain a controlled atmosphere or a vacuum narrows down to quartz or a dense ceramic, such as sillimanite. The compacts are usually embedded in an inert insulating powder, such as alumina, magnesia, or zirconia. Only if the container is placed outside of the high-frequency coil can it be constructed of metal. Bells made of copper sheet, having a diameter about 5 to 10 times that of the coil, are used in carbide manufacture. The fact that, on heating, the resistance of green compacts decreases drastically with the formation of effective sinter bonds can be utilized for the regulation of the high-frequency current and for the automatic control and maintenance of a specific sintering temperature.[16]

By adopting the previously described method of placing the compacts on separator trees (which must be of nonconducting materials if direct induction heating of the compacts is to be accomplished), a fairly large-size charge may be heated in one batch. For example, it is believed possible that a 50-kw., high-frequency converter would be capable of heating a 50-pound net charge of iron compacts to 1300°C. (2370°F.) in approximately 1 hour; cooling would take about the same time. The crucible necessary to handle 50 one-pound iron compacts would have to be about 9 inches in diameter and 15 inches high. The compacts would have to be arranged concentrically in the field, each piece equidistant from the high-frequency coil, with no specimen touching its neighbors.

Experiments with chromium and stainless steel mixtures have shown that direct, high-frequency induction sintering may materially accelerate sintering and diffusion. Sintering times of only one-tenth and less are

[14] U. S. Pat. 2,228,600.
[15] Brit. Pat. 405,983.
[16] U. S. Pats. 2,235,835 and 2,247,370.

necessary as compared with sintering in an electric muffle furnace under otherwise identical conditions.[17]   The reason for this rather significant phenomenon is not easily understood. The explanation that a higher temperature or that more intense heat is vested in the compact heated by high frequency is not entirely satisfactory, since the techniques for temperature measurement have been greatly improved in recent years.

Summary

Several methods are available for carrying out the sintering operation by indirect heating of the charge. Sintering by submersion in a liquid salt or metal bath causes uniform heating but is not generally used on account of the corrosive effect of the salt or the difficulties encountered in removing a residue of the bath left in the pores. Sintering by embedding the compact in a heated powder pack of an inert or partially reducing material is a simple method, but requires careful control of the composition of the pack in order to prevent undesired reactions with the compact. Sintering in heated ceramic or graphite molds that closely fit the contours of the compact may be advantageous where complexity of the shape would cause deformation of unsupported sections of the part. The most conventional method, by far, is sintering by placing the compact freely on a base or tray within a box or muffle which is filled with a controlled atmosphere and heated externally. For large-scale operations continuous transport of the trays through the muffle is preferable. The atmosphere protecting the compact from oxidation may be above or below normal atmospheric pressure. Special attention must be given to the tight seal of the doors if a vacuum or an atmosphere under pressure is to be used. Pressure sintering may also be effected by subjecting the individual compact to a static load.

There exist two methods of generating heat directly in the compacts. Direct resistance sintering is universally employed in the manufacture of refractory metal bars and ingots for wires and strips. A conventional design provides for the clamping of the prismatic shapes inside of a bell, retaining the protective atmosphere. Extremely high temperatures (exceeding 3000°C., 5430°F.) can be reached effectively in this manner with regulation of the current input controlling the heat treatment. The other method, generating heat in the compact by high-frequency induction heating, is less commonly used. Although rates of heating and cooling are comparable with those obtainable in resistance sintering—provided that the high-frequency furnace is sufficiently powerful—the maximum temperatures rarely exceed 2000°C. (3630°F.). But this method does not restrict

[17] C. Hardy, *private communication,* based on Laboratory Report No. 107 of Chas. Hardy Inc., New York.

the form of the compact to a cylindrical or prismatic bar, and makes possible the simultaneous sintering of many compacts of identical shape. Thus, high-frequency induction sintering, though a batch-type operation. is in its technical and economical aspects more comparable to the methods of sintering by indirect heating. In view of the wide temperature range possible by high-frequency sintering, coupled with the advantages of quick rates of heating and cooling, effective introduction of the heat into the compact itself, and clean operating conditions all around, it appears likely that this method of sintering will emerge in the near future as one of the most important improvements in powder metallurgy practice.

CHAPTER XVI

Sintering Furnaces

Unlike the situation existing for presses—where special machines have been developed and standardized for powder metallurgy purposes—there exists at present no standardization of furnaces and furnace atmospheres for sintering. For this process the industry is entirely dependent on the controlled-atmosphere furnaces which have been developed for other applications, such as furnace brazing, hardening, and bright annealing. Fortunately, a wide variety of such furnace equipment can be used for sintering—usually requiring only minor modifications. The purpose of this chapter is to survey briefly the various types of sintering furnaces used here and abroad.

In monometallic systems the sintering temperature is commonly fixed between $2/3$ and $7/8$ of the absolute melting temperature ($a = 0.67$ to 0.87), the notable exception being the refractory metals which are sintered at temperatures very close to their melting points ($a = 0.90$ to 0.95). In polymetallic systems consisting of components with widely differing melting points sintering is usually carried out above the melting point of the lowest melting component. Typical sintering temperatures are:

Material	Temperature	
	°C.	°F.
Brasses and bronzes	650–850	1200–1560
Copper and high melting copper alloys	750–1050	1380–2550
Iron, nickel, and cobalt alloys	1000–1400	1830–2550
Refractory composites	1200–1500	2190–2730
Hard metals	1400–1600	2550–2910
Refractory metals	2000–3000	3630–5430

Sintering time, like sintering temperature, is dependent upon the type of material being processed. Sometimes a sintering time of ten minutes is sufficient, as in the case of porous bearings. In other instances, such as for refractory metals, the sintering time is about $1/2$ hour. For many applications involving iron-base alloys, sintering time is at least one hour, and hard metals, permanent magnets, or stainless steel alloys usually require several hours of sintering at the maximum temperature. Generally speaking, sintering time is inversely proportional to sintering temperature.

TABLE 57

Temperature Range and Applications of Different Sintering Furnace Constructions

Furnace type	Method of heating	Temperature range, °C. (°F.)	Application
Crucible furnace	Gas Nichrome	Up to 1100 (2000)	Batch sintering of nonferrous parts where uniform heating and slow cooling are essential
	Molybdenum	700–1400 (1300–2550)	Batch sintering of ferrous parts
Bell-type furnace	Nichrome Kanthal	Up to 1100 (2000)	Batch sintering of larger charges of small pieces, especially of nonferrous alloys
Box furnace	Nichrome Kanthal	Up to 1100 (2000)	Batch sintering: nonferrous parts, porous bearings, and filters
	Globar	700–1350 (1300–2450)	Batch sintering of ferrous alloys, machine parts, soft-magnetic materials
	Molybdenum	700–1500 (1300–2750)	
Tube furnace	Gas Nichrome Kanthal	Up to 1100 (2000)	Batch or continuous sintering of nonferrous alloy parts, porous bearings, filters
	Globar	700–1350 (1300–2450)	Batch or continuous sintering of ferrous alloys, machine parts, soft magnets, permanent magnets, composite metals, hard metals
	Molybdenum	700–1500 (1300–2750)	
Continuous production furnace	Gas Nichrome	Up to 1100 (2000)	Continuous sintering of nonferrous articles on a mass production scale
	Globar	700–1200 (1300–2200)	Continuous sintering of ferrous articles on a mass production scale
Vertical furnace	Molybdenum	700–1400 (1300–2550)	Continuous sintering of small parts where uniform heating is essential
Carbon short-circuiting furnace	Carbon	Up to 2000 (3600)	Sintering of hard metals and composite metals
High-frequency induction furnace	Induction	Up to 3000 (5400)	Sintering of hard metals and composite metals
Sinter bell jar	Resistance	Up to 3200 (5800)	Sintering of refractory metals

In Table 57 a number of industrial furnaces suitable for sintering are given, with operating temperature ranges and fields of application. For sintering operations up to a temperature of about 1050°C., (1920°F.), either gas-fired furnaces or electrically heated furnaces with nickel–chromium (Nichrome) or iron–chromium–aluminum (Kanthal) elements are satisfactory. Up to about 1350°C. (2460°F.), electric furnaces equipped with silicon carbide (Globar) elements are usable, while for a sintering temperature range of 1000–1600°C. (1830–2910°F.), furnaces with molybdenum windings or hairpin elements are in use. High-frequency furnaces are useful up to 1800–2000°C. (3270–3630°F.). Where a carbon monoxide-containing furnace atmosphere is not objectionable (*e.g.*, in sintering cemented carbides) a carbon-tube, short-circuit furnace can be used. This type of furnace, more favored in Europe than in this country, can also produce temperatures up to 2000°C. (3630°F.), although such high temperatures would not be permissible in carbide production. For the sintering of refractory metals where temperatures up to 3000°C. (5430°F.) are required, special "sinter bell jars" are used in which the current in a protective atmosphere passes directly through the bar (see page 586, Chapter XV).

It appears appropriate to distinguish between batch-type and continuous furnaces. Batch-type furnaces are convenient for smaller scale operations and for diversified applications because of the simplicity and flexibility of the equipment. Ordinary crucible furnaces, box furnaces, and muffle and retort furnaces belong to this class; gas, as well as electricity, may provide the heat. Although of specialized construction, modern high-frequency furnaces and resistance-heated sinter bell jars also belong to this group. In large-scale manufacture of parts, such as porous metals bearings, ferrous machine parts, etc., the problem arises of sintering in quantity a large variety of sizes and shapes. For such applications the continuous furnace type is more suitable, both from a technical and an economic point of view. Continuous furnaces, too, may vary in construction, and conveyor-belt, roller hearth, and pusher-type furnaces are the principal variations.

BATCH-TYPE FURNACES

Crucible Furnaces

Probably the simplest type of furnace adaptable for sintering is the ordinary crucible or pot furnace, either fired by gas or heated by electrical resistor elements, as shown in Figure 240. Although the furnace may be built as an inexpensive and handy laboratory tool, it also permits small-scale production (up to 50-pound charges) as in the type shown in Figure 241.[1] All liquid-bath sintering must be performed in this type of furnace, with refractories or graphite most suitable as crucible materials. The normal operating temperature ranges from 500° to 1100°C. (930° to 2010°F.), and if this range is not exceeded, cast iron crucibles can also be used. Tight-fitting covers permit sintering in a pack of charcoal or inert substances in gas-tight conditions. If the cover is fitted with

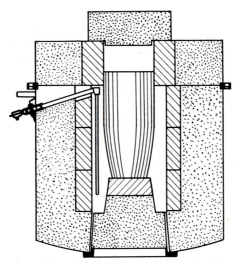

Fig. 240. Diagrammatic sketch of cross section through crucible furnace suitable for batch sintering.

inlet and outlet tubings the use of a controlled atmosphere is possible, so that such a set-up is not restricted to sintering in a liquid bath (Fig. 240). At the end of the heating period, the crucible containing the charge can be removed from the furnace to cool in air and a new crucible with another charge can be inserted in the furnace. During heating of the second crucible the first cools to room temperature, the charge remaining protected from contact with oxygen. Two to four charged crucibles—depending on the details of the sintering and cooling cycle—permit consecutive, batch-type operation.

If the heating elements are made of molybdenum, and the entire furnace is built gas-tight to retain the protective atmosphere around the

[1] R. Kieffer and W. Hotop, *Pulvermetallurgie und Sinterwerkstoffe.* Springer, Berlin, 1943, pp. 50, 51.

elements (as shown in Figure 241), temperatures up to 1400°C. (2550°F.) are obtainable. Such construction, however, does not permit removal of the hot crucible after completion of sintering, and it becomes necessary to cool the charge in the furnace or cooling chamber. If the crucible is completely closed and furnished with its own protective atmosphere it may be removed while still hot, resulting, however, in considerable loss of the heat stored in the furnace. On the other hand, furnace cooling is slow and can only be accelerated slightly by cooling the protective gas before its entry into the furnace chamber. Another disadvantage of this

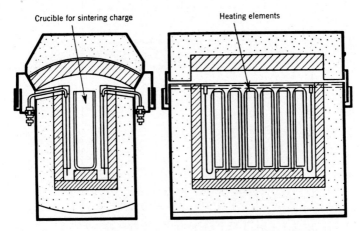

Fig. 241.   Diagrammatic cross section of crucible furnace for sintering of large blanks and forms (according to Kieffer and Hotop[1]).

construction lies in the fact that the heating elements must undergo continuous heat changes during heating and cooling. This usually tends to shorten the life of the resistor elements, particularly in the case of molybdenum.

Bell-Type Furnaces

A furnace somewhat different in construction from the one previously described is the bell-type furnace. Its principal temperature range also lies between 600° and 1100°C. (1110° and 2010°F.). A diagrammatic cross section of a typical design is shown in Figure 242 according to a description by Webber.[2]   In such a furnace the work can be stacked and, if necessary, considerable pressure can be applied by means of a downward-moving hydraulic ram. The work is loaded on a support that rests on

[2] H. M. Webber, in J. Wulff, *Powder Metallurgy*. Am. Soc. Metals, Cleveland, 1942, p. 292 ff.

a rigid base. A retort from alloy sheet is placed over the charge, its lower lip being engaged with a liquid seal provided for in the base. This seal permits the retention of a protective atmosphere within the retort and surrounding the charge. The atmosphere is introduced through an inlet at the base and circulated through the charge by a fan in the base. This

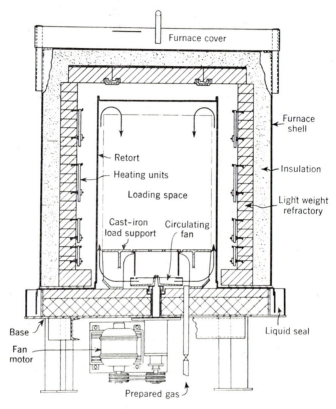

Fig. 242.   Diagrammatic cross section of typical bell-type furnace in which atmosphere is circulated through charge by a fan below the load (according to Webber[2]).

arrangement accelerates both the heating and cooling rates and creates a more uniform temperature distribution. The heating elements, usually made of Nichrome wire, are built into the furnace bell, and heat the charge after the bell is lowered over the retort. At the end of the heating cycle the bell is raised and the retort containing the charge may cool in air unless accelerated cooling is effected by a water spray. Immediately after completion of the heating cycle the furnace bell can be placed over another charged retort, and the heating cycle started anew. The loss of heat energy stored in the bell is considerably reduced by this method.

Box-Type Furnaces

The common box-type furnace is probably the most popular construction for small-scale and diversified operations carried out at a temperature range of 600–1050°C. (1110–1920°F.). A typical design is

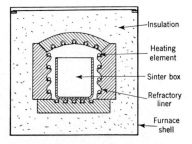

Fig. 243. Diagrammatic cross section of box furnace with Nichrome resistor heating elements suitable for sintering of powder metallurgy products (according to Kieffer and Hotop[1]).

Fig. 244. Atmosphere-retaining sinter boxes with covers, trays, locking cross bars and wedges. Covers are sealed gas tight to boxes with refractory cement (in back), and gas inlets and outlets are connected to copper tubing (courtesy of Sintercast Corporation of America).

shown in cross section in Figure 243. The electrical heating elements are fastened to the top, bottom, and sides of the chamber so that the charge contained in a box is heated uniformly from all sides. Since heat transfer is obtained principally by radiation, heating must occur slowly—especially in the case of heavy loads. Since sintering must always take place in the absence of oxygen, the sinter boxes must close tightly. The charge may either be submerged in an inert or slightly reducing pack, or a protective atmosphere must be introduced into the box through its cover,

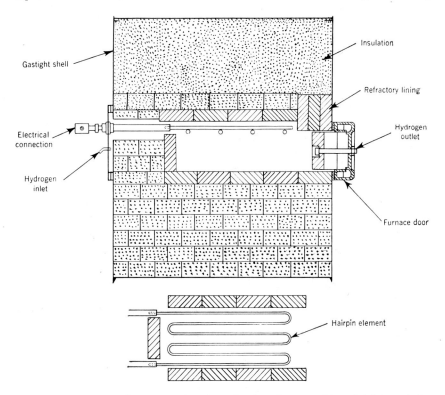

Fig. 245. Diagrammatic cross section through a box furnace with molybdenum hairpin-heating element on top of chamber (according to Kieffer and Krall[2a]).

as shown in the photograph of Figure 244. After the box is charged, the cover is closed and sealed with refractory cement which is baked during the initial stage of heating. The box is placed in the chamber and the furnace door closed to avoid unnecessary heat losses. After the sintering cycle is completed, the box may either cool in the furnace or be removed

[2a] R. Kieffer and F. Krall, *VDE Fachber., 11,* 107 (1939).

for cooling in air or under a water spray. When the furnace has re-covered its heat losses, it is ready to receive a new box. The temperature is controlled by a thermocouple either placed between the box and the chamber walls or extended into the sinter box within a metallic protecting tube that is sealed toward the inside to prevent loss of atmosphere.

A different construction of box furnace provides a gas-tight furnace shell. No cover is necessary for the box, and the protective atmosphere fills the entire furnace. If the heating elements are made of molybdenum,

Fig. 246. 20-kw. laboratory box-type electric furnace (left) with molybdenum heating elements and connected cooling chamber; for operating temperature up to 1700°C. (3100°F.) and for use with protective atmosphere and "Reactrol" control (right) (courtesy of General Electric Co.).

hydrogen or dissociated ammonia must be used as protective atmosphere, and temperatures up to 1500°C. (2730°F.) can then be reached. A typical construction of such a furnace is shown in Figure 245 and in the photograph of Figure 246. A gas-tight sliding door retains the hydrogen in the furnace. The danger of explosion during opening of the door (by a mixing of air with the furnace hydrogen) is prevented by a dense flame curtain.

Gas-heated box furnaces are also adaptable to batch sintering, but temperatures are lower (not exceeding 1100°C.; 2010°F.). Sinter boxes to be used in gas-fired furnaces must be constructed of high-temperature-resistant alloys, and will tend to deteriorate faster than if used in electric furnaces because of the contaminating effects of the furnace gases. Care must be taken that the furnace gases do not come into contact with the charge through leaks in the sinter box, since unwanted chemical reactions may otherwise follow.

Tube and Muffle Furnaces

One of the most useful furnaces for sintering is that having an Alundum tube muffle, around which an electric resistor wire is wound.

Fig. 247. Tubular laboratory furnaces with Nichrome heating elements and quartz retorts. Furnaces are of split-design type, permitting lifting of hot retorts into water shower; retorts can also be rotated while being heated or quenched. (Courtesy of American Electro Metal Corp.)

This type of furnace was one of the first developed especially for sintering purposes. It has been and still is being used almost exclusively for the sintering of cemented carbide tool materials and permanent magnets, and for presintering tungsten and molybdenum bars.[2] The furnace type is particularly favored on account of its simple construction, efficient heating, and adaptability to a continuous transporting of the charge.

For low-temperature work (up to 1050°C.; 1920°F.) Nichrome or Kanthal wire is used; the furnace housing simply consists of a shell hold-

ing the brickwork or powdered furnace insulation in place. The furnace
retorts are made of stainless steel, Nichrome, or, as shown in Figure 247,
of quartz. For higher temperatures, molybdenum or silicon carbide is

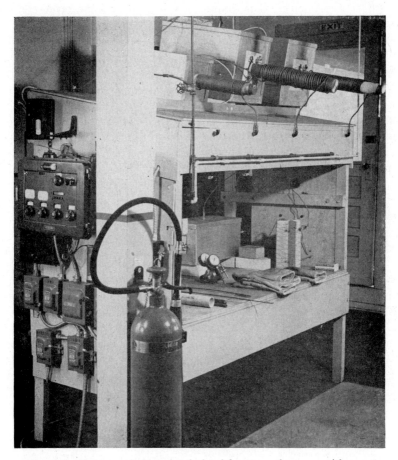

Fig. 248. Battery of slanted tubular laboratory furnaces with quartz
muffles and Globar resistor elements for reduction of powders and sintering
of ferrous and nonferrous compacts (courtesy of Ekstrand and Tholand).

used as resistor, and the furnace housing must be made gas-tight to retain
the hydrogen protecting the wire. The furnace may be closed on one end,
the other end serving for charging and discharging purposes. A cooling
chamber can be attached to the open end of the furnace so that the work
can cool under protective atmospheric conditions. The most favored
construction, as shown in Figure 248, has an opening on each end of the

tube. The work is charged on boats at one end, and placed in the middle of the furnace. After completion of the heating cycle the charge is pulled into the cooling chamber at the other end, and the muffle is ready to receive another charge. The charge end is usually kept open (or only partially closed by a brick) to permit unobstructed escape of the atmosphere. A flame curtain at the discharge end prevents the formation of an explosive mixture of the hydrogen with air entering through the draft created when the discharge door is opened.

The chief disadvantage of this type of furnace construction lies in the immediate contact of the heating wire with the refractory muffle. Contamination of the wire leads to frequent "break-throughs," and once the wire has been disrupted, arcing results in local fusion of the muffle. Increasing embrittlement of the wire is another cause for failure. Replacement of the wire on the same Alundum tube is rarely possible, and a failure usually requires a newly wound tube. This replacement necessitates the dismantling of the brickwork and constitutes a more cumbersome task than the simple replacement of resistor elements in a box furnace.

Carbon Short-Circuiting Furnaces

To assure better service and longer furnace life, carbon tubes can be used instead of Alundum, provided that a carbon monoxide-containing atmosphere is not objectionable. By connecting the carbon tube directly with the current, heat is generated by short circuiting—no metallic heating element being used. This furnace has been used extensively in Germany and England for the sintering of hard metals in a temperature range of 1350–1550°C. (2460–2820°F.).[3,4] In this country it is used mainly in the manufacture of graphite. The construction of the furnace as a whole is similar to that of the Alundum tube furnace.

The carbon-tube furnace is also adaptable for vacuum sintering as, for instance, is required in the production of tantalum carbide. The schematic design of such a furnace is shown in Figure 249. A uniform furnace temperature is obtainable over most of the length of the carbon tube by preventing undue radiation losses at the ends by means of tungsten or molybdenum shields. The temperature must be controlled optically through a window in the upper cover. All joints and current connections through the housing must, of course, be vacuum sealed. The covers and housing are water cooled.

[3] K. Becker, *Hochschmelzende Hartstoffe und Ihre Technische Anwendung.* Verlag Chemie, Berlin, 1933, p. 32.
[4] W. D. Jones, *Principles of Powder Metallurgy.* Arnold, London, 1937, p. 164.
[4a] R. Kieffer and W. Hotop, *Pulvermetallurgie und Sinterwerkstoffe.* Springer, Berlin, 1943, pp. 55, 57.

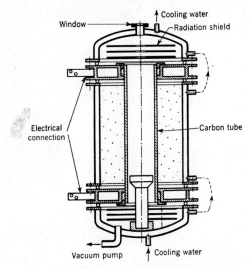

Fig. 249. A diagrammatic cross section through a carbon-tube, short-circuiting furnace for vacuum sintering of hard metals (according to Kieffer and Hotop[4a]).

High-Frequency Furnaces

In recent years high-frequency induction furnaces have been used increasingly in the sintering of hard metals, composite metals, and in other applications where temperatures between 1300° and 2000°C. (2370° and 3630°F.) were needed. Various frequencies have proved suitable for this purpose, ranging all the way from about 10,000 cycles (as produced by a motor generator) to more than 1,000,000 cycles (as produced by a vacuum tube-type converter). Figure 250 shows a 50-kva machine of the latter type for the sintering of large compacts under hydrogen gas in sealed Nichrome retorts.

A high-frequency furnace arrangement suitable for sintering of hard metals and refractory metal bits in a protective atmosphere is represented diagrammatically in Figure 251. Graphite serves as the crucible material, and heat is generated in the crucible and transferred to the charge by radiation. A high-frequency furnace operating in vacuum is shown in Figure 252. The furnace is surrounded by a vacuum-sealed bell; it can operate at temperatures ranging from 1000° to 1800°C. (1830° to 3270°F.), the temperature being measured (optically) through a window on top of the bell.

Fig. 250. 50-kva vacuum-tube-type, high-frequency converter and furnace equipment for induction sintering (courtesy of American Electro Metal Corp, and Federal Telephone and Radio Corp.).

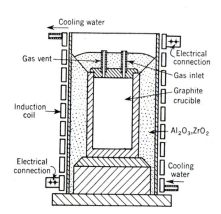

Fig. 251. Diagrammatic cross section through a high-frequency induction furnace for sintering of hard metals and other compositions at temperatures from 1500–2000°C. (2730–3630°F.) (according to Kieffer and Hotop[4a]).

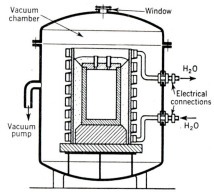

Fig. 252. Diagrammatic cross section through a high-frequency induction furnace for vacuum sintering at temperatures from 1000–1800°C. (1830–3270°F.) (according to Kieffer and Hotop[4a]).

If the graphite crucible is replaced by a refractory crucible, direct induction heating of the compacts is possible. This may be particularly desirable where carbon would lead to undesirable reactions with the charge. Uniform heating of the charge depends on the type of insulation and on the pattern in which the compacts are placed in the high-frequency field.

Sinter Bell Jars

The present-day production of the refractory metals, molybdenum (melting point 2620°C.; 4750°F.), tantalum (melting point 2850°C.;

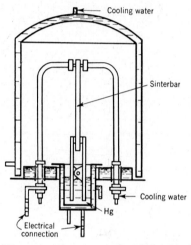

Fig. 253. Diagrammatic cross section through a bell jar for sintering tungsten bars (according to Kieffer and Hotop[4b]).

5150°F.), and tungsten (melting point 3370°C.; 6100°F.) is based entirely on the sintering of compacted powder bars by passing an electric current through them. Figure 253 shows a diagrammatic sketch of the cross section of a bell jar; the apparatus has already been described in Chapter XV.

Tilting Furnaces

For the production of composite metals, such as tungsten–copper, tungsten–silver, and molybdenum–silver contact materials, special furnaces have been developed in Germany.[5,6] Due consideration has been given to the fact that the production technique embraces both powder

[4b] R. Kieffer and W. Hotop, loc. cit., p. 230.
[5] R. Kieffer, Z. tech. Physik, 21, 35 (1940).
[6] R. Kieffer and F. Krall, VDE Fachber., 11, 107 (1939).

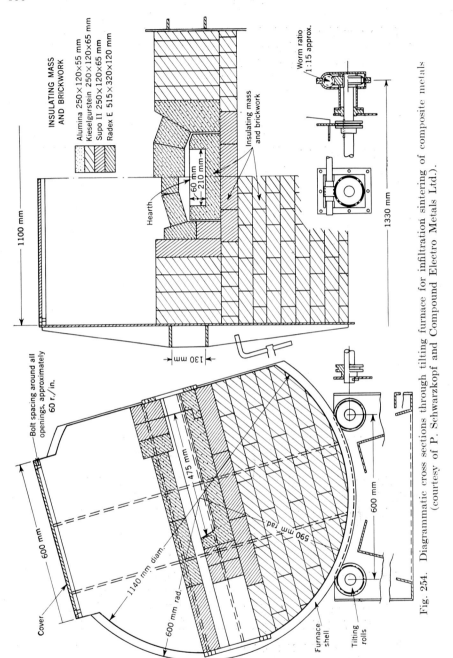

INSULATING MASS
AND BRICKWORK

Alumina 250×120×55 mm
Kieselgurstein 250×120×65 mm
Supo II 250×120×65 mm
Radex E 515×320×120 mm

1100 mm

Hearth

Insulating mass
and brickwork

60 mm
210 mm

130 mm

Worm ratio
1:15 approx.

1330 mm

Bolt spacing around all
openings, approximately
60 r./in.

Cover

600 mm

600 mm

475 mm

590 mm rad.

1140 mm diam.

600 mm rad.

600 mm

Furnace
shell

Tilting
rolls

Fig. 254. Diagrammatic cross sections through tilting furnace for infiltration sintering of composite metals
(courtesy of P. Schwarzkopf and Compound Electro Metals Ltd.).

metallurgy and fusion metallurgy. Composite bodies are obtained by the impregnation of a presintered porous skeleton body of tungsten or molybdenum with fused copper or silver. This impregnation is accomplished advantageously with tilting crucible furnaces, a typical design of which is sketched in Figure 254. The porous skeleton compacts and the metal to be impregnated are placed on the same tray, and the latter is first slanted at an angle which permits its melting without touching the compacts. After sufficient time has been allowed for degassing of the melt, the furnace is tilted, so that the tray slants at an opposite angle, thereby causing a contacting of the melt with the porous compacts. In order to prevent segregation or the formation of pipes and blowholes in the solidifying metal the furnace is tilted back into its original position after completion of the impregnation cycle. The charge can then be removed from the furnace in a customary manner. The impregnation operation is usually carried out in a protective atmosphere, such as hydrogen, at temperatures between 1100° and 1300°C. (2010° and 2370°F.).

CONTINUOUS FURNACES

The chief distinction between a batch-type and a continuous sintering furnace is the manner in which the work enters, passes through, and leaves the furnace. Among the designs discussed so far there were several which lent themselves to continuous transportation of the work from one end of the furnace to the other. All tube and muffle furnaces can be used for both batch and continuous sintering, the constructions varying only in detail for either application.

Several prerequisites must be fulfilled to permit continuous transport of the work. The furnace must have charge and discharge openings at opposite ends, so that the charge can travel in one direction. A means of transport must be available to convey the charge smoothly without mechanical interference. A cooling chamber must be attached to the discharge end of the heating chamber to enable the charge to cool in a protective atmosphere. This atmosphere must flush the entire furnace and cooling chamber to prevent possible oxidation of the compacts. A counter-stream flow of a cooled atmosphere, moreover, aids cooling of the charge before discharge. Flame curtain and exhaust arrangements or internal sliding doors are apt to eliminate the danger of explosion which may be caused by an undesirable influx of air.

Aside from the favorable economics of a continuous sintering furnace, there are several technical advantages worthy of mention. The continuous furnace demonstrates excellent heat distribution and heat utilization, with remarkably uniform heatings of consecutive charges being possible. Large quantities of material of different sizes can be sintered efficiently and under controlled and easily adjusted conditions of temperature, rates of heating and cooling, etc.

Pusher-Type Furnaces

The simplest way of changing a batch-type muffle or tube furnace into a continuous sintering furnace is by pushing the work-containing trays ahead of each other through the tube. Provided the tray material is sufficiently rigid and temperature-resistant, considerable loads can be transported in this manner. Pushing may be accomplished by hand or by a mechanical or hydraulic power drive. The push may be either gradual or instantaneous, but the latter method has the disadvantage of introducing cold trays and charges into the heating chamber suddenly, causing undesirable temperature fluctuations.

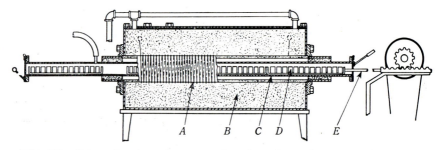

Fig. 255. Schematic view of cross section through continuous tube furnace for sintering small Alnico magnets (according to Howe[7]): A, Alundum tube wound with molybdenum wire; B, alumina insulation; C, the iron tube containing D—the magnets in an appropriate boat; and E, the pusher mechanism.

In Figure 255 a schematic view is shown of a continuous tube furnace as it is used for sintering small Alnico magnets.[7] It is very important that adverse effects of temperature gradients be eliminated in this operation and that all magnets are subjected to a substantially uniform sintering treatment. The furnace consists of a cylindrical Alundum tube wound with molybdenum wire. Through the Alundum tube is placed an iron tube which extends beyond each end of the furnace for a charging and discharging zone. Purified hydrogen is used in the furnace housing and in the iron tube, which is closed by doors and furnished with flame curtains to prevent an inrush of air while the doors are open. The boats containing the charges are slowly pushed through the hot zone with a mechanical pusher device, whose push rod leads through the charging door.

[7] G. H. Howe, in J. Wulff, *Powder Metallurgy*. Am. Soc. Metals, Cleveland, 1942, p. 532.

The carbon-tube, short-circuiting furnace (page 602) can also be constructed as a continuous, pusher-type furnace. As such, it has won favor in the European production of hard metal alloys.[8] A typical construction is shown in Figure 256. Instead of a protective flame curtain arrangement, duplex gates are provided at the charge and discharge ends of the furnace. The boats are also made of graphite.

Pusher-type furnaces have also found use in the sintering of iron and steel parts. The lower sintering temperature makes it possible to use cast or fabricated retorts from heat-resistant alloys (Nichrome, Inconel, Rezistal) in gas-heated or electrically heated furnaces. Trays are usually fabricated from ordinary rolled steel, stainless steel, or nickel–chromium

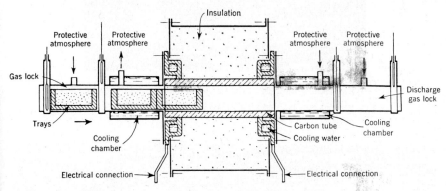

Fig. 256. Diagrammatic cross section through a carbon-tube, short-circuiting furnace for the continuous sintering of hard metal compositions (according to Kieffer and Hotop[8]).

alloys (Nichrome, Inconel, etc.); they may also be cast in a honeycomb pattern, on top of which a wire mesh is fastened to carry the charge. Where contamination of the charge by carbon is not possible, trays pressed or machined from graphite are advantageous because of their self-lubricating properties. The charging of a "double-decker" electric push furnace with fabricated steel trays is shown in Figure 257, while Figure 258 shows a two-ton-a-day, gas-heated, continuous sintering furnace with hydraulic push system.

The latter furnace is known to have had a remarkable service record, since it has operated for a number of years at a temperature range of 1100–1200°C. (2000–2200°F.) with varying loads and transport rates, without serious breakdowns or interferences. At times, prolonged periods

[8] R. Kieffer and W. Hotop, *Pulvermetallurgie und Sinterwerkstoffe.* Springer. Berlin, 1943, pp. 52, 55.

Fig. 257. Charging of double-deck electric push furnace heated with molybdenum wire elements (courtesy of American Electro Metal Corp.).

Fig. 258. Two-ton-a-day continuous sintering furnace with hydraulic push system (courtesy of American Electro Metal Corp.).

of service of close to a year were obtained without need of shutdown for repair. This may be attributed chiefly to the excellent performance of the fabricated Inconel muffle construction used, which stood up remarkably well in spite of the contaminating effects of the sulfur-containing city gas used as fuel source. The D-shaped muffle was constructed from $^5/_{16}$-inch-thick annealed Inconel sheet, and gas-tight, durable welds could be readily produced with a new, improved type of Inconel welding rod. To counteract sagging and warpage of the muffle, the construction incorporated reinforcement braces and ribs transverse to the muffle axis at 12-inch intervals, and longitudinal rails at the bottom of the muffle, both inside and outside. The life of the welds was much prolonged by placing the joint of the top and bottom sections of the D-muffle about one inch above the corner and by introducing a liberal radius at the bend. The service life of the fabricated Inconel trays used in this furnace was extended to about 500 alternate heating and cooling cycles by water quenching and straightening the trays periodically. This record is even more remarkable, if it is realized that the weight of the empty trays alone was about 25 pounds per piece.

Roller Hearth Furnace

One of the decided disadvantages of the ordinary push furnace is that trays must be constructed sufficiently rigid and large enough in cross section to withstand deformation in the heat zone caused by the friction

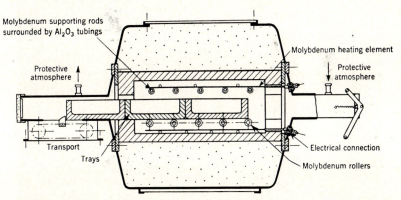

Fig. 259. Diagrammatic cross section through a roller hearth push furnace with molybdenum heating elements for sintering ferrous and hard metal alloys (according to Kieffer and Hotop[8]).

against the driving force. Sheet iron boats, for example, tend to become shorter and wider in their middle with each successive cycle. Their center sections will soon wedge inside the muffle, resulting in serious interferences.

Fig. 260. Roller hearth sintering furnace, charging end. Door opening 20 in. wide, 8 in. high; heating chamber 10 ft. long; cooling chamber 30 ft. long; rated 110 kw. (Courtesy of General Electric Co.)

Fig. 261. Roller hearth sintering furnace, discharge end. Gas equipment in background. Reactors for "Reactrol" control, and transformers, at right center. (Courtesy of General Electric Co.)

In order to prevent this occurrence, the tray material must be very thick in cross section ($1/4$ inch and more), and complex re-enforcements become necessary. As a result, trays are heavy and the ratio between tray weight and charge weight becomes unfavorable, necessitating the needless heating and cooling of a large mass of ineffective load.

Installation of rolls in the hearth of the furnace or muffle reduces friction considerably, permitting the use of light-weight trays. A furnace construction providing molybdenum rolls in the heating zone, but depending on a push mechanism for transport of the trays, is shown in Figure 259. If the rolls are power driven they can convey the trays without the need of a pusher mechanism. By individual drives for the rolls in these zones, trays can travel through different sections of the furnace at different speeds. A roller hearth furnace is particularly suitable for high temperature work (1100–1200°C.; 2000–2200°F.), and for large-scale production because its trays are not subjected to pushing or pulling stresses. Work of sizable dimensions may be carried on light-weight trays with attached mesh grids.

The modern furnace of this type has a series of driven rolls running throughout the length of the furnace. In the heating zone the rolls are constructed from a cast alloy, such as Nichrome, and are water cooled. The stresses in the rolls are redistributed continuously as they turn. Finally, to permit passage of a protective atmosphere through the furnace chamber, the roller hearth drive mechanism alongside the furnace is contained in a gas-tight housing. A modern roller hearth sintering furnace is shown from the charging end in Figure 260 and from the discharge end in Figure 261.

Mesh-Belt Conveyor Furnace

The most widely used continuous sintering furnace is the conveyor-type furnace. It has proved its value in the sintering of copper–tin–graphite compacts for self-lubricating bearings at about 800°C. (1470°F.), and iron and steel compacts at about 1100°C. (2000°F.). A furnace of this type is described in detail by Koehring,[9] and a schematic view of a typical construction is reproduced in Figure 262. In Figure 263 a photograph of a modern mesh-belt conveyor furnace is shown. The furnace has connected heating and cooling chambers, through which runs an endless mesh belt which is fabricated from high-alloy (Nichrome) wire. The belt is supported by a cast-alloy hearth plate in the heating chamber, and by the bottom of the liner in the cooling chamber. The conveyor is constantly kept under tension by a suitable belt-tightening mechanism. If the belt

[9] R. P. Koehring, in J. Wulff, *Powder Metallurgy*. Am. Soc. Metals, Cleveland, 1942, p. 287.

stretches beyond limits it can be shortened by removing one link at a time. The drive mechanism usually provides various belt speeds to accommodate the individual sintering cycles of the different materials. The work is either placed directly on the belt or carried through on trays.

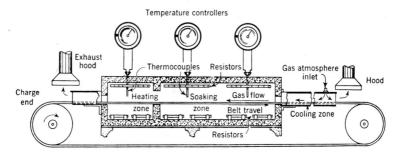

Fig. 262.   Schematic cross section through continuous conveyor-type sintering furnace (according to Koehring[9]).

The furnace shown in Figure 262 consists of three heating zones (each having its own temperature control) in addition to the cooling chamber. The first zone on the charging end is further divided by a wall, whose purpose is to facilitate a temperature difference between the zones—if so desired. In order to bring the work up to the sintering temperature quickly

Fig. 263.   Mesh-belt, conveyor-type continuous sintering furnace. Belt 12 in. wide, door opening 8 in. high, heating chamber 10 ft. long, cooling chamber 20 ft. long, rating 102 kw., operating temperature 1150°C. (2100°F.) max. (Courtesy of General Electric Co.)

and at a controlled rate, a large part of the heating capacity of the furnace is stored in the first zone. If a heavy part is being sintered, the temperature of the first zone is increased so that the rate of heating is comparable to that of a lighter part heated at lower temperature. If the furnace is used near its maximum temperature, as in the sintering of iron parts, all three zones are usually kept at the same high temperature. The large heating capacity of the first zone allows speedy heating of the work to maximum (soaking) temperature, thereby reducing the over-all time of the sintering cycle—even in a relatively short furnace. Suitable disposition of the heating elements and the use of trays with side baffles eliminate excessive radiation heating of the outer rows of work on each side, and therefore aid in the maintenance of a uniform temperature and heating rate from one side of the furnace to the other.

The controlled sintering atmosphere is conducted into the cooling chamber near the discharge end. Some of the gas escapes at the discharge end to maintain a gas seal, but most of it passes through the furnace in counterstream to the work—maintaining a gas seal at the charge end. If the gas is previously cooled it serves as an additional cooling medium while passing over the hot charge leaving the furnace.

Lift Beam Furnace

An ingenious method of conveying heavy furnace charges through horizontal or inclined furnaces was recently developed by the Metallwerk Plansee in Reutte, Austria, and may be designated as the lift beam ("Hubbalken") transport system.[9a] Again, the work to be sintered is contained in suitable trays which pass through an entrance chamber, the heating zone, and a cooling tunnel. The bed of all three zones is constructed as a multiple-beam system, with some transverse beams stationary and others movable. Movement of the trays is caused by first raising some of the movable beams, whereby all trays are lifted from the stationary beams; then the movable beams are pushed a short distance in the direction of the furnace axis carrying the trays along; finally, the beams are again lowered and the trays rested on the stationary beams. The movable beams are then withdrawn and brought into their original position, ready for a repeat performance. Chief advantages of the furnace are the possibilty of transporting single trays (without need of running chains of empty trays), a minimum of wear and tear on the trays and furnace bed, and the comparatively safe mechanical performance of the system. On the other hand, the lift beam system is more complicated than the pusher system, and sintering in reducing atmospheres requires its complete encasing within a gas-tight furnace housing. Atmospheres offering little explosion hazard are preferred for the furnace.

Vertical Lowering Furnace

The development of a European furnace construction that provides transport of the charge in a vertical direction has attracted particular

---

[9a] R. Kieffer and W. Hotop, *Sintereisen und Sinterstahl.* Springer, Vienna, 1948, p. 299.

interest recently.[10]   Figure **264** presents a view of the entire furnace, while a cross section of it is shown in Figure 265. As in pusher-type and roller hearth furnaces, the work has to travel in boats and is surrounded by a protective atmosphere. A cooling chamber which is also filled with the atmosphere is attached to the bottom of the furnace. The boats have the form of crucibles and are made of heat-resistant cast alloys, ceramics,

Fig. 264. View of continuous vertical lowering furnace (according to Kieffer and Krall[10]).

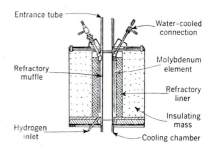

Fig. 265. Diagrammatic cross section through lowering furnace heated by molybdenum elements (according to Kieffer and Krall[10]).

or graphite. They are lowered through a circular vertical tube by means of three feeding rolls at the bottom of the tube, one of the rolls being power driven. The crucibles are squeezed between these rolls and moved slowly downward until they drop onto a chute or conveyor. The rolls are placed at a slight angle so that the crucibles can enter and pass through the opening left by them on the top. To provide sufficient friction between the crucible and the rolls, the latter are rubber-lined. Details of the feeding-roll arrangement are sketched in Figure 266.

The revolution of the power-driven roll is translated to the other

[10] R. Kieffer and F. Krall, *VDE Fachber.*, *11*, 107 (1939).

roll through the crucible, caught by friction. This force can be regulated by adjusting the bearings of the rolls. As a consequence, the crucible revolves during lowering, and with it all others placed above—provided

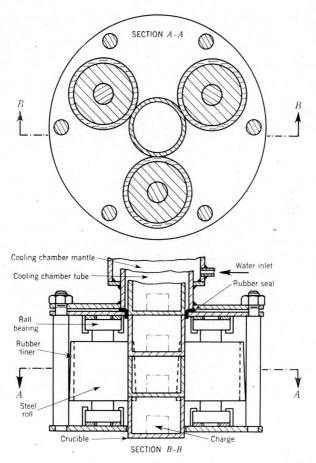

SECTION A-A

B
B

Cooling chamber mantle
Cooling chamber tube
Water inlet
Rubber seal
Ball bearing
Rubber liner
A
A
Steel roll
Crucible
Charge
SECTION B-B

Fig. 266. Diagrammatic sketch of transport mechanism for lowering furnace.

their weight is considerable and the bearing surface between the walls and bottoms of the crucibles causes sufficient friction. Thus, the entire column of crucibles revolves, and a very uniform heat distribution is obtained in the furnace.

There are other advantages in this furnace construction. The helpful action of gravity reduces the power input necessary for driving the rolls to a mere fraction of that needed to push a similar load in a horizontal furnace. A long furnace makes it possible to charge and discharge crucibles at a rapid rate, *e.g.*, one per minute. The construction would be particularly advantageous for plants operating on two floors, with powders prepared and compacts pressed and charged into the vertical sintering furnace on the upper floor, and the sintered metal discharged and processed further on the lower floor. The vertical furnace appears most suitable for high-temperature work of limited volume. Suspended molybdenum heating elements make possible attainment of temperatures up to 2000°C. (3630°F.),[10] provided the crucibles are made of graphite and are not over 3 inches in diameter. For larger crucibles (up to 6 inches in diameter) lower temperatures and transport rates are necessary to assure uniform temperature distribution. In this country so far, only smaller size furnaces have been used on an industrial scale, but the experience with them has been very good. A furnace built for 3-inch crucibles and transporting a column of 40 crucibles, each 2 inches high, can sinter a maximum of one-pound net load every two minutes, *e.g.*, approximately 700 pounds per day. Larger furnaces are estimated to handle as much as a ton a day.

The vertical furnace has been particularly favored in the continuous sintering of composite alloys where one constituent is liquid at the highest temperature and is incorporated into the compact by impregnation of a skeleton body.[11] Extremely dense and sound composite bodies can apparently be obtained by solidifying the impregnant while in vertical motion. The impregnated body undergoes a minute temperature gradient while being lowered from the heat zone to the cooling chamber, and the impregnant freezes from the bottom upward. All harmful gases are pushed upward through the constantly diminishing liquid phase until they are completely pressed into the open.

The latest devlopment in vertical furnace construction, as recently reported by Kieffer,[12] differs from the aforementioned mainly by its mode of heating. Extremely high temperatures (up to 3000°C.; 5400°F.) result from a multitude of arcs caused by the passing of a three-phase current through a pack of coarse granular carbon which backs the central graphite furnace tube. A 300-kw. furnace of this type is now effectively used by the Metallwerk Plansee in Reutte for the continuous production of multicarbides from their powdered ingredients.

[11] U. S. Pat. 2,422,439.
[12] R. Kieffer, Communication to New York Regional Group, Powder Metallurgy Committee, *Inst. Metals. Div., Am. Inst. Mining Met. Engrs.*, Dec 15, 1948.

Summary

Batch-type furnaces of conventional design can be adapted for sintering purposes. Gas-heated crucible or box furnaces are available for low-temperature work, but the charge must be protected from the furnace gases by a gas-tight crucible or retort. Electric furnaces with Nichrome or Kanthal resistor elements can be used up to 1050°C. (1920°F.). Tightly sealed boxes or retorts in which the compacts are surrounded by inert packs or protective atmospheres can be heated in bell-type or box-type furnaces. Electric muffle furnaces require a retort if heated by a Nichrome or Kanthal wire spiral wound on the refractory muffle. For higher temperatures, up to about 1500°C. (2730°F.), muffle furnaces must be wound with molybdenum wire. Since the wire must be surrounded by hydrogen, the furnace housing must be of a gas-tight construction; no box or retort is necessary in this case. Box furnaces may also be heated with molybdenum resistors (usually of hair needle shape). Electric box and muffle furnaces may also be heated with Globar elements up to an operating temperature of 1350°C. (2460°F.). Since these elements burn best in air, no gas-tight housing is necessary, but the charge must be contained in a closed retort or box to retain the protective atmosphere or pack.

Where higher temperatures are needed, sintering may be carried out in a carbon-tube, short-circuiting furnace or in graphite crucibles heated by high-frequency current, provided the carbon does not cause harmful reactions with the compacts.

The sintering of refractory metals requires specialized equipment, so that compacted bars can be heated to temperatures between 2000° and 3000°C. (3630° and 5430°F.) by passing a current through them. Composite, refractory-metal-base materials are also treated in specially constructed furnaces, in which sintering, fusion, and impregnation are carried out in protective atmospheres at temperatures up to 1500°C. (2730°F.). Furnaces of this type can be tilted and permit degassing of the molten phase before impregnating.

The continuous furnace has attained widespread industrial use for the large-scale sintering of a large variety of sizes and shapes. Three types of continuous horizontal furnaces are employed predominantly for sintering temperatures ranging from 800 to 1150°C. (1470 to 2100°F.). The pusher furnace is the simplest type, being used with metallic or refractory muffles. The furnace can be heated by gas or electricity, and the work must be placed in rigid trays of cast or fabricated alloy, or of graphite. Either mechanical or hydraulic pusher mechanisms may be used, and the push may be either gradual or sudden.

Ineffective tray load can be reduced considerably if friction is eliminated through the incorporation of rolls in the muffle bed. High-
capacity, roller hearth furnaces have rolls in the heating and cooling
zones, and permit flexible transport of light-weight trays by individual
driving mechanisms. Internal gates may subdivide the entrance and
cooling chambers from the hot zone—preventing the inrush of air during
opening of the charging or discharging doors. The mesh-belt conveyor
furnace is most popular for sintering below 1100°C. (2000°F.). Although
the work must travel through the entire furnace at the same speed, rapid
heating of the work is possible by proper distribution of the heat input.
If the furnace is divided into several zones, a large part of the heating
capacity can be stored in the first zone. The charge can be placed directly
on the conveyor, or can be contained in light-weight trays provided with
shields to eliminate excessive side radiation from the heating elements.

A novel construction is the vertical, continuous sintering furnace. The
work is lowered in crucibles through the furnace and cooling chamber by
means of power-driven, rubberized feeding rolls. Rotation of the entire
crucible column makes possible a very uniform heat distribution. Freely
suspended molybdenum heating elements permit temperatures up to
2000°C. (3630°F.) for small units. The capacity of the vertical furnace
is less than that of the other types discussed, but larger furnaces capable
of sintering up to one ton of material per day are conceivable. The vertical
furnace appears to have its particular merits in the sintering and impregnation of composite metals.

CHAPTER XVII

Sintering Atmospheres

In contrast to the sintering of ceramic materials, metal powder compacts as a rule can be sintered only in a protective atmosphere that excludes a reaction with oxygen. An exception occurs in the sintering of precious metals or metals forming irreducible oxide films, where heating can be carried out in air and where, as a matter of fact, gases normally employed as protective atmospheres, such as hydrogen, may prove to be harmful.

The importance of the proper choice of the protective atmosphere may be fully understood if one realizes that any specific effect that a harmful gas, such as oxygen, may have on the surface of a solid piece of metal is increased tremendously in the case of a compacted metal powder, because of the large surface area of the particles and the interlocking porosity of powder compacts. Therefore, the principal purpose of any sintering atmosphere must be to prevent the access of gases which form undesirable reaction products, such as oxides or carbides, with the particle surfaces. This purpose is achieved satisfactorily by sintering in a neutral atmosphere of controlled analysis and pressure, such as a liberal flow of pure nitrogen above atmospheric pressure. If undesirable surface oxide films are already present on the powders as the compact enters the sintering furnace, a mere neutral atmosphere is not sufficient, since the oxide films must be at least partially removed to permit action of the attractive forces between the metal surfaces. Under these conditions, which incidentally prevail in the powder metallurgy of copper, nickel, iron, tungsten, and many other metals and alloys sensitive to oxidation, the sintering atmosphere must be reducing in character. Hydrogen is the most effective gas for this purpose, and it is therefore widely used where its beneficial properties overbalance its high cost.

In the powder metallurgy of iron and steels, reactions involving carbon are of great significance. Although desirable for the reduction of oxide films around the steel particles, hydrogen reacts with the carbon combined in the structure, so that decarburization of the steel compact occurs, especially if the atmosphere is not absolutely dry. Similar to the scale-free heat treating of steels in general, sintering of powdered steel compacts must also be conducted in an atmosphere inert to the carbon contained in the structure. Atmospheres consisting of mixtures of carbon monoxide, hydrocarbons, hydrogen, and nitrogen are typical examples, but the atmosphere must be capable of reducing the oxide films or inclusions so effectively that no internal reaction between the oxygen and carbon combined in the structure is possible. Where this is beyond control, the atmosphere must restore all lost carbon, but must not otherwise directly affect the character (structure of pearlite) or distribution (diffusion) of the carbon in the steel compact. The successful production of sintered steel parts depends largely on the selection of an atmosphere that fulfills these requirements as closely as possible.

The general subject of controlled atmospheres for the heat treatment of metals has recently been presented in a comprehensive book by Jenkins.[1]

NEUTRAL ATMOSPHERES

Vacuum

Among the sintering atmospheres or media inert to a metal powder compact, a vacuum is the foremost. Although it is less practical than an inert gas for application to large-scale production, vacuum sintering offers the only possibility for the study of sintering in the absence of any gaseous reactions or interferences. As such, it is an essential practice in any powder metallurgy laboratory involved in fundamental research problems.

The furnace arrangement necessary for vacuum sintering on a laboratory scale in simple and flexible. In Figure 236 (Chapter XV) a sketch shows a vacuum sintering furnace suitable for temperatures up to 1000°C. (1830°F.) and a vacuum of 1-mm. pressure and less—as it was used by the author in sintering experiments with copper.[1a] The compact resting on a metal tray is placed in a hermetically sealed quartz tube that is heated externally by an electric, Nichrome resistor furnace. While the quartz tube is evacuated at the open end through a glass tube sealed into a rubber stopper, hydrogen or any other gas can be introduced into the tube from the other end which is reduced in cross section to nipple size. The arrangement thus permits purging with a reducing atmosphere during the early stages of heating. Similarly, cooling can be accelerated by a rapid flow of a cool gas. Alternate sintering in vacuum and hydrogen is also possible. For temperatures from 1000° to 1200°C. (1830° to 2200°F.),

[1] I. Jenkins, *Controlled Atmospheres for Heat Treatment of Metals.* Chapman & Hall, London, 1946.

[1a] C. G. Goetzel, *The Influence of Processing Methods on the Structure and Properties of Compressed and Heat-Treated Copper Powders. Dissertation.* Columbia University, New York, 1939.

quartz tubes are no longer practicable as furnace muffles, and high-temperature-resistant ceramics impervious to gases must be used; sillimanite tubes are most suitable, but must be protected carefully from thermal shock or sharp temperature gradients in the medium temperature zones. Vacuum sintering at temperatures above 1200°C. (2200°F.) can be carried out in the carbon-tube, short-circuiting furnace described in Chapter XVI and shown in Figure 249. It can also be done in a high-frequency, bell-type furnace if the vacuum required is not much below 1-mm. pressure. (At lower pressures, down to 1×10^{-6} mm., ionization and flash-overs between the turns of the induction coil may occur for frequencies exceeding 20,000 cycles.) A typical design of such a furnace has also been described in Chapter XVI (Fig. 252).

High-vacuum sintering is applied industrially in the sintering of tantalum, columbium, and certain types of cemented carbides. Compressed bars of the refractory metals are sintered in a bell jar at about 2700°C. (4900°F.) by passing a current through them, as in sintering of tungsten and molybdenum. Sintering in a gaseous atmosphere, such as hydrogen, is out of the question because of the great solubility of tantalum and columbium for all gases, causing undesirable embrittlement of the metals. The bell jars used for vacuum sintering are distinguished from their counterparts for hydrogen sintering solely by a slightly changed design of the mercury seal forming contact with the lower clamp.

Dean and co-workers[2,3] recently described the preparation of ductile titanium by sintering in a vacuum of about 1×10^{-4} mm. at 1000°C. (1830°F.). This high vacuum was required for the removal of residual magnesium and hydrogen from the compacts which were pressed from magnesium-reduced powder. Porcelain tubes were found to be satisfactory for the sintering chambers.

Argon

Argon has been used experimentally for the sintering of brittle metals and alloys. Kroll[4] sintered alloys of titanium with small percentages of other metals in argon at 50-mm. mercury pressure, in order to render them ductile. The same investigator[5] reduced chromium powder with calcium under argon, and sintered chromium compacts at about 1700°C. (3100°F.) in argon at 100-mm. mercury pressure, prior to hot rolling into sheets. In the same manner Kroll also sintered chromium alloys containing small percentages of iron, nickel, cobalt, and other metals. Argon was used for

[2] R. S. Dean, J. R. Long, F. S. Wartman, and E. L. Anderson, *Trans. Am. Inst. Mining Met. Engrs.*, *166*, 369 (1946).

[3] R. S. Dean, J. R. Long, F. S. Wartman, and E. T. Hayes, *Trans. Am. Inst., Mining Met. Engrs.*, *166*, 382 (1946).

[4] W. Kroll, *Z. Metallkunde*, *29*, 189 (1937).

[5] W. Kroll, *Z. anorg. allgem. Chem.*, *226*, 23 (1935).

the high sintering operation after presintering in high vacuum at a sufficiently low temperature to prevent sublimation of the chromium.

Helium

Helium has been mentioned most recently by Duwez and Martens[5a] in connection with the sintering of special, highly permeable, stainless steel compacts in experiments with power engine components having a controlled porosity. It was found that sintering in dried helium has a noticeable effect in improving the physical properties of compacts from 18–8 stainless steel alloy powders. An average increase of 20% in tensile strength and 50% in elongation of specimens sintered in an atmosphere of helium instead of hydrogen could not be attributed to a decrease in porosity, since the density remained practically unchanged; rather, it is believed likely that sintering in helium instead of in hydrogen produces a better bond between the particles of the 18–8 stainless steel.

Nitrogen

Nitrogen has been used successfully as a protective atmosphere in a number of applications where other gases such as hydrogen would cause reactions with the metal, resulting in harmful by-products or in undesirable gasification phenomena. It has also been used occasionally where inherent oxide films could not be removed by a reducing gas, while additional oxidation was to be avoided. Although maintenance of the atmosphere in a sintering furnace is considerably less complicated than maintaining a vacuum, and the costs of the gas are much lower than that of argon or helium, the use of nitrogen has been reported only in laboratory experiments. No production method involving sintering in nitrogen is known to the author.

Sintering of copper and brass may be cited as examples of the successful use of a nitrogen atmosphere in the elimination of degasification difficulties. If copper is sintered in hydrogen or carbon monoxide, an embrittlement of the copper may be caused by the products from the reaction of the hydrogen or carbon monoxide with the cuprous oxide films surrounding the copper particles. While the small molecules of hydrogen can enter the most minute passages between the particles, the larger molecules of water vapor or carbon dioxide can escape only by forcibly widening these passages, causing strains, distortions, or ruptures in the metal. This effect is particularly serious in compacts pressed to high green density. Additional complications arise from the possibility that hydrogen dissolved in copper at high temperatures is trapped in the metal upon

[5a] P. Duwez and H. E. Martens, *Metals Technol.*, *15*, No. 3, T.P. 2343 (1948).

cooling, and builds up pressure. Having a larger molecular size, and being inert to oxygen, nitrogen cannot cause this embrittlement. In the case of brass it appears also that the gas has a retarding influence on dezincification.

If silver compacts are sintered in nitrogen they retain their pressed dimensions closely, or may even shrink if their green density is moderate, while the same compacts sintered in hydrogen under otherwise equal conditions will show a tendency to grow in size.

The sintering of aluminum and aluminum alloys in nitrogen has been reported by Cremer and Cordiano,[6] who compared the effect of various sintering atmospheres on these materials. The investigators found that for compacts formed at higher pressure, highest density and ductility values were obtained for nitrogen-sintered material, with the tensile strength being about equal to that of hydrogen-sintered compacts. No apparent advantage was found in using nitrogen for more porous compacts.

One of the difficulties in using nitrogen for sintering of oxidizing metals is the danger of oxidation by air leaks. Large gas flow and high pressure maintained in the system reduce this danger, however, particularly if sintering is conducted in sealed boxes. A small percentage of hydrogen blended with oil-pumped desiccated nitrogen has proved to be an excellent atmosphere for sintering copper and copper-base alloys. The small amount of hydrogen reacts slowly with the cuprous oxide film, and sufficient time is allowed for the diffusion of the water vapor into the open. Appreciable deformation or growth of the compacts does not occur.

Air

Sintering without controlled atmospheric conditions is practical where oxygen cannot react with the metal powder surfaces. All precious metals —including gold—are best sintered in air, since no protection against oxygen is necessary. Other gases, such as hydrogen, may cause a loosening of the structure because of solubility changes during heating and cooling. Gold, however, was sintered effectively in hydrogen (by Trzebiatowski[7]).

Sintering in air is also feasible where solid oxide films surround the metal particles and the oxide is not reducible by common gases. Sintering of these metals compares with that of ceramics, except for the lower temperatures imposed by the lower melting points of the encased metals. Aluminum, magnesium, and alloys of these metals are typical examples of powder metal products that can simply be sintered in air in conventional

[6] G. D. Cremer and J. J. Cordiano, *Trans. Am. Inst. Mining Met. Engrs.*, *152*, 152 (1943).
[7] W. Trzebiatowski, *Z. physik. Chem.*, *B24*, 75, 87 (1934).

metallurgy furnace equipment. Cremer and Cordiano[6] showed that air treatment results in inferior physical properties only for very porous compacts. Higher compacting pressures applied to these light metal alloys during compaction apparently rupture the oxide casings at a sufficient number of points to form the genuine metal-to-metal sintering contacts that account for the amazing cohesive strength of these compacts. Because of the high plasticity of the metal cores, severe deformation aids in puncturing or dislodging the oxide films; at the same time the entire structure is densified to such an extent that during the sintering treatment the metal cores are well protected from further oxidation by contact with outside air.

Steam

According to Koehring,[8] some sintering operations on copper and its usual alloys might be conducted in steam; for production purposes this is far more economical than nitrogen. But again, precautions are necessary to prevent the inrush of air that would oxidize the work—a difficulty for continuous production furnaces. In order to neutralize the ill effects a considerable amount of hydrogen would have to be added—a procedure which would make the proposition too costly for most commercial operations.

Sintering of ferrous alloys in steam, of course, is possible only if decarburization is desired. It may, for example, be expedient to treat iron compacts containing combined carbon in this manner in order to obtain a pure iron which is more susceptible to cold deformation in coining.

Carbon Dioxide

In connection with nonferrous metals such as copper and some of its alloys, pure carbon dioxide may also be used as a sintering atmosphere. It has the disadvantage of being more costly, but can be blended with smaller percentages of hydrogen in order to gain some reducing properties.

Combusted Hydrocarbon Gases

Sintering atmospheres inert to certain metals are obtainable by mixing completely burned hydrocarbon gases with premixed air in suitable atmosphere converters. A typical composition of a furnace atmosphere, according to Webber,[9] from complete combustion of the hydrocarbon is as follows:

[8] R. P. Koehring, in J. Wulff, *Powder Metallurgy*. Am. Soc. Metals, Cleveland, 1942, p. 278.
[9] H. M. Webber, in J. Wulff, *Powder Metallurgy*. Am. Soc. Metals, Cleveland, 1942, p. 292.

Gas	Composition, %	Gas	Composition, %
Carbon dioxide	11.0	Hydrogen	0.7
Illuminants	0.0	Methane	0.0
Oxygen	0.0	Nitrogen	84.6
Carbon monoxide	0.7	Water vapor	3.0

This gas mixture is inert to copper and is sometimes used for the sintering of this metal. Iron cannot be treated in this atmosphere, since it contains a sufficient amount of water vapor and carbon dioxide to oxidize the metal. A more complete discussion of this type of sintering atmosphere will follow under carbon-affecting atmospheres.

REDUCING ATMOSPHERES

Hydrogen

Electrolytic hydrogen is a reliable and convenient atmosphere for the sintering of most metal powders. It is widely used in small laboratory furnaces, and is used commercially in the sintering of tungsten, molybdenum, cemented carbides, Alnico, iron–nickel alloys, and certain types of steels. In all these applications advantage is taken of the reducing power of the gas, compensating for the high cost. Commercial hydrogen always contains small amounts of oxygen and water vapor; it is not sufficiently dry and pure for the sintering of all those metals that have a strong affinity to oxygen. The sole exception is in the sintering of cemented carbides, where graphite carriers or containers convert any traces of these impurities to carbon monoxide.

In general, a two-step purification of the hydrogen is required before it reaches the furnace: (*1*) the conversion of all oxygen into water with the aid of a catalyst and (*2*) condensation of the water and removal with drying agents. A very good method of purifying hydrogen for laboratory work has been suggested by Grube and Schlecht[10]; the apparatus is reproduced in Figure 267. After leaving the cylinder, the hydrogen is conducted through a catalyst furnace over copper chips and heated to 800°C. (1470° F.), a process by which all oxygen is combined with the hydrogen. The gas is then passed through a calcium chloride drying bottle where part of the water is adsorbed. Further removal of the water takes place while the gas is led over magnesium chips heated to 600°C. (1110°F.); this is followed by passage through a chromium sulfate bath and a potassium hydroxide drying tower which remove traces of carbon dioxide and water. The last traces of moisture are withdrawn from the gas when it is passed through a phosphorus pentoxide drying tower immediately before entering the sintering furnace.

[10] G. Grube and H. Schlecht, *Z. Elektrochem.*, *44*, 367, 413 (1938).

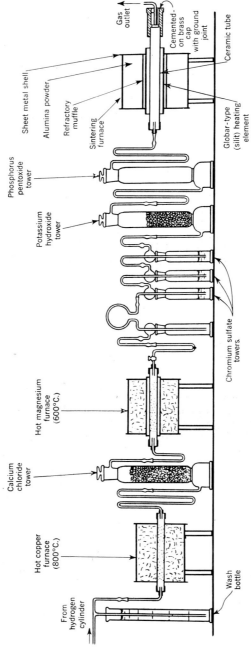

Fig. 267. Flow diagram of hydrogen purification train for laboratory-scale sintering (according to Grube and Schlecht[10]).

A purification system of similar efficiency, giving extremely dry hydrogen (dew point better than −70°C.; −94°F.), is shown in Figure 268 The hydrogen is pressed with 2 pounds pressure first through a fully automatic self-regenerating "Puridryer"[10a] (top) consisting of a palladium catalyst and an activated alumina dryer. Thereafter the gas passes over a hot copper catalyst (furnace 1), then through a calcium chloride dryer (last glass tower to the right), over hot magnesium chips (furnace 2), and then, successively, through glass towers containing (from right to left) potassium hydroxide, magnesium perchlorate, and finally phosphorous pentoxide. The adjacent dew point determinator (between gas cylinders) permits constant testing of the moisture content of the atmosphere.

Fig. 268. Hydrogen purification system for laboratory-scale sintering, consisting of automatic self-regenerating palladium catalyst plus activated alumina unit; hot copper furnace (*1*), calcium chloride drying tower (furthest to the right); hot magnesium furnace (*2*); and potassium hydroxide, magnesium perchlorate and phosphorous pentoxide drying towers (second, third and fourth from right, respectively). Dew point determinator is located between hydrogen cylinders. (Courtesy of Sintercast Corporation of America.)

[10a] Product of Baker & Co., Inc., Newark, N. J.

A less elaborate purification system is employed in industrial sintering practices. The oxygen-hydrogen reaction can be effected with less danger if palladium is used as the catalyst, since the reaction can then take place at room temperature. Commercial palladium catalysts of 200-ft.³-per-hour

Fig. 269. Palladium catalyst purifier (right center—black tower) for complete removal of oxygen; and dual chamber, activated alumina dryer (left) for removal of moisture from hydrogen. (Courtesy of General Electric Co.)

capacity are now available.[11] Various ways of drying the gas effectively for large flow rates are used in the industry. Drying towers may be filled with caustic potash which must be replenished when saturated with water. Activated alumina contained in self-regenerating, commercial gas-drying units[12] of capacities up to 1000 ft.³ per hour has won great favor in the powder metallurgy industry. Finally, moisture may be removed effectively

[11] Type *Deoxo-Purifier,* product of Baker & Co., Inc., Newark, N. J.
[12] Type *Lectrodryer,* product of the Pittsburgh Lectrodryer Corp., Pittsburgh, Pa.

by condensation either in cold water condensers or in liquid air traps. Figure 269 shows a palladium-catalyst purifier in series with a dual chamber activated alumina dryer.

Moisture content in hydrogen (or other reducing gases) is usually measured by determining the dew point, *i.e.*, the temperature at which condensation occurs. The dew point of dry hydrogen varies with decreasing moisture content between about $-40°$ and $-70°C$. ($-40°$ and $-94°F$.), and is generally difficult to determine. In a new device recently developed by Westinghouse,[13] water vapor, as well as oxygen, is determined by electron bombardment, which ionizes these two gases while not affecting other constituents of the furnace atmosphere. The pressure of water vapor or oxygen is thus indicated by changes in the electron current. The electron tube used is essentially a diode tube with both ends open operating in a stream of hot furnace gases. This electronic dew point indicator eliminates the difficulty of judging dew points in the usual way, a method subject to considerable variation. Furthermore, it provides a continuous indication of the quality of the gas.

The relationship between the dew-point temperature, as determined by the electronic indicator or by a dew-point potentiometer, and the moisture content of air or other gases at a pressure of one atmosphere (14.7 psi) is shown graphically in Figure 270 and is also given in Table 58 together with moisture weight data of Webber and Hotchkiss.[14] The extent to which various methods are able to remove the moisture from the sintering atmospheres is also indicated in Figure 270.

Where material of particularly great affinity to oxygen is sintered in hydrogen, the ordinary purification methods are inadequate. Even if the gas is dried at the source to a very low dew point (*e.g.*, $-60°C$.; $-76°F$.), it may pick up moisture and oxygen in the piping system or in the furnace chamber because of air seepage through the charging and discharging doors. This experience is especially persistent where sintering is carried out in continuous furnace operation. In order to protect the charge properly, it becomes necessary to desiccate the gas before it touches the compact. This can best be accomplished by passing the hydrogen over a getter which is placed in immediate proximity to the charge. Various getters are suitable, such as metallic calcium or lithium. In some instances, it is best to imbed completely the charge in the getter material; it is necessary to choose (as pack) a material that does not react with the

[13] F. W. Lucht, *Iron Age, 153,* No. 11, 56 (1944).
[14] H. M. Webber and A. G. Hotchkiss, *Proc. Second Annual Spring Meeting of Metal Powder Association,* New York, June 13, 1946, p. 13.

compact. For nonferrous metals lampblack may serve as pack, while iron-base compacts have been successfully imbedded in a mixture of nine parts powdered beryllium oxide and one part molybdenum or ferromolybdenum powder; other suitable packs contain from two to four parts alumina or magnesia and one part chromium, silicon, ferrochromium, or ferrosilicon powder.

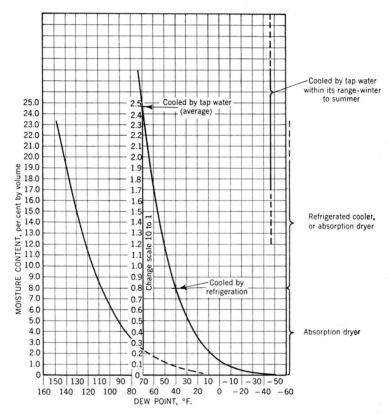

Fig. 270. Relation between dew point and moisture content of gases and the extent of moisture removal by various methods (according to Webber and Hotchkiss[14]).

Where alloys are extremely sensitive to oxygen, metal hydrides have been used successfully as "activators." Wulff[15] has thus produced a very dry atmosphere for sintering stainless steel powder compacts of high ductility. When he used ordinary tank hydrogen after passing it through

[15] J. Wulff, in J. Wulff, *Powder Metallurgy*. Am. Soc. Metals, Cleveland, 1942, p. 137.

TABLE 58

Relation between Dew Point and Moisture Content of Gases[a]
(Webber and Hotchkiss[14])

Dew-point temp. °C	°F	Pounds per 1000 ft.³	Milligrams per liter	Per cent by volume	Dew-point temp. °C	°F	Pounds per 1000 ft.³	Milligrams per liter	Per cent by volume
43.3	110	3.77	60.5	8.70	− 8.9	16	0.160	2.56	0.308
42.2	108	3.57	57.0	8.20	−10.0	14	0.147	2.35	0.282
41.1	106	3.38	54.0	7.75	−11.1	12	0.135	2.16	0.258
40.0	104	3.20	51.0	7.30	−12.2	10	0.124	1.99	0.236
38.9	102	3.02	48.5	6.90	−13.3	8	0.114	1.83	0.216
37.8	100	2.85	45.5	6.45	−14.4	6	0.105	1.68	0.198
36.7	98	2.70	43.2	6.10	−15.6	4	0.096	1.54	0.180
35.6	96	2.55	40.8	5.75	−16.7	2	0.088	1.41	0.165
34.4	94	2.40	38.4	5.40	−17.8	0	0.081	1.30	0.150
33.3	92	2.26	36.2	5.05	−18.9	− 2	0.074	1.18	0.137
32.2	90	2.13	34.2	4.75	−20	− 4	0.0676	1.08	0.125
31.1	88	2.01	32.2	4.46	−21.1	− 6	0.0617	0.990	0.113
30.0	86	1.89	30.2	4.18	−22.2	− 8	0.0562	0.900	0.103
28.9	84	1.78	28.5	3.92	−23.3	−10	0.0511	0.820	0.093
27.8	82	1.68	26.9	3.68	−24.4	−12	0.0464	0.745	0.084
26.7	80	1.58	25.3	3.46	−25.6	−14	0.0421	0.675	0.076
25.6	78	1.48	23.7	3.22	−26.7	−16	0.0381	0.610	0.0685
24.4	76	1.39	22.2	3.02	−27.8	−18	0.0346	0.555	0.0620
23.3	74	1.31	21.0	2.84	−28.9	−20	0.0314	0.505	0.0560
22.2	72	1.23	19.7	2.65	−30.0	−22	0.0284	0.455	0.0503
21.1	70	1.15	18.4	2.47	−31.1	−24	0.0257	0.410	0.0453
20.0	68	1.08	17.3	2.31	−32.2	−26	0.0232	0.372	0.0407
18.9	66	1.01	16.2	2.16	−33.3	−28	0.0210	0.336	0.0368
17.8	64	0.95	15.2	2.02	−34.4	−30	0.0189	0.303	0.0329
16.7	62	0.89	14.2	1.88	−35.6	−32	0.0170	0.272	0.0294
15.6	60	0.83	13.3	1.75	−36.7	−34	0.0153	0.245	0.0264
14.4	58	0.775	12.4	1.63	−37.8	−36	0.0137	0.220	0.0235
13.3	56	0.724	11.6	1.51	−38.9	−38	0.0123	0.197	0.0210
12.2	54	0.676	10.8	1.40	−40.0	−40	0.0111	0.178	0.0189
11.1	52	0.630	10.1	1.30	−41.1	−42	0.0101	0.162	0.0171
10.0	50	0.587	9.4	1.21	−42.2	−44	0.0092	0.147	0.0155
8.9	48	0.546	8.75	1.12	−43.3	−46	0.0083	0.133	0.0139
7.8	46	0.508	8.15	1.04	−44.4	−48	0.0075	0.120	0.0125
6.7	44	0.473	7.55	0.96	−45.6	−50	0.0067	0.107	0.0112
5.6	42	0.440	7.05	0.895	−46.7	−52	0.0060	0.096	0.0099
4.4	40	0.409	6.55	0.825	−47.8	−54	0.0053	0.085	0.0087
3.3	38	0.379	6.05	0.765	−48.9	−56	0.0046	0.075	0.0076
2.2	36	0.351	5.60	0.705	−50.0	−58	0.0040	0.064	0.0065
1.1	34	0.325	5.20	0.650	−51.1	−60	0.0034	0.054	0.0055
0	32	0.301	4.80	0.600	−53.9	−65	0.0025	0.040	0.0040
−1.1	30	0.279	4.45	0.553	−56.7	−70	0.0018	0.029	0.0028
−2.2	28	0.259	4.15	0.511	−59.4	−75	0.0013	0.021	0.0020
−3.3	26	0.240	3.84	0.472	−62.2	−80	0.0009	0.014	0.0014
−4.4	24	0.222	3.56	0.435	−65.0	−85	0.0006	0.009	0.0008
−5.6	22	0.205	3.28	0.400	−67.8	−90	0.0004	0.006	0.0005
−6.7	20	0.189	3.02	0.367	−70.6	−95	0.0002	0.003	0.0003
−7.8	18	0.174	2.79	0.337	−73.3	−100	0.0001	0.002	0.0002

[a] With a known dew-point temperature (indicated by the dew-point potentiometer), the moisture content can be read directly from the table. The table shows the amount of water vapor in air or other gas at various dew-point temperatures at a pressure of one atmosphere (14.7 psi).

a drying tower of activated alumina, the elongation of the sintered compacts was insignificant. But when he used an "activator," consisting of a furnace containing a hydride-forming metal powder, alloys of appreciable ductility were obtained. A sketch of Wulff's hydrogen purification system is reproduced in Figure 271. In the first tower the oxygen is

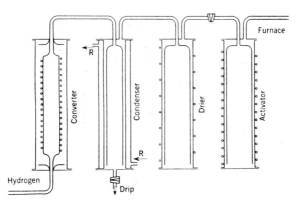

Fig. 271. Flow diagram of purifying system used for hydrogen in sintering stainless steel (according to Wulff[15]). The converter changes contained oxygen to water vapor; the condenser removes most of this water vapor by condensation; the drier is in two sections, both containing activated alumina—one is at heat while the other is in use. The activator is a simple furnace containing a hydride-forming metal powder.

converted into water vapor, either by passing the hydrogen over hot copper or by leading the gas over a catalyst, such as palladium. The second tower is a refrigerant condenser. The third tower is filled with activated alumina; an alternate tower is reactivated continuously while the original tower is in use, and is put into service when the other one is exhausted and requires reactivation. The last tower contains the activator, a finely divided elemental or alloy powder that will combine readily with the last traces of water vapor, oxygen, hydrocarbons, etc. Heated titanium hydride, as well as certain active metals, or alloys, such as pure titanium, calcium, iron–titanium, or alloys of magnesium or cerium may serve as activators. It is claimed that hydrogen treated in this manner is even pure enough to bright-anneal stainless steel which has been oxidized previously.

Another way of securing extremely pure atmospheric conditions has been described by Kalischer[16] in connection with the sintering of Alnico

[16] P. R. Kalischer, *Trans. Am. Inst. Mining Met. Engrs.*, *145*, 369 (1941).

magnets. Complete diffusion of the aluminum into the alloy was obtained by reducing the aluminum oxide films with atomic hydrogen produced by decomposition of metal hydrides at elevated temperatures (above 450°C.; 840°F. for titanium and zirconium hydrides). Atomic hydrogen is so unstable that its powerful reducing qualities can be utilized only if hydrogen reacts with the aluminum oxide at the instant of generation; this requirement necessitates such immediate proximity of the powdered metal hydride to the aluminum oxide, as can be achieved only by a blend of the powdered hydride with the alloy mixture.

The principal objection to a hydrogen reducing atmosphere is its high cost, which is prohibitive even for large-scale consumption, except where high-priced materials are to be sintered. In the refractory metal industry, where hydrogen is essential, recirculation of the gas not consumed in the reaction has generally been accepted for the sake of greater economy. Figure 272 shows the diagram of a typical hydrogen recirculation system.

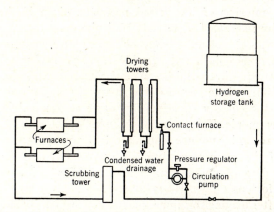

Fig. 272. Flow diagram of typical hydrogen recirculation system for reduction and sintering furnaces (according to Kieffer and Hotop[16a]).

The recovered gas is first washed in a scrubber and then dried by cooling in a condenser or by passing over caustic potash, silica gel, or activated alumina. Recirculation is effected with the aid of a suction pump placed in the return line. The entrance of air into the system during opening of the furnace doors is prevented by maintaining a considerable overatmospheric pressure of the gas in the furnace tubes (2 to 4 inches water at the charge end of the furnace tube). Nevertheless, it is impossible to prevent

[16a] R. Kieffer and W. Hotop, *Pulvermetallurgie und Sinterwerkstoffe*. Springer, Berlin, 1943, p. 53.

small amounts of air from mixing with the hydrogen, and conversion of the oxygen to water must be accomplished with catalysts. A certain quantity of fresh hydrogen must be added constantly to the system. This is necessary to compensate for losses caused, first, by a reducing reaction with oxides present in the compacts and, second, by escape of the gas during opening of the furnace doors when the return line must be shut automatically (*e.g.*, by the pressure drop in the furnace tube) to prevent an intake of air through the suction pump. A properly functioning recirculation system recovers at least 50% of the hydrogen flow passing over the charge. If an appreciable amount is not required for reduction purposes and the sintering cycle is slow, requiring infrequent opening of the furnace door, the hydrogen recovery may reach as high as 75%.

Dissociated Ammonia

A pure hydrogen atmosphere has proved too costly for large-scale commercial sintering of low-priced products. Even if the recovery of unused gas by recirculation exceeds 50% the costs rarely go below one-half of the original costs on account of expenses connected with the maintenance of the recirculation and purification system. In addition, precautions against explosion require constant vigilance and close engineering control. It is for these reasons that hydrogen has given way in most instances to the equally effective, but more economical and convenient dissociated ammonia. This atmosphere, which consists of 75% hydrogen and 25% nitrogen by volume, can almost always replace pure hydrogen. Formation of nitrides from mixtures of hydrogen and nitrogen is unlikely, so that dissociated ammonia can even be used for such metals as molybdenum[17] or ferrous alloys. Its great technical advantage lies in its low dew point. After leaving the ammonia cracker (which consists of a furnace box filled with a catalytic reagent), the gas is completely free of oxygen and water vapor and often has a dew point as low as $-70°C.$ ($-94°F.$). Thus, no further purification is necessary and the gas can be used directly as it leaves the cracking unit. An activated alumina drying tower directly in front of the sintering furnace may, however, be needed to eliminate any traces of moisture that may seep into the atmosphere through the piping system. Modern ammonia dissociators are self-containing units which operate fully automatically and require practically no maintenance whatsoever. The low dew point makes the atmosphere particularly suitable for the bright-sintering of iron, steel, and stainless steel compacts. Figure 273 is a picture of an ammonia dissociator of 1000-ft.³-per-hour capacity.

Table 59 gives a comparison of the costs of hydrogen and cracked

[17] R. Kieffer and F. Krall, *VDE Fachber.* 11, 107 (1939).

Fig. 273. 1000-cubic-foot-per-hour ammonia dissociator with control panel (courtesy of the Drever Company).

TABLE 59
Cost of Dissociated Ammonia Versus Purchased Hydrogen[18]

Consumption, ft.³				
per hour	100	250	500	1,000
per year	720,000	1,800,000	3,600,000	7,200,000
Cost of dissociated ammonia per year,[a] dollars				
Cylinder ammonia (15c./lb.)	2,400	6,000	12,000	24,000
Power (1c./kw.-hr.)	216	360	720	1,368
Maintenance	125	200	300	400
Total	2,741	6,560	13,020	25,768
Cost of equivalent quantity of hydrogen per year, dollars				
Cylinder hydrogen (1c./ft.³)	7,220	18,050	36,100	72,200
Recirculated hydrogen (0.5c./ft.³)	3,610	9,025	18,050	36,100
Maintenance	2,000	3,000	4,000	5,000
Total recirculated hydrogen	5,610	12,025	22,050	41,100
Comparative cost, dissociated ammonia				
vs. cylinder hydrogen, %	38.0	36.4	36.1	35.7
vs. recirculated hydrogen, %	48.9	54.5	59.1	62.7
Savings by using dissociated ammonia per year, dollars				
vs. cylinder hydrogen	4,479	11,490	23,080	46,432
vs. recirculated hydrogen	2,869	5,465	9,030	15,362

[a] Based on 7200 operating hours per year.

[18] The Drever Co., Philadelphia, Pa. *Bull.* No. 51.

ammonia atmospheres on a yearly basis, as tabulated from available information by one manufacturer of ammonia dissociators.[18] The cost of the atmosphere is reduced to nearly one-third if compared with cylinder hydrogen, and is still nearly one-half less than that of recirculated hydrogen. All these figures are based on cylinder ammonia; users of large quantities would purchase the ammonia in tank cars, in which case the price would drop to about 3.5 cents per pound. The cost of the dissociated ammonia atmosphere is thus lowered to less than one quarter, and would amount to only 8.5% of that for cylinder hydrogen and 15% of that for recirculated hydrogen. The gas volume from one standard cylinder of ammonia equals that of about 30 standard cylinders of hydrogen.

Carbon Monoxide

Pure carbon monoxide, because of its toxic character, is not suitable for a sintering atmosphere. Besides, its price as cylinder gas is prohibitively high. It would, finally, drastically affect the carbon content in ferrous compositions.

Hydrocarbons

Pure hydrocarbon gases, such as propane or methane, are equally objectionable on account of high cost and carburizing tendencies.

TABLE 60
Composition of Manufactured and Natural Gases Possible for Sintering Atmospheres[a]

Gas	Percentage of								
	N_2	H_2	CO	CO_2	O_2	CH_4	C_2H_4	C_2H_6	C_2H_8
Blast-furnace gas	57.6	3.2	26.2	13.0	—	—	—	—	—
Producer gas	58.8	10.5	22.0	5.7	—	2.6	0.4	—	—
Water gas, blue	1.3	51.8	43.4	3.5	—	—	—	—	—
carburetted	1.8	35.2	33.9	1.5	—	14.8	12.8	—	—
Coal gas	2.3	47.0	9.0	1.1	—	34.0	6.6	—	—
Coke-oven gas	4.2	57.4	5.1	1.4	0.5	28.5	2.9	—	—
Oil gas	3.4	53.5	10.6	2.8	—	27.0	2.7	—	—
Natural gas,[b] Type I	0.5	—	—	—	—	—	—	31.8	67.7
Type Ia	0.6	—	—	—	—	—	—	79.4	20.0
Type II	0.7	—	—	—	—	32.2	—	67.0	—
Type III	3.8	—	—	0.2	—	83.5	—	12.5	—

[a] Data from Handbook of Chemistry and Physics, 30th Ed., 1947, p. 1508.
[b] Type I, at Follansbee, W. Va.; Type Ia, same, residual; Type II, at McKean County, Pa.; and Type III, at Sandusky, O.

Manufactured and Natural Hydrocarbon Gases

The manufactured and natural gases offer an inexpensive source of gas and can be used as protective atmospheres in the sintering of nonferrous products, provided such harmful impurities as water vapor, sulfur dioxide, and hydrogen sulfide are removed. Sintering of ferrous products in these atmospheres is generally connected with difficulties, since the carbon content of the metal is hard to control, especially if the gas consists of a mixture of pure hydrocarbons, as in natural gas. Table 60 lists the compositions of some of these manufactured and natural gases that have been used as sintering atmospheres.

Partially Combusted Hydrocarbon Gases

Partially combusted hydrocarbon gases have found widespread use in the sintering of powdered metals, both as reducing gases and as neutral atmospheres for steel compositions of definite carbon content. Their chief advantage lies in their extremely low costs which, depending on the gas source, varies from about 5 to 40 cents per 1000 ft.^{3}, *i.e.*, 0.5 to 4% of the cost of hydrogen. Table 61 compares the costs of partially combusted gases with fully combusted fuel gases and also with hydrogen and dissociated ammonia (according to Webber and Hotchkiss[14]). Natural gas or other hydrocarbon gases are premixed with air in definite ratios and are burned in a gas converter. This generator consists essentially of a combustion chamber in which the reactions take place over a catalyst bed at carefully controlled temperatures between 900° and 1100°C. (1650° and 2010°F.). The catalyst is responsible for the high hydrogen content in the effluent gas when operating at partial combustion.[18a] A typical gas converter is shown in Figure 274, and its flow diagram is illustrated in Figure 275. After leaving the cracker, the gas is usually dried by condensing the water vapor; additional thorough drying by means of chemical moisture absorbents is often necessary to make the atmosphere reducing and non-decarburizing.

In Figure 276 a typical analysis chart gives the variation in composition of the gas that can be obtained from such a gas converter—operating over the entire range of air–gas ratios from partial to complete combustion. For complete combustion of air and natural gas, for example, the carbon dioxide content is at a maximum of about 11%, the actual value depending on the composition and specific gravity of the natural gas. With increasing proportion of gas in the mixture, the carbon dioxide content decreases to a minimum of about 4% with corresponding increases of hydrogen to about 20% and carbon monoxide to about 12%. The limited ratio for operating this type of atmosphere generator is about 5

[18a] H. M. Webber, in J. Wulff, *Powder Metallurgy*. Am. Soc. Metals, Cleveland 1942, pp. 299, 300.

TABLE 61

Approximate Analyses and Costs of Sintering Furnace Atmospheres (Webber and Hotchkiss[14])

Furnace atmosphere	Air/gas ratio	Percentage of							Water vapor dew-point, °F.	Cost[a] per 1000 ft.³, dollars
		CO_2	Illumi-nants	O_2	CO	H_2	CH_4	N_2		
Electrolytic hydrogen										
From cells..........	—	0.0	0.0	0.2	0.0	99.8	0.0	0.0	Saturated	1.50–3.00
From bottles.........	—	0.0	0.0	0.2	0.0	99.8	0.0	0.0	−30 to −10	5.00–10.00
Dissociated ammonia		0.0	0.0	0.0	0.0	75.0	0.0	25.0	−60	1.25–4.00
Partially burned......... 1/1 (low)	1/1 (low)	0.0	0.0	0.0	0.0	24.0	0.0	76.0	Saturated	0.75–2.45
Completely burned.........1.75/1 (high)	1.75/1 (high)	0.0	0.0	0.0	0.0	1.0	0.0	99.0	Saturated	0.95–2.95
Combusted fuel gas										
Partially burned......... 6/1 (low)	6/1 (low)	5.0	0.0	0.0	10.0	14.0	1.5	69.5	Saturated	0.08–0.40
Partially burned......... 8/1 (medium)	8/1 (medium)	8.0	0.0	0.0	6.0	7.0	0.0	79.0	Saturated	0.06–0.37
Completely burned.........10.5/1 (high)	10.5/1 (high)	11.5	0.0	0.0	0.7	0.7	0.0	87.1	Saturated	0.05–0.31
Carbon dioxide-free and dry..... 6/1 (low)	6/1 (low)	0.0	0.0	0.0	10.5	14.8	0.0	73.1	−40	0.12–0.44
Drycolene ....10.5/1 (high)	10.5/1 (high)	0.0	0.0	0.0	30.0	16.0	1.6	54.0	−15	0.32–0.72

[a] Costs include only materials consumed, such as power, gas, and water. They do not include factors for obsolescence, installation, maintenance, etc. The estimates have been based on the following unit costs:

Power...............$0.01 per kw-hr. Charcoal ....$0.03 per pound
Natural gas..............0.50 per 1000 ft.³ Propane ....0.25 per gallon
Water................0.05 per 1000 gallons Ammonia ....0.035–0.15 per pound

Fig. 274. Atmosphere gas converter (right), sulfur-removing iron oxide tower (left), and water scrubber (center) supplying protective atmosphere (courtesy of General Electric Co.).

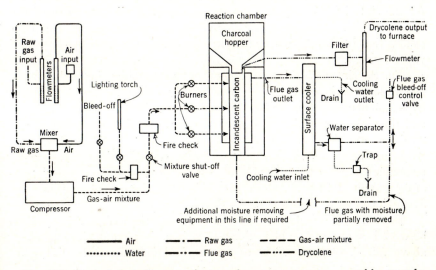

Fig. 275. Typical flow diagram of atmosphere gas converter, operable on coke oven gas, natural gas, or propane (according to Webber[18a]):

642   SINTERING ATMOSPHERES

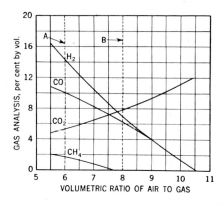

Fig. 276. Diagram indicating approxi mate analysis of gas obtained from combustion-type atmosphere generator fed mixed natural gas and air (according to Webber and Hotchkiss[14]): A indicates atmosphere suitable for sintering iron parts, B for sintering bronze parts.

parts of air to 1 part of gas. A typical analysis of an atmosphere obtained with such a ration is, according to Webber:[18a]

Gas	Composition, %	Gas	Composition, %
Carbon dioxide	4.0	Hydrogen	17.0
Illuminants	0.0	Methane	2.0
Oxygen	0.0	Nitrogen	63.0
Carbon monoxide	11.0	Water vapor	3.0

If no free oxygen is present, any atmosphere obtained from different air–gas ratios up to nearly complete combustion is sufficiently reducing for the sintering of copper and many of its usual alloys. For economic reasons the air–gas ratio is kept fairly lean. A gas obtained by complete combustion is inert to copper in spite of high carbon dioxide and water vapor content.

For the sintering of iron and steel compacts a more thoroughly purified atmosphere is required, since pure iron will oxidize in the gas obtained by complete combustion. Only if the water vapor is substantially removed and the carbon monoxide–carbon dioxide ratio has increased to at least 2 : 1 can iron be satisfactorily sintered at about 1100°C. (2000°F.). Koehring[8] recommends an atmosphere with approximately 10% carbon monoxide and 5% carbon dioxide, as obtained by combustion of a mixture of 6 parts air and 1 part natural gas. This atmosphere must be dried to a dew point of at least –40°C. (–40°F.).

In view of the fact that austenitic iron absorbs carbon, the reaction between iron and carbon monoxide and carbon dioxide must be controlled above 750°C. (1380°F.). The diagrams of Figure 277[19] show the equi-

[19] *Metal Progress, 46*, No. 4, 876 (1944).

librium conditions for gas-steel reactions and the relation between various gas compositions, temperature, and the carbon content of the steel. In

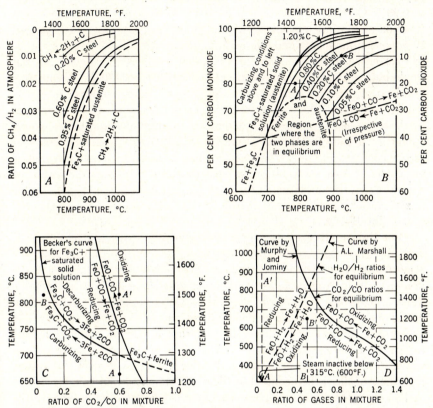

Fig. 277. Equilibrium diagrams for gas–steel reactions: *A*, relation between temperature, methane–hydrogen ratio in atmosphere, and carbon content in steel; *B*, relation between temperature, carbon monoxide in atmosphere, and carbon and oxygen content in steel (carburizing curves are for pressure of 1 atmosphere; 760 mm. Hg); *C*, relation between temperature and carbon dioxide–carbon monoxide ratio in mixture, and carbon and oxygen content of steel; and *D*, relation between temperature, ratio of gases in atmosphere, and carbon and oxygen content in steel. (Courtesy of Metal Progress.)

Figure 277*A* it is shown that almost pure hydrogen decarburizes hot steel (upper and left of diagram) while methane decomposes almost completely into hydrogen and carbon, thus carburizing steel and depositing excess soot (lower right of diagram). It is apparent from Figure 277*B* that the carburizing reactions depend on the carbon content of the steel, indicating that, for instance, a proportion of 80% carbon monoxide, 20% carbon dioxide at a temperature of 800°C. (1470°F.) will carburize a low-carbon steel, but decarburize a high-carbon steel, while 90% carbon monoxide, 10% carbon dioxide at 920°C. (1670°F.) (point B) is relatively inert to 0.20% C steels. Figure 277C shows the permissible proportions as a function of the temperature. For example, a dried atmosphere containing 10% carbon monoxide and 6% carbon dioxide would tend both to reduce and carburize an iron compact at about 660°C. (1220°F.) (point A), but would both decarburize and oxidize at about 800°C. (1470°F.) (point A'). For heating without a trace of oxidation, which is essential for iron-powder compacts, the carbon dioxide content must be well below one-tenth of the carbon monoxide content for normal sintering temperatures (point B). Figure 277D shows that the oxidizing action of steam on iron can be counteracted only if at least 20 times as much hydrogen as steam is present (line A–A'). Oxidation and decarburization through the moist gas may likewise be counteracted by carrying excess carbon monoxide in the mixed carbon oxides present in the furnace atmosphere (e.g., $CO_2 : CO = 0.5$, line B–B').

For the sintering of medium-carbon and high-carbon steel compositions, it is necessary to remove water vapor and carbon dioxide from the atmosphere, since these constituents are active decarburizing agents in the presence of hydrogen and carbon monoxide (Fig. 277C). A graphic illustration of the need of dry combusted gas, completely free of carbon dioxide, to prevent surface decarburization, is shown in Figure 278 (according to Gonser[20]). From this diagram it will be noted that best results were obtained with 0% carbon dioxide (curves 11 and 12), in which cases no decarburization took place and parts were carburized regardless of the carbon content of the steel. Activated alumina dryers and carbon dioxide solvents can be used to prevent decarburization. If the removal of the undesired constituents is complete, and the carbon monoxide content of the gas is high, the atmosphere will not decarburize the steel compacts. Sulfur dioxide and hydrogen sulfide are other constituents that must be removed. They are present in the cracked atmosphere when coke oven gas is burned, and attack copper, nickel, and other nonferrous metals. Their removal is effected by passing the gas through a water-spray tower and a tower filled with wood shavings coated with iron oxide.[18a]

The production of a furnace atmosphere that is virtually free of carbon dioxide and water vapor without the use of refrigeration or drying has recently been reported by Koebel.[20a] This gas is produced by pass-

[20] B. W. Gonser, *Ind. Heating, 6,* No. 12, 1123 (1939).

[20a] N. K. Koebel, *Proc. Fourth Annual Spring Meeting of Metal Powder Association,* Chicago, April 15–16, 1948, p. 6.

ing rich gas-air mixtures over an endothermically heated catalyst at about 1200°C. (2200°F.) thereby splitting the hydrocarbon constituent with air into hydrogen and carbon monoxide directly without side reactions to form carbon dioxide and water vapor. A gas containing an analysis of about 40% maximum hydrogen, 22% maximum carbon monoxide, 0.5–1.5% methane, and balance nitrogen with a dew point of below —18°C. (0°F.) can thus be produced. This analysis will not change during the passage of the gas into the sintering furnace chamber if it is drastically cooled immediately after leaving the catalyst section of the gas generator.

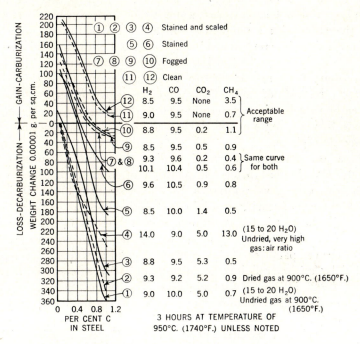

Fig. 278. Effect on carbon steels of using partly combusted generator gases with and without removal of water or of carbon dioxide or addition of methane (according to Gonser[20]).

A slightly reducing atmosphere of pure nitrogen with only small percentages of hydrogen can be obtained by completely removing all carbon dioxide from the dried products of complete combustion with the aid of regenerative absorption equipment. The resulting gas analyzes about 98% nitrogen and not more than 1% each of hydrogen and carbon monoxide. Commercially pure nitrogen can also be produced by burning dissociated ammonia to complete combustion in a similar atmosphere converter. Any desired ratio of hydrogen and nitrogen—from 75% to about 0.5% hydrogen—may be obtained by partial combustion of the dissociated ammonia. The water vapor, always present in these mixtures, can be removed by drying agents.

Lithium-Containing Atmospheres

Metallic lithium vapors have been used successfully for a reducing medium in the sintering atmosphere. The affinity of lithium to oxygen is so great that it reacts with water to produce hydrogen and lithium oxide. Thus, lithium is an extremely active moisture absorbent. Since its action takes place at low temperatures, water vapor and oxygen are removed before the gas reaches a temperature at which it could decarburize steel. Moreover, oxides inherent in steel compacts react with the lithium vapor during the early stages of heating where the temperature is still below that necessary for a reaction between the oxygen and the carbon of the steel structure. Both of these effects are of great advantage in the sintering of carbon steels, which generally tend to decarburize more by internal reactions between oxygen and carbon than by an inadequately dried sintering atmosphere.

There are numerous ways of leading the lithium vapors to the compacts, the simplest being the addition of a small percentage of a lithium salt to the powder mixture. If lithium stearate is used it has the added advantage of serving as a lubricant during pressing of the compact. Another simple way is to use lithium salts as getters in the immediate vicinity of the sinter charge. The salt may either be placed in containers ahead of the compacts, or it may serve as a mold in which the compacts, protected by an inner lining, are placed and stay submerged after the salt has fused at higher temperatures. The protective coating (e.g., aluminum oxide or zirconium oxide) is necessary as the lithium salts are very corrosive.

Perfect drying of the atmosphere and an effective reaction with the oxygen in the compact can also be achieved by a thorough blending of the lithium vapor with the furnace atmosphere. One furnace manufacturer generates metallic lithium vapors by fusing and decomposing lithium salts at a controlled temperature immediately above the sintering furnace muffle. A partially combusted hydrocarbon (propane) is very suitable as a furnace atmosphere, since—with this gas—lithium enters in a self-regenerating cycle which assures greatest economy.

According to Thomas,[21] the reactions of lithium with water vapor, oxygen, and the carbon oxides are as follows:

[21] C. E. Thomas, *Ind. Heating, 11,* No. 9, 1405 (1944).

$$2\,Li + H_2O = Li_2O + H_2 \qquad (1)$$
$$4\,Li + O_2 = 2\,Li_2O \qquad (2)$$

It would normally be expected that the active lithium in the furnace atmosphere would reduce the carbon dioxide to carbon monoxide, but actually the opposite occurs. The oxygen, taken from the reduction of water vapor by lithium, reacts with the carbon monoxide to produce carbon dioxide; lithium is freed to react with more water vapor. This is represented by the following equations:

$$2\,Li_2O + CO = Li_2CO_3 + 2\,Li \qquad (3)$$
$$Li_2CO_3 = Li_2O + CO_2 \qquad (4)$$

This reduction of lithium oxide by carbon monoxide is important, and essential to the economy of the process. Were it not for this reaction, it would be necessary to introduce large quantities of lithium continuously to take care of the large amount of oxygen and water vapor passing through the furnace chamber, since only a portion of the water is condensed from the combustion gases used. It will be noted below that all of the lithium is recovered during each cycle, as shown by equations (5) and (6), etc.:

$$8\,Li_2O + 4\,CO = 4\,Li_2CO_3 + 8\,Li \qquad (5)$$
$$4\,Li_2CO_3 = 4\,Li_2O + 4\,CO_2 \qquad (6)$$
$$4\,Li_2O + 2\,CO = 2\,Li_2CO_3 + 4\,Li \qquad (7)$$
$$2\,Li_2CO_3 = 2\,Li_2O + 2\,CO_2 \qquad (8)$$
$$2\,Li_2O + CO = Li_2CO_3 + 2\,Li \qquad (3)$$
$$Li_2CO_3 = Li_2O + CO_2 \qquad (4)$$

The lithium metal recovered in the above sequence is again available to eliminate water vapor and oxygen from the furnace atmosphere, according to equations (1) and (2) above.

The furnace atmosphere passes over the crucible in which the lithium salt is fused, carrying the lithium vapor into the adjacent furnace chamber, as shown in Figure 279—illustrating the cross section of a batch furnace designed for the lithiated atmosphere. Excessive cooling of the atmosphere must be avoided, since this would lead to a heavy deposit of lithium oxide in the connecting pipes. Therefore, the atmosphere is best introduced into the hot furnace muffle. In a continuous furnace a small portion of the gas can be branched off after passing the lithium generator and can be introduced at the discharge end of the cooling chamber for the sake of more effective cooling by counterflow of the gas. The exterior

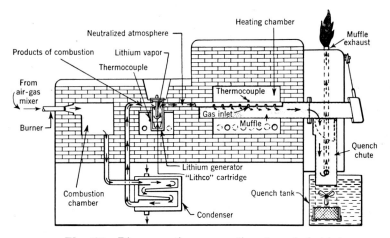

Fig. 279. Diagrammatic cross section through batch-type, heat-treating furnace operating with lithiated atmosphere (according to Thomas[21]).

view of a combined continuous sintering furnace and lithium generator is shown in Figure 258 (Chapter XVI).

Summary

Sintering atmospheres may be classified into inert and reducing gases. They may also be grouped into (1) gases ineffective toward carbon and (2) gases that carburize or decarburize steels. Among the atmospheres that are inert toward oxygen as well as carbon, vacuum is the most favorable. Vacuum sintering is used for laboratory work, where it is desirable to eliminate the gas problem entirely. It is, however, a difficult and costly practice for production operations. Pure nitrogen is used as an inert atmosphere, both for laboratory and production work, but infiltration of oxygen into the furnace chamber must be compensated for by small additions of hydrogen or carbon monoxide in order to preserve a truly neutral atmosphere. Argon has been used occasionally in research work involving expensive materials. Helium has lately been used to sinter stainless steel with good results. Steam and carbon dioxide are sometimes used in the sintering of nonferrous metals. Sintering in air is possible with precious metals and light metals whose powders already contain irreducible oxide films.

Hydrogen is widely used as a reducing atmosphere in laboratory and production work; if carefully purified, it is a suitable atmosphere for most sinter metals. Its high cost can, moreover, be halved by recirculating that portion of the gas unused in the reaction with oxygen and not lost by leakage or through the opening of the furnace doors. Dissociated ammonia can be substituted for hydrogen in practically all applications. The cost is considerably less than, and, indeed, for large-scale consumption, may amount to only one-tenth that of recirculated hydrogen.

In the sintering of ferrous metals, as in the heat treatment of steels, the sintering atmosphere must be in equilibrium with the carbon contained in the compacts. Partially combusted hydrocarbons have proved particularly suitable for this purpose and, on account of their low costs and closely controllable contents, are preferred over manufactured and natural gases. By premixing natural gas or other hydrocarbons with air in definite ratios, and burning the mixture in gas converters, an atmosphere of variable composition is obtained. Depending on the ratio, the atmosphere can be varied from oxidizing toward ferrous metals and neutral toward nonferrous metals, to neutral or reducing toward ferrous metals.

The action of metallic lithium vapors in conjunction with partially combusted gases is particularly beneficial in the sintering of steels. Lithium has a strongly desiccating effect on the atmosphere and, furthermore, tends to react with the oxygen present in the compact before a reaction can take place with the carbon combined in the structure.

CHAPTER XVIII

Subsequent Working

The sintering operation concludes the true synthesis of the powder metal. We deal now with a material that closely resembles an annealed casting in many respects; the principal difference lies in a finer structure of randomly oriented grains, and in the abundance of pores that may vary in size from ultramicroscopic cavities in the boundary areas to holes of a magnitude similar to that of the original particles (*e.g.,* 10 to 50 μ). Because of this unrefined grain structure, the physical properties of the material as it leaves the sintering furnace are usually unsatisfactory. Some of these properties, such as tensile, compressive, and fatigue strengths, as well as hardness and electrical conductivity, depend largely upon the character of the porosity inherent in the sinter metal. They tend to increase linearly with the density of the compact, as indicated schematically in Figure 280*A*; they reach normal values if density closely approaches the theoretical density of the particular metal or alloy. Other physical properties, such as tensile elongation, reduction in area, bending and torsional properties, and especially impact strength, are generally low; they are affected not only by percentage porosity, but also by additional factors, such as shape of the pores or grain boundary conditions. Usually, only a limited improvement in ductility and impact values can be obtained with increasing density, and even substantially densified compacts show a deficiency in these properties as compared to those obtainable with fusion products. This is schematically indicated in Figure 280*B*.

The only logical explanation of this phenomenon is that stress-raisers are incorporated within the structure to prevent optimum deformation under external stresses. The stress-raising effect can be attributed to impurity films and inclusions in the grain boudary areas only to a minor extent. Its main origin appears to be the notch effect produced by the residual pores. An exact stress analysis involving a multitude of pores of different shades is not possible at present, and the mechanism of the notch effect in sintered products is still far from being fully understood. Qualitatively, however, it can be expected that the stress concentrations and triaxial stresses produced by notches will affect ductility and shock resistance to a larger extent than strength. Since this agrees with experimental observation, the assumption that the poor ductility and impact values of sintered products are due mainly to residual porosity appears to be justified. This subject is more fully discussed in Chapter XXXIII, Volume II.

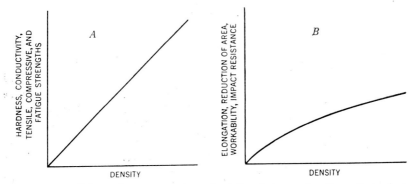

Fig. 280. Schematic representation of relation between density and physical properties of sintered metals: *A*, hardness, conductivity, tensile, compressive, and fatigue strengths; and *B*, elongation, reduction of area, workability, and impact resistance.

On the basis of this assumption, attempts to improve the mechanical characteristics of sintered products would have to be concentrated on methods to eliminate entirely—if possible—residual porosity along two main lines, *viz.*, (1) increasing the purity of the powders, and, therefore, the resulting grain boundaries, and (2) increasing the density to the limit. In the author's opinion, the problem of the effects of residual porosity represents one of the most important unsolved problems in the powder metallurgy process. After fifty years of industrial development of the art, there is now a definite demand for scientific research on this problem. An earnest attempt toward solution, perhaps with the aid of such modern tools as x-rays and the electron microscope, should go a long way in advancing our knowledge about the fundamentals of the sintering process, and research concerned solely with the problem of pore elimination might well become the basis for building a science of powder metallurgy.

PLASTIC DEFORMATION

Principles for Metals in General

The deficiency of sinter metals in ductility and shock resistance has been recognized by powder metallurgists from the earliest development phases of the art. Then, it was only natural to compare sinter metals with castings, and to apply the same recipe that transforms a brittle casting into a tough solid metal, *i.e.*, metal working. Amazing changes in the properties of fused metals can be produced by mechanical deformation and annealing of metals, effected by industrial forming operations, such as forging, swaging, rolling, or drawing. Of course, the magnitude of change differs with different properties and with the extent of working to which the metal has been subjected. The same essential procedures are indicated by the modern viewpoints on the role of residual porosity. Working methods which increase the density can be expected to increase the mechanical performance of sintered products to an even larger extent than in the case of cast materials.

According to accepted theory the deformation of metals takes place principally by slipping along certain crystallographic planes. Because of atomic rearrangements in the slip planes, an increase in frictional resistance stops this slip after it has proceeded to a certain extent. An increase in applied stress causes continuation of the slip phenomenon. As cold deformation of the metal proceeds, the grains, orginally more or less randomly oriented, gradually elongate and orient themselves with respect to the direction of flow. The grain structure transforms into a "fiber" structure, and the metal is regarded as being in a cold-worked state. A marked increase in hardness and tensile strength, and a substantial drop in ductility accompany the change in structure. At low temperatures the cold-worked condition in most metals persists permanently, but if the temperature is raised during an annealing process, recovery from the unstable condition begins, and the physical properties begin to return to "normal." At higher temperatures—actual values depending principally on the metal and amount of grain fragmentation or internal stress—the formation of new minute crystals occurs, a phenomenon known as recrystallization. With rising temperature new grains grow gradually in favored positions at the expense of their neighbors; such grain growth occurs even without previous recrystallization if the degree of deformation has only been slight. In case plastic deformation is carried out at temperatures above the recrystallization range, as in hot rolling or forging, the rate of hardening by working is slower than the rate of softening by annealing, and the metal does not harden during the working. However, appreciable refinement of grain size (with simultaneous elimination of a cast structure), severe plastic deformation, and simultaneous annealing break up the shells of impurities usually surrounding the individual grains and permit homogenization of the structure by means of diffusion. Generally speaking, a metal is in the most desirable condition (*i.e.*, highly plastic and, at the same time, tough, strong, and hard) when recrystallization is just complete and when a fine polygonal grain structure has developed. The coarse grain size, the envelopes of impurities around the grains, a certain amount of porosity, and the relative brittleness of cast metal, as well as the internal stresses and the distorted fibrous structure of a cold-worked material, have then been replacd by an ideal homogeneous, stress-free state which accounts for the superior performance of wrought material over castings.

Application to Powder Compacts after Sintering

NATURE OF SINTER METALS

To determine how principles of plastic deformation apply to "synthetic" metals obtained by processes of powder metallurgy, we should recall our previous analysis of the structure and the conditions as they exist in a simple metal. In Chapter VIII it was shown that in the formation of the compact, the metallic powder particles varying in size from fractions of a micron to several hundred microns are brought into close contact by high external force, and the surface adhesion forces between individual units give the compressed powder briquette a certain stability. The material undergoes a marked structural change during subsequent sintering (Chapter XIV). Plasticity increases with rising temperature and new crystallites form at temperatures in the recrystallization range of the particular metal, first at particle borders where the highest concentration of energy can be expected and then inside the powder particles. With additional temperature rises, atomic mobility becomes pronounced, and the new minute crystals tend to grow inward, and out toward the neighboring particles, producing new, "artificial" bonds between the different constituents of the compact. With continuation of the heat treatment the old particle contact areas gradually give way to new grain boundaries. In contrast to normal metals, where the structure is crystallized out of the liquid state and the grain boundaries generate from the residual melt, the grain boundaries of a sintered metal are formed in the solid state. These artificial grain boundaries are usually weakened by microscopic cavities, and only a prolonged "high" sintering operation at temperatures close to the melting point brings substantial consolidation by grain growth. The total grain boundary area is then diminished, the boundaries seem to be reinforced, and the cavities are incased by the expanded grains or completely closed up.

The sintered metal is in a peculiar state after this heat treatment. Its new crystals give it characteristics similar to those of the regular metal. In copper, nickel, and iron, for instance, a certain degree of ductility and malleability, corresponding to the plastic nature of these metals, can be obtained. On the other hand, no ductility can be expected in sintered chromium or molybdenum, for instance, since these metals are extremely brittle. In any case, cavities present inside the crystals or in the boundary areas, and impurities present as isolated inclusions, finely dispersed in the grain boundary regions, or as films surrounding the grains, render a sintered metal inhomogeneous and weakened in structure with a tendency to produce intercrystalline fractures predominantly. If, therefore, sintered metals are considered to be similar in behavior and qualities to castings subjected to a certain limited degree of homogenizing treatment, their specific properties, such as lack of ductility and resistance to sudden shock, can be well understood.

SUSCEPTIBILITY OF SINTERED COMPACTS TO DEFORMATION

It is only natural that attempts to refine the structure of sintered metals should follow the conventional lines so well established in the metallurgy of wrought materials. That this can be successfully accomplished, though perhaps by a more careful and therefore more costly technique, can be seen from the production of the refractory metals. The manufacture of wires from tungsten, tantalum, and molybdenum is the classic example of how the application of work can transform a fundamentally brittle material into a ductile wire or sheet of physical and chemical perfection. Only the laborious procedure of alternate working and annealing of the brittle sinter bars finally creates wire spun out to fine and strong filaments with great resistance to shock and creep. The example of the refractory metals clearly points the way by which most of the lower melting metals and many of the common alloys can be processed from powders to the desired degree of consolidation or metal working, provided the necessary processing steps and the proper care are taken during the manufacturing operations.

Application of such metal-working procedures, however, is quite impossible in the production of "custom-molded" metal powder parts having accurate dimensions, since gross deformation is incompatible with the purpose of the molding process. There must be introduced in this case a consolidating operation existing within the framework of common powder processing practices. Repressing of the sintered compact—perhaps in the same die used for the original molding of the powder—fulfills such a requirement, but of course is less effective than a true metal-working operation, as only limited plastic deformation can be effected during compression of the porous structure. While in such compression the axial deformation will be more or less drastic in accordance with the degree of porosity in the compact, the lateral deformation is usually only of minor order, its extent depending on the fit of the part into the die.

In analyzing the possibilities of consolidating the structure of sinter metals by subsequent mechanical operations we can distinguish between operations of limited and of unlimited deformation. To the former belong repressing, coining, and sizing; to the latter forging, hammering, swaging, rolling, and drawing. While these operations can theoretically be applied to hot compacts as well as to those kept at normal temperature, hot working is used mostly where extreme deformation is desirable (*e.g.,* in the manufacture of refractory metal products). Hot coining in a confined die space, although resembling conventional drop forging, has not yet become an operation that can be applied generally and on an industrial scale.

CONSOLIDATION BY RESTRICTED DEFORMATION

Repressing at Room Temperature

Repressing is today quite generally applied to powder metal parts that require accurate contours. Although this processing step is mostly a sizing operation in producing porous bearings, the structure is thereby somewhat condensed and hardened. Special sizing presses, sometimes similar in construction to the powder presses, and sizing dies giving the final accurate dimensions, are used for this purpose. In the manufacture of parts, repressing—usually called coining—serves the primary purpose of condensing the sintered compact. If the sinter bonds are sufficiently strong, considerable plastic deformation within the die is possible and, in this way, shapes of greater complexity are possible than can be molded directly from powder. The pressure necessary for this work is dependent upon the particular metal and the required properties. Although the recompression pressure is usually somewhere in the neighborhood of the compacting pressure, it sometimes exceeds the latter by as much as 100% and more.

The extent of deformation attainable by repressing a compact in a confined die cavity is chiefly dependent upon four factors: (*1*) the nature of the metal under consideration (its hardness and plasticity); (*2*) the density and grain structure of the sintered compact (effectiveness of sinter treatment); (*3*) the numerical specific pressure applied during repressing; and (*4*) the specific shape and die design under consideration. As far as consolidation is concerned lubrication between the compact and the die walls plays a less important role than it does in molding the powder. If customary compression dies with a unidirectional application of pressure are used, compression of the compact will also be in the direction of the axis of pressure. Consequently, the metal object will flatten out and the grains will deform in the plane perpendicular to the axis of pressure.

The effect of this limited degree of plastic deformation on the structure and physical properties of the compact is not insignificant. The grain structure of vacuum-sintered copper before and after recompression is shown in Figure 281. Although preferred orientation of the grains— evidence of severe plastic deformation—is hardly noticeable, the densification of the structure and the reduction in size and quantity of the

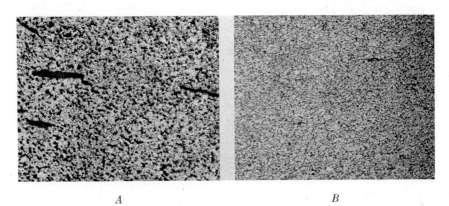

A   B

Fig. 281. Effect of recompressing on the microstructure of vacuum-sintered copper produced from ultrafine electrolytic powder ($\times$ 110): A, compact compressed at 45 tsi and sintered at 750°C. (1380°F.) for 16 hours; and B, same compact after recompressing at 45 tsi.

oriented cracks and many cavities are apparent. An indication of the effect of the recompression pressure on density and hardness is given in the diagrams of Figure 282 for copper and Figure 283 for iron. In the softer copper the increase in density and hardness for a given recompression pressure is greater than in the harder and tougher iron. The absolute values of the curves indicate that only extremely high recompression pressures (far in excess of 100 tsi) could yield a density comparable to that of the wrought metal. With the increased resilience of the compact caused by work-hardening, it will be impossible to close the last remainders of the pores and, consequently, the density as well as the physical properties will fall short of normal. Only an alternate repressing and annealing cycle, employing moderately high pressure (e.g., 50 tsi) and heating to sintering

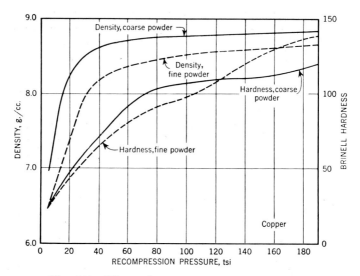

Fig. 282. Effect of recompression pressure on the density and hardness of sintered copper compacts from fine and coarse electrolytic powders.

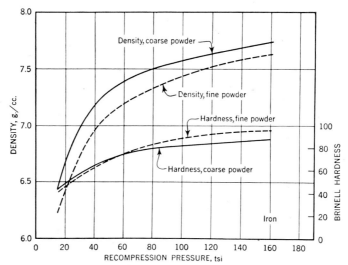

Fig. 283. Effect of recompression pressure on the density and hardness of sintered iron compacts from fine and coarse electrolytic powders

TABLE 62

Physical Properties of Sintered Iron before and after Cold
Coining and after Subsequent Annealing

Type of powder	State	Density, absolute vs. theoretical		Brinell hardness	Tensile strength, psi	Elongation, %
		g./cc.	%			
Hydrogen-reduced iron powder, all passing through 100 mesh	Compacted at 25 tsi, sintered at 1200°C. (2190°F.), 1 hr.	5.86	74.5	36	19,600	7.0
	Same, after coining at 25 tsi and annealing at 1200°C. (2190°F.), 1 hr.	6.88	87.4	66	26,700	10.0
	Compacted at 50 tsi, sintered at 1000°C. (1830°F.), 8 hrs.	6.68	85.0	47	27,000	10.0
	Same, after coining at 50 tsi, and annealing at 1000°C. (1830°F.), 8 hrs.	7.23	92.0	63	34,900	20.5
	Compacted at 50 tsi, sintered at 1200°C. (2190°F.), 1 hr.	6.93	88.0	57	29,300	11.0
	Same, after coining at 50 tsi	7.60	96.6	102	55,500	2.0
	Same, after annealing at 1200°C. (2190°F), 1 hr.	7.61	96.7	76	38,100	26.0
Coarse, electrolytic iron powder, all passing through 100 mesh	Compacted at 25 tsi, sintered at 1200°C. (2190°F.), 1 hr.	6.75	85.7	40	20,000	12.0
	Same, after coining at 25 tsi and annealing at 1200°C. (2190°F.), 1 hr.	7.40	94.0	58	30,900	19.5
	Compacted at 50 tsi, sintered at 1200°C. (2190°F.), 1 hr.	7.53	95.6	61	30,800	15.0
	Same, after coining at 50 tsi	7.62	96.8	88	44,000	7.5
	Same, after annealing at 1200°C. (2190°F.), 1 hr.	7.75	98.5	66	33,000	27.5
Fine, electrolytic iron powder, all passing through 325 mesh	Compacted at 25 tsi, sintered at 1200°C. (2190°F.), 1 hr.	6.66	84.6	44	23,900	12.0
	Same, after coining at 25 tsi and annealing at 1200°C. (2190°F.), 1 hr.	7.23	92.0	59	29,500	16.0
	Compacted at 50 tsi, sintered at 1200°C. (2190°F.), 1 hr.	7.33	93.3	63	30,300	15.0
	Same, after coining at 50 tsi	7.59	96.5	93	51,100	5.0
	Same, after annealing at 1200°C. (2190°F.), 1 hr.	7.69	97.7	69	31,700	23.0

temperatures, will accomplish a gradual consolidation that may reach theoretical densities after a number of repeat performances. It need hardly be said that such a procedure is uneconomical. In practice, a single coining and annealing operation must suffice in most instances and, accordingly, the improvement of density and physical properties remains limited. In Table 62 the physical properties of sintered iron are given before and after cold coining for different powders and processing conditions.

The coining operation often encounters severe mechanical difficulties, caused either by insufficient lubrication at the die walls or by the shape of the compact. Hard or abrasive powder compacts wear the walls of the die cavities appreciably, especially during the knock-out in conventional ejector-type press dies. This effect has been found in certain iron and steel compacts, and also in sintered aluminum, magnesium, and refractory metals. The effects of the shape factor are varied. Irregular cross sections or nonuniform density cause distortion or localized shrinkage or growth during sintering, and careful refitting of the compacts into the sizing or coining die becomes necessary. Sharp-angular or steep-curved contours, or shapes having lateral projections or re-entrant angles, often cannot be fitted into the coining die by simply inserting them into the cavity; parts of the die may first have to be removed, and then be replaced before compression. The sintered body will of course deform and flow to a much lesser degree than the original powder mass, and certain shapes that may form to precise dimensions in the molding operation will not materialize in coining a sintered blank. There remains also the difficulty of preventing seizure (cold welding), or physical adherence of the repressed compact to parts of the die or punches. This is particularly troublesome if the upper coining punch must be recessed to form a projection in the compact's upper surface. Upon withdrawal of the punch the compact may be pulled out of the die and will remain squeezed into the recess of the upper tool, thus requiring mechanical separation—which will impede an automatic coining operation. Consequently allowances must often be made in the design of the part to assure the maximum benefit of a subsequent repressing step.

Repressing at Elevated Temperatures

Repressing in hot-coining dies resembles hot-pressing, and has already been discussed as such in Chapter XII. One of the most practical methods of hot-pressing is by inserting preformed and preheated compacts into a hot die. If these compacts are considered to be in the sintered condition, the hot-pressing operation really becomes a hot-coining operation. Since the accomplishments of such a step have been analyzed fully in Chapter XII, it may suffice to add in Table 63 some properties of hot-coined iron and steel compacts for the sake of comparison with the properties of ordinarily coined material as given in Table 62. The great improvement in density and tensile values, a result of the better plastic deformability of the heated compacts, is quite impressive.

TABLE 63

Effect of Hot-Coining on the Physical Properties of Cold-Pressed and Sintered Iron and Steel (0.33% C) Compacts

Properties	After cold pressing at 25 tsi and sintering at 1100° C. (2010°F.) for ¼ hr.	After cold pressing, sintering, and hot coining at 1100° C. (2010°F.) and 2 tsi	After cold pressing, sintering, hot coining and cold coining at 50 tsi
Electrolytic Iron Powder, −100 Mesh			
Density, g./cc	5.88	7.47	7.77
Brinell hardness	37	87	111
Yield point. psi	—	39,000	55,000
Tensile strength, psi	20,200	40,800	59,500
Elongation in ½ inch, %	8	15	12
Reduction in area, %	6	14	12
Izod impact strength, ft.-lb	2	6	6
Electrolytic Iron, −100 Mesh Plus 0.5% Graphite Powder			
Density, g./cc	6.06	7.44	7.75
Brinell hardness	44	99	123
Yield point, psi	—	39,000	55,000
Tensile strength, psi	27,500	50,500	66,000
Elongation in ½ inch, %	3	14	12
Reduction of area, %	2	14	12
Izod impact strength	2	6	5

With the advancement of the hot-pressing technique, hot-coined powder metallurgy parts are bound to replace some of the cold-coined compacts, especially where the material offers great resistance to cold deformation, as in alloy steels. In this connection the reader is referred to Koehring's[1] results, obtained by hot die forging of various iron and steel compositions (page 463).

CONSOLIDATION BY UNRESTRICTED DEFORMATION

Deformation by Cold Working

Sintered metals cannot be cold deformed as extensively as fused metals. The inherent porosity and cold shortness of the structure impose severe restrictions even on such plastic metals as pure iron or copper. As a rule, the reduction in cross section between individual passes in the roll-

[1] R. P. Koehring, in J. Wulff, *Powder Metallurgy*. Am. Soc. Metals, Cleveland, 1942, p. 304.

ing mill or between individual blows of a hammer must be appreciably smaller than is conventional in metal-working practices. At the same time, in-between annealings of the structure must take place more frequently, and must be conducted under carefully controlled conditions that may

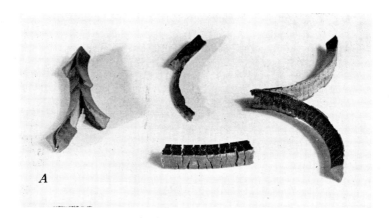

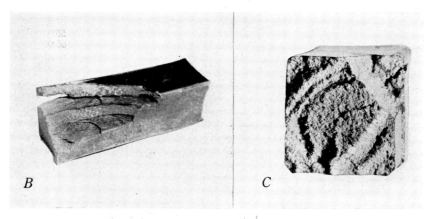

Fig. 284. Typical failures of sintered copper bars during cold rolling in a square-wire mill: A, specimens with different types of failures; B, specimen with laminated cracks; and C, appearance of fracture of bottom specimen in A, with cracks perpendicular to the direction of rolling.

require heating in a protective atmosphere. If these precautions are not observed, the sintered metal will soon fail, as shown in the photographs of Figure 284.

Various types of cracks may occur during cold rolling. The three top specimens in Figure 284A and the one shown in Figure 284B indicate cracks that are typical of too rapid and too severe reduction. Under these conditions the cracks run parallel to the axis of the bar and in the direction of rolling. By the use of an extremely low rate of reduction, failures of this kind can be avoided in very soft metals, such as copper, even if no intermediate anneals take place. In that case the rate of reduction must be approximately 0.001 in. per pass.[2]

The lower specimen in Figure 284A displays cracks perpendicular to the direction of rolling. This failure is typical of sintered metals, whose structure is insufficiently consolidated, and therefore cannot withstand severe plastic deformation; a typical fracture of this kind is shown in Figure 284C. Apparently, the specimen becomes more compressed in the

TABLE 64

Physical Properties of Sintered and Cold-Worked Copper

State	Density, g./cc.	Brinell hardness	Tensile strength, psi	Elongation in 2 in., %
Compacted at 50 tsi.............................7.47		73	970	—
Sintered at 800°C. (1470°F.), 8 hrs...............7.90		34	16,000	9.5
Coined at 50 tsi.................................8.39		70	22,200	4.0
Coined and annealed at 800°C. (1470°F.), 8 hrs...8.37		39	25,500	17.0
After 25% cold-rolling reduction; hard..........8.33		97	37,300	4.0
After 50% cold-rolling reduction; hard..........8.57		109	44,400	2.5
After 75% cold-rolling reduction; hard..........8.80		117	49,000	1.0
After 25% cold-rolling reduction; annealed......8.35		39	17,000	16.5
After 50% cold-rolling reduction; annealed......8.59		41	24,600	22.0
After 75% cold-rolling reduction; annealed......8.82		44	32,700	27.5

outer zones, especially at the corners, but the pressure exerted by a square-wire mill is not sufficiently effective to densify the cross section uniformly. The center section retains its original porosity until the reduction has advanced appreciably.

The effect on the physical properties of an unrestricted plastic deformation through cold working is best demonstrated if compared with that caused by a restricted deformation, such as achieved by coining in a die. In Tables 64 and 65, a comparison is made between electrolytic copper and iron compacts during various phases of processing, after cold rolling and annealing. It is apparent from these figures that both for copper and iron substantial plastic deformation in the rolling mill yields properties that are decidedly superior to those obtainable by merely re-pressing and resintering the compacts. The change of the physical prop-

[2] C. G. Goetzel, *The Influence of Processing Methods on the Structure and Properties of Compressed and Heat Treated Copper Powders. Dissertation.* Columbia University, New York, 1939.

erties of the sinter metal with increasingly severe deformation is clearly seen if the values are plotted as a function of the degree of work applied, the latter measured by the reduction in thickness of the sintered plates. In Figure 285 such curves are shown for copper, and in Figures 286 and 287 for iron. In all cases the metals were subjected to intermediate annealings after each 25% reduction in thickness and a final annealing before test specimens were machined from the plates. The values for 0% working reduction are those of the material as sintered. All properties rise continuously to highest values for a 75% reduction, and the figures probably could be increased slightly by an additional reduction in thickness. Values for rolled copper of 8.81 g./cc. for density and 32,700 psi for tensile

TABLE 65

Physical Properties of Sintered and Worked Iron

State	Density, g./cc.	Brinell hardness	Tensile strength, psi	Elongation in 2 in., %
Compacted at 50 psi.	6.23	69	470	—
Sintered at 1000°C. (1830°F.), 8 hrs.	6.68	47	27,000	10.0
Coined at 50 tsi.	7.27	67	30,500	4.0
Coined and annealed at 1000°C. (1830°F.), 8 hrs.	7.23	63	34,900	20.5
After 25% cold-rolling reduction; hard.	7.39	107	50,500	2.0
After 50% cold-rolling reduction; hard.	7.67	133	63,000	1.0
After 75% cold-rolling reduction; hard.	7.74	161	77,700	0
After 25% cold-rolling reduction; annealed.	7.40	63.5	30,600	15.5
After 50% cold-rolling reduction; annealed.	7.69	68.5	32,800	21.5
After 75% cold-rolling reduction; annealed.	7.75	68.5	33,800	26.0

strength are very close to those of the ordinary wrought metal, but an elongation of 27.5% is still considerably below normal. The situation is much the same for iron; sheets rolled from electrolytic powder compacts again reach their highest density and tensile strength after a 75% reduction (Fig. 286); the actual values of 7.74 g./cc. for density and 31,000 psi for tensile strength are not very high. The elongation reaches a peak value of 30% after 50% reduction and drops to 25% after 75% reduction. For cold-rolled, hydrogen-reduced, sponge iron compacts the trend of the curves is again the same (Fig. 287), but tensile values are generally higher (reaching 43,000 psi for 75% reduction), and the elongation figures are somewhat lower (22%). The modulus of elasticity lies appreciably below normal, even for 75% reduced sheets. In the cold-rolled condition, the 75% reduced sponge iron has a tensile strength of 63,000 psi at a hardness of 120 Brinell, as compared to 77,700 psi and 161 Brinell for electrolytic iron reduced to the same degree. These figures

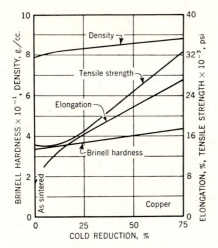

Fig. 285. Effect of cold-rolling reduction on the physical properties of electrolytic copper sintered at 800°C. (1470°F.) for 8 hours; material annealed before testing.

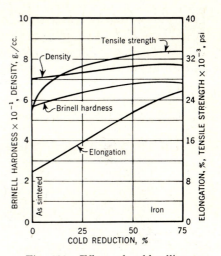

Fig. 286. Effect of cold-rolling reduction on the physical properties of electrolytic iron sintered at 1000°C. (1830°F.) for 8 hours; material annealed before testing.

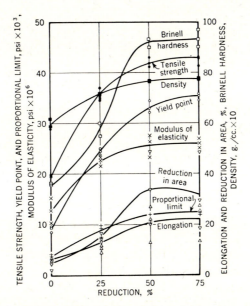

Fig. 287. Effect of cold-rolling reduction on the physical properties of hydrogen-reduced iron powder compacts sintered at 1200°C. (2190°F.) for 1 hour; material annealed before testing.

give evidence of the severe cold deformation to which the metal is subjected by such a procedure.

The relative weakness of these cold-worked sinter metals must be explained by the fact that even a 75% reduction is not sufficient to close all
pores and disrupt impurity films entirely. This can only be accomplished
by a working procedure which frequently alternates working and annealing, and reduces the metal above 90%. Figure 288 shows the perfect

Fig. 288. Fibrous microstructure of 0.060-in. diameter hard
wire from cold-rolled and drawn sintered copper (× 165).

microstructure of a fine copper wire in the hard-drawn state. The wire
was obtained by cold rolling and drawing a compact from electrolytic
copper powder; the total reduction was 97%. The electrical conductivity
of this wire is 99% of I.A.C.S.; its tensile strength is 67,000 psi, accompanied by an elongation of 5%.

Deformation by Hot Working

It is quite evident that the method of working a sintered metal by cold
deformation, such as rolling, is too one-sided and, at the same time, too
energy-absorbing to close all cavities, to crack all films of impurities, and
to condense the structure to the limit. Only hot working can fulfill these
requirements, since it takes advantage of the increased plasticity of most
metals at higher temperatures. Therefore, practically all industrial metal
working of powder metals is done hot. The reduction of tungsten and
molybdenum to wires and thin sheets is a typical example.

Tungsten and molybdenum bars are worked down into wires and filaments by many tedious operations. Hot swaging by hand and in machines down to about 0.080-in. wires gives the metal sufficient ductility to permit hot drawing, first through alloy dies, then through cemented carbide dies, and finally through diamond dies. These operations are frequently interrupted for intermediate annealing in hydrogen above the recrystallization temperature (approximately 1200° C. (2200°F.) for tungsten and

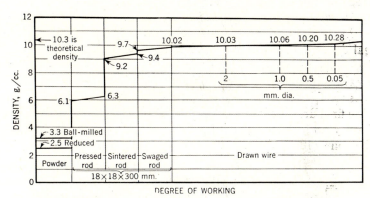

Fig. 289. Density of molybdenum powder, rod, and wire as affected by pressing, sintering, and degree of hot working.

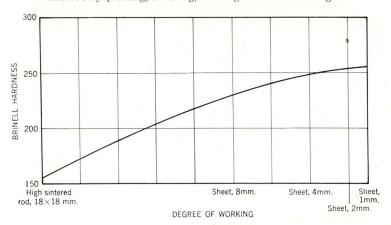

Fig. 290. Brinell hardness of molybdenum as affected by degree of hot working.

1000°C. (1830°F.) for molybdenum). In sheet manufacture the working methods are modified, forging replacing swaging, and final hot-rolling substituting for hot-drawing.

The effect of hot working on the properties of molybdenum is demonstrated in Figures **289** and **290**. Figure **289** shows the density of molybdenum powder after the various processing steps. The apparent density of reduced molybdenum powder is **2.5** g./cc., or approximately **24%** of the theoretical density of molybdenum. Ball milling increases the apparent density of the powder to **3.3** g./cc., or approximately **32%** of the theoretical density. Conventional compression of the powder at **30** tsi brings the density of an **18**-mm.-square rod up to **6.2** g./cc., or **61%**. Moreover, shrinkage during electric resistance sintering of the bar-shaped compact at a temperature approximately **250°**C. (**500°**F.) below the melting point causes the density to rise to a value between **9.2** and **9.4** g./cc., or **90–92%** of ideal density. Bars sintered even as high as **25°**C. (**50°**F.) below the melting point still show a porosity of **7** to **8%**, and are extremely brittle. During the subsequent working, swaging of the bars down to 3-mm. diameter lifts the density to values between **9.7** and **10.02** g./cc., or **95** to **98%**. Only the final drawing into wire size as fine as **0.05** mm. in diameter brings the density up to **10.20** g./cc., that is, **99%** of the theoretical density of molybdenum (**10.3** g./cc.).

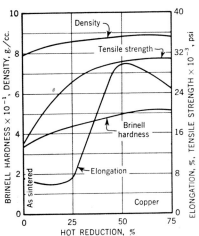

Fig. 291. Effect of hot-forging reduction on physical properties of electrolytic copper sintered at 800°C. (1470°F.) for 8 hours.

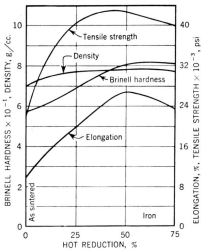

Fig. 292. Effect of hot-forging reduction on physical properties of electrolytic iron sintered at 1000°C (1830°F.) for 8 hours.

Similarly, as shown in Figure 290, hardness increases with sintering and with working. The hardness of a high-sintered molybdenum bar, 10 × 10 mm. in cross-sectional area, is 156 Brinell. Hot-rolling the material to sheet of 8-mm. thickness increases the hardness of the deformed metal to

229 Brinell, and this figure is further increased to 255 Brinell by hot-rolling to 1 mm.

Hot-forging experiments on copper and iron yield results of a similar nature to that seen in the cold-rolled metals. Although in this case too, work is applied only in one direction (perpendicular to the plane of the plates), consolidation and marked improvement of physical properties can be obtained at a lower percentage of reduction because of the increased plasticity of the metal, as shown in Figure 291 for electrolytic copper, and in Figure 292 for electrolytic iron. The apparent peak values for all properties at approximately 50% reduction are quite interesting; they indi-

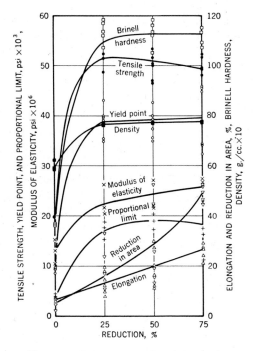

Fig. 293. Effect of hot-forging reduction on physical properties of hydrogen-reduced iron powder compacts sintered at 1200°C. (2190°F.) for 1 hour.

cate that complete consolidation is achieved for this reduction. The drop of the properties for the highest degree of reduction tested may be attributed to the beginning of overstraining of the boundary areas, and could possibly be avoided by a lengthy annealing. Tests conducted on compacts pressed from hydrogen-reduced sponge iron and sintered at a higher temperature (1200°C.; 2190°F.) than the electrolytic powder compacts (1000°C.; 1830°F.) produce a similar trend in the curves, as shown in Figure 293, although there is no marked decline in most values for the 75% reduction. The density value of 7.85 g./cc. for sintered electrolytic iron reduced 50% by forging closely approaches theoretical density, but tensile strength (42,000 psi average) and especially elongation (27%) are still rather low in comparison with wrought pure iron. The yield point and tensile strength values obtained by hot forging hydrogen-reduced sponge iron compacts (38,000 and 50,000 psi, respectively) are more comparable to those of carbon-free wrought iron. However, the modulus of elasticity, as well as elongation and reduction in area, though improved by hot forging, remains considerably short of standard. If the material is completely released of all stresses by lengthy intermediate annealings, or by an effective final annealing, values for elongation and reduction in areas can be still further improved. After a final annealing of the forgings at 1000°C. (1830°F.) for 1 hour, the hardness was lowered 20 Brinell units and, while yield point and ultimate strength figures remained practically at the same high level, elongation and reduction in area were incrased to maximum values of 35 and 60%, respectively, for a forging reduction of 50%.[3]

It is apparent from the foregoing examples that where strength and ductility are important, and high machining cost can be tolerated to get close dimensions, some form of hot forging could be used on the more plastic metals and alloys. There are many modifications of the hot-working technique that lend themselves to improvement of the physical properties. On the other hand, where close tolerances are just as important as strength and ductility, and where the metal belongs to the low-priced category, unrestricted deformation is not permissible; in such event, hot-pressing in a confined die space must replace forging.

A hot-working operation of somewhat intermediate character between those causing restricted and unrestricted deformation is extrusion. Here, the metal is confined in its lateral dimensions by the extrusion die, but can deform freely in the longitudinal direction. Under favorable circumstances (sufficient plasticity of the metal, endurable die material, good lubrication) the cross-sectional dimensions of the work can be held very closely (within 0.0005 in.) for one pass, and dies may withstand a number of passes before the extruded metal exceeds its cross-sectional tolerance.

[3] C. G. Goetzel, *Iron Age, 150,* No. 14, 82 (1942).

The extrusion of metals prepared from powders has already been discussed in Chapter XIII, and the reader is again referred to the pioneering work by Stout[4] and Tyssowski[5] on "coalesced" copper. Sintered iron and iron–lead alloys containing 3% lead are apparently also extruded readily, as reported by Kieffer and Hotop.[6] According to the same source, the extrusion of composite refractory metal-base contact materials has been used extensively abroad.[7] For this operation, temperatures of about 700°C. (1300°F.) and extrusion pressures of about 100 tsi were found necessary, and the copper or silver content had to be at least 20% by weight. The tungsten or molybdenum powder is obtained by pulverizing the sintered metal, and the low-melting constituent can be added either by mixing the powders or by impregnation of a sponge body pressed from the refractory metal. A thorough sintering treatment at temperatures above the melting point of the silver, or copper, respectively, is required in order to render the metal susceptible to the plastic deformation during extrusion.

Summary

Most metal powder objects are not usable after sintering. The demand for precise dimensions or high physical properties makes a subsequent operation necessary. This operation may cause a consolidation of the structure either by a restricted deformation of the compact, controlling the cross-sectional dimension in a die, or by an unrestricted deformation through conventional metal-working techniques. In both cases the plastic deformation may be accomplished at ordinary or at elevated temperatures—the nature of the metal permitting.

Restricted deformation of the compact by repressing at room temperature in a sizing or coining die is a widely applied operation in the production of parts from such plastic metals as copper, iron, and some of their alloys. Since this method does not constitute a very drastic deformation, the consolidation of the compact is only partially accomplished, and the resulting physical properties fall below normal values, especially with regard to ductility and toughness. On the other hand, closer dimensional tolerances can be held without the necessity of costly machining.

If the same restricted deformation is carried out at higher temperatures the physical properties can be improved without losing the advan-

[4] H. H. Stout, *Trans. Am. Inst. Mining Met. Engrs.*, *143*, 326 (1941).
[5] J. Tyssowski, *Trans. Am. Inst. Mining Met. Engrs.*, *143*, 335 (1941).
[6] R. Kieffer and W. Hotop, *Pulvermetallurgie und Sinterwerkstoffe*. Springer, Berlin, 1943, p. 205.
[7] R. Kieffer and W. Hotop, *loc. cit.*, p. 321.

tage of dimensional accuracy; but difficulties inherent in the process of hot-pressing have not yet been overcome sufficiently to make a hot-coining operation of sintered metals practical for mass production. The somewhat related hot-extrusion process, however, has found several noteworthy applications in powder metallurgy.

If sintered metals are sufficiently plastic they can be subjected to unrestricted deformation by working, like their counterparts derived from the fusion process; but in the case of cold working, more care is necessary, since the sintered structure tends to fail readily under adverse stress applications.

Hot working of sintered metals represents an important step in the manufacture of refractory metal products. It is practical where the metal is not hot-short. Considerable improvement of the physical characteristics of sintered metals is possible by hot forging, swaging, rolling, and the like, because of the unrestricted deformation in all directions. Where definite shapes must be produced, machining is mandatory, and the procedure is then limited to the higher-priced metals and alloys.

CHAPTER XIX

Subsequent Heat Treating

The quality of metal powder products can often be improved by heat treatment following the sintering or subsequent working process. Although sinter metals do not necessarily respond to such treatment as fully as fusion metals, ordinary heat-treating or annealing practices are quite generally adaptable for use with these materials. As a matter of fact, this circumstance alone is responsible for the successful manufacture of refractory metal products through mechanical working as indicated in the previous chapter.

The purpose behind all heat treatment in powder metallurgy is to improve or stabilize certain desirable properties by heating and cooling the material under controlled conditions of temperature, time, and atmosphere as long as a substantial part of the material remains in the solid state. The rates of heating and cooling are governed by the metallurgy laws peculiar to the composition. They are also affected by phenomena specific to sintered metals: internal porosity, for example, lowers the heat conductivity and thus retards a drastic quench treatment; surface porosity, on the other hand, requires heat treating in a protective atmosphere as a precaution against oxidation.

The heat treating of metal powder compacts after completion of the sintering cycle may have several objectives. If consolidation or homogenization of the structure is intended primarily, the heat treatment may be properly designated as "resintering," since this implies a direct sequel to the sintering and repressing operation. Other treatments affecting the structure and physical properties are analogous to those used conventionally. They include annealing, quench hardening, solution and precipitation treatments, and case hardening.

RESINTERING

Cold-Coined Compacts

In its true sense resintering is a continuation of the sintering process. In the first sintering operation the compact undergoes a fundamental structural change, the initial particle agglomerates, giving way to a metallic grain structure of abnormal texture. While recrystallization proceeds unobstructed within the particles and transgressing their boundaries at the points of contact, grain growth may be inhibited to a considerable degree by the interstices between the particles and by interfering foreign phases at the particle surfaces. Cold deformation after completion of sintering treatment, *e.g.*, by coining, tends to minimize both of these obstacles: the cavities are reduced in size and amount, and films of foreign substances are disrupted. If sintering treatment is now resumed, diffusion can proceed with fewer obstructions and grain growth will generally commence at a lower temperature and be more effective. This has been shown by Sauerwald,[1] who experimented with iron and copper. Iron,

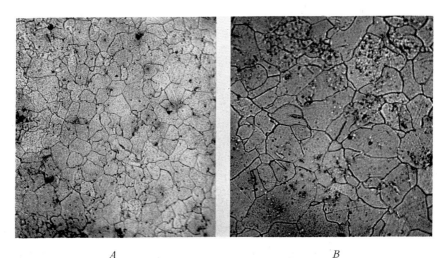

A   *B*

Fig. 294. Microstructures of sintered, and coined and resintered iron. Compact compressed from decarburized steel powder at 50 tsi, sintered at 1200°C. (2190°F.) for 2 hours, coined at 50 tsi and resintered at 1200°C. (2190°F.) for 2 hours (×145): *A*, compact after pressing and sintering; *B*, same, after coining and resintering.

after repressing at 65 tsi (representing 60% reduction), underwent grain growth at 750°C. (1380°F.); copper, after cold rolling (97% reduction), showed first indications of grain growth at 420°C. (790°F.).

    If resintering is carried out at relatively high temperatures, equal to those employed during the first sintering, grain growth may become very pronounced. This is indicated by the photomicrographs of Figure 294, where a compact from reduced iron powder is shown after sintering (*A*), and after coining and resintering (*B*). In both treatments the compact was kept at 1200°C. (2190°F.) for 2 hours. The effect of resintering at high temperatures on the physical properties of cold-coined copper and

[1] F. Sauerwald, *Z. Elektrochem., 29,* 79 (1923).

iron can be noted from the data compiled in Tables 64 and 65 (Chapter XVIII). As compared to straight sintered metal, the tensile strength can be increased through repressing and resintering by about 60% in the case of copper, and by about 30% in the case of iron. For both metals, the increase in elongation by repressing and resintering can amount to 100%. Even more impressive is a comparison of the same properties on the sintered and repressed metals before and after resintering. With the particular heat-treating cycles of these experiments[2] (800°C. (1470°F.), 8 hours for copper; 1000°C. (1830°F.), 8 hours for iron), resintering effects an increase in tensile strength of approximately 15% for both metals, in spite of an appreciable drop in hardness. At the same time the elongation increases more than fourfold for copper and more than fivefold for iron.

Although the physical properties of the compact can thus be improved markedly, other factors may make the resintering operation less desirable. Effective diffusion causes not only an increase in grain size, but frequently also further consolidation of the structure, accompanied by additional densification (shrinkage). This, however, results in a loss in the closeness of the part's dimensions and undoes one of the most important accomplishments of the coining or sizing operation. If, moreover, high coining pressures are used, gas expulsion and stress relief take place even at relatively low temperatures, resulting in expansion and warpage of the part. These phenomena, superimposed on the shrinkage effects at higher temperatures, may cause complex deformations which cannot be controlled easily, especially in parts with complicated contours. Consequently, it often becomes necessary, in practice, to strike a balance between that which is desirable in physical properties, and that which is permissible in dimensional accuracy. Where both must be of the highest quality, a second sizing must follow the resintering operation. This, of course, constitutes an additional operation, and may also require an additional die—both unfavorable economic factors.

Hot-Pressed Compacts

Resintering has produced very interesting results in the case of hot-pressed metal powders. In an investigation concerned with copper[3] it was found that grain structure could be affected materially by such a treatment. This becomes apparent from the photomicrographs of Figure 156 (Chapter XII). In Figure 156A the structure of an electrolytic copper

[2] C. G. Goetzel, in J. Wulff, *Powder Metallurgy*. Am. Soc. Metals, Cleveland. 1942, p. 97.
[3] C. G. Goetzel, *The Influence of Processing Methods on the Structure and Properties of Compressed and Heat Treated Copper Powders. Dissertation.* Columbia University, New York, 1939.

powder, hot-pressed at 400°C. (750°F.), displays a very dense agglomerate of the powder particles. The original shape and size distribution of the particles are still retained, but they have already undergone some internal recrystallization. During resintering at 750°C. (1380°F.) for one hour, a complete crystalline metamorphosis has taken place (Fig. 156B), and the structure has become homogeneous and finely crystalline, in addition to being virtually fully dense. The physical properties of such a material were found to be equal to those of wrought copper:

Density ....8.90 g./cc.	Elongation in 1 inch..............44%
Brinell hardness..............38.0 g./cc.	Reduction of area................66%
Tensile strength..............33,300 psi	Izod impact strength..........14 ft.-lbs.

Encouraging results can also be obtained if hot-pressed iron or steel powders are resintered. Using a method described elsewhere (Chapter XIII), pure iron (electrolytic, −100 mesh) and iron compacts containing 0.5% graphite were first preformed and then hot-pressed at 1125°C. (2060°F.) at a pressure of $1^1/_2$ tsi. In Table 66 the physical properties of

TABLE 66

Effect of Resintering on the Physical Properties of Hot-Pressed
Iron and Steel (0.33% C) Compacts

Properties	After hot pressing at 1100°C. (2010° F.) and 2 tsi	After hot pressing, and resintering at 1100°C. (2010° F.) for 1 hr.	After hot pressing, resintering, and coining at 50 tsi	After hot pressing, resintering, coining and second resintering at 1000°C. (1830° F.) for 1 hr.
Electrolytic Iron Powder, −100 Mesh, Pre-pressed at 25 tsi				
Density, g./cc................	7.39	7.53	7.80	7.80
Brinell hardness..............	88	69	114	77
Yield point, psi..............	30,000	29,000	53,000	33,000
Tensile strength, psi..........	39,000	37,000	57,000	47,000
Elongation in $^1/_2$ inch, %.....	14	30	12	47
Reduction of area, %.........	14	37	14	63
Izod impact strength, ft.-ib....	5	10	6	15
Electrolytic Iron, −100 Mesh Plus 0.5% Graphite Powder, Pre-pressed at 25 tsi				
Density, g./cc................	7.40	7.55	7.75	7.77
Brinell hardness..............	97	89	117	97
Yield point, psi	37,000	33,000	57,000	44,000
Tensile strength, psi..........	47,000	44,000	63,000	60,000
Elongation in $^1/_2$ inch, %.....	10	20	8	39
Reduction of area, %........	12	23	10	44
Izod impact strength, ft.-ib....	4	8	5	14

these materials are compared with the properties obtained after the hot-pressed compacts were resintered at 1100°C. (2010°F.) for 1 hour, and then cold coined at 50 tsi and annealed at 800°C. (1470°F.). The improvement in ductility of the resintered materials is particularly impressive. The microstructures of hot-pressed, hydrogen-reduced, iron powder compacts after resintering under different conditions are shown in Figure 295.

SOFTENING

Annealing Treatments

The comprehensive term "annealing" is defined by the metallurgist as a heating and cooling operation performed for one of the following purposes:[4] (1) to remove stress; (2) to induce softness; (3) to alter ductility, toughness, or other physical properties; (4) to refine the grain structure; (5) to remove gases; and (6) to produce a definite microstructure.

The temperature of the annealing operation and the rate of cooling depend upon the material being heat-treated and the purpose of the treatment. In ferrous metallurgy, certain specific heat treatments coming under the comprehensive term "annealing" include full annealing, process annealing, normalizing, spheroidizing, tempering, malleablizing, and graphitizing. All of these treatments usually imply a relatively slow cooling rate. On the other hand, many nonferrous alloys, especially of the copper-base type, are rendered "dead-soft" by a very rapid cooling (quench).

With certain modifications in technique the same treatments can also be applied to sinter-metals, the results varying according to the nature and history of the material and the deviation from normal of the heat-treating technique. As mentioned before, the inherent porosity of the sintered compact reduces the heat conductivity of the metal and consequently requires a more prolonged heating period. Oxidation of the surface and interior of the compact must be prevented by heating and cooling in a protective atmosphere.

The effectiveness of annealing is perhaps best demonstrated by several examples of sintered metals that were worked afterward. In Figure 296 microstructures of an annealed copper wire 0.060 in. in diameter are shown. The wire was obtained by cold rolling and drawing a sintered compact prepared from electrolytic powder.[3] After annealing at 600°C. (1110°F.) for 1 hour, the change of the fibrous structure of the hard-drawn wire (see Fig. 288 of Chapter XVIII) into a homogeneous structure containing very small uniform equiaxed grains (Fig. 296) evidences the

[4] *Metals Handbook*. Am. Soc. Metals, Cleveland, 1948, p. 2.

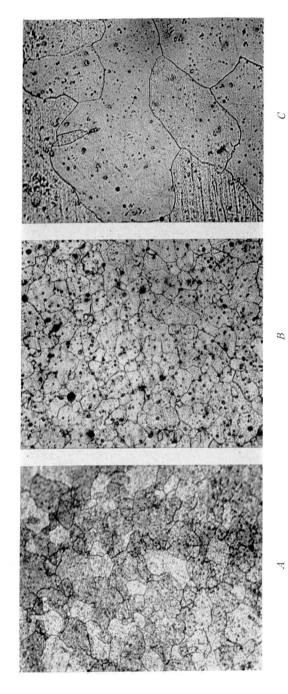

Fig. 295. Microstructures of hot-pressed, hydrogen-reduced iron powder compacts after resintering under different conditions (×150): A, hot-pressed at 500°C. (930°F.) and 50 tsi—resintered at 1000°C. (1830°F.) for two hours; B, hot-pressed at 500°C. (930°F.) and 50 tsi—resintered at 1300°C. (2370°F.) for 1 hour; and C, hot-pressed at 900°C. (1650°F.) and 2.5 tsi—resintered at 1300°C. (2370°F.) for 1 hour.

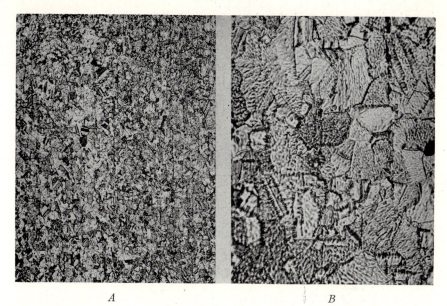

Fig. 296. Microstructure of annealed 0.060-in. diameter wire produced by cold rolling and drawing sintered copper: A (×135); B (×675).

normal behavior of the severely cold-worked (97% reduced) sinter-metal. The physical properties of the wire before and after annealing are given in Table 67. The effectiveness of the annealing treatment is indicated by the change of each of these properties.

TABLE 67

Physical Properties of Wire from Sintered Copper before and after Annealing

Property[a]	Hard drawn	After annealing[b]
Electrical conductivity, × 10^2, megmho (cm.).........	53.9	55.5
Electrical conductivity, % vs. standard...............	93.0	95.7
Tensile strength, psi.................................	66,000	36,600
Elongation in 2 inches, %............................	4.5	41.0
Number of bends to failure..........................	11	19
Number of twists to failure in 3 inches...............	20	57

[a] Wire, 0.060-inch ϕ, rolled and drawn from double-pressed and sintered copper.
[b] At 600°C. (1110°F). for 1 hour.

Table 68 illustrates that annealing can also markedly alter the properties of hot-worked sinter-metals; data of physical properties are given for sintered iron that has been hot-forged to different percentages of reduction.[5] Apart from the natural decrease in hardness through anneal-

[5] C. G. Goetzel, *Iron Age, 150,* No. 14, 82 (1942).

ing, the improvement of the elongation and reduction of area are most distinct for a cross-sectional reduction in forging of 25 and 50%. While the tensile strength is only slightly lowered for the lowest reduction, the yield point is markedly decreased for all reductions. Evidence of an effective softening and increase in ductility through the annealing treatment is therefore produced.

TABLE 68

Effect of Annealing on Physical Properties of Hot-Forged Hydrogen-Reduced Iron[a]

Percentage reduction	Property	After forging	After forging and annealing at 950°C. (1740°F.) for 1 hr.
25	Brinell hardness	112	101
	Yield point, psi	38,550	33,800
	Tensile strength, psi	51,550	48,900
	Elongation in 2 inches,%	13.5	22.2
	Reduction of area, %	19.2	39.4
50	Brinell hardness	110	89
	Yield point, psi	38,150	33,570
	Tensile strength, psi	49,700	49,430
	Elongation in 2 inches, %	19.2	30.3
	Reduction of area, %	28.1	51.5
75	Brinell hardness	111	93
	Yield point, psi	39,410	30,930
	Tensile strength, psi	49,980	49,500
	Elongation in 2 inches, %	26.1	27.8
	Reduction of area, %	49.4	55.1

[a] Compacts compressed at 25 tsi; sintered at 1200°C. (2190°F.) for one hour; forged at 1000°C. (1830°F.).

In the production of parts requiring close dimensional tolerances, a high-temperature annealing after coining may cause intolerable deformations, especially where the metal undergoes an allotropic change during heating, such as the alpha–gamma inversion in ferrous materials. If the annealing temperature is kept below the transformation temperature, the deformation and shrinkage can generally be expected to remain negligible, even for complex parts with differences in cross section. Of course, heat treatment of this type accomplishes merely the recovery of the work-hardened structure and removal of internal stresses, the grain size and texture having essentially the same appearance as after sintering. The ductilizing effect of such treatment becomes evident if the bending properties are studied. Data for sintered and coined, pure iron plates are given in Table 69.

TABLE 69

Bending Properties for Sintered Iron

Operation	Brinell hardness	Bending angle, degrees
Compaction at 25 tsi, sintering for 1 hr. at 1100°C. (2010°F.).	47	12
Coining at 50 tsi..	123	6
Annealing for 1 hr. at 600°C. (1110°F.)......................	97	14
700°C. (1290°F.)...	89	33
800°C. (1470°F.)...	77	63
900°C. (1650°F.)...	69	97
1000°C. (1830°F.)..	66	180+

Quench Treatments

A number of technical metals and alloys can be rendered "dead-soft" by a quench treatment. Pure copper or silver, for example, is best annealed if drastically cooled from a temperature above recrystallization. On the other hand, certain copper base alloys susceptible to precipitation hardening (copper–beryllium, copper–nickel–beryllium, copper–chromium–cobalt–beryllium, copper–iron, copper–aluminum–iron, etc.) are made quite ductile by forming a maximum solid solution phase through quenching.

If these treatments are to be applied to powder metallurgy products, the porosity again demands certain extraordinary precautions. The pores, acting as stress raisers, may cause severe distortion or fracturing during the quench treatment, especially if the part is complex in shape, or if it has sharp transitions from one cross section to another. The reduced heat conductivity may also require a more drastic quench in order to be effective. On the other hand, if porosity is large and of the interconnected pore type, the quenching medium would readily penetrate the structure, and the quench may be somewhat more rapid. In this case, of course, precautions must be taken so that the quenching medium does not cause internal corrosion. Moreover, in order to prevent excessive oxidation, it may even be necessary to surround the part by a protective atmosphere while it is dropped into the quench tank.

The softening effect of a water quench treatment on sintered copper is demonstrated by the following comparison with coined and furnace annealed compacts in Table 70.

TABLE 70

Effect of Quench Treatment on the Physical Properties of Sintered Copper

Property	Sintered copper, after coining at 50 tsi	Sintered and coined copper, after annealing at 750°C. (1380° F.) for 1 hour, furnace cooled	Sintered and coined copper, after annealing at 750°C. (1380° F.) for 1 hour, water quenched
Brinell hardness........	72	40	37
Tensile strength, psi.....	44,000	25,500	23,500
Elongation in 2 in., %...	4.0	17.0	20.5
Bending angle, degrees..	5	123	180+

The effect is most distinct for the bending angle, where a 50% improvement over furnace-cooled material can be obtained.

An increase in ductility through enlargement of the solid solution in an alloy is illustrated in Table 71 by a comparison of an annealed and a water-quenched composition containing 85% iron and 15% copper.

TABLE 71

Effect of Quench Treatment on the Physical Properties of an 85% Iron, 15% Copper Alloy

Property	85-15 Fe-Cu, after sintering	85-15 Fe-Cu, after reheat to 850°C. (1560°F.) for 1 hour, furnace cooled	85-15 Fe-Cu, after reheat to 850°C. (1560°F.) for 1 hour, water quenched
Brinell hardness...........	97	113	95
Tensile strength, psi........	47,700	54,000	53,000
Elongation in 1 in., %......	4	4	7
Electrical conductivity, $\times 10^2$, megmho (cm.).......	9.49	8.52	6.62

HARDENING

In applying the conventional hardening practices to powder metals, the same precautions are necessary as already described for solution treating by quenching. This applies equally to quench hardening and to case hardening. Where precipitation or age hardening is the purpose, a solution treatment (which again involves quenching) must precede.

Age and Precipitation Hardening

ALUMINUM-BASE ALLOYS

Age hardening of sintered aluminum alloys has led to interesting results. According to Kikuchi,[6] the hardness of compacts prepared from dur-

[6] R. Kikuchi, Science Repts. Tôhoku Imp. Univ. First Ser. (in English), 26, No. 1, 125 (1937).

alumin powder under a pressure of 95 tsi and after sintering and fully annealing at 570°C. (1060°F.) changed as follows:

Duralumin compact, completely soft annealed..........................77 Brinell
   immediately after quenching...94 Brinell
   after aging at room temperature for 80 hours.....................136 Brinell

These increases in hardness are quite remarkable, since they exceed those customarily experienced in massive alloys of this type. Using a similar

TABLE 72

Physical Properties of Precipitation-Hardened Aluminum Alloys after Sintering for 16 Hours at 500°C.(930°F.) in Hydrogen

Alloy[a]	Type 1		Type 2		Type 3	
Composition, %	Al, 94.0 Cu, 4.5 Si, 1.5		Al, 94.5 Cu, 4.5 Mg, 0.5 Mn, 0.5		Al, 92.5 Cu, 4.0 Ni, 2.0 Mg, 1.5	
Pressure, tsi						
Molding	35	70	70	35	30	70
Coining	35	70	70	35	30	70
Reheating cycle						
Temperature, °C.	530	530	530	530	510	510
°F.	986	986	986	986	950	950
Time, hr.	1	1	1	1	1	1
Atmosphere	CO_2	CO_2	CO_2	CO_2	CO_2	CO_2
Quenching operation[b]						
Temperature, °C.	520	520	500	500	500	500
°F.	968	968	932	932	932	932
Properties after quenching						
Brinell hardness	71	77	50	73	70	90
Density, g./cc.	2.44	2.63	2.39	2.69	2.37	2.55
Porosity, %	10.9	4.0	14.3	3.5	14.5	7.9
Precipitation treatment						
Temperature, °C.	120	120	120	120	120	120
°F.	248	248	248	248	248	248
Time, hr.	2	2	2	2	2	2
Hardness after precipitation						
Brinell hardness	102	110	112	140	129	147

[a] All alloys were prepared by mixing the powders in elemental form, compacting the mixtures, and diffusing the metals during sintering.
[b] Quenching medium, water.

treatment for a variety of aluminum alloys, the author was able to substantiate these results.[7] The remarkable response of some of the aluminum alloys to the age-hardening treatment becomes apparent from Table 72 and the curves in Figure 297. All compositions investigated were ob-

[7] U. S. Pat. 2,155,651.

tained by compacting powder mixtures containing the alloying constituents in virgin form. A long period of sintering (16 hours) was found necessary to obtain complete diffusion. With all alloys an aging treatment at 120°C. (250°F.) for a short time (2 hours) was found considerably more effective than aging at room temperature for longer periods. After a 4-day aging of duralumin compacts that had been pressed and repressed at 70 tsi, and sintered and resintered at 500°C. (930°F.) for 16 hours, the Brinell hardness had reached only 104. During aging for 10 days, no saturation values could be obtained, and the hardness of the compacts had only increased to 114 Brinell.

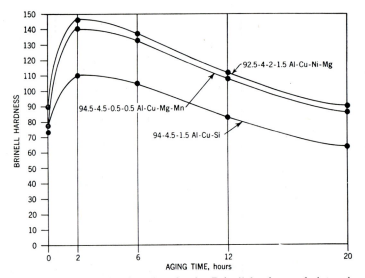

Fig. 397.   Eﬀect of aging time in the Brinell hardness of sintered aluminum alloys (courtesy of Charles Hardy, Inc.).

Cremer and Cordiano,[8] by producing a duralumin compact that had a Brinell hardness of 92 after aging at room temperature for 4 days, confirmed these findings. The powder mixture was compacted at 50 tsi, sintered in dry nitrogen for 30 minutes at 580°C. (1075°F.), cooled to 510°C. (950°F.) in 15 minutes, and quenched in cold water. By employing the unusually high heat-treating temperature of 580°C. (1075°F.), i.e., above the eutectic temperature of the alloy, remarkable physical properties were obtained as shown in Table 73.

    [8] G. D. Cremer and J. J. Cordiano, *Trans. Am. Inst. Mining Met. Engrs., 152,* 152 (1943).

TABLE 73
Tensile Properties of Sintered Duralumin (Cremer and Cordiano[8])

Methods	Tensile strength, psi	Elongation in 1 in., %
Compact pressed from duralumin powder mixture at 50 tsi, sintered at 580°C. (1075°F.) for 30 min. in nitrogen...	33,000	10
Compact treated identically, aged at room temperature for 4 days.................................	50,000	10

COPPER-BASE ALLOYS

Precipitation hardening of copper alloys can be achieved with equal success; many compositions are susceptible to the treatment. A dense copper–iron composition, containing 75% iron, balance copper, changes its properties through precipitation hardening as shown in Table 74.

TABLE 74
Effect of Precipitation Treatment on the Physical Properties on an 85% Iron, 15% Copper Alloy

Property	75-25 Fe–Cu, after sintering	75-25 Fe–Cu, after reheat to 850°C. (1560°F.) for 1 hour, water quenched	75-25 Fe–Cu, after reheat, quench, and precipitation treatment at 500°C. (930°F.) for 1 hour
Brinell hardness............	94	91	118
Tensile strength, psi........	46,900	53,500	58,000
Elongation in in., %........	5	7	4
Electrical conductivity, $\times$ 10^2, megmho (cm.).......	9.94	7.13	11.66

An extensive investigation of copper-base compositions susceptible to precipitation hardening has been reported by Hensel, Larsen, and Swazy,[9] with results given in part in Tables 75 and 76. They indicate that very good physical and electrical properties can be obtained with precipitation hardening of compositions containing chromium, cobalt, and beryllium, or nickel and beryllium. The addition of fractions of 1% of phosphorus or titanium hydride served as precaution against oxidation, and had beneficial effects.

A sintered copper–base bearing material, developed more recently by the same authors,[10] contains nickel silicide as the precipitation-hard-

[9] F. R. Hensel, E. I. Larsen, and E. F. Swazy, *Trans. Am. Inst. Mining Met. Engrs., 166*, 533 (1946).

[10] F. R. Hensel, E. I. Larsen, and E. F. Swazy, *Trans. Am. Inst. Mining Met. Engrs., 166*, 548 (1946).

TABLE 75

Composition, Processing, and Physical Properties of Precipitation-Hardened
Copper Alloys (Hensel, Larsen, and Swazy[9])

Properties	Type 1 (L-1511)	Type 2 (L-1512)	Type 3 (L-1271)	Type 4 (L-1293)
Composition, %				
Chromium	3.00	—	—	—
Cobalt	—	2.00	—	2.50
Nickel	—	—	2.50	—
Beryllium	—	—	0.50	0.50
Phosphorus	0.10	0.50	—	—
Titanium hydride	0.50	—	—	—
Copper	Balance	Balance	Balance	Balance
Processing				
Compacting pressure, tsi	30	30	40	40
Sintering temperature, °C	975	975	975	975
°F	1785	1785	1785	1785
Sintering time, min	60	60	30	30
Coining pressure, tsi	40	40	60	60
Quenching temperature, °C	975	975	950	975
°F	1785	1785	1740	1785
Precipitation temperature, °C	450	450	450	450
°F	840	840	840	840
Precipitation time, hrs	16	16	16	16
Physical properties (after complete heat treatment)				
Tensile strength, psi	52,000	36,500	52,000	78,200
Elongation in 2 inches, %	8	15	7	3
Fatigue strength, psi	20.000	—	—	—
Brinell hardness				
As coined	73	67	71	71
As solution treated	43	37	—	—
As precipitation treated	107	74	114	142
Electrical conductivity, per cent I.A.C.S.				
As solution treated	33.5	28.5	—	—
As precipitation treated	69.1	43.4	61.0	41.0

ening agent. The nominal composition of this product is 2.4% nickel, 0.8% silicon, and 0.3% phosphorus, and the remainder copper. Preferential silicon oxidation in surface layers reduces the efficiency of the precipitation hardening, providing the soft shell which is essential for bearing applications. The phosphorus addition controls the rate of preferential oxidation.

TABLE 76

Composition, Density, and Surface Hardness of Precipitation-Hardened
Copper–Beryllium Alloys Containing Phosphorus (Hensel, Larsen, and Swazy[9])

Properties	Alloy			
	A	B	C	D
Composition, %				
Cobalt............................	2.50	2.50	—	—
Beryllium.........................	0.72	0.72	0.32	0.32
Nickel............................	—	—	1.80	1.80
Phosphorus	0.19	0.27	0.22	0.30
Copper ....Balance		Balance	Balance	Balance
After compacting at 30 tsi				
Density, g./cc....................	6.7	6.7	6.8	6.8
Sintered, 1 hr. in hydrogen at 975°C. (1785°F.) and quenched				
Density, g./cc....................	7.1	7.1	7.2	7.2
Brinell hardness..................	65	58.5	36	40
Coined at 60 tsi				
Density, g./cc....................	8.6	8.2	8.5	8.5
Brinell hardness..................	102	106	83	86
After precipitation hardening for 8 hours at 450°C. (840°F.)				
Brinell hardness at surface........	160	167	122	140
0.015 in. below surface...........	175	175	133	145
0.030 in. below surface...........	179	171	130	145
0.060 in. below surface...........	179	175	133	145

Quench Hardening

If a carbon steel is prepared from powders, it can be heat-treated exactly as an ordinary carbon steel. Because of the porosity of the compact and the greater surface exposed to the action of the atmosphere, great care must be taken that the heat-treating atmosphere is truly neutral and dry.

By heat treating, the properties of sintered steels can be varied in much the same way as with ordinary steels. In Table 77 the properties of sintered and heat-treated steels of different carbon content, obtained by diffusion of graphite into electrolytic iron, are compared with similarly treated SAE steels, according to Stern.[11]

It is evident from these figures that remarkably good physical properties are obtainable by hardening and heat-treating sintered steels. The values for yield point and tensile strength are close to those of SAE plain

[11] G. Stern, *Trans. Am. Inst. Mining Met. Engrs., 166,* 556 (1946).

TABLE 77

Effect of Heat Treatment on Physical Properties of Sintered Carbon Steels[a] (Stern[11])

Carbon content of steel	Treatment	Ultimate tensile strength, psi	Yield strength, psi	Elongation, %	Reduction in area, %	Brinell hardness	Izod impact, ft.-lb.
0.28	Sintered and furnace cooled from 1095°C. (2000°F.)	56,050	38,350	23.5	25.0	124	9
	Furnace cooled from 1095°C. (2000°F.), oil quenched 860°C. (1575°F.), drawn 315°C. (600°F.)	66,550	51,000	22.5	22.9	134	8.5
	Furnace cooled from 1095°C. (2000°F.), oil quenched 860°C. (1575°F.), drawn 705°C. (1300°F.)	52,450	36,025	34.0	38.8	120	10.5
	Furnace cooled from 1095°C. (2000°F.), water quenched 830°C. (1525°F.), drawn 315°C. (600°F.)	82,500	65,800	13.0	20.2	158	—
0.52	Sintered and furnace cooled from 1095°C. (2000°F.)	66,150	41,000	17.0	16.5	137	4
	Furnace cooled from 1095°C. (2000°F.), oil quenched 830°C. (1525°F.), drawn 315°C. (600°F.)	84,000	57,700	14.5	14.7	146	5
	Furnace cooled from 1095°C. (2000°F.), oil quenched 830°C. (1525°F.), drawn 705°C. (1300°F.)	58,800	41,700	25.0	30.0	126	5
	Furnace cooled from 1095°C. (2000°F.), water quenched 830°C. (1525°F.), drawn 315°C. (600°F.)	122,250	102,100	8.0	8.5	203	—
0.64	Sintered and furnace cooled from 1095°C. (2000°F.)	69,900	42,975	11.5	10.7	132	2.5
	Furnace cooled from 1095°C. (2000°F.), oil quenched 830°C. (1525°F.), drawn 425°C. (800°F.)	97,050	72,200	13.0	11.2	161	4
	Furnace cooled from 1095°C. (2000°F.), oil quenched 830°C. (1525°F.), drawn 705°C. (1300°F.)	65,725	46,650	16.0	18.6	132	2
	Furnace cooled from 1095°C. (2000°F.), water quenched 830°C. (1525°F.), drawn 315°C. (600°F.)	136,800	109,500	8.0	8.5	218	—
0.87	Sintered and furnace cooled from 1095°C. (2000°F.)	84,075	63,150	7.0	4.1	158	2
	Oil quenched 830°C. (1525°F.), drawn 425°C. (800°F.)	109,400	86,500	5.5	5.9	177	2.5
	Oil quenched 830°C. (1525°F.), drawn 705°C. (1300°F.)	69,550	53,850	9.5	9.1	135	1
	Water quenched 830°C. (1525°F.), drawn 315°C. (600°F.)	152,500	132,000	4.0	3.3	245	—

[a] Prepared from mixtures of electrolytic iron and graphite by pressing and repressing at 50 tsi, sintering, and resintering in cracked NH_3 for one hour at 1095°C. (2000°F.).

carbon steels; for elongation, reduction in area, and impact strength they are, however, considerably lower. The poor ductility and toughness must be attributed to a deficiency in the density of the sintered steels; the pores act as stress risers and slight elongation causes rupture. The relatively low hardness values are partially due to the method of measuring, the pores in the structure causing a deeper penetration of the hardened steel ball of a Rockwell tester. In addition, a high purity of the sintered steel retards formation of a very hard structure.

The microstructure of sintered carbon steels is equivalent to that of similarly heat-treated ordinary carbon steels, with the exception of the presence of pores in the sintered structure. An abnormality in the form of coarse pearlite can usually be found in steel prepared by diffusion of free carbon (graphite) in high-purity iron particles during sintering, followed by slow cooling. It is difficult to obtain the formation of a martensitic structure, even in the case of a high carbon content, with a more drastic quench required than in the case of ordinary carbon steels. An ordinary water quench generally produces only very fine pearlite. This phenomenon can be explained by the lower heat conductivity caused by the pores, and also by the high purity of the steel—if a pure iron (*e.g.*, electrolytic) is used as raw material. The complete absence of such elements as silicon and manganese in such a sintered steel makes the material considerably more inactive in transforming to a harder structure upon rapid cooling. The abnormalities of the sintered and heat-treated structures are less pronounced if steel powder (*e.g.*, crushed steel), in which the required amount of carbon is combined with the iron, is used as raw material (see Figure 224*B* and *C*, Chapter XIV). The photomicrograph of Figure 298*A* displays the structure of a sintered 0.5% carbon steel obtained from a mixture of pure electrolytic iron and graphite, and Figure 298*B* of the same steel after furnace cooling, reheating to 830°C. (1525°F.), water quenching and tempering at 315°C. (600°F.). The water quench was not drastic enough to form martensite, and instead the microstructure displayed very fine pearlite.

Surface Hardening

Surface hardening includes a number of heat treatments in which the surface layer (case) of an iron-base alloy is made substantially harder than the interior (core) by altering its composition. Included in this kind of treatment are such well-known operations as carburizing or cyaniding, both of which are ordinarily followed by quenching and nitriding. In responding to any of these treatments, sintered steels do not vary materially from ordinary steels, except possibly in lower hardness values. This

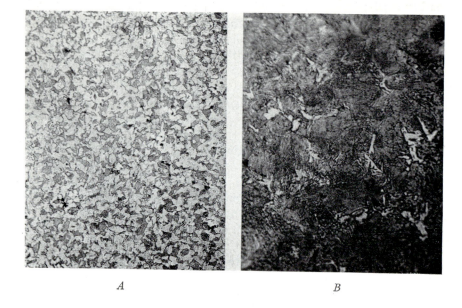

A   B

Fig. 298. Microstructures of sintered and heat-treated steels (according to Stern,[12] A and B; and Stern and Greenberg[12] C–F): A, sintered 0.52% carbon steel from mixture of electrolytic iron and graphite powders, compressed at 50 tsi, sintered at 1100°C. (2010°F.) for 15 minutes, repressed at 50 tsi, resintered at 1100°C. (2010°F.) for 1 hour, furnace cooled (×150); B, same, after reheating to 830°C. (1525°F.) water quenching and tempering at 315°C. (600°F.) (×750); C, sintered carbon steel from mixture of electrolytic iron and 0.4% graphite powders produced as in A and gas carburized in "Vapocarb"[12a] furnace at 900°C. (1650°F.) for 1 hour, lime cooled (×150); D, same as C, but steel produced from mixture containing 1% manganese in form of ferromanganese (×150); E, case of same material as C, but after water quenching from 850°C. (1560°F.) (×750); and F, case of same material as D, but after water quenching from 850°C. (1560°F.) (×750).

[12] G. Stern and J. Greenberg, Iron Age, 157, No. 17, 56 (1946).
[12a] Product of Leeds and Northrup Co., Philadelphia, Pa.

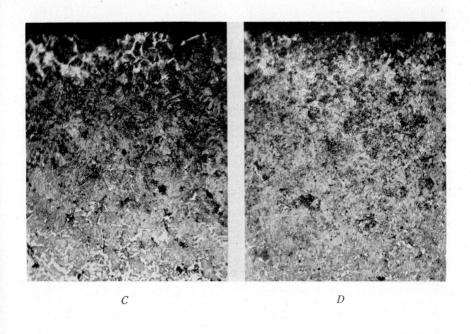

C   D

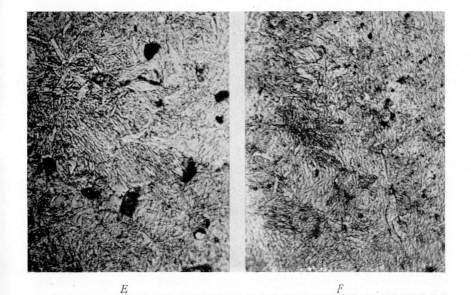

E   F

can again be explained by the residual porosity in the sintered steel (permitting the hardened steel ball to penetrate to a greater depth).

CARBURIZING

The carburization or cementation of a sintered steel compact can be accomplished by adding carbon to the compact's surface while heating slightly above the critical temperature range in the presence of carbonaceous solids, liquids, or gases; the depth of the case thus formed depends on the medium and on the time allowed for the reaction. It has generally been found that porosity in the compact tends to facilitate diffusion of the carbonaceous gases into the core, equalizing the carbon concentration

TABLE 78

Composition, Processing, Heat Treatment, and Surface Hardness of Alloy Steels

Properties	Alloy			
	Type 1	Type 2	Type 3	Type 4
Composition, %				
Carbon	0.50	0.10	0.20	0.20
Nickel	3.50	3.50	3.50	3.50
Chromium	—	—	—	1.50
Iron	Balance	Balance	Balance	Balance
Processing				
Compacting pressure, tsi	65	65	65	65
Sintering temperature, °C	950	950	950	950
°F	1740	1740	1740	1740
Sintering time, hrs	8	8	8	8
Coining pressure, tsi	65	65	65	65
Carburizing temperature, °C	925	925	925	925
°F	1700	1700	1700	1700
Carburizing time, hrs	20	20	20	20
Reheating temperature, °C	820	820	820	820
°F	1510	1510	1510	1510
Reheating time, hr	1	1	1	1
Quenching temperature, °C	800	800	800	800
°F	1470	1470	1470	1470
Surface hardness				
As sintered, Brinell	89	91	83	88
As coined, Brinell	143	143	134	140
As carburized, Rockwell	A-58	A-57	A-55	A-60
Brinell	210	200	180	228
As quenched in oil, Rockwell	C-32	C-31	C-33	C-38
Brinell	297	290	305	352
As quenched in water, Rockwell	C-57	C-55	C-48	C-52
Brinell	573	547	460	509
Case depth, in	0.069	0.066	0.059	0.063

between surface and interior; the control of the case depth and uniformity is therefore more difficult in sintered steels, especially if the porosity is appreciable. In Table 78 hardness values are given for sintered steel composition prepared from powder mixtures and case hardened in a commercial carburizing pack. A Rockwell hardness of C57 obtained by water quenching a 0.5% carbon-containing nickel steel is not much below the hardness of a solid steel treated in the same manner. By quenching in brine the hardness of a plain 0.5% carbon steel case hardened in the same manner can reach Rockwell C60–62.

TABLE 79

Depth of Case of Sintered Steels for Various Methods of Carburizing at 900°C, (1650°F.) for 1 Hour (Stern and Greenberg[12])

| Material composition | Carbon content, % | Density | | Pack carburized | | Gas carburized,[c] in. |
		Actual g./cc.	vs. Theoretical, %	Coarse compound,[a] in.	Fine compound,[b] in.	
Reduced iron + 0.4% graphite + 2% copper....	0.16	7.02	89.4	0.009	0.008	0.012
Electrolytic iron + 0.4% graphite	0.26	7.46	95.0	0.012	0.012	0.014
Electrolytic iron + 0.4% graphite + 1% manganese[d]	0.49	7.27	92.8	0.015	0.015	0.013
Cold-rolled SAE 1015 steel..	—	—	—	0.013	0.012	0.013

[a] Houghton Nucarb No. 3 Mesh.
[b] Houghton High-Carbon Bone Black No. 10 Mesh.
[c] In Leeds and Northrup "Vapocarb" Furnace.
[d] Added as ferromanganese.

A systematic comparison of different carburizing methods was recently carried out by Stern and Greenberg.[12] The three types of steel studied are characterized in Table 79, and typical microstructures of electrolytic iron base compacts with and without manganese are shown in Figures 298C to F. The sintered compacts were pack carburized either with a coarse carburizing compound (Houghton Nucarb No. 3 Mesh) or with a fine compound (Houghton High Carbon Boneblack No. 10 Mesh), or they were gas carburized in a Leeds and Northrup Vapocarb furnace (a photograph of this furnace is shown in Figure 234, Chapter XV). Houghton No. 2 Soluble Oil was employed as quenching oil. The results are summarized in Table 79, which gives comparative values for depth of case, and in Table 80, which compares the average surface hardness obtained by the different carburizing methods. The microstructures of water-quenched pieces have fully martensitic cases, while those of oil-

TABLE 80. Average Surface Hardness of Pack- and Gas-Carburized Steel Compacts—Carburized for 1 Hour (Stern and Greenberg[12])

Material	Hardness scale	Water quenched			Oil quenched		
		Nucarb	H.C. boneblack	Vapocarb	Nucarb	H.C. boneblack	Vapocarb
Reduced iron + 0.4% graphite + 2% Cu (R242)	R15N	88.5	88	86.5	87.5	87.5	77
	RC	56	55	50.5	54	54	34
Electrolytic iron + 0.4% graphite (S34) + 1% Mn	R15N	88.5	89	85.5	80	82.5	69
	RC	57	57	53	39	44	20
Electrolytic iron + 0.4% graphite (S34)	R15N	89	89.5	87	79	77	71
	RC[a]	56	59	25	37	34	23
Cold-rolled SAE 1015 steel	R15N	93	93	92	90	91	92
	RC	67	67	65	60	62	65
Reduced iron + 0.4% graphite + 2% Cu (R242)	R30N	73.5	73.5	—	71	69.5	—
	RC	55.5	55.5	—	53	51	—
Electrolytic iron + 0.4% graphite (S34) + 1% Mn	R30N	74	74	—	53	55	—
	RC	56	56	—	33	35	—
Electrolytic iron + 0.4% graphite (S34)	R30N	76	76.5	—	51	51	—
	RC	58.5	59	—	30.5	30.5	—
Cold-rolled SAE 1015 steel	R30N	83	84	82	77	77	82
	RC	66	67.5	65	59.5	59.5	65
Reduced iron + 0.4% graphite + 2% Cu (R242)	RA	78.5	78.5	—	74	76	—
	RC	55	55	—	47	50	—
Electrolytic iron + 0.4% graphite (S34) + 1% Mn	RA	80	80.5	—	66.5	69	—
	RC	58	59	—	32	37	—
Electrolytic iron + 0.4% graphite (S34)	RA	81	81.5	—	65	64.5	—
	RC	60	61	—	29	28	—
Cold-rolled SAE 1015 steel	RA	84	86	84	83	83	84
	RC	65	69	65	63	63	65
Reduced iron + 0.4% graphite + 2% Cu (R242)	Vhm[b]	885	—	—	—	—	—
	RC	66.5	—	—	—	—	—
Electrolytic iron + 0.4% graphite (S34) + 1% Mn	Vhm	945	863	—	—	—	—
	RC	68	66	—	—	—	—
Electrolytic iron + 0.4% graphite (S34)	Vhm	868	720	—	—	—	—
	RC	66.5	61	—	—	—	—
Cold-rolled SAE 1015 steel	Vhm	958	—	—	—	—	—
	RC	68	—	—	—	—	—

a All Rockwell C readings converted from corresponding scales. b Vickers readings converted from microhardness readings.

quenched pieces varied from troosto-martensite to very fine or emulsified pearlite. The results yield the conclusion that either of the investigated methods can be used to obtain a satisfactory carburized case for sintered steel. Stern and Greenberg point out that for most reliable surface-hardness values of sintered products, the Vickers or microhardness tester should be used. For practical purposes, however, the superficial Rockwell "15N" scale may be used.

The possibility of combining pack-carburizing and sintering in a single operation has been investigated by Margolies.[13] In these experiments, compacts were pressed at 25, 50, or 75 tsi, packed in the carburizing compound, and sintered at 925°C. (1700°F.) for 2, 4, or 8 hours. After "carbusintering," which does not require a protective gas atmosphere in the furnace, the pieces were water quenched and tested for tensile properties, Rockwell hardness, depth of case, and distribution of combined carbon between case and core. Maximum surface hardness and depth of case were obtained at a carbusintering period of 4 hours. Compacts pressed at 75 tsi, carbusintered at 925°C. (1700°F.), and water quenched from this temperature gave a hardness (as quenched) of Rockwell C52, a case depth of 0.04 in., a case carbon content of 1.25%, and a tensile strength of 66,000 psi; the same iron powder, processed by conventional sintering technique, had a hardness of Rockwell B13, and tensile strength of 31,000 psi. A practical application of the carbusintering process does not appear feasible, however, since the reported elongation values are zero.

CYANIDING

Surface hardening by carbon and nitrogen absorption of sintered steels through heating at a suitable temperature in contact with cyanide salt, followed by quenching, is possible only if the compacts are very dense. As in the case of sintering in a liquid bath, the salt readily penetrates into the compact through the interconnected pores, corroding the entire cross section badly. The only application where cyanide hardening appears to be practical for sintered steels is for high-density parts requiring a high wear resistance and file hardness at the surface. The case of a compact hardened in sodium cyanide is usually not as deep as that of a carburized compact, rarely exceeding 0.030 in. Iron nitride needles formed at the surface at the lower cyaniding temperature obstruct rapid carbon diffusion at the maximum heat-treating temperature. This could be noticed on compacts from electrolytic iron powder that were either hardened in sodium cyanide or in a so-called "activated" liquid carburizing

[13] A. S. Margolies, *Iron Age, 157,* No. 9, 60 (1946).

bath (calcium cyanamid plus sodium cyanide). 0.2% graphite was added to the iron in powder form, and the compacts were produced by pressing and coining, each at 50 tsi, and by sintering and resintering, each at 1100°C. (2010°F.) for 1 hour in hydrogen. The carbon content before the surface-hardening treatment was 0.14%. Microscopic examination after the treatment disclosed the following case depths and carbon contents at the surface listed in Table 81.

TABLE 81

Effect of Cyanide Treatment on Sintered Steel

Property	Compact treated in cyaniding bath, 97% NaCN	Compact treated in "activated" carburizing bath
Temperature....................	870°C. (1600°F.)	(870°C. 1600°F.)
Time of immersion, hours........	3	3
Case depth, inch................	0.020	0.027
Maximum carbon content at surface, %....................	0.75	0.87

After water quenching, the cyanide-treated compact had a surface hardness of 75 on the Rockwell Superficial N30 scale (Rockwell C57) as compared with 77 (Rockwell C59) for the compact hardened by the liquid carburizing method.

Stern and Greenberg[14] recently extended their surface-hardening experiments with sintered steels by including tests with liquid carburizing compounds. In this work two different types of sintered steel were used: (1) made of reduced iron powder plus graphite and about 2% copper, and (2) made of electrolytic iron plus graphite without copper addition. The first material had a density of 7.02 g./cc. (89.4% of theoretical) and analyzed 0.16% carbon; the second had a density of 7.46 g./cc. (95%) and a final carbon content of 0.26%. Two commercial liquid carburizing salts were investigated, the first being a low cyanide carburizer having carbon in suspension and being generally recommended for producing thick cases; the second contained about 30% cyanide and was of the type generally recommended for producing light cases. During the treatment at 790–850°C. (1450–1550°F.) the samples were suspended in the molten salts. They were removed rapidly upon completion of the treatment and quenched immediately in a commercial grade of quenching oil. This was followed by a thorough rinsing procedure involving a 15-minute rinse in a boiling solution of commercial salt cleaner followed by several rinses in boiling water, with fresh water used for each rinse. The degree of salt removal by this procedure was tested by precipitation of residual salt by

---

[14] G. Stern and J. Greenberg, *Powder Met. Bull.*, *2*, No. 4, 85 (1947).

means of silver nitrate. Stern and Greenberg found that sintered steels of the type employed in the investigation can be satisfactorily cyanide-hardened to yield file-hard cases which displayed a microhardness equivalent to Rockwell C62–70. If the Superficial Rockwell 15N scale is to be used for measuring the surface hardness, the case must be at least 0.008 in. thick. The method permits production of very shallow cases as well as a bright and comparatively corrosion-resistant surface finish. The possibility of residual salt oozing out during subsequent use of the sintered steel part is eliminated by the above-described rinsing technique.

NITRIDING

Surface hardening through the action of nitrogen is also possible in powder compacts, but the necessity of heat treating for very long times (up to 3 days) to obtain a reasonably deep case has made the process rarely competitive with straight pack or gas carburizing. Moreover, the nitrides formed are sufficiently stable only if aluminum is added to the iron in amounts between 1.00 and 1.50%. This addition requires careful manipulations, since the element can become effective only if aluminum nitrides can be allowed to form; it may be necessary to add a prealloyed iron–aluminum powder to the mixture to eliminate obstruction of diffusion by oxide films.

TABLE 82

Effects of Aluminum and Chromium Additions on Hardness of Nitrided Sintered Steel

Composition	0.4% C steel compact	0.4% C, 1% Al steel compact	0.4% C, 1% Al, 1% Cr steel compact
Powdered raw material.....	Electrolytic iron, graphite	Electrolytic iron, graphite, 50–50 Fe–Al pre-alloy	Electrolytic iron, graphite, 50–50 Fe–Al pre-alloy, high-carbon ferrochrome
Nitriding temperature, °C...	525	525	525
°F...	977	977	977
Nitriding time, hrs..........	48	48	48
Case depth, in..............	0.008	0.014	0.018
Surface hardness (Rockwell superficial 30-N).........	71	75	79
Equivalent Rockwell.......	C-53	C-57	C-62

The nitriding temperature is relatively low, and decomposition of the nitrides takes place at higher temperatures. As a consequence, only completely sintered or resintered iron or steel compacts can be nitrided. The treatment is best carried out in ammonia gas, and an ordinary batch-type furnace—equipped with a nitriding box—can be used.[15] The hardness, wear resistance, and corrosion resistance of the surface regions are retained without the necessity of further treatment.

The beneficial effect of alloying additions of aluminum and chromium on the stabilization of the nitride structure and on the depth of case is best illustrated by comparing (Table 82) the data of sintered, coined, and repressed compacts constituting a pure iron–carbon composition (0.4% carbon) with compositions containing 1% aluminum and 1% chromium. All steels were quenched in water from 940°C. (1725°F.) and tempered at 540°C. (1005°F.) for 1 hour prior to nitriding.

Summary

After completion of the sintering cycle, metal powder compacts can be subjected in principle to the same kind of heat treatments employed in metallurgy. However, complexity in shape or drastic changes in cross section may cause severe deformation during the treatment and may make additional forming necessary; the response of the sintered metal is affected by its unique structure. Inherent porosity, impurity films around the particle, or high-purity compositions may cause less pronounced structural changes than would ordinarily be expected. This is particularly true for quench-hardened steels, where only extremely drastic quenching, or alloying additions would produce a martensitic structure.

Heat treating of powder compacts may serve a variety of purposes. It may simply be a stress relief or annealing in the case of coined or otherwise cold-worked compacts of pure metals. Quenching will render certain metals dead-soft, and will also produce a maximum solid solution in precipitation treatable alloys. Age and precipitation hardening of nonferrous compositions are equally feasible. Steels can be quench hardened and surface hardened by carburizing, cyaniding, or nitriding. Of these processes, quench hardening and carburizing are most effective and have found industrial applications in the production of steel parts.

[15] V. O. Homerberg, in *Metals Handbook*. Am. Soc. Metals, Cleveland, 1948, p. 697.

CHAPTER XX

Finishing Treatments

The mechanical and metallurgical operations of molding, sintering, coining, working, annealing, and heat treating comprise the true steps of the technology of sintered metals. In the manufacture of ingots, wire bars, or other intermediate forms—as used, for example, in the refractory metal industry—the powder metallurgy process may be considered to be completed after these operations; but where parts are to be "custom-molded" to great accuracy and high surface finish, the powder-metal technology must be supplemented by specific finishing treatments. Although these treatments are analogous to those conventionally used in engineering industries, the unique character of powder metallurgy structures requires definite modifications in mechanical as well as in chemical or electrochemical operations.

It may be said without exaggeration that finishing of metal powder products ranks in importance with the other fundamental processing steps of powder preparation, molding, and sintering. No part can be considered ready for use unless it has passed through all four stages of manufacture. It is amazing that the voluminous literature on powder metallurgy contains virtually no reference to finishing of the parts; even the most recent texts and symposiums fail to deal with this subject.

Here, the survey of the many finishing operations adaptable to metal powder parts will be restricted to mechanical, surface, and impregnation treatments, and a short reference to joining operations. Elsewhere (Chapter XXXIV, Volume II), in connection with a detailed discussion of the properties of sintered metals, the problem of testing the physical properties of finished powder metallurgy products will be treated.

MECHANICAL OPERATIONS

Shaping

It is not always possible to achieve the exact shape and dimensional accuracy of a part in the molding or coining die. Certain mechanical forming operations may become indispensable especially in the production of some of the more complicated parts. This final shaping step may involve only a minor operation such as straightening or sizing, or may involve one or several cutting operations. Some basic shapes that require final machining have already been indicated in Chapter IX, *e.g.*, parts with undercuts and threads, steep angles in the direction of pressing, greatly diverse cross sections, and holes that are not strictly parallel to the pressure axis.

MACHINING

Machining is probably most important of the mechanical finishing operations. Because it is fully realized that this operation is incompatible to a certain extent with the principal objective of powder metallurgy, namely, to mold metal powders *directly* into solid shapes of *accurate* dimensions, the problem of machinability of sintered metals has been given only passing notice, and has been described only in a few specific instances (cemented carbide tool bits, composite contact metals). Considering the present-day shortcomings in the quality of powders and molding techniques, it is unavoidable that the more complicated shapes from steels or nonferrous alloys should require some kind of machining operation at the end of the manufacturing process.

Sintered metals machine poorly because the inherent porosity of the metal—causing discontinuous chip formation—continuously subjects the cutting edge to shock and impact when passing over the pores and re-entering the metal, and so causes premature dullness of the tool. Once the cutting edge is blunted, the metal particles of the surface layers are easily torn out, and a rough finish results. The use of ordinary coolants (soap, water, cutting oil) is prohibitive where the pores form an interconnected pattern (*i.e.*, where the density is below 90%) because of internal corrosion or other undesirable aftereffects.

There are a number of expedients that may be used to relieve this condition—at least in part. A change in the tool angle usually helps to prolong the tool life. In tapping, the relief of the tap should be at least twice that ordinarily used for ferrous metals. If the use of a coolant on the work and cutter is indispensable for cooling, washing away of chips, or protection from oxidation or other corrosion, a volatile fluid, such as carbon tetrachloride should be used. Finally, the addition of small percentages (2% maximum) of lead or copper–lead to iron or steel aids machinability considerably, and is the best means known to date for achieving consistent results.

SHEARING AND BROACHING

A shearing or broaching operation may be used instead of cutting. Gears, for example, can be molded and sintered to slightly oversize dimensions, and can then be forced through a hardened steel die (in a press) that shears off the excess amount at the surface of the teeth. The same method is also adaptable for bushings or other cylindrical parts. Since the operation usually requires additional finishing to remove burs, it is not often used.

SIZING

This operation has already been discussed in the sections on molding and coining. In contrast to the preceding two methods, sizing is not a metal-removing, but a metal-displacing process. It may simply involve reshaping of the part to comply with very close demands on dimensional tolerances, or it may accomplish a densification and strengthening of the structure at the same time. Sizing is usually conducted in power presses, using hardened punches and dies. The operation is employed chiefly in the manufacture of porous bearings, bushings, seal rings, etc., where dimensional accuracy is of prime importance. Other products that are sized include certain types of gears, cams, and other machine parts.

BURNISHING

This operation has found widespread use where internal bores are to be finished, e.g., in the manufacture of bushings. It consists essentially of a displacement of some metal by the forcing of one or several hardened steel balls through the bored hole. In the most common form of application, the balls are successively larger in diameter, exceeding the diameter of the hole, so that the bore is enlarged and sized, and the metal work hardened and made more wear resistant at the same time. Special ball burnishing tools are required to match the final permissible tolerances.

STRAIGHTENING

Certain shapes have the tendency to deform or warp during sintering or annealing. The effect is particularly pronounced in long and flat forms, or in parts having large curved faces or varying cross sections. Deformation occurs when uneven stresses imposed on the part during molding or coining are released during the treatment at higher temperatures. In order to meet the required dimensional accuracy, the part must be reshaped in a straightening operation. This process, too, can be applied with the aid of hardened steel dies placed in a mechanical press. If the metal is in the annealed condition, no elastic spring-back can be expected and very close tolerances can be met.

Surface Cleaning

DEBURRING

The removal of burs is a most essential and most commonly used mechanical finishing operation. Because of the necessary clearance between the stationary molding or coining die and the moving punch sections, there is a tendency for the metal to extrude into the free space. With increasing tool wear, burs and flash will become more pronounced, and their removal will become a necessity. Where small quantities are involved, deburring may be carried out by manual operation with files. scrapers, or sand stones; in large-scale production work, deburring by machining—most suitably by filing, milling, and grinding—is more economical. Very large quantities can be deburred advantageously with a machine of the double disk-grinding type. If compatible with the production cycle, deburring is best done while the metal is in a strain-hardened condition, *e.g.*, after coining and before annealing. In this state the extruded metal particles are very brittle and shear off readily. In the soft-annealed state, the extruded particles form a solid unity with the rest of the surface, and more pressure and a slower motion of the tool are necessary for the dislodging of the burs.

BELT GRINDING

An effective method of surface cleaning and deburring of flat or rectangular shapes is by sanding the parts with a belt-grinding machine. Although this operation usually requires individual handling of the part, high production rates can be achieved. It is possible to obtain good surface finishes if fine-grit grinding belts are used. Special feeding attachments may permit automatic transport over the belt for simple shapes, with handling confined to a turning over of the parts. Depending on the construction of the machine, the endless belt operates horizontally, or vertically, or at an angle; the shape of the part decides which type of machine is preferable. Figure 299 is a general view of a finishing department showing belt-grinding and deburring operations on powder metallurgy parts.

TUMBLING

If the burs are very small (not exceeding 0.0005 in.), they can be removed readily by tumbling. This operation, however, is more suitable for solid, round or cylindrical parts than for cored parts, or for square shapes with sharp corners. A number of parts are placed in a drum together with a medium; when the drum is rotated in an automatic tumbling machine, the parts are thrown against each other and against the walls of the drum; the burs break off by impact, and a good finish all around the parts is obtained. The tumbling must be timed accurately, since too long a period may cause unwanted deformations of the parts. Sand or sawdust are commonly used for polishing the surface—if the

Fig. 299. Belt grinding and deburring of sintered iron products (courtesy of American Electro Metal Corp.).

operation is done dry. Certain types of powdered soaps and hard waxes are also very effective and yield excellent surface finishes. Wet tumbling is rarely done, and never with water, unless the parts have been treated previously with a water-repellent impregnant. Liquids suitable for wet tumbling include chlorinated solvents (perchloroethylene, carbon tetrachloride, etc.).

ABRADING

Surface cleaning by blasting with sand or other abrasives is usually too rough a treatment for soft powder metallurgy parts. Among a number of well-known sand-blasting methods, the Vaporblast process[1] is con-

---

[1] *Anonym., Metals Review, 18,* No. 3, 1 (1945); *Anonym., Steel, 115,* No. 22, 100 (1944); A. H. Eppler, *Product Eng., 15,* No. 7, 468 (1944); F. M. Reck, *Aero Digest, 44,* 5, 103, 200 (1944).

sidered best for a good finish on any part. A minimum of metal is removed in this process. The method, for example, is used before polishing to prefinish dies used in the plastics industry. Other suitable methods include the Rotoblast operation with airless blast cleaning tables or rocker-barrel-type machines,[2] and the Tumblast operation with Wheelabrator-type machines,[3] that operate fully automatically. The ordinary sandblasting method involving individual handling has become obsolete.

Ferrous grits are the preferred abrasives, since fine size and close size distribution—both important for a good finish—are more easily maintained in ferrous grits than in other abrasives. Nonmetallic abrasives such as sand, alundum, silicon, carbide, etc. are used only for nonferrous metal parts where ferrous abrasives would cause contamination of the surface. Diamond crushed steel grits[4] varying between 150 and 200 mesh are used for very hard alloy steel parts. Cast iron grits, especially the finer sizes (150 and 200 mesh), are satisfactory for all other iron and steel products, giving an excellent surface finish.

BUFFING

Buffing and surface polishing are not generally employed in the production of powder metallurgy parts. Sole exceptions are coins and medallions, and salesman's samples. Rubber-bonded abrasive wheels have lately been used successfully for high-polishing of flat surfaces and a variety of polishes exist which are useful for buffing, such as rouges and mixtures of abrasives and grease.

SURFACE PROTECTION TREATMENTS

Electrochemical Treatments

ELECTROPLATING

The electrodeposition of an adherent metallic coating upon another metal functioning as electrode is commonly practiced for the purpose of securing a surface with properties or dimensions different from those of the base metal. A complete treatise on the subject can be found in *Modern Electroplating*.[5] Among the objectives that can be attained by applying

---

[2] Products of the Pangborn Corp., Hagerstown, Md. (Bull. 211A. 213). See also A. L. Gardner, *Metal Progress*, *48*, No. 4, 959 (1945).
[3] Products of the American Foundry Equipment Co., Mishawaka, Ind. (Catalog No. 212).
[4] Products of Pittsburgh Crushed Steel Co., Pittsburgh, Pa. (See also Chapt. VI).
[5] *Modern Electroplating*. Symposium, Electrochemical Society, New York, 1942.

electroplated coatings are the following:[6] (*1*) to produce and retain a desired color, luster, or resistance to tarnish (*e.g.*, plating with platinum, gold, chromium, nickel) ; (*2*) to protect the base metal against corrosion or rust (*e.g.*, plating with tin, cadmium, zinc, silver, copper, nickel, chromium) ; (*3*) to protect the base metal against some specific chemical reaction (*e.g.*, plating steel with copper) to prevent carburization; (*4*) to protect the base metal against wear caused by either abrasion or erosion, *i.e.*, a combination of corrosion and abrasion (*e.g.*, hard chromium plating) ; (*5*) to increase the dimensions of parts accidentally made undersize or worn in service with coatings that must possess same or greater hardness than the original metal surface (*e.g.*, chromium, nickel, and iron plating).

Were it not for the porosity inherent in most sintered metals, desired effects could be obtained on powder metallurgy products with the same bath compositions and plating techniques used for ordinary solid metals. Unfortunately, however, the presence of interconnected pores, evident in most products made from powders—including the so-called "high density" parts—causes complete penetration of the core by the plating solutions. After the part is removed from the plating bath the solution remains in the pores, and the chemicals cannot be removed by ordinary washing techniques. This condition causes internal corrosion, discoloration of the plating, or oozing of some salt through the porous plating over the surfaces of the part. The latter effect, known as "flowering," is especially pronounced in parts of appreciable porosity, and has been the main reason that little progress has been made in the electroplating of metal powder parts. Another disadvantage due to interconnected porosity is that deposition may take place on internal rather than on external surfaces.

Only recently some important practices were developed that promise ultimate success in at least partially solving these problems: (*1*) production of high density parts without interconnected pores; (*2*) change of plating techniques; (*3*) coverage of the exterior and interior surfaces with a very thin film of nonmetallic water-repellent protective that allows sufficient penetration of the ions to form the electrodeposit; and (*4*) plating of *green* pressings and removal of residual salts during sintering.

At present, only the first and last methods form platings that are effective as corrosion inhibitors. Where parts of appreciable porosity are involved, the plated surface is usually as porous as the underlying material, and the plate can serve only for decorative purposes. Even if water repellents are used, the surface voids are not covered by the plate, and the surface of the part remains porous. Moreover, the water repellents themselves are not corrosion-proof.

[6] W. Blum, in *Metals Handbook*. Am. Soc. Metals, Cleveland, 1948, p. 716.

High Density Parts. Careful study of the proportion and character of the porosity in sintered, coined, or otherwise worked parts from different metals has established the fact that up to approximately 90% of the true density the pores are generally interlocked, and fluids of low viscosity, such as water or aqueous solutions, will penetrate readily through the skeletonlike metal. If the density exceeds 90%, the extent of absorption of the fluid begins to depend more and more on the pore distribution, as affected by the method of manufacture. For example, a part of 92% density produced simply by molding at a high pressure followed by sintering may have a fairly even distribution of pores throughout the cross section, with an appreciable proportion of the pores still interconnected. On the other hand, a part of the same density produced by molding, sintering, and coining operations, implying severe deformation, may display a thin surface case practically free of pores, and without any channels leading into the more porous interior of the metal. Thus, it can be expected that although the part made by the first method may still absorb some electrolyte, the part produced by the second method may not do so, and can therefore be plated successfully. Products with a density of 95% and above generally contain only isolated pores which will not cause a deep penetration of the electrolyte, and so permit any kind of plating. But since parts of a high density usually require many alternate pressing and heating operations, as well as very high pressures, the process is rendered rather uneconomical, and is restricted to the more ductile metals, such as pure iron or low carbon steels.

Special Plating Techniques. Direct electroplating of parts of substantial porosity is possible from both alkaline and acid baths, provided due attention is paid to the fact that the conditions on the inside walls of the pores will vary according to the type of ions present. The two types of baths must therefore be handled differently. In order to avoid discoloration of the plate and oozing of the salt in plating from alkaline baths, a neutralizing aftertreatment is used, *e.g.*, an anodic treatment at reduced current density after the plating has reached the required thickness. One specific plating technique provides heating of the parts to slightly above room temperature, followed by placing in cold water for half an hour. This is followed by an acid pickle,—concentration and pickling time depending upon the metal and the condition of the surface. For the sake of good plating, the time should be as short as possible; the parts are then washed and plated. Without delay, after plating, the parts are again washed, given the anodic treatment in a cold water bath that contains a small proportion of sulfuric acid, then washed again in cold water, and finally dried.

Acid plating baths are considerably less troublesome—possibly due to less throwing power in the plating of porous parts—than are the cyanide type baths. The chemicals are easily washed from the pores, and possible residues are not detrimental to plating or to internal structure, since the acid concentration is not strong enough to harm the exposed surfaces. However, it is advisable to use an acid dip or pickling treatment immediately before plating, and to wash and neutralize the part after plating. It is inadvisable to use strongly alkaline solutions in cleaning the part because of the corrosive effects of the caustic. The best method of cleaning the parts before the acid dip and plating is vapor degreasing.

In the plating of very porous parts, where the plating solution is allowed to run through the pores of the interior of the part, a uniform plate can be secured only if a high current density is used. In experimenting with chromium plating it has been found that very porous pieces with their greater microscopic surface area require about four times as much current as solid parts. However, porous and nonporous sections which lie close together tend to plate well with a high current density and show little evidence of burning. Porosity and required current density vary approximately proportional to each other; as the density of the part nears the ideal value, the current density also nears a normal value.

Water-Repellent Surface Films. A promising way of stopping the flow of the plating bath solution into the pores is to impregnate the part with a water-repellent material. If the film is very thin, anchorage of the electrodeposit through the film onto the surface particles of the porous compact is possible with standard plating techniques. Some of the recently developed liquid organo-silicon oxide polymers (silicones[7]) are ideal for this purpose; they tend to repel water while allowing electrodeposition to proceed without necessitating the removal of the thin surface film. By presenting a surface which, although being discontinuous, is at least sufficiently repellent to prevent harmful liquids of the plating baths from entering the porous part, the impregnation with silicones affords the best answer so far to the plating problem in powder metallurgy. The method is especially suitable for electroplating from alkaline baths; but even in acid baths the parts can be silicone-impregnated in order to permit the use of a normal plating current.

A technique that has proved very successful in many applications involves heating the cleaned part to approximately 200°C. (390°F.), and then quenching it in a 4% solution of Silicone Fluid[8] in perchloro-

---

[7] W. R. Collings, *Chem. Eng. News, 23,* No. 18, 1616 (1945).

[8] *Silicone, Type 200 Fluid* (viscosity 350 centistokes at 25°C.), product of Dow Corning Corp., Midland, Mich.

ethylene; a photograph of this operation is shown in Figure 300. The part must remain submerged until effervescence ceases (30 seconds to 2 minutes—depending on the porosity). The part is then removed and dried,

Fig. 300. Silicone impregnation of porous iron parts in batch-type heat-treating furnace in preparation of electroplating (courtesy of American Electro Metal Corp.).

and is followed by a baking treatment, again at 200°C. (390°F.), for 1/2 to 2 hours—depending on the size of part. Air cooling completes the treatment, and plating can now be carried out in the conventional manner.

An oxide tarnish at the surface caused by the cooling in air has no detrimental effect on the plating as long as it is only superficial. By using this method, parts having up to 25% volumetric porosity have been plated successfully with cadmium and nickel to a plate thickness of 0.010 in. and the water-repellent capacity has remained noticeable even after plating. After the silicone impregnation, an acid pickle (*e.g.*, 2% sulfuric acid solution) is necessary if any discoloration results. After plating, the parts must be thoroughly washed and dried.

Plating of Green Pressings. This process, recently disclosed by Ekstrand and Tholand,[8a] and apparently especially suitable for nickel-plating of iron–graphite parts or silver-plating of iron-base bearings, assures a complete removal of residual salts during the subsequent sintering treatment, and provides at the same time alloying of the coating with the base compact or with a second coating. In the case of a cyanide bath, treatment of the plated green compact with oxalic acid or dilute sulfuric acid results in more easily volatilizable salts.

Conversion Coatings. It is advisable to form a protective conversion coating in the case of very porous parts that have been plated previously, because in most cases the plate does not cover the surface voids, and the parts remain—in effect—porous. For example, cadmium-plated porous parts are not corrosion resistant, although after plating they retain some of the water-repellent property of the silicone and display an appealing surface. The corrosion resistance of parts plated with zinc or cadmium, however, may be improved by subjecting the surfaces to further chemical treatment. Iriditing and Chromaking are treatments in which the plated parts are dipped into chromium salt solutions, and a protective layer is obtained by formation of a complex chromate. The treatments are possible only if the surfaces are plated with cadmium or zinc; they cannot be applied directly to iron and steel.

The Iridite process, recently described by Albin,[9] permits the deposition of a sufficiently thick and solid surface film to withstand the action of wear, salt spray, liquids, condensed moisture, etc. The process requires close temperature control (25° to 40°C.; 75° to 105°F.) in order to start a chemical reaction between the Iridite solution and the plated metal prior to the formation of the protective coat. The plated surface must be clean, and oxide layers must first be removed by a bright dip in 2% sulfuric acid or $1/2$% nitric acid solution. The Iridite chemicals available[10]

[8a] Brit. Pat. Appl. 3836/48 and 3837/48, Feb. 11, 1947; see also *Metal Powder Rept., 3,* No. 3, 40 (1948).

[9] J. Albin, *Iron Age, 155,* No. 20, 44 (1945).

[10] Products of the Rheem Research Products, Inc., Baltimore, Md.

produce varicolored coatings, including olive-drab, green, blue, black, and bronze. The method is of particular interest since the protective coating formed appears to be stable and impervious to moist air, even for relatively porous parts. In one instance, where iron parts of approximately 15% porosity were zinc plated and afterward Iridited, the surfaces of the parts were found to be still perfect after a four months' exposure to humid industrial air during the summer season, while the untreated material subjected to the same test showed definite signs of corrosion.

DISPLACEMENT

According to the relative position of the metals in the electromotive series, the replacement of one metal in a salt by one above it in the series is more or less actively effected. Thus, *e.g.* the copper in a copper sulfate solution is replaced by the iron if the solution is painted on steel. At the same time, the copper plates out and forms a film on the surface. Because of the corrosive action of such salts, the method is not generally used for powder metallurgy parts, except where the density of the part is very high. Copper sulfate dipping is sometimes used to facilitate brazing, and the same solution may also serve to mark the part. A more important application of the displacement method can be found in powder production where solid particle shapes are involved (see Chapter III).

ELECTROPOLISHING

This operation is used in some instances where highly polished surfaces are desired. One example where very hard surfaces must have a perfect finish is the tungsten carbide shaft for high-speed internal grinding quirls running at speeds up to 100,000 rpm. In order to operate satisfactorily at this speed, the surface of the shaft in contact with the bearing surface must be perfectly smooth, and the maximum tolerance permitted cannot exceed 0.0002 in.

ELECTROLYTIC COLORING

Electrolysis can also be applied for the creation of interference colors. The colors produced have a metallic luster and are determined by the thickness of the deposit. Practically all colors of the spectrum can be produced in a single bath. The baths usually contain the metal to be deposited as organic salts. Typical compositions are solutions containing cupric acetate, or ammonium molybdate and ammonium hydroxide. In the case of the cupric acetate bath, a ferrous part must first be electroplated with copper; copper is used as anode material and the part acts as cathode. The finish of the plate is controlled by the surface polish of the part. In order to prevent undue penetration of the bath into the pores of the metal, a high-density part is preferred for this treatment.

Production of Oxide Layers by Chemical Treatment

Controlled chemical oxidation of metallic surfaces serves, in general, the dual purpose of protection against corrosion and coloring. Many treatments have been worked out for iron and steels, and some also for nonferrous metals. Those readers especially interested in the subject of protective coatings for metals are referred to the book by Burns and Schuh.[11]

COLORING BY TEMPERING IN AIR

By this method, one of the oldest ways of producing a desired color on the surface of steels, the metal is heated gradually over a range of temperatures, each degree producing a corresponding shade which varies from straw yellow (200°C.; 390°F.), through brown (250°C.; 480°F.), purple (300°C.; 570°F.), to blue (350°C.; 660°F.). The colors produced are generally not uniform because of fluctuations in temperature. Very porous powder metallurgy parts cannot be subjected to this kind of treatment, since they would oxidize throughout and become brittle.

COLORING IN SALT BATHS

Hot salt solutions or molten salt baths produce colors on polished iron and steel surfaces. Although in view of the corrosive effects of the salts, porous powder metallurgy parts cannot be handled in this way, sufficiently dense parts could be treated by such methods. For example, a black color is produced in a solution of 400 g. sodium hydroxide, 10 g. sodium nitrate, 10 g. potassium nitrate dissolved in 600 cc. water and heated to 140°C. (285°F.) for about 20 minutes. Niter baths consisting of a molten mixture of half sodium nitrate and half potassium nitrate to which 2% by weight of manganese dioxide is added, produce a blue color at about 330°C. (625°F.) within a few minutes. The same effect can be obtained with the less expensive sodium nitrate or Chile saltpeter alone, but higher bath temperatures are then necessary. The reaction caused by these baths is the production of a uniformly colored oxide film, such as is obtained when tempering steel to a certain color.

[11] R. M. Burns and A. E. Schuh, *Protective Coatings for Metals*. Reinhold, New York, 1939.

COLORING THROUGH THE ACTION OF CARBON

Surfaces of ferrous powder metallurgy parts can be effectively colored with carbonaceous matter. The porosity of the part is then of little importance since the treatments are conducted at subcarburization temperatures. For example, a blue surface color can be obtained by introducing a highly polished part into a bed of hot charcoal. The bed must be relatively deep, the lower regions being in a state of incandescence, while the upper layer is held at a temperature suitable for the development of blue oxide colors. A black color can be obtained by oil quenching the polished steel parts, provided impregnation of the pores is not likely or objectionable. The part may be packed in a spent carburizing compound and heated to about 650°C. (1200°F.); the oil quench must follow immediately after removal of the part from the pack. Equally good results can be obtained without embedding by heating the part to about 600°C. (1110°F.) prior to oil quenching. A durable black color can also be obtained by wiping the surface of the parts with oil and then heating in a closed rotary drum retort to about 300°C. (570°F.) for at least one hour.

In a method especially suitable for small parts known as Carbonia (gun metal) finishing, the parts are placed loosely in a retort with some charred bone; heated to about 400°C. (750°F.); then, after some time is allowed for surface oxidation, the temperature is dropped to about 350°C. (660°F.), and a mixture of charred bone and a few spoonfuls of carbonia oil are added. Heating is continued for several hours, and is followed by cooling and tumbling in oily cork to obtain a uniform black finish.

STEAM TREATMENTS

Durable and inert surfaces can be produced on iron parts by forming a magnetic iron oxide coating (Fe_3O_4) through the action of superheated steam at a pressure of about 60 psi. The parts are heated to a temperature of about 600°C. (1110°F.), before steam is introduced; the oxide layers thus produced are of appreciable depth. According to Lenel,[12] porous steel parts treated in steam show increased resistance to corrosion and rust formation, in addition to increased rigidity, hardness, and wear resistance.

Another treatment (Bower-Barff method) involves heating of the parts in a closed retort to 875°C. (1610°F.); superheated steam is injected and forms coatings of both Fe_2O_3 to Fe_3O_4; then the alternate

---

[12] F. V. Lenel, in J. Wulff, *Powder Metallurgy*. Am. Soc. Metals Cleveland 1942, p. 512.

injection of steam and carbon monoxide is continued until a sufficient depth of the magnetic iron oxide surface is obtained.

PHOSPHATE TREATMENTS

Steel parts can be given a light gray coating by placing them in a ferrous phosphate solution (made from iron filings placed in a 1% phosphoric acid solution), which is then boiled for about 2 hours. The treatment, known as Coslettizing, is particularly suitable for hardened and tempered steel parts, since the temperature is too low to affect the temper. Another method, known as Parkerizing, includes the use of an oxidizing agent such as manganese oxide to the 1% ferrous phosphate solution. Upon immersion of the polished parts into boiling manganese dihydrogen phosphate, a considerable amount of hydrogen is evolved, and when the effervescence eventually ceases, the process is completed with the removal and drying of the parts. The gray color can be changed to deep black by dipping the parts into paraffin oil. More recently the process has been accelerated by conveyerizing and spray application.[13] The Parkerizing process has been used effectively as a rust preventive on large quantities of soft magnetic parts whose density was in excess of 85%; their resistance to corrosion by atmospheric humidity was found to be remarkably improved.

Metal Coatings

In addition to electroplating procedures, many other methods can provide protective metal coatings, but few of these are adaptable for powder metallurgy products. Most of these processes involve operation at higher temperatures and the continuous porosity of the part provides an extensive surface area that can react with the metal or salt to be used for the coat. Metal coatings can be applied to the surface of the base metal in three different ways: (*1*) the part may be brought in contact with a fine metal powder, and a surface film obtained by diffusion at the contact points; (*2*) the part may be dipped into a liquid, lower melting metal, and form a protective film; and (*3*) the metal coating can be sprayed onto the surface. In the case of the powder and hot-dip processes, the film may have a simple mechanical bond or may form compounds and alloys with the base metal.

COATINGS BY BURYING IN METAL POWDER PACKS

Sherardizing. A typical metal powder coating method is the zinc "impregnation" or Sherardizing process. The parts are put in a tightly

---

[13] R. M. Burns and A. E. Schuh, *Protective Coatings for Metals* Reinhold, New York, 1939, p. 374.

sealed revolving drum or retort filled with zinc dust; oxidation of the zinc powder is prevented by the addition of small quantities of powdered charcoal. The retort is heated to about 350° to 375°C. (660° to 710°F.); the length of time required for Sherardizing varies from 3 to 12 hours, depending on the thickness of the coat desired. The gray coating (the color of zinc after cleaning and brushing) is much affected by temperature changes. If the treatment is carried out much below 350°C. (660°F.), the coating remains relatively thin, soft, and porous; if the operating temperature is materially above 375°C. (710°F.), the coating becomes coarse, crystalline, and richer in iron content—with correspondingly reduced resistance to atmospheric action. Sherardized coatings are filled with minute cracks, probably caused by the difference in expansion between the zinc and the iron base. If the porosity of the part is large, the coating will also contain pores, and its effectiveness will be diminished. Since the method produces a coating of uniform thickness, it is generally preferred for threaded solid parts, the threads not being filled up as is the case of parts dipped in a liquid. Powder metallurgy parts having several projections or complicated contours are well suited for the process.

Calorizing. There are several processes that provide a coating of aluminum on iron or steel by diffusion. One method is the packing of the part with a calorizing compound which consists mainly of aluminum powder in a hermetically sealed revolving retort. The charge is heated to a temperature above 800°C. (1470°F.) for 4 to 6 hours, and the liquid aluminum droplets are allowed to diffuse into the surfaces of the parts. The surface layer of iron–aluminum alloy formed contains approximately 25% aluminum. In order to increase the depth of the aluminum–iron alloy case, the parts can be removed from the rotating retort and heated to 1000°C. (1830°F.) under stationary conditions. This treatment causes the aluminum to penetrate by diffusion to depths reaching 0.040 in. Another method consists of packing the parts in hermetically sealed stationary boxes with the powdered aluminum compound. This process requires at least 6 hours at temperatures ranging from 800° to 900°C. (1470° to 1650°F.), but it is preferable for parts of irregular shape which might be damaged by processing in rotating retorts. The aluminum–iron surface alloy is desirable for its properties of good heat endurance and resistance to sulfur corrosion. These properties, however, become effective only if the part, and therefore the coat, are comparatively dense.

Chromizing. Chromium can be diffused into the surface layers of parts by packing the material to be treated in a mixture of chromium and alumina powders (each about 50% by weight)[13a]; the alumina prevents the sintering of the chromium into a solid mass. The parts are packed with the mix in a container and heated to 1300–1400°C. (2370–2550°F.) in pure

[13a] F. C. Kelley, in *Metals Handbook*. Am. Soc. Metals, Cleveland, 1948, p. 705.

hydrogen for 3 to 4 hours, depending on the penetration and chromium concentration desired; the chromium content of the diffused surface layer may reach 20%. Chromized iron parts have been found to be highly resistant to corrosion of atmosphere, moisture, steam, etc., but the surface as well as the interior of the part must be sufficiently dense to provide effective corrosion resistance.

A more effective process, developed by Becker, Daeves, and Steinberg[13b] in Germany during the war, involves chromizing at about 1040°C. (1900°F.) by exchange action with the surface of the iron part of chromous chloride vapors which are formed by reaction of hydrogen chloride with the chromium of the pack. After 8 hours of diffusion of the deposit, the chromium content at the surface is 30–35%, and 12–13% at a depth of 0.004 inch.

Siliconizing. Iron and steel parts can also be contacted with silicon, and a case containing about 14% silicon can be produced. The process may consist essentially of a reaction between the surface of the part and powdered silicon carbide and chlorine at temperatures near 1000°C. (1830°F.). The chloride apparently liberates the silicon from the carbide in the nascent form, the silicon diffusing immediately into the iron surface. It is essential that the silicon carbide particles be in close contact with the surface during decomposition. Instead of silicon carbide, ferrosilocon or mixtures of the two can be used. The parts may be treated in a rotary or pot furnace and the chlorine added when the parts have reached the operating temperature; ordinary carburizing equipment can be adapted for the purpose. The fact that no inert or reducing atmosphere is used during the treatment, coupled with the highly corrosive action of the chlorine gas, makes the process unsuitable for any powder metallurgy part that contains more than a few isolated pores. Very dense iron parts, on the other hand, can be furnished with a surface that not only possesses improved heat and corrosion resistance, but also displays appreciable wear resistance.

COATINGS BY DIPPING INTO LIQUID METALS

Among the metal coats obtained by dipping into a liquid, tin and zinc coatings are most frequently used in the metal industry, followed in frequency of use by aluminum and low melting alloy coatings. The same processes may, of course, be used for powder metallurgy products, but again, it is advisable to employ the method only on dense parts. Parts with interconnected porosity are infiltrated readily with the liquid metal and, as a result, brittle constituents may form throughout the structure by diffusion of the low melting metal into the iron or other base metal.

[13b] G. Becker, K. Daeves, and F. Steinberg, *Stahl u. Eisen, 61,* 189 (1941); *Z. VDI., 85,* 127 (1941); see also *Metallwirtschaft, 20,* 217 (1941).

Tinning. Ferrous metal powder products can be tinned by hot dipping or immersing the part to be coated in a molten tin bath, a method generally preferred to electroplating, spraying, or fusing a tin powder coating over the surfaces of the part. Before dipping, the parts are cleaned by vapor degreasing and then pickled in a 3 to 6% sulfuric acid solution. This is followed by washing and wetting by water that is made slightly acid with hydrochloric acid; the wet base metal is then introduced into the tin bath at a bath temperature ranging between 300° and 350°C. (570° to 660°F.). Small parts cannot be tinned in the automatic machines used for sheet tinning and must be coated in batch-type tin baths. Usually the parts are immersed first in a preliminary tinning bath, and then in a second tinning pot to secure a better coating.[14]   In this way much of the iron-tin dross that forms remains in the first bath, and the second pot contains very pure tin. If the contours are not too complex, the surfaces of the parts may be wiped while the coating is still liquid to assure complete coverage; a smooth surface is then obtained by briefly redipping the part. Finally, the parts may be dipped in a separate pot that allows the tin coating to drain, or they may simply be withdrawn through a palm oil or molten tallow division of the final tinning pot, the latter serving simultaneously as a flux; zinc chloride is generally used for fluxing in the first pot. A comparatively large amount of tin adheres to the surface—the amount depending on the temperature of the finishing bath, the time permitted for drainage, and the configuration of the part. Where excess tin has drained to an edge or bottom section of the part, redipping of this portion in a hot tin bath is helpful in removing the heavy tin concentration.

Galvanizing. Hot dipping, applicable to parts having relatively simple contours, is the most widely used method of zinc coating of ferrous products. The chief advantage of the zinc coating lies in the sacrificial corrosion of zinc which forms one or more alloy layers with the iron between the base metal and the substantially pure zinc coating at the surface. The character and thickness of these alloy layers depend on the bath temperature, time of immersion, and composition of the base metal. The coating can be treated further to render it more resistant to corrosion, as by chromate surface treatments, developed especially for this purpose. The base metal must first be cleaned perfectly by pickling in sulfuric acid or hydrochloric acid. All scale and free iron particles must be removed by scrubbing, since they would form dross with the galvanizing bath, or render impossible the formation of a satisfactory coating. The parts are then conditioned by wetting with hydrochloric acid or zinc chloride solution, and are introduced into the molten zinc bath while wet. Ammonium

    [14] R. J. Nekervis and B. W. Gonser, in *Metals Handbook*. Am. Soc. Metals, Cleveland, 1948, p. 709.

zinc chloride flux protects the surface of the bath. In order to distribute the coating evenly and remove excess zinc, the parts may be jarred, rotated, or even centrifuged; moving of the parts should be continued until the coating has frozen. The surface appearance of the coating is controlled by the rate of cooling, with air or ammonium chloride being applied to the still molten coating or providing for local chilling.

Alloy Coatings. The coating of low-melting alloys onto ferrous parts can be accomplished with practically the same methods as used in tinning or galvanizing. Coatings of tin–lead alloys (containing about 20% tin), known as "terne plates," are made by dipping in a bath kept at a temperature slightly above that for tin plates. A flux of zinc ammonium chloride is sometimes used instead of zinc chloride. Other alloys used for coatings may contain zinc in addition to tin and lead. If the base metal is sufficiently dense, it may first be cleaned by the Lohmanizing method. The parts are first immersed in a bath containing amalgamating salt, then pickled, and dipped into two baths of molten alloys, the second bath containing the final composition to be used for the surface coating.

Dip Calorizing. Parts may also be dipped into molten aluminum in a manner similar to that employed for hot-dip galvanizing, although a protective atmosphere may be required; the surfaces must again be carefully degreased and pickled. Dip calorizing produces a coat of practically pure aluminum with an iron–aluminum compound at the surface of the base metal. This alloy layer can be increased in thickness by a prolonged diffusion treatment.

SPRAYED METAL COATINGS

The process of coating metal parts by spraying molten metal against the surface to be coated is also of some interest for powder metallurgy products; the act of spraying does not affect the internal surface areas of porous parts. On the other hand, such coatings require rough surfaces or preheating to provide for sufficient adherence; they are relatively thick and do not lend themselves to the preservation of close tolerances unless they are subsequently machined. Moreover, sprayed surfaces in general are quite porous, and such coatings are therefore not well suited for corrosion-resistant uses when relatively noble metals are used. The spraying process makes it possible, by means of simple apparatus, to apply a durable coating of many commercial metals and alloys upon the surfaces of metal parts (e.g., nonferrous coatings of tin, lead, zinc, aluminum, copper, brass, or bronze can be sprayed onto ferrous base metals, or iron or nickel coatings can be applied to nonferrous metal surfaces). The thickness of the coating may be varied according to requirements.

The metal-spraying process involves melting of the metal which is to be deposited, atomizing the molten metal by means of a blast of air or inert gas, and depositing the atomized metal upon the surface to be coated. The standard apparatus used for metal spraying today is of the portable pistol type, weighing about 4 pounds. For the spraying of small parts, however, mass coating machines have been constructed, usually consisting of rotating drums and spray guns operating at both ends. In this equipment the metal to be deposited is used in the form of wire or powder. Liquid metal can also be used to feed the spray gun, but is limited to low-melting metals. Wire is fed automatically from a reel through the nozzle and into contact with an oxygen-fuel gas flame, which melts the wire, and the atomizing gas blows the metal against the surface. If a powder is used, it must be heated to the melting point by passing the particles through the flame zone, and projecting the resulting spray by an air blast. As the metal particles impinge against the surface at high velocity, they are flattened into relatively flat scales which become interlocked and form a closely knit unit as the coating is built up. Cavities of microscopic size remain between the scales at numerous points, however, and the coating is accordingly somewhat porous. Moreover, during spraying, superficial oxide films form on the minute particles and become incorporated in the coating.

Mechanically Applied Nonmetallic Coatings

Besides nonmetallic compound coatings obtained via chemical reactions involving the base metal, nonmetallic protective films entirely foreign to the base metal can be applied to the surface by mechanical means. Two basic types of such nonmetallic coatings (apart from vitreous types) are used in the metal industry, namely organic finishes (including paints, varnishes, lacquers, etc.) and slushing compounds. The methods of applying these coatings include brushing, dipping, spraying, tumbling, roller-coating, etc.

ORGANIC FINISHES

Organic finishes, including paints, varnishes, lacquers, etc., fall into three major component groups: pigments, binders, and thinners. The pigment particles ordinarily produce the desired color and are cemented in place by the binders which constitute the film-forming materials. The thinners serve to convert the whole mass into a liquid so that it can be applied easily. Binders and thinners, serving together to carry the pigments, are generally referred to as vehicles. Some organic finishes, for instance, protect the surface primarily by the formation of a film, which acts as a mechanical barrier. Other types depend on specific chemical reactions to provide protection, *e.g.*, the use of zinc chromate and red lead pigments as corrosion inhibitors. For further details the reader is referred to the book by Burns and Schuh.[11]

SLUSHING COMPOUNDS

Unlike the aforementioned organic finishes that are converted after application into relatively dry films by chemical and physical changes, oils and greaselike materials are in general chemically inactive and are affected solely by physical conditions. Their protective quality depends entirely on the providing of a continuous film which excludes air or moisture from the metal surface. The chief advantages in using these protective agents—commonly referred to as "slushing compounds"—are the relative ease of application and removal, and their great economy over paint, since a very thin coat usually is sufficient to provide adequate protection. The coat may be applied either cold or hot, by brushing, dipping, or spraying. The slushing compounds in general use comprise a number of organic substances that vary widely in their viscosity. For powder metallurgy products, however, only the more viscous and less penetrating compounds are effective. Slushing compounds containing a corrosion inhibitor have been applied with considerable success to surfaces of ferrous and nonferrous powder metallurgy products of various densities; for further details see Burns and Schuh.[15]

IMPREGNATION TREATMENTS

The continuous porosity in sintered bronzes or steels facilitates certain impregnation treatments whose purpose it is to incorporate into the structure elements which cannot suitably be mixed initially. The impregnants, possibly including a number of metallic or nonmetallic materials, may serve to improve (among other things) such different properties as specific gravity, corrosion resistance, or resistance to friction. The impregnation treatment is carried out at temperature and pressure conditions that vary according to the individual physical and chemical properties of the liquid impregnant. Where very high melting metals are to be infiltrated, the porous base metal need not be sintered before the treatment, since sintering will precede the infiltration of the liquid metal during heating of the charge. For the sake of dimensional accuracy it may also become necessary to size the sintered parts before impregnation,[16] although care must be taken that drastic densification of the surface regions

[15] R. M. Burns and A. E. Schuh, *Protective Coatings for Metals*. Reinhold, New York, 1939, p. 337.
[16] W. D. Jones, *Can. Metals Met. Ind., 6,* 10, 31, 53, (1943).

does not take place, since this would tend to impede impregnation. In the manufacture of self-lubricating bearings, sizing is frequently performed after oil-impregnation, with the oil serving as lubricant during pressing in the sizing dies.

Nonmetallic Impregnants

OILS

The same characteristic of penetrating throughout the network of pores that makes the low-viscosity oils of the mineral, fatty, or petroleum type unsuitable for corrosion-inhibiting surface protection, is utilized where it is desirable to retain a lubricant inside the structure. An entire industry—the manufacture of self-lubricating bearings—is built on the principle of impregnating sintered bronzes or iron-base compositions with lubricating oils through the capillary action of the intercommunicating pores. Porosity of the metal can be closely controlled within a wide percentage range up to 40% by volume. The same capillary action rapidly transmits the oil held in the pores to the bearing surface and shaft, when the pressure or temperature of the oil is increased by the friction during initial operation. The oil apparently most favored for impregnating porous bronze bearings is a high-grade, nongumming petroleum oil.[17] The impregnation may be done in two ways. (1) The simpler but slower method is to dip a basketful of parts into a container of hot oil, 10 to 15 minutes of soaking time being necessary at a temperature of 110°C. (230°F.) to assure complete impregnation. (2) The faster and more efficient method is to use a somewhat lower temperature and to draw the air out of the pores by vacuum, then force the oil into the cavities under pressure. Sometimes, the parts are preheated to the temperature of the oil in order to accelerate the process. Capillary action retains the oil in the pores until external pressure or heat will cause the lubricant to ooze out.

WAXES AND GREASES

The same impregnation principle is also applicable to more viscous hydrocarbons, such as waxes or greases, the only difference being in the need for a somewhat higher impregnation temperature. Wax impregnation can be combined with tumbling—necessitating operation at particular temperatures. Waxlike impregnants would be preferable where a continuous protective surface film is required in service, without pressure or frictional heat being available as a force for driving the impregnant to

[17] U. S. Army-Navy Aeronautical Spec. AN-B-7, Aug. 26, 1942.

the surface or where corrosion effects of the conventionally used oils are objectionable. In the case of the waxes that are solid at room temperature, a worn-off coating must be regenerated by a special baking treatment; high-viscosity greases that are still somewhat fluid at room temperature may be forced to the wearing surface by the intermittent application of injection pressure methods.

PAINTS AND INKS

In recent years, numerous attempts have been made to produce porous typewriter and printing types impregnated with paints and inks, and so to establish a self-inking process analogous to the principle of self-lubricating bearings. As far as is known, these attempts have not been very successful, partly because most inks have a tendency to dry quickly—sealing the surface cavities. A pressure-feed arrangement would overcome this difficulty, but might lead to serious complications for small types.

Metallic Impregnants

LOW-MELTING METALS

Low-melting metals are infiltrated into high-melting porous metal structures for two principal reasons: (*1*) to increase the weight of the part and (*2*) to improve antifriction properties. Although such infiltration processes, in their true sense, constitute an integral part of the processing operations and cannot be considered to be finishing treatments, they will be mentioned here.

In recent years iron-base parts of high specific gravity have been produced by impregnating a ferrous skeleton metal with liquid lead.[18] The sintered iron or steel briquettes are immersed in the molten lead bath, and are kept in the liquid for a sufficient time to assure complete penetration of the lead throughout the interconnected pores. A simplification of the procedure consists of combining the lead impregnation with the sintering operation, *i.e.*, by introducing the molded parts into a lead bath kept at about 1000° to 1100°C. (1830° to 2010°F.). Careful control of the shape and size distribution of the iron particles permits close regulation of the infiltration process, and makes possible a uniform dispersal of the lead throughout the structure in spite of the fact that there is no practical solubility or surface affinity between the two metals.

The infiltration of tin and lead babbitt alloys into a spongy matrix of nonferrous alloys tends to improve the bearing properties of the metal.

[18] U. S. Pat. 2,192,792.

The process—developed by Boegehold for General Motors,[19,20] and used in the manufacture of main and connecting rod bearings—involves the impregnation of a sintered, porous, copper–nickel layer. Although various babbitt and lead alloys can be used as impregnants, an alloy containing 94% lead, 3% tin, and 3% antimony appears to be the most suitable for strength, deformability, and corrosion resistance. The matrix is formed by pouring the copper–nickel powder on a copper-plated, continuous steel strip that moves through a sintering furnace (at the rate of about 25 feet per minute). Impregnation with the babbitt is carried out in a special furnace in a vacuum. First, the strip is heated, then the air is drawn from the pores by the vacuum, and finally the matrix is impregnated with the liquid metal. The desired thickness of the surface coat is obtained by regulating the speed at which the strip passes through the furnace. (Additional information on the subject is given in Chapter XXVI, Volume II.)

Another recently developed process involves the impregnation of lead and lead alloys into porous silver bearings.[20a]

HIGH-MELTING METALS

The impregnation of porous matrices of high-melting metals with silver or copper to form composite electrical contact or structural materials constitutes one of the major processes in powder metallurgy. Some of the implications of this process have already been discussed in the section on sintering (see page 568). Since the infiltration of these higher melting metals is always an integral part of the synthesis of the composite metal, and not a subsequent finishing operation, the subject will not be developed here, but will be elaborated in connection with the manufacture of composite metals (Chapters XXIII and XXV, Volume II).

JOINING

As in the case of surface protection coatings, the porosity of the sintered metal is an important consideration when metal-joining operations are contemplated. If the porosity is appreciable and most of the pores are interconnected, the liquid soldering, brazing, or welding metals will tend to penetrate through the pores, and the joints will become imperfect; at the same time they will require excessive quantities of joining metal. The absorption of the liquid metal is small where relatively dense parts are involved and the joining operation will consequently be successful in this case.

[19] U. S. Pat. 2,198,240.
[20] A. L. Boegehold, in J. Wulff, *Powder Metallurgy.* Am. Soc. Metals, Cleveland, 1942, p. 520.
[20a] U. S. Pat. 2,294,404.

Soldering

There are only rare instances in which sintered metals must be soldered onto another base. Parts pressed from aluminum and zinc powder have been soldered onto steel strips for display purposes; and brass parts of complicated shapes have experimentally been produced by soldering two less complex sections. Ordinary soldering techniques must be modified to a certain extent if a perfect joint is to result. A noncorrosive flux must be applied liberally since superficial pores make the surface of the sinter-metal considerably larger than the comparable surface of ordinary metals. In addition, more pressure than usual may have to be applied to the soldering iron in order to produce a solid bond.

Brazing

Brazing is the one joining operation that is used extensively in connection with powder metallurgy products. Cemented tungsten carbide tool tips and die nibs are brazed onto steel shanks or into steel rings, and tungsten-base composite contact tips and electrodes are brazed onto copper or other metal supports. In the latter instance, brazing is facilitated by the considerable proportions of copper or silver always present in the material. In fact, one manufacturer attaches the refractory-base contact metal to the backing material by direct fusion if the low-melting metal content of the composite is sufficiently high.[21]

Tools, drawing dies, and other wear-resistant products are generally so constructed that the cemented carbides form only the tip or insert. They are brazed to the tool shank or die ring by a straight copper braze, by silver solder, or by other suitable brazing alloys, such as "Eutectic low-temperature welding" alloys.[22] For a high-strength braze, it is essential that the cemented carbide face be highly finished. Most production brazing of cemented carbides is carried out in a hydrogen-brazing, muffle-type furnace that is heated either by electricity or by gas. Brazing under hydrogen in a high-frequency furnace is the method used when tungsten carbide precision tools are produced. The great difference in thermal expansion between the cemented carbide and the steel support calls for great caution in the brazing operation. Aside from the careful selection of the shank material, it may be necessary to incorporate a cushioning shim between the steel and carbide[23] for larger or more intricately shaped

[21] F. R. Hensel, E. I. Larsen, and E. Szwazy, in J. Wulff, *Powder Metallurgy.* Am. Soc. Metals, Cleveland, 1942, p. 490.

[22] Products of Eutectic Welding Alloys, Inc., New York, N. Y. (*e.g.*, type No. 16).

[23] E. W. Engle, in J. Wulff, *Powder Metallurgy.* Am. Soc. Metals, Cleveland, 1942, p. 450.

pieces. One way of providing such a cushion is by nickel plating the pocket of the steel shank.

Welding

Direct welding of sintered metals to other metals is not employed extensively. Welding is feasible only in joining refractory-metal-base contacts or cemented carbides to supporting materials. However, when copper supports are to be faced with tungsten–copper compositions it is more practical to copper-braze or to cast the copper support on the refactory-base composite; when silver-containing compositions are to be attached to supporting materials, silver soldering is usually satisfactory. Welding of cemented carbides to steel supports gives a strong joint; because of the large differences in thermal expansion of the two materials, however, precautions must be taken in the design, in the manipulations during welding, and especially during cooling. (Engle[23] has reported that a special process makes possible the welding of cemented carbides to steel, but the process apparently cannot be adapted to a great variety of shapes.)

Summary

The finishing treatments of powder metallurgy products include operations designed to perfect the metal in a number of ways. Mechanical operations may have the dual purpose of giving the product its final shape to the desired degree of dimensional accuracy and of producing a satisfactory surface finish. Machining, sizing, and straightening perform the shaping operation, whereas grinding, tumbling, or sand blasting impart the desired surface appearance to the metal.

Special surface treatments may be required where the metal is to be protected against corrosion or wear. Electroplating of sintered metals is possible, but as a corrosion inhibitor, it is effective only if the metal is sufficiently dense, if the coating metal is capable of affording sacrificial protection, or if the plated surface is afterwards treated with a corrosion-inhibiting agent. Chemical treatments based on controlled surface oxidation are applicable for certain metals to obtain a desired color effect. Metal coatings for the purpose of increasing corrosion or wear resistance can be obtained by contact with metal powder packs, by dipping into molten metal baths, or by spraying the coating on the surface of the base metal. Nonmetallic coatings consisting of paints, lacquers, or slushing compounds can be applied mechanically by brushing, dipping, spraying, tumbling, etc. Sintered parts containing a large number of intercommunicating pores are capable of absorbing nometallic as well as metallic liquids. To provide automatic lubrication during service, oils and greases can be impregnated by simple means into porous bronze and iron-base bearings and other parts subjected to wear. Low-melting metals can be infiltrated into copper- or iron-base porous structures to improve anti-friction properties or to increase the specific gravity of the part by dipping them into the low-melting metals. Higher melting metals, finally, are infiltrated into refractory metal skeletons as an integral part of the processing of heavy-duty contact metals or structural parts.

Brazing is the only joining operation used extensively in connection with powder metallurgy products. Refractory-base contact metals and cemented carbides are attached to supporting materials by copper brazing, silver soldering, or other commercial brazing methods. Soldering and welding have so far found few applications in powder metallurgy.

AUTHOR INDEX

SUBJECT INDEX

A

Abrading of sinter metals, 703–704

Abrasive(s), in metal–nonmetal combinations, 8

use in friction elements, 7, 252

Abrasiveness, of aluminum oxide surface layers, 264

of powders, and die construction, 350, 367

Abrasion, effect on molding tools and dies, 363–364

of particle surfaces, in cold-pressed compacts, 306

Absolute size values, correlation of screen analysis figures with, 131

Absorbed moisture and flow, 108

Acicular particles, factors affecting compressibility, 269, 272

Acrawax, as lubricant, 256

Activated alumina, use as atmosphere drying agent, 629, 630, 634

Activating agents in precipitation of powders, 61

Actual density of individual powder particles, determination, 107

Adaptor, cast steel, in die construction, 377

Addition agents. See also *Crystallization centers.*

function, 231, 251–257

Adhesion. See also *Surface adhesion, Interparticle cohesion,* etc.

and size of powder particle, 88, 89

Adhesive forces, and applied pressure on powder, 274, 275

and impure surface films, 118

and powder dispersion, 275

during compression, 274, 297

influence of gases, 518

in sintering, 105

of metal particles during compression, 266

Adsorptive capacity. See *Sorptive capacity.*

Adsorption, in determination of range of particle size, 129

in measurement of surface area, 130

Adsorption method(s), application to porous materials, 146

for particle determination, application, 145

Adsorptive capacity of copper powder, 121, 122

Age and precipitation hardening, 682–686

Agglomerates, causing error in sizing, 132, 137

Agglomeration(s), effect on blending, 246

effect on powder flow, 241

Aggregation of particles, and applied pressure on powder, 274

by sliding and rotation, 101–102

Air, as sintering atmosphere, 625–626

entrapped, causing lamination during compression, 90

Air conditioning, in powder metallurgy technique, 15

in powder production, 240–241

Allotropic changes during sintering, 497

Alloy(s), by electrolytic process, 69

eutectic, use of hydride in production, 59–60

gold–silver, as binder in production of platinum, 18

homogeneous, sintering, 562–564

iron–nickel, in electric circuit cores, 33

production by pressure, 27

tin–silver, early use, 20

Alloy coatings of sinter–metals, 717

Alloy disintegration, of powdered products, 75, 78, 99

in powder production, 74–76

Alloying, in production of composite metals, 7

nonmetallic coating in, 249

Alloying elements, metals used as, 159–160

Alloying powders, by diffusion treatment, 235

Alloy powder(s), 203–217

by rotating disk method, 45

coating in manufacture of magnetic cores, 250